Wireless Mobile Internet Security

Wireless Mobile Internet Security

Second Edition

Man Young Rhee

Endowed Chair Professor, Kyung Hee University
Professor Emeritus, Hanyang University, Republic of Korea

WILEY

A John Wiley & Sons, Ltd., Publication

Library of Congress Cataloging-in-Publication Data

Rhee, Man Young.
  Wireless mobile internet security / Man Young Rhee. – Second edition.
 pages cm
 Includes bibliographical references and index.
  ISBN 978-1-118-49653-4 (cloth)
  1. Wireless Internet – Security measures. I. Title.
  TK5103.4885.R49 2013
  004.67′8 – dc23

                                        2012040165

A catalogue record for this book is available from the British Library.

ISBN: 9781118496534

Set in 10/12pt Times by Laserwords Private Limited, Chennai, India
Printed and bound in Singapore by Markono Print Media Pte Ltd

Contents

Preface			**xiii**
About the Author			**xxi**
Acknowledgments			**xxiii**

1	**Internetworking and Layered Models**		**1**
	1.1	Networking Technology	2
		1.1.1 Local Area Networks (LANs)	2
		1.1.2 Wide Area Networks (WANs)	3
	1.2	Connecting Devices	5
		1.2.1 Switches	5
		1.2.2 Repeaters	6
		1.2.3 Bridges	7
		1.2.4 Routers	7
		1.2.5 Gateways	8
	1.3	The OSI Model	8
	1.4	TCP/IP Model	12
		1.4.1 Network Access Layer	13
		1.4.2 Internet Layer	14
		1.4.3 Transport Layer	14
		1.4.4 Application Layer	14

2	**TCP/IP Suite and Internet Stack Protocols**		**15**
	2.1	Network Layer Protocols	15
		2.1.1 Internet Protocol (IP)	15
		2.1.2 Address Resolution Protocol (ARP)	28
		2.1.3 Reverse Address Resolution Protocol (RARP)	31
		2.1.4 Classless Interdomain Routing (CIDR)	31
		2.1.5 IP Version 6 (IPv6 or IPng)	32
		2.1.6 Internet Control Message Protocol (ICMP)	40
		2.1.7 Internet Group Management Protocol (IGMP)	41
	2.2	Transport Layer Protocols	41
		2.2.1 Transmission Control Protocol (TCP)	41
		2.2.2 User Datagram Protocol (UDP)	44

2.3	World Wide Web	47
	2.3.1 Hypertext Transfer Protocol (HTTP)	47
	2.3.2 Hypertext Markup Language (HTML)	47
	2.3.3 Common Gateway Interface (CGI)	48
	2.3.4 Java	49
2.4	File Transfer	49
	2.4.1 File Transfer Protocol (FTP)	49
	2.4.2 Trivial File Transfer Protocol (TFTP)	49
	2.4.3 Network File System (NFS)	50
2.5	E-Mail	50
	2.5.1 Simple Mail Transfer Protocol (SMTP)	50
	2.5.2 Post Office Protocol Version 3 (POP3)	51
	2.5.3 Internet Message Access Protocol (IMAP)	51
	2.5.4 Multipurpose Internet Mail Extension (MIME)	52
2.6	Network Management Service	52
	2.6.1 Simple Network Management Protocol (SNMP)	52
2.7	Converting IP Addresses	53
	2.7.1 Domain Name System (DNS)	53
2.8	Routing Protocols	54
	2.8.1 Routing Information Protocol (RIP)	54
	2.8.2 Open Shortest Path First (OSPF)	54
	2.8.3 Border Gateway Protocol (BGP)	55
2.9	Remote System Programs	55
	2.9.1 TELNET	55
	2.9.2 Remote Login (Rlogin)	56
2.10	Social Networking Services	56
	2.10.1 Facebook	56
	2.10.2 Twitter	56
	2.10.3 Linkedin	57
	2.10.4 Groupon	57
2.11	Smart IT Devices	57
	2.11.1 Smartphones	57
	2.11.2 Smart TV	57
	2.11.3 Video Game Console	58
2.12	Network Security Threats	58
	2.12.1 Worm	58
	2.12.2 Virus	58
	2.12.3 DDoS	58
2.13	Internet Security Threats	58
	2.13.1 Phishing	58
	2.13.2 SNS Security Threats	59
2.14	Computer Security Threats	59
	2.14.1 Exploit	59
	2.14.2 Password Cracking	60
	2.14.3 Rootkit	60

	2.14.4	Trojan Horse	60
	2.14.5	Keylogging	61
	2.14.6	Spoofing Attack	61
	2.14.7	Packet Sniffer	62
	2.14.8	Session Hijacking	62

3 | **Global Trend of Mobile Wireless Technology** | | **63** |
3.1	1G Cellular Technology		63
	3.1.1	AMPS (Advanced Mobile Phone System)	64
	3.1.2	NMT (Nordic Mobile Telephone)	64
	3.1.3	TACS (Total Access Communications System)	64
3.2	2G Mobile Radio Technology		64
	3.2.1	CDPD (Cellular Digital Packet Data), North American Protocol	65
	3.2.2	GSM (Global System for Mobile Communications)	65
	3.2.3	TDMA-136 or IS-54	66
	3.2.4	iDEN (Integrated Digital Enhanced Network)	66
	3.2.5	cdmaOne IS-95A	67
	3.2.6	PDC (Personal Digital Cellular)	67
	3.2.7	i-mode	67
	3.2.8	WAP (Wireless Application Protocol)	67
3.3	2.5G Mobile Radio Technology		67
	3.3.1	ECSD (Enhanced Circuit-Switched Data)	69
	3.3.2	HSCSD (High-Speed Circuit-Switched Data)	69
	3.3.3	GPRS (General Packet Radio Service)	69
	3.3.4	EDGE (Enhanced Data rate for GSM Evolution)	69
	3.3.5	cdmaOne IS-95B	69
3.4	3G Mobile Radio Technology (Situation and Status of 3G)		70
	3.4.1	UMTS (Universal Mobile Telecommunication System)	73
	3.4.2	HSDPA (High-Speed Downlink Packet Access)	73
	3.4.3	CDMA2000 1x	74
	3.4.4	CDMA2000 1xEV (1x Evolution)	74
	3.4.5	CDMA2000 1xEV-DO (1x Evolution Data Only)	74
	3.4.6	CDMA2000 1xEV-DV (1x Evolution Data Voice)	74
3.5	3G UMTS Security-Related Encryption Algorithm		75
	3.5.1	KASUMI Encryption Function	75

4 | **Symmetric Block Ciphers** | | **81** |
4.1	Data Encryption Standard (DES)		81
	4.1.1	Description of the Algorithm	82
	4.1.2	Key Schedule	84
	4.1.3	DES Encryption	86
	4.1.4	DES Decryption	91
	4.1.5	Triple DES	95
	4.1.6	DES-CBC Cipher Algorithm with IV	97

	4.2	International Data Encryption Algorithm (IDEA)	99
		4.2.1 Subkey Generation and Assignment	100
		4.2.2 IDEA Encryption	101
		4.2.3 IDEA Decryption	106
	4.3	RC5 Algorithm	108
		4.3.1 Description of RC5	109
		4.3.2 Key Expansion	110
		4.3.3 Encryption	114
		4.3.4 Decryption	117
	4.4	RC6 Algorithm	123
		4.4.1 Description of RC6	123
		4.4.2 Key Schedule	124
		4.4.3 Encryption	125
		4.4.4 Decryption	128
	4.5	AES (Rijndael) Algorithm	135
		4.5.1 Notational Conventions	135
		4.5.2 Mathematical Operations	137
		4.5.3 AES Algorithm Specification	140

5 Hash Function, Message Digest, and Message Authentication Code   161

	5.1	DMDC Algorithm	161
		5.1.1 Key Schedule	162
		5.1.2 Computation of Message Digests	166
	5.2	Advanced DMDC Algorithm	171
		5.2.1 Key Schedule	171
		5.2.2 Computation of Message Digests	173
	5.3	MD5 Message-Digest Algorithm	176
		5.3.1 Append Padding Bits	176
		5.3.2 Append Length	177
		5.3.3 Initialize MD Buffer	177
		5.3.4 Define Four Auxiliary Functions (F, G, H, I)	177
		5.3.5 FF, GG, HH, and II Transformations for	
		Rounds 1, 2, 3, and 4	178
		5.3.6 Computation of Four Rounds (64 Steps)	178
	5.4	Secure Hash Algorithm (SHA-1)	188
		5.4.1 Message Padding	188
		5.4.2 Initialize 160-bit Buffer	189
		5.4.3 Functions Used	189
		5.4.4 Constants Used	190
		5.4.5 Computing the Message Digest	191
	5.5	Hashed Message Authentication Codes (HMAC)	195

6 Asymmetric Public-Key Cryptosystems   203

| | 6.1 | Diffie–Hellman Exponential Key Exchange | 203 |
| | 6.2 | RSA Public-Key Cryptosystem | 207 |

	6.2.1	RSA Encryption Algorithm	208
	6.2.2	RSA Signature Scheme	212
6.3	ElGamal's Public-Key Cryptosystem		215
	6.3.1	ElGamal Encryption	215
	6.3.2	ElGamal Signatures	217
	6.3.3	ElGamal Authentication Scheme	219
6.4	Schnorr's Public-Key Cryptosystem		222
	6.4.1	Schnorr's Authentication Algorithm	222
	6.4.2	Schnorr's Signature Algorithm	224
6.5	Digital Signature Algorithm		227
6.6	The Elliptic Curve Cryptosystem (ECC)		230
	6.6.1	Elliptic Curves	230
	6.6.2	Elliptic Curve Cryptosystem Applied to the ElGamal Algorithm	239
	6.6.3	Elliptic Curve Digital Signature Algorithm	240
	6.6.4	ECDSA Signature Computation	244

7 Public-Key Infrastructure   **249**

7.1	Internet Publications for Standards		250
7.2	Digital Signing Techniques		251
7.3	Functional Roles of PKI Entities		258
	7.3.1	Policy Approval Authority	258
	7.3.2	Policy Certification Authority	260
	7.3.3	Certification Authority	261
	7.3.4	Organizational Registration Authority	262
7.4	Key Elements for PKI Operations		263
	7.4.1	Hierarchical Tree Structures	264
	7.4.2	Policy-Making Authority	265
	7.4.3	Cross-Certification	266
	7.4.4	X.500 Distinguished Naming	269
	7.4.5	Secure Key Generation and Distribution	270
7.5	X.509 Certificate Formats		271
	7.5.1	X.509 v1 Certificate Format	271
	7.5.2	X.509 v2 Certificate Format	273
	7.5.3	X.509 v3 Certificate Format	274
7.6	Certificate Revocation List		282
	7.6.1	CRL Fields	282
	7.6.2	CRL Extensions	284
	7.6.3	CRL Entry Extensions	285
7.7	Certification Path Validation		287
	7.7.1	Basic Path Validation	287
	7.7.2	Extending Path Validation	289

8 Network Layer Security   **291**

| 8.1 | IPsec Protocol | | 291 |

		8.1.1	IPsec Protocol Documents	292
		8.1.2	Security Associations (SAs)	294
		8.1.3	Hashed Message Authentication Code (HMAC)	296
	8.2	IP Authentication Header		299
		8.2.1	AH Format	300
		8.2.2	AH Location	301
	8.3	IP ESP		301
		8.3.1	ESP Packet Format	303
		8.3.2	ESP Header Location	304
		8.3.3	Encryption and Authentication Algorithms	306
	8.4	Key Management Protocol for IPsec		308
		8.4.1	OAKLEY Key Determination Protocol	308
		8.4.2	ISAKMP	309

9   Transport Layer Security: SSLv3 and TLSv1   **325**
	9.1	SSL Protocol		325
		9.1.1	Session and Connection States	326
		9.1.2	SSL Record Protocol	327
		9.1.3	SSL Change Cipher Spec Protocol	331
		9.1.4	SSL Alert Protocol	331
		9.1.5	SSL Handshake Protocol	332
	9.2	Cryptographic Computations		338
		9.2.1	Computing the Master Secret	338
		9.2.2	Converting the Master Secret into Cryptographic Parameters	339
	9.3	TLS Protocol		339
		9.3.1	HMAC Algorithm	340
		9.3.2	Pseudo-random Function	344
		9.3.3	Error Alerts	349
		9.3.4	Certificate Verify Message	350
		9.3.5	Finished Message	351
		9.3.6	Cryptographic Computations (for TLS)	351

10 Electronic Mail Security: PGP, S/MIME   **353**
	10.1	PGP		353
		10.1.1	Confidentiality via Encryption	354
		10.1.2	Authentication via Digital Signature	355
		10.1.3	Compression	356
		10.1.4	Radix-64 Conversion	357
		10.1.5	Packet Headers	361
		10.1.6	PGP Packet Structure	363
		10.1.7	Key Material Packet	367
		10.1.8	Algorithms for PGP 5.x	371
	10.2	S/MIME		372
		10.2.1	MIME	372
		10.2.2	S/MIME	379

10.2.3 Enhanced Security Services for S/MIME 382

11 Internet Firewalls for Trusted Systems **387**
11.1 Role of Firewalls 387
11.2 Firewall-Related Terminology 388
11.2.1 Bastion Host 389
11.2.2 Proxy Server 389
11.2.3 SOCKS 390
11.2.4 Choke Point 391
11.2.5 Demilitarized Zone (DMZ) 391
11.2.6 Logging and Alarms 391
11.2.7 VPN 392
11.3 Types of Firewalls 392
11.3.1 Packet Filters 392
11.3.2 Circuit-Level Gateways 397
11.3.3 Application-Level Gateways 397
11.4 Firewall Designs 398
11.4.1 Screened Host Firewall (Single-Homed Bastion Host) 399
11.4.2 Screened Host Firewall (Dual-Homed Bastion Host) 400
11.4.3 Screened Subnet Firewall 400
11.5 IDS Against Cyber Attacks 401
11.5.1 Internet Worm Detection 401
11.5.2 Computer Virus 402
11.5.3 Special Kind of Viruses 403
11.6 Intrusion Detections Systems 404
11.6.1 Network-Based Intrusion Detection System (NIDS) 404
11.6.2 Wireless Intrusion Detection System (WIDS) 406
11.6.3 Network Behavior Analysis System (NBAS) 408
11.6.4 Host-Based Intrusion Detection System (HIDS) 409
11.6.5 Signature-Based Systems 410
11.6.6 Anomaly-Based Systems 411
11.6.7 Evasion Techniques of IDS Systems 412

12 SET for E-Commerce Transactions **415**
12.1 Business Requirements for SET 415
12.2 SET System Participants 417
12.3 Cryptographic Operation Principles 418
12.4 Dual Signature and Signature Verification 420
12.5 Authentication and Message Integrity 424
12.6 Payment Processing 427
12.6.1 Cardholder Registration 427
12.6.2 Merchant Registration 433
12.6.3 Purchase Request 434
12.6.4 Payment Authorization 435
12.6.5 Payment Capture 437

13 **4G Wireless Internet Communication Technology** **439**
   13.1   Mobile WiMAX   440
       13.1.1   Mobile WiMAX Network Architecture   440
       13.1.2   Reference Points in WiMAX Network
                  Reference Model (NRM)   442
       13.1.3   Key Supporting Technologies   444
       13.1.4   Comparison between Mobile WiMAX Network and Cellular
                  Wireless Network   447
   13.2   WiBro (Wireless Broadband)   448
       13.2.1   WiBro Network Architecture   448
       13.2.2   Key Elements in WiBro System Configuration   449
       13.2.3   System Comparison between HSDPA and WiBro   451
       13.2.4   Key Features on WiBro Operation   451
   13.3   UMB (Ultra Mobile Broadband)   452
       13.3.1   Design Objectives of UMB   453
       13.3.2   Key Technologies Applicable to UMB   453
       13.3.3   UMB IP-Based Network Architecture   455
       13.3.4   Conclusive Remarks   456
   13.4   LTE (Long Term Evolution)   457
       13.4.1   LTE Features and Capabilities   457
       13.4.2   LTE Frame Structure   458
       13.4.3   LTE Time-Frequency Structure for Downlink   458
       13.4.4   LTE SC-FDMA on Uplink   460
       13.4.5   LTE Network Architecture   461
       13.4.6   Key Components Supporting LTE Design   463
       13.4.7   Concluding Remarks   464

Acronyms   467

Bibliography   473

Index   481

Preface

The mobile industry for wireless cellular services has grown at a rapid pace over the past decade. Similarly, Internet service technology has also made dramatic growth through the World Wide Web with a wire line infrastructure. Realization for complete mobile Internet technologies will become the future objectives for convergence of these technologies through multiple enhancements of both cellular mobile systems and Internet interoperability.

Flawless integration between these two wired/wireless networks will enable subscribers to not only roam worldwide but also solve the ever increasing demand for data/Internet services. However, the new technology development and service perspective of 4G systems will take many years to come. In order to keep up with this noteworthy growth in the demand for wireless broadband, new technologies and structural architectures are needed to improve system performance and network scalability greatly, while significantly reducing the cost of equipment and deployment. The present concept of P2P networking to exchange information needs to be extended to implement intelligent appliances such as a ubiquitous connectivity to the Internet services, the provision of fast broadband access technologies at more than 50 Mbps data rate, seamless global roaming, and Internet data/voice multimedia services.

The 4G system is a development initiative based on the currently deployed 2G/3G infrastructure, enabling seamless integration to emerging 4G access technologies. For successful interoperability, the path toward 4G networks should be incorporated with a number of critical trends to network integration. MIMO/OFDMA-based air interface for beyond 3G systems are called *4G systems* such as Long Term Evolution (LTE), Ultra Mobile Broadband (UMB), Mobile WiMAX (Worldwide Interoperability for Microwave Access) or Wireless Broadband (WiBro).

Chapter 1 begins with a brief history of the Internet and describes topics covering (i) networking fundamentals such as LANs (Ethernet, Token Ring, FDDI), WANs (Frame Relay, X.25, PPP), and ATM; (ii) connecting devices such as circuit- and packet-switches, repeaters, bridges, routers, and gateways; (iii) the OSI model that specifies the functionality of its seven layers; and finally, (iv) a TCP/IP five-layer suite providing a hierarchical protocol made up of physical standards, a network interface, and internetworking.

Chapter 2 presents a state-of-the-art survey of the TCP/IP suite. Topics covered include (i) TCP/IP network layer protocols such as ICMP, IP version 4, and IP version 6 relating to the IP packet format, addressing (including ARP, RARP, and CIDR), and routing; (ii) transport layer protocols such as TCP and UDP; (iii) HTTP for the World Wide Web; (iv) FTP, TFTP, and NFS protocols for file transfer; (v) SMTP, POP3, IMAP, and MIME for e-mail; and (vi) SNMP for network management. This chapter also introduces latest Social Network Services and smart IT devices. With the introduction of smart services and devices, security problems became an issue. This chapter introduces security threats such as (i) Worm, Virus, and DDoS for network security; (ii) Phishing and SNS security for Internet security; (iii) Exploit, password cracking, Rootkit, Trojan Horse, and so on for computer security.

Chapter 3 presents the evolution and migration of mobile radio technologies from first generation (1G) to third generation (3G). 1G, or circuit-switched analog systems, consist of voice-only communications; 2G and beyond systems, comprising both voice and data communications, largely rely on packet-switched wireless mobile technologies. This chapter covers the technological development of mobile radio communications in compliance with each iterative generation over the past decade. At present, mobile data services have been rapidly transforming to facilitate and ultimately profit from the increased demand for nonvoice services. Through aggressive 3G deployment plans, the world's major operators boast attractive and homogeneous portal offerings in all of their markets, notably in music and video multimedia services. Despite the improbability of any major changes in the next 4–5 years, rapid technological advances have already bolstered talks for 3.5G and even 4G systems. For each generation, the following technologies are introduced:

1. 1G Cellular Technology
   - AMPS (Advanced Mobile Phone System)
   - NMT (Nordic Mobile Telephone)
   - TACS (Total Access Communications System)
2. 2G Mobile Radio Technology
   - CDPD (Cellular Digital Packet Data), North American protocol
   - GSM (Global System for Mobile Communications)
   - TDMA-136 or IS-54
   - iDEN (Integrated Digital Enhanced Network)
   - cdmaOne IS-95A
   - PDC (Personal Digital Cellular)
   - i-mode
   - WAP (Wireless Application Protocol)
3. 2.5G Mobile Radio Technology
   - ECSD (Enhanced Circuit-Switched Data)
   - HSCSD (High-Speed Circuit-Switched Data)
   - GPRS (General Packet Radio Service)
   - EDGE (Enhanced Data rates for GSM Evolution)
   - cdmaOne IS-95B
4. 3G Mobile Radio Technology
   - UMTS (Universal Mobile Telecommunication System)
   - HSDPA (High-Speed Downlink Packet Access)
   - FOMA
   - CDMA2000 1x
   - CDMA2000 1xEV (1x Evolution)

- CDMA2000 1xEV-DO (1x Evolution Data Only)
- CDMA2000 1xEV-DV (1x Evolution Data Voice)
- KASUMI Encryption Function

Chapter 4 deals with some of the important contemporary block cipher algorithms that have been developed over recent years with an emphasis on the most widely used encryption techniques such as Data Encryption Standard (DES), the International Data Encryption Algorithm (IDEA), the RC5 and RC6 encryption algorithms, and the Advanced Encryption Standard (AES). AES specifies an FIPS-approved Rijndael algorithm (2001) that can process data blocks of 128 bits, using cipher keys with lengths of 128, 192, and 256 bits. DES is not new, but it has survived remarkably well over 20 years of intense cryptanalysis. The complete analysis of triple DES-EDE in CBC mode is also included. Pretty Good Privacy (PGP) used for e-mail and file storage applications utilizes IDEA for conventional block encryption, along with RSA for public-key encryption and MD5 for hash coding. RC5 and RC6 are both parameterized block algorithms of variable size, variable number of rounds, and a variable-length key. They are designed for great flexibility in both performance and level of security.

Chapter 5 covers the various authentication techniques based on digital signatures. It is often necessary for communication parties to verify each other's identity. One practical way to do this is with the use of cryptographic authentication protocols employing a one-way hash function. Several contemporary hash functions (such as DMDC, MD5, and SHA-1) are introduced to compute message digests or hash codes for providing a systematic approach to authentication. This chapter also extends the discussion to include the Internet standard HMAC, which is a secure digest of protected data. HMAC is used with a variety of different hash algorithms, including MD5 and SHA-1. Transport Layer Security (TLS) also makes use of the HMAC algorithm.

Chapter 6 describes several public-key cryptosystems brought in after conventional encryption. This chapter concentrates on their use in providing techniques for public-key encryption, digital signature, and authentication. This chapter covers in detail the widely used Diffie–Hellman key exchange technique (1976), the Rivest-Schamir-Adleman (RSA) algorithm (1978), the ElGamal algorithm (1985), the Schnorr algorithm (1990), the Digital Signature Algorithm (DSA, 1991), and the Elliptic Curve Cryptosystem (ECC, 1985) and Elliptic Curve Digital Signature Algorithm (ECDSA, 1999).

Chapter 7 presents profiles related to a public-key infrastructure (PKI) for the Internet. The PKI automatically manages public keys through the use of public-key certificates. The Policy Approval Authority (PAA) is the root of the certificate management infrastructure. This authority is known to all entities at entire levels in the PKI and creates guidelines that all users, CAs, and subordinate policy-making authorities must follow. Policy Certificate Authorities (PCAs) are formed by all entities at the second level of the infrastructure. PCAs must publish their security policies, procedures, legal issues, fees, and any other subjects they may consider necessary. Certification Authorities (CAs) form the next level below the PCAs. The PKI contains many CAs that have no policy-making responsibilities. A CA has any combination of users and RAs whom it certifies. The primary function of the CA is to generate and manage the public-key certificates that bind the user's identity with the user's public key. The Registration Authority (RA) is the interface between a user and a CA. The primary function of the RA is user identification and authentication on behalf of a CA. It also delivers the CA-generated certificate to the end user. X.500 specifies the directory service. X.509 describes the authentication service using the X.500 directory. X.509 certificates have evolved through three versions: version 1 in 1988, version 2 in 1993, and version 3 in 1996. X.509 v3 is now found in numerous products and Internet standards. These three versions are explained in turn. Finally, Certificate Revocation Lists (CRLs) are used to list unexpired certificates that have been revoked. CRLs may be revoked for a variety of reasons, ranging from routine administrative revocations to situations where private keys are compromised. This chapter also includes the certification path validation procedure for the Internet PKI and architectural structures for the PKI certificate management infrastructure.

Chapter 8 describes the IPsec protocol for network layer security. IPsec provides the capability to secure communications across a LAN, across a virtual private network (VPN) over the Internet, or over a public WAN. Provision of IPsec enables a business to rely heavily on the Internet. The IPsec protocol is a set of security extensions developed by IETF to provide privacy and authentication services at the IP layer using crypto-graphic algorithms and protocols. To protect the contents of an IP datagram, there are two main transformation types: the Authentication Header (AH) and the Encapsulating Security Payload (ESP). These are protocols to provide connectionless integrity, data origin authentication, confidentiality, and an antireplay service. A Security Association (SA) is fundamental to IPsec. Both AH and ESP make use of an SA that is a simple connection between a sender and receiver, providing security services to the traffic carried on it. This chapter also includes the OAKLEY key determination protocol and ISAKMP.

Chapter 9 discusses Secure Socket Layer version 3 (SSLv3) and TLS version 1 (TLSv1). The TLSv1 protocol itself is based on the SSLv3 protocol specification. Many of the algorithm-dependent data structures and rules are very similar, so the differences between TLSv1 and SSLv3 are not dramatic. The TLSv1 protocol provides communications privacy and data integrity between two communicating parties over the Internet. Both protocols allow client/server applications to communicate in a way that is designed to prevent eavesdropping, tampering, or message forgery. The SSL or TLS protocol is composed of two layers: Record Protocol and Handshake Protocol. The Record Protocol takes an upper-layer application message to be transmitted, fragments the data into manageable blocks, optionally compresses the data, applies a MAC, encrypts it, adds a header, and transmits the result to TCP. Received data is decrypted to higher-level clients. The Handshake Protocol operated on top of the Record Layer is the most important part of SSL or TLS. The Handshake Protocol consists of a series of messages exchanged by client and server. This protocol provides three services between the server and client. The Handshake Protocol allows the client/server to agree on a protocol version, to authenticate each other by forming a MAC, and to negotiate an encryption algorithm and cryptographic keys for protecting data sent in an SSL record before the application protocol transmits or receives its first byte of data.

A keyed hashing message authentication code (HMAC) is a secure digest of some protected data. Forging an HMAC is impossible without knowledge of the MAC secret. HMAC can be used with a variety of different hash algorithms: MD5 and SHA-1, denoting these as HMAC-MD5 (secret, data) and SHA-1 (secret, data). There are two differences between the SSLv3 scheme and the TLS MAC scheme: TLS makes use of the HMAC algorithm defined in RFC 2104 and the TLS master-secret computation is also different from that of SSLv3.

Chapter 10 describes e-mail security. PGP, invented by Philip Zimmermann, is widely used in both individual and commercial versions that run on a variety of platforms throughout the global computer community. PGP uses a combination of symmetric secret-key and asymmetric public-key encryption to provide security services for e-mail and data files. PGP also provides data integrity services for messages and data files using digital signatures, encryption, compression (ZIP), and radix-64 conversion (ASCII Armor). With growing reliance on e-mail and file storage, authentication and confidentiality services are becoming increasingly important. Multipurpose Internet Mail Extension (MIME) is an extension to the RFC 822 framework that defines a format for text messages sent using e-mail. MIME is actually intended to address some of the problems and limitations of the use of SMTP. S/MIME is a security enhancement to the MIME Internet e-mail format standard, based on the technology from RSA Data Security. Although both PGP and S/MIME are on an IETF standards track, it appears likely that PGP will remain the choice for personal e-mail security for many users, while S/MIME will emerge as the industry standard for commercial and organizational use. The two PGP and S/MIME schemes are covered in this chapter.

Chapter 11 discusses the topic of firewalls and intrusion detection systems (IDSs) as an effective means of protecting an internal system from Internet-based security threats: Internet Worm, Computer Virus, and Special Kinds of Viruses. The Internet Worm is a standalone program that can replicate itself through the network to spread, so it does not need to be attached. It makes the network performance weak by consuming bandwidth, increasing network traffic, or causing the Denial of Service (DoS). Morris worm, Blaster worm, Sasser worm, and Mydoom worm are some examples of the most notorious worms. The Computer Virus is a kind of malicious program that can damage the victim computer and spread itself to another computer. The word "Virus" is used for most of malicious programs. There are special kind of viruses such as Trojan horse, Botnet, and Key Logger. Trojan horse (or Trojan) is made to steal some information by social engineering. The term *Trojan horse* is derived from Greek mythology. The Trojan gives a cracker remote access permission, like the Greek soldiers, avoiding detection of their user. It looks like some useful or helpful program, or a legitimate access process, but it just steals password, card number, or other useful information. The popular Trojan horses are Netbus, Back Orifice, and Zeus. Botnet is a set of zombie computers connected to the Internet. Each compromised zombie computer is called as *bot*, and the botmaster, called as *C&C* (*Command & Control server*), controls these bots. Key logger program monitors the action of the key inputs. The key logger is of two types: software and hardware. This chapter is concerned with the software type only. It gets installed in the victim computers and logs all the strokes of keys. The logs are saved in some files or sent to the hacker by network. Key logger can steal the action of key input by kernel level, memory level, API level, packet level, and so on.

A firewall is a security gateway that controls access between the public Internet and a private internal network (or intranet). A firewall is an agent that screens network traffic in some way, blocking traffic it believes to be inappropriate, dangerous, or both. The security concerns that inevitably arise between the sometimes hostile Internet and secure intranets are often dealt with by inserting one or more firewalls on the path between the Internet and the internal network. In reality, Internet access provides benefits to individual users, government agencies, and most organizations. But this access often creates a security threat. Firewalls act as an intermediate server in handling SMTP and HTTP connections in either direction. Firewalls also require the use of an access negotiation and encapsulation protocol such as SOCKS to gain access to the Internet, intranet, or both. Many firewalls support tri-homing, allowing the use of a DMZ network. To design and configure a firewall, it needs to be familiar with some basic terminology such as a bastion host, proxy server, SOCKS, choke point, DMZ, logging and alarming, and VPN. Firewalls are classified into three main categories: packet filters, circuit-level gateways, and application-level gateways. In this chapter, each of these firewalls is examined in turn. Finally, this chapter discusses screened host firewalls and how to implement a firewall strategy. To provide a certain level of security, the three basic firewall designs are considered: a single-homed bastion host, a dual-homed bastion host, and a screened subnet firewall.

An IDS is a device or software application that monitors network or system activities for malicious activities or policy violations and produces reports to a Management Station. Intrusion detection and systems are primarily focused on identifying possible incidents, logging information about them, and reporting attempts. In addition, organizations use IDSs for other purposes, such as identifying problems with security policies, documenting existing threats, and deterring individuals from violating security policies. IDSs have become a necessary addition to the security infrastructure of nearly every organization. Regarding IDS, this chapter presents a survey and comparison of various IDSs including Internet Worm/Virus detection.

IDSs are categorized as Network-Based Intrusion Detection System (NIDS), Wireless Intrusion Detection System (WIDS), Network Behavior Analysis System (NBAS), Host-Based Intrusion Detection System (HIDS), Signature-Based Systems and Anomaly-Based Systems. An NIDS monitors network traffic for particular network segments or devices and analyzes network, transport, and application protocols to identify suspicious activity.

NIDSs typically perform most of their analysis at the application layer, such as HTTP, DNS, FTP, SMTP, and SNMP. They also analyze activity at the transport and network layers both to identify attacks at those layers and to facilitate the analysis of the application layer activity (e.g., a TCP port number may indicate which application is being used). Some NIDSs also perform limited analysis at the hardware layer. A WIDS monitors wireless network traffic and analyzes its wireless networking protocols to identify suspicious activity. The typical components in a WIDS are the same as an NIDS: consoles, database servers (optional), management servers, and sensors. However, unlike an NIDS sensor, which can see all packets on the networks it monitors, a WIDS sensor works by sampling traffic because it can only monitor a single channel at a time. An NBAS examines network traffic or statistics on network traffic to identify unusual traffic flows. NBA solutions usually have sensors and consoles, with some products also offering management servers. Some sensors are similar to NIDS sensors in that they sniff packets to monitor network activity on one or a few network segments. Other NBA sensors do not monitor the networks directly, and instead rely on network flow information provided by routers and other networking devices. HIDS monitors the characteristics of a single host and the events occurring within that host for suspicious activity. Examples of the types of characteristics an HIDS might monitor are wired and wireless network traffic, system logs, running processes, file access and modification, and system and application configuration changes. Most HIDSs have detection software known as *agents* installed on the hosts of interest. Each agent monitors activity on a single host and if prevention capabilities are enabled, also performs prevention actions. The agents transmit data to management servers. Each agent is typically designed to protect a server, a desktop or laptop, or an application service. A signature-based IDS is based on pattern matching techniques. The IDS contains a database of patterns. Some patterns are well known by public program or domain, for example, Snort (http://www.snort.org/), and some are found by signature-based IDS companies. Using database of already found signature is much like antivirus software. The IDS tries to match these signatures with the analyzed data. If a match is found, an alert is raised. An Anomaly-Based IDS is a system for detecting computer intrusions and misuse by monitoring system activity and classifying it as either normal or anomalous. The classification is based on heuristics or rules, rather than patterns or signatures, and will detect any type of misuse that falls out of normal system operation. This is as opposed to signature-based systems that can only detect attacks for which a signature has previously been created.

Chapter 12 covers the SET protocol designed for protecting credit card transactions over the Internet. The recent explosion in e-commerce has created huge opportunities for consumers, retailers, and financial institutions alike. SET relies on cryptography and X.509 v3 digital certificates to ensure message confidentiality, payment integrity, and identity authentication. Using SET, consumers and merchants are protected by ensuring that payment information is safe and can only be accessed by the intended recipient. SET combats the risk of transaction information being altered in transit by keeping information securely encrypted at all times and by using digital certificates to verify the identity of those accessing payment details. SET is the only Internet transaction protocol to provide security through authentication. Message data is encrypted with a random symmetric key that is then encrypted using the recipient's public key. The encrypted message, along with this digital envelope, is sent to the recipient. The recipient decrypts the digital envelope with a private key and then uses the symmetric key to recover the original message. SET addresses the anonymity of Internet shopping by using digital signatures and digital certificates to authenticate the banking relationships of cardholders and merchants. The process of ensuring secure payment card transactions on the Internet is fully explored in this chapter.

Chapter 13 deals with 4G Wireless Internet Communications Technology including Mobile WiMAX, WiBro, UMB, and LTE. WiMAX is a wireless communications standard designed to provide high-speed data communications for fixed and mobile stations. WiMAX far surpasses the 30-m wireless range of a conventional Wi-Fi LAN, offering a metropolitan area network with a signal radius of about 50 km. The name *WiMAX* was created by the WiMAX Forum, which was formed in June 2001 to promote conformity and interoperability of the standard. Mobile WiMAX (originally based on 802.16e-2005) is the revision that was deployed in many countries and is the basis of future revisions such as 802.16m-2011.

WiBro is a wireless broadband Internet technology developed by the South Korean telecoms industry. WiBro is the South Korean service name for IEEE 802.16e (mobile WiMAX) international standard. WiBro adopts TDD for duplexing, OFDMA for multiple access, and 8.75/10.00 MHz as a channel bandwidth. WiBro was devised to overcome the data rate limitation of mobile phones (for example, CDMA 1x) and to add mobility to broadband Internet access (for example, ADSL or WLAN). WiBro base stations will offer an aggregate data throughput of 30–50 Mbps per carrier and cover a radius of 1–5 km allowing for the use of portable Internet usage.

UMB was the brand name for a project within 3GPP2 (3rd Generation Partnership Project) to improve the CDMA2000 mobile phone standard for next-generation applications and requirements. In November 2008, Qualcomm, UMB's lead sponsor, announced it was ending development of the technology, favoring LTE instead. Like LTE, the UMB system was to be based on Internet (TCP/IP) networking technologies running over a next-generation radio system, with peak rates of up to 280 Mbps. Its designers intended for the system to be more efficient and capable of providing more services than the technologies it was intended to replace. To provide compatibility with the systems it was intended to replace, UMB was to support handoffs with other technologies including existing CDMA2000 1x and 1xEV-DO systems. However, 3GPP added this functionality to LTE, allowing LTE to become the single upgrade path for all wireless networks. No carrier had announced plans to adopt UMB, and most CDMA carriers in Australia, the United States, Canada, China, Japan, and South Korea have already announced plans to adopt either WiMAX or LTE as their 4G technology.

LTE, marketed as 4G LTE, is a standard for wireless communication of high-speed data for mobile phones and data terminals. It is based on the GSM/EDGE and UMTS/HSPA network technologies, increasing the capacity and speed using new modulation techniques. The standard is developed by the 3GPP. The world's first publicly available LTE service was launched by TeliaSonera in Oslo and Stockholm on 14 December 2009. LTE is the natural upgrade path for carriers with GSM/UMTS networks, but even CDMA holdouts such as Verizon Wireless, which launched the first large-scale LTE network in North America in 2010, and au by KDDI in Japan have announced they will migrate to LTE. LTE is, therefore, anticipated to become the first truly global mobile phone standard, although the use of different frequency bands in different countries will mean that only multiband phones will be able to utilize LTE in all countries where it is supported.

The scope of this book is adequate to span a one- or two-semester course at a senior or first-year graduate level. As a reference book, it will be useful to computer engineers, communications engineers, and system engineers. It is also suitable for self-study. The book is intended for use in both academic and professional circles, and it is also suitable for corporate training programs or seminars for industrial organizations as well as in research institutes. At the end of the book, there is a list of frequently used acronyms and a bibliography section.

About the Author

Man Young Rhee is an Endowed Chair Professor at the Kyung Hee University and has over 50 years of research and teaching experience in the field of communication technologies, coding theory, cryptography, and information security. His career in academia includes professorships at the Hanyang University (he also held the position of Vice President at this university), the Virginia Polytechnic Institute and State University, the Seoul National University, and the University of Tokyo. Dr. Rhee has held a number of high-level positions in both government and corporate sectors: President of Samsung Semiconductor Communications (currently, Samsung Electronics), President of Korea Telecommunications Company, Chairman of the Korea Information Security Agency at the Ministry of Information and Communication, President of the Korea Institute of Information Security & Cryptology (founding President), and Vice President of the Agency for Defense Development at the Ministry of National Defense. He is a Member of the National Academy of Sciences, a Senior Fellow at the Korea Academy of Science and Technology, and an Honorary Member of the National Academy of Engineering of Korea. His awards include the "Dongbaek" Order of National Service Merit and the "Mugunghwa" Order of National Service Merit, the highest grade honor for a scientist in Korea; the NAS Prize, the National Academy of Sciences; the NAEK Grand Prize, the National Academy of Engineering of Korea; and Information Security Grand Prize, KIISC. Dr. Rhee is the author of six books: *Error Correcting Coding Theory* (McGraw-Hill, 1989), *Cryptography and Secure Communications* (McGraw-Hill, 1994), *CDMA Cellular Mobile Communications and Network Security* (Prentice Hall, 1998), *Internet Security* (John Wiley, 2003), *Mobile Communication Systems and Security* (John Wiley, 2009), and *Wireless Mobile Internet Security*, Second Edition (John Wiley, 2013). Dr. Rhee has a B.S. in Electrical Engineering from the Seoul National University, as well as an M.S. in Electrical Engineering and a Ph.D. from the University of Colorado.

Acknowledgments

This book is the outgrowth of my teaching and research efforts in information security over the past 20 years at the Seoul National University and the Kyung Hee University. I thank all my graduate students, even if not by name. Special thanks go to Yoon Il Choi, Ho Cheol Lee, Ju Young Kim, and others of Samsung Electronics for collecting materials related to this book. Finally, I am grateful to my son Dr. Frank Chung-Hoon Rhee for editing and organizing the manuscript during my illness throughout the production process.

1

Internetworking and Layered Models

The Internet today is a widespread information infrastructure, but it is inherently an insecure channel for sending messages. When a message (or packet) is sent from one web site to another, the data contained in the message are routed through a number of intermediate sites before reaching their destination. The Internet was designed to accommodate heterogeneous platforms so that people who are using different computers and operating systems can communicate. The history of the Internet is complex and involves many aspects – technological, organizational, and community. The Internet concept has been a big step along the path toward electronic commerce, information acquisition, and community operations.

Early ARPANET researchers accomplished the initial demonstrations of packet-switching technology. In the late 1970s, the growth of the Internet was recognized and subsequently a growth in the size of the interested research community was accompanied by an increased need for a coordination mechanism. The Defense Advanced Research Projects Agency (DARPA) then formed an International Cooperation Board (ICB) to coordinate activities with some European countries centered on packet satellite research, while the Internet Configuration Control Board (ICCB) assisted DARPA in managing Internet activity. In 1983, DARPA recognized that the continuing growth of the Internet community demanded a restructuring of coordination mechanisms. The ICCB was disbanded and in its place the Internet Activities Board (IAB) was formed from the chairs of the Task Forces. The IAB revitalized the Internet Engineering Task Force (IETF) as a member of the IAB. By 1985, there was a tremendous growth in the more practical engineering side of the Internet. This growth resulted in the creation of a substructure to the IETF in the form of working groups. DARPA was no longer the major player in the funding of the Internet. Since then, there has been a significant decrease in Internet activity at DARPA. The IAB recognized the increasing importance of IETF, and restructured to recognize the Internet Engineering Steering Group (IESG) as the major standards review body. The IAB also restructured to create the Internet Research Task Force (IRTF) along with the IETF.

Wireless Mobile Internet Security, Second Edition. Man Young Rhee.
© 2013 John Wiley & Sons, Ltd. Published 2013 by John Wiley & Sons, Ltd.

Since the early 1980s, the Internet has grown beyond its primarily research roots, to include both a broad user community and increased commercial activity. This growth in the commercial sector brought increasing concern regarding the standards process. Increased attention was paid to making progress, eventually leading to the formation of the Internet Society in 1991. In 1992, the Internet Activities Board was reorganized and renamed the Internet Architecture Board (IAB) operating under the auspices of the Internet Society. The mutually supportive relationship between the new IAB, IESG, and IETF led to them taking more responsibility for the approval of standards, along with the provision of services and other measures which would facilitate the work of the IETF.

1.1   Networking Technology

Data signals are transmitted from one device to another using one or more types of transmission media, including twisted-pair cable, coaxial cable, and fiber-optic cable. A message to be transmitted is the basic unit of network communications. A message may consist of one or more cells, frames, or packets which are the elemental units for network communications. Networking technology includes everything from local area networks (LANs) in a limited geographic area such as a single building, department, or campus to wide area networks (WANs) over large geographical areas that may comprise a country, a continent, or even the whole world.

1.1.1   Local Area Networks (LANs)

A LAN is a communication system that allows a number of independent devices to communicate directly with each other in a limited geographic area such as a single office building, a warehouse, or a campus. LANs are standardized by three architectural structures: Ethernet, token ring, and fiber distributed data interface (FDDI).

Ethernet

Ethernet is a LAN standard originally developed by Xerox and later extended by a joint venture between Digital Equipment Corporation (DEC), Intel Corporation, and Xerox. The access mechanism used in an Ethernet is called *Carrier Sense Multiple Access with Collision Detection* (*CSMA/CD*). In CSMA/CD, before a station transmits data, it must check the medium where any other station is currently using the medium. If no other station is transmitting, the station can send its data. If two or more stations send data at the same time, it may result in a collision. Therefore, all stations should continuously check the medium to detect any collision. If a collision occurs, all stations ignore the data received. The sending stations wait for a period of time before resending the data. To reduce the possibility of a second collision, the sending stations individually generate a random number that determinates how long the station should wait before resending data.

Token Ring

Token ring, a LAN standard originally developed by IBM, uses a logical ring topology. The access method used by CSMA/CD may result in collisions. Therefore, stations may attempt to send data many times before a transmission captures a perfect link. This redundancy can create delays of indeterminable length if traffic is heavy. There is no way to predict either the occurrence of collisions or the delays produced by multiple stations attempting to capture the link at the same time. Token ring resolves this uncertainty by making stations take turns in sending data.

As an access method, the token is passed from station to station in sequence until it encounters a station with data to send. The station to be sent data waits for the token. The station then captures the token and sends its data frame. This data frame proceeds around the ring and each station regenerates the frame. Each intermediate station examines the destination address, finds that the frame is addressed to another station, and relays it to its neighboring station. The intended recipient recognizes its own address, copies the message, checks for errors, and changes four bits in the last byte of the frame to indicate that the address has been recognized and the frame copied. The full packet then continues around the ring until it returns to the station that sent it.

Fiber Distributed Data Interface (FDDI)

FDDI is a LAN protocol standardized by the ANSI (American National Standards Institute) and the ITU-T. It supports data rates of 100 Mbps and provides a high-speed alternative to Ethernet and token ring. When FDDI was designed, the data rate of 100 Mbps required fiber-optic cable.

The access method in FDDI is also called *token passing*. In a token ring network, a station can send only one frame each time it captures the token. In FDDI, the token passing mechanism is slightly different in that access is limited by time. Each station keeps a timer which shows when the token should leave the station. If a station receives the token earlier than the designated time, it can keep the token and send data until the scheduled leaving time. On the other hand, if a station receives the token at the designated time or later than this time, it should let the token pass to the next station and wait for its next turn.

FDDI is implemented as a dual ring. In most cases, data transmission is confined to the primary ring. The secondary ring is provided in case of the primary ring's failure. When a problem occurs on the primary ring, the secondary ring can be activated to complete data circuits and maintain service.

1.1.2 Wide Area Networks (WANs)

A WAN provides long-distance transmission of data, voice, image, and video information over large geographical areas that may comprise a country, a continent, or even the world. In contrast to LANs (which depend on their own hardware for transmission), WANs can utilize public, leased, or private communication devices, usually in combination.

PPP

The Point-to-Point Protocol (PPP) is designed to handle the transfer of data using either asynchronous modem links or high-speed synchronous leased lines. The PPP frame uses the following format:

- *Flag field.* Each frame starts with a 1-byte flag whose value is 7E(0111 1110). The flag is used for synchronization at the bit level between the sender and receiver.

- *Address field.* This field has the value of FF(1111 1111).

- *Control field.* This field has the value of 03(0000 0011).

- *Protocol field.* This is a 2-byte field whose value is 0021(0000 0000 0010 0001) for TCP/IP (Transmission Control Protocol/Internet Protocol).

- *Data field.* The data field ranges up to 1500 bytes.

- *CRC.* This is a 2-byte cyclic redundancy check (CRC). CRC is implemented in the physical layer for use in the data link layer. A sequence of redundant bits (CRC) is appended to the end of a data unit so that the resulting data unit becomes exactly divisible by a predetermined binary number. At its destination, the incoming data unit is divided by the same number. If there is no remainder, the data unit is accepted. If a remainder exists, the data unit has been damaged in transit and therefore must be rejected.

X.25

X.25 is widely used, as the packet-switching protocol provided for use in a WAN. It was developed by the ITU-T in 1976. X.25 is an interface between data terminal equipment and data circuit terminating equipment for terminal operations at the packet mode on a public data network.

X.25 defines how a packet mode terminal can be connected to a packet network for the exchange of data. It describes the procedures necessary for establishing connection, data exchange, acknowledgment, flow control, and data control.

Frame Relay

Frame relay is a WAN protocol designed in response to X.25 deficiencies. X.25 provides extensive error-checking and flow control. Packets are checked for accuracy at each station to which they are routed. Each station keeps a copy of the original frame until it receives confirmation from the next station that the frame has arrived intact. Such station-to-station checking is implemented at the data link layer of the Open Systems Interconnect (OSI) model, but X.25 only checks for errors from source to receiver at the network layer. The source keeps a copy of the original packet until it receives confirmation from the final destination. Much of the traffic on an X.25 network is devoted to error-checking to ensure reliability of service. Frame relay does not provide error-checking or require acknowledgment in the data link layer. Instead, all error-checking is left to the protocols at the network and transport layers, which use the frame relay service. Frame relay only operates at the physical and data link layer.

Asynchronous Transfer Mode (ATM)

Asynchronous transfer mode (ATM) is a revolutionary idea for restructuring the infrastructure of data communication. It is designed to support the transmission of data, voice, and video through a high-data-rate transmission medium such as fiber-optic cable. ATM is a protocol for transferring cells. A cell is a small data unit of 53 bytes long, made of a 5-byte header and a 48-byte payload. The header contains a virtual path identifier (VPI) and a virtual channel identifier (VCI). These two identifiers are used to route the cell through the network to the final destination.

An ATM network is a connection-oriented cell switching network. This means that the unit of data is not a packet as in a packet-switching network, or a frame as in a frame relay, but a cell. However, ATM, like X.25 and frame relay, is a connection-oriented network, which means that before two systems can communicate, they must make a connection. To startup a connection, a system uses a 20-byte address. After the connection is established, the combination of VPI/VCI leads a cell from its source to its final destination.

1.2   Connecting Devices

Connecting devices are used to connect the segments of a network together or to connect networks to create an internetwork. These devices are classified into five categories: switches, repeaters, bridges, routers, and gateways. Each of these devices except the first one (switches) interacts with protocols at different layers of the OSI model.

Repeaters forward all electrical signals and are active only at the physical layer. Bridges store and forward complete packets and affect the flow control of a single LAN. Bridges are active at the physical and data link layers. Routers provide links between two separate LANs and are active in the physical, data link, and network layers. Finally, gateways provide translation services between incompatible LANs or applications, and are active in all layers.

Connection devices that interact with protocols at different layers of the OSI model are shown in Figure 1.1.

1.2.1   Switches

A switched network consists of a series of interlinked switches. Switches are hardware/ software devices capable of creating temporary connections between two or more devices to the switch but not to each other. Switching mechanisms are generally classified into three methods: circuit switching, packet switching, and message switching.

- Circuit switching creates a direct physical connection between two devices such as telephones or computers. Once a connection is made between two systems, circuit switching creates a dedicated path between two end users. The end users can use the path for as long as they want.

- Packet switching is one way to provide a reasonable solution for data transmission. In a packet-switched network, data are transmitted in discrete units of variable-length blocks called *packets*. Each packet contains not only data, but also a header with control information. The packets are sent over the network node to node. At each node, the packet is stored briefly before being routed according to the information in its header.

- In the datagram approach to packet switching, each packet is treated independently of all others as though it exists alone. In the virtual circuit approach to packet switching, if a single route is chosen between sender and receiver at the beginning of the session, all packets travel one after another along that route. Although these two approaches seem the same, there exists a fundamental difference between them. In circuit switching, the path between the two end users consists of only one channel. In the virtual circuit, the line is not dedicated to two users. The line is divided into channels and each channel can use one of the channels in a link.

- Message switching is known as the *store and forwarding method*. In this approach, a computer (or a node) receives a message, stores it until the appropriate route is free, then sends it out. This method has now been phased out.

1.2.2 Repeaters

A repeater is an electronic device that operates on the physical layer only of the OSI model. A repeater boosts the transmission signal from one segment and continues the signal to another segment. Thus, a repeater allows us to extend the physical length of a network. Signals that carry information can travel a limited distance within a network before degradation of the data integrity due to noise. A repeater receives the signal before attenuation, regenerates the original bit pattern, and puts the restored copy back on to the link.

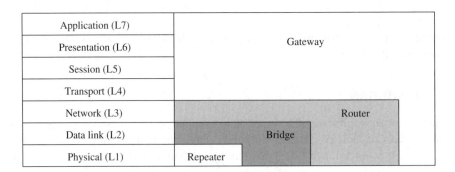

Figure 1.1 Connecting devices.

1.2.3 Bridges

Bridges operate in both the physical and the data link layers of the OSI model. A single bridge connects different types of networks together and promotes interconnectivity between networks. Bridges divide a large network into smaller segments. Unlike repeaters, bridges contain logic that allows them to keep separate the traffic for each segment. Bridges are smart enough to relay a frame toward the intended recipient so that traffic can be filtered. In fact, this filtering operation makes bridges useful for controlling congestion, isolating problem links, and promoting security through this partitioning of traffic.

A bridge can access the physical addresses of all stations connected to it. When a frame enters a bridge, the bridge not only regenerates the signal but also checks the address of the destination and forwards the new copy to the segment to which the address belongs. When a bridge encounters a packet, it reads the address contained in the frame and compares that address with a table of all the stations on both segments. When it finds a match, it discovers to which segment the station belongs and relays the packet to that segment only.

1.2.4 Routers

Routers operate in the physical, data link, and network layers of the OSI model. The Internet is a combination of networks connected by routers. When a datagram goes from a source to a destination, it will probably pass through many routers until it reaches the router attached to the destination network. Routers determine the path a packet should take. Routers relay packets among multiple interconnected networks. In particular, an IP router forwards IP datagrams among the networks to which it connects. A router uses the destination address on a datagram to choose a next-hop to which it forwards the datagram. A packet sent from a station on one network to a station on a neighboring network goes first to a jointly held router, which switches it over the destination network. In fact, the easiest way to build the Internet is to connect two or more networks with a router. Routers provide connections to many different types of physical networks: Ethernet, token ring, point-to-point links, FDDI, and so on.

- The routing module receives an IP packet from the processing module. If the packet is to be forwarded, it should be passed to the routing module. It finds the IP address of the next station along with the interface number from which the packet should be sent. It then sends the packet with information to the fragmentation module. The fragmentation module consults the MTU table to find the maximum transfer unit (MTU) for the specific interface number.

- The routing table is used by the routing module to determine the next-hop address of the packet. Every router keeps a routing table that has one entry for each destination network. The entry consists of the destination network IP address, the shortest distance to reach the destination in hop count, and the next router (next-hop) to which the packet should be delivered to reach its final destination. The hop count is the number of networks a packet enters to reach its final destination. A router should have a routing table to consult when a packet is ready to be forwarded. The routing table should specify the optimum path for the packet. The table can be either static or dynamic. A static table is one that is not changed frequently, but a dynamic table is one that is updated automatically when there is a change somewhere in the Internet. Today, the Internet needs dynamic routing tables.

- A metric is a cost assigned for passing through a network. The total metric of a particular router is equal to the sum of the metrics of networks that comprise the route. A router chooses the route with the shortest (smallest value) metric. The metric assigned to each network depends on the type of protocol. The Routing Information Protocol (RIP) treats each network as 1 hop count. So if a packet passes through 10 networks to reach the destination, the total cost is 10 hop counts. The Open Shortest Path First (OSPF) protocol allows the administrator to assign a cost for passing through a network based on the type of service required. A route through a network can have different metrics (costs). OSPF allows each router to have several routing tables based on the required type of service. The Border Gateway Protocol (BGP) defines the metric totally differently. The policy criterion in BGP is set by the administrator. The policy defines the paths that should be chosen.

1.2.5　Gateways

Gateways operate over the entire range in all seven layers of the OSI model. Internet routing devices have traditionally been called *gateways*. A gateway is a protocol converter which connects two or more heterogeneous systems and translates among them. The gateway thus refers to a device that performs protocol translation between devices. A gateway can accept a packet formatted for one protocol and convert it to a packet formatted for another protocol before forwarding it. The gateway understands the protocol used by each network linked into the router and is therefore able to translate from one to another.

1.3　The OSI Model

The Ethernet, originally called the *Alto Aloha network*, was designed by the Xerox Palo Alto Research Center in 1973 to provide communication for research and development CP/M computers. When in 1976 Xerox started to develop the Ethernet as a 20 Mbps product, the network prototype was called the *Xerox Wire*. In 1980, when the Digital, Intel, and Xerox standard was published to make it a LAN standard at 10 Mbps, Xerox Wire changed its name back to Ethernet. Ethernet became a commercial product in 1980 at 10 Mbps. The IEEE called its Ethernet 802.3 standard CSMA/CD. As the 802.3 standard evolved, it has acquired such names as Thicknet (IEEE 10Base-5), Thinnet or Cheapernet (10Base-2), Twisted Ethernet (10Base-T), and Fast Ethernet (100Base-T).

The design of Ethernet preceded the development of the seven-layer OSI model. The OSI model was developed and published in 1982 by the International Organization for

Standardization (ISO) as a generic model for data communication. The OSI model is useful because it is a broadly based document, widely available and often referenced. Since modularity of communication functions is a key design criterion in the OSI model, vendors who adhere to the standards and guidelines of this model can supply Ethernet-compatible devices, alternative Ethernet channels, higher-performance Ethernet networks, and bridging protocols that easily and reliably connect other types of data network to Ethernet.

Since the OSI model was developed after Ethernet and Signaling System #7 (SS7), there are obviously some discrepancies between these three protocols. Yet the functions and processes outlined in the OSI model were already in practice when Ethernet or SS7 was developed. In fact, SS7 networks use point-to-point configurations between signaling points. Due to the point-to-point configurations and the nature of the transmissions, the simple data link layer does not require much complexity.

The OSI reference model specifies the seven layers of functionality, as shown in Figure 1.2. It defines the seven layers from the physical layer (which includes the network adapters), up to the application layer, where application programs can access network services. However, the OSI model does not define the protocols that implement the functions at each layer. The OSI model is still important for compatibility, protocol independence, and the future growth of network technology. Implementations of the OSI model stipulate communication between layers on two processors and an interface for interlayer communication on one processor. Physical communication occurs only at layer 1. All other layers communicate downward (or upward) to lower (or higher) levels in steps through protocol stacks.

The following briefly describes the seven layers of the OSI model:

1. *Physical layer.* The physical layer provides the interface with physical media. The interface itself is a mechanical connection from the device to the physical medium used to transmit the digital bit stream. The mechanical specifications do not specify the electrical characteristics of the interface, which will depend on the medium being used and the type of interface. This layer is responsible for converting the digital data into a bit stream for transmission over the network. The physical layer includes the method of connection used between the network cable and the network adapter, as well as the basic communication stream of data bits over the network cable. The physical layer is responsible for the conversion of the digital data into a bit stream for transmission when using a device such as a modem, and even light, as in fiber optics. For example, when using a modem, digital signals are converted into analog-audible tones which are then transmitted at varying frequencies over the telephone line. The OSI model does not specify the medium, only the operative functionality for a standardized communication protocol. The transmission media layer specifies the physical medium used in constructing the network, including size, thickness, and other characteristics.

2. *Data link layer.* The data link layer represents the basic communication link that exists between computers and is responsible for sending frames or packets of data without errors. The software in this layer manages transmissions, error acknowledgment, and recovery. The transceivers are mapped data units to data units to provide physical error

Layer No.	OSI layer	Functionality
7	Application	• Provides user interface • System computing and user application process • Of the many application services, this layer provides support for services such as e-mail, remote file access and transfer, message handling services (X.400) to send an e-mail message, directory services (X.500) for distributed database sources, and access for global information about various objects and services
6	Presentation	• Data interpretation (compression, encryption, formatting, and syntax selection) and code transformations
5	Session	• Administrative control of transmissions and transfers between nodes • Dialog control between two systems • Synchronization process by inserting checkpoints into data stream
4	Transport	• Source-to-destination delivery of entire message • Message segmentation at the sending layer and reassembling at the receiving layer • Transfer control by either connectionless or connection-oriented mechanism for delivering packets • Flow control for end-to-end services • Error control based on performing end-to-end rather than a single link
3	Network	• Source-to-destination delivery of individual packets • Routing or switching packets to final destination • Logical addressing to help distinguish the source/destination systems
2	Data Link	• Framing, physical addressing, data flow control, access control, and error control
1	Physical	• Physical control of the actual data circuit (electrical, mechanical, and optical)

Figure 1.2   ISO/OSI model.

detection and notification and link activation/deactivation of a logical communication connection. Error control refers to mechanisms to detect and correct errors that occur in the transmission of data frames. Therefore, this layer includes error correction, so when a packet of data is received incorrectly, the data link layer makes system send the data again. The data link layer is also defined in the IEEE 802.2 logical link control specifications.

Data link control protocols are designed to satisfy a wide variety of data link requirements:

- High-Level Data Link Control (HDLC) developed by the ISO (ISO 3309, ISO 4335),
- Advanced Data Communication Control Procedures (ADCCP) developed by the ANSI (ANSI X3.66),

- Link Access Procedure, Balanced (LAP-B) adopted by the CCITT as part of its X.25 packet-switched network standard,
- Synchronous Data Link Control (SDLC) is not a standard, but is in widespread use. There is practically no difference between HDLC and ADCCP. Both LAP-B and SDLC are subsets of HDLC, but they include several additional features.

3. *Network layer.* The network layer is responsible for data transmission across networks. This layer handles the routing of data between computers. Routing requires some complex and crucial techniques for a packet-switched network design. To accomplish the routing of packets sending from a source and delivering to a destination, a path or route through the network must be selected. This layer translates logical network addressing into physical addresses and manages issues such as frame fragmentation and traffic control. The network layer examines the destination address and determines the link to be used to reach that destination. It is the borderline between hardware and software. At this layer, protocol mechanisms activate data routing by providing network address resolution, flow control in terms of segmentation, and blocking and collision control (Ethernet). The network layer also provides service selection, connection resets, and expedited data transfers. The IP runs at this layer.

   The IP was originally designed simply to interconnect as many sites as possible without undue burdens on the type of hardware and software at different sites. To address the shortcomings of the IP and to provide a more reliable service, the TCP is stacked on top of the IP to provide end-to-end service. This combination is known as *TCP/IP* and is used by most Internet sites today to provide a reliable service.

4. *Transport layer.* The transport layer is responsible for ensuring that messages are delivered error-free and in the correct sequence. This layer splits messages into smaller segments if necessary, and provides network traffic control of messages. Traffic control is a technique for ensuring that a source does not overwhelm a destination with data. When data is received, a certain amount of processing must take place before the buffer is clear and ready to receive more data. In the absence of flow control, the receiver's buffer may overflow while it is processing old data. The transport layer, therefore, controls data transfer and transmission. This software is called *TCP*, common on most Ethernet networks, or System Packet Exchange (SPE), a corresponding Novell specification for data exchange. Today, most Internet sites use the TCP/IP protocol along with the Internet Control Message Protocol (ICMP) to provide a reliable service.

5. *Session layer.* The session layer controls the network connections between the computers in the network. The session layer recognizes nodes on the LAN and sets up tables of source and destination addresses. It establishes a handshake for each session between different nodes. Technically, this layer is responsible for session connection (i.e., for creating, terminating, and maintaining network sessions), exception reporting, coordination of send/receive modes, and data exchange.

6. *Presentation layer.* The presentation layer is responsible for the data format, which includes the task of hashing the data to reduce the number of bits (hash code) that will be transferred. This layer transfers information from the application software to the

network session layer to the operating system. The interface at this layer performs data transformations, data compression, data encryption, data formatting, syntax selection (i.e., ASCII, EBCDIC, or other numeric or graphic formats), and device selection and control. It actually translates data from the application layer into the format used when transmitting across the network. On the receiving end, this layer translates the data back into a format that the application layer can understand.

7. *Application layer.* The application layer is the highest layer defined in the OSI model and is responsible for providing user-layer applications and network management functions. This layer supports identification of communicating partners, establishes authority to communicate, transfers information, and applies privacy mechanisms and cost allocations. It is usually a complex layer with a client/server, a distributed database, data replication, and synchronization. The application layer supports file services, print services, remote login, and e-mail. The application layer is the network system software that supports user-layer applications, such as Word or data processing, CAD/CAM, document storage, and retrieval and image scanning.

1.4 TCP/IP Model

A protocol is a set of rules governing the way data will be transmitted and received over data communication networks. Protocols are then the rules that determine everything about the way a network operates. Protocols must provide reliable, error-free communication of user data as well as a network management function. Therefore, protocols govern how applications access the network, the way that data from an application is divided into packets for transmission through cable, and which electrical signals represent data on a network cable.

The OSI model, defined by a seven-layer architecture, is partitioned into a vertical set of layers, as illustrated in Figure 1.2. The OSI model is based on open systems and peer-to-peer communications. Each layer performs a related subset of the functions required to communicate with another system. Each system contains seven layers. If a user or application entity A wishes to send a message to another user or application entity B, it invokes the application layer (layer 7). Layer 7 (corresponding to application A) establishes a peer relationship with layer 7 of the target machine (application B), using a layer 7 protocol.

In an effort to standardize a way of looking at network protocols, the TCP/IP four-layer model is created with reference to the seven-layer OSI model, as shown in Figure 1.3. The protocol suite is designed in distinct layers to make it easier to substitute one protocol for another. The protocol suite governs how data is exchanged above and below each protocol layer. When protocols are designed, specifications set out how a protocol exchanges data with a protocol layered above or below it.

Both the OSI model and the TCP/IP layered model are based on many similarities, but there are philosophical and practical differences between the two models. However, they both deal with communications among heterogeneous computers.

Electronic payment system	Internet security
E-cash, Mondex, Proton, Visa Cash, SET, CyberCash, CyberCoin, E-check, First Virtual	SSL, TLS, S/HTTP, IPsec, SOCKS V5, PEM, PGP, S/MIME

OSI model (seven layers)	TCP/IP model (four layers)	Internet protocol suite
Application Presentation	Application	HTTP, FTP, TFTP, NFS, RPC, XDR, SMTP, POP, IMAP, MIME, SNMP, DNS, RIP, OSPF, BGP, TELNET, Rlogin
Session Transport	Transport	TCP, UDP
Network	Internet	IP, ICMP, IGMP, ARP, RARP
Data link Physical	Network access	Ethernet, token ring, FDDI, PPP, X.25, frame replay, ATM

Figure 1.3 The TCP/IP model and Internet protocol suite.

Since TCP was developed before the OSI model, the layers in the TCP/IP model do not exactly match those in the OSI model. The important fact is the hierarchical ordering of protocols. The TCP/IP model is made up of four layers: application layer, transport layer, Internet layer, and network access layer. These will be discussed below.

1.4.1 Network Access Layer

The network access layer contains protocols that provide access to a communication network. At this layer, systems are interfaced to a variety of networks. One function of this layer is to route data between hosts attached to the same network. The services to be provided are flow control and error control between hosts. The network access layer is invoked either by the Internet layer or the application layer. This layer provides the device drivers that support interactions with communications hardware such as the token ring or Ethernet. The IEEE token ring, referred to as the *Newhall ring*, is probably the oldest ring control technique and has become the most popular ring access technique in the United States. The FDDI is a standard for a high-speed ring LAN. Like the IEEE 802 standard, FDDI employs the token ring algorithm.

1.4.2   Internet Layer

The Internet layer provides a routing function. Therefore, this layer consists of the procedures required within hosts and gateways to allow data to traverse multiple networks. A gateway connecting two networks relays data between networks using an internetwork protocol. This layer consists of the IP and the ICMP.

1.4.3   Transport Layer

The transport layer delivers data between two processes on different host computers. A protocol entity at this level provides a logical connection between higher-level entities. Possible services include error and flow controls and the ability to deal with control signals not associated with a logical data connection. This layer contains the TCP and the User Datagram Protocol (UDP).

1.4.4   Application Layer

This layer contains protocols for resource sharing and remote access. The application layer actually represents the higher-level protocols that are used to provide a direct interface with users or applications. Some of the important application protocols are File Transfer Protocol (FTP) for file transfers, Hypertext Transfer Protocol (HTTP) for the World Wide Web, and Simple Network Management Protocol (SNMP) for controlling network devices. The Domain Naming Service (DNS) is also useful because it is responsible for converting numeric IP addresses into names that can be more easily remembered by users. Many other protocols dealing with the finer details of applications are included in this application layer. These include Simple Mail Transport Protocol (SMTP), Post Office Protocol (POP), Internet Message Access Protocol (IMAP), ICMP for e-mail, Privacy Enhanced Mail (PEM), Pretty Good Privacy (PGP), and Secure/Multipurpose Internet Mail Extension (S/MIME) for e-mail security. All protocols contained in the TCP/IP suite are fully described in Chapter 2.

2

TCP/IP Suite and Internet Stack Protocols

The Internet protocols consist of a suite of communication protocols, of which the two best known are the Transmission Control Protocol (TCP) and the Internet Protocol (IP). The TCP/IP suite includes not only lower-layer protocols (TCP, UDP, IP, ARP (Address Resolution Protocol), RARP (Reverse Address Resolution Protocol), ICMP, and IGMP (Internet Group Management Protocol)), but also specifies common applications such as the World Wide Web, e-mail, domain naming service, login, and file transfer. Figure 1.3 depicts many of the protocols of the TCP/IP suite and their corresponding OSI layer.

It may not be important for the novice to understand the details of all protocols, but it is important to know which protocols exist, how they can be used, and where they belong in the TCP/IP suite.

This chapter addresses various layered protocols in relation to Internet security, and shows which are available for use with which applications.

2.1 Network Layer Protocols

At the network layer in the OSI model, TCP/IP supports the IP. IP contains four supporting protocols: ARP, RARP, ICMP, and IGMP. Each of these protocols is described below.

2.1.1 Internet Protocol (IP)

The IP is a network layer (layer 3 in the OSI model or the Internet layer in the TCP/IP model) protocol which contains addressing information and some control information to enable packets to be controlled. IP is well documented in RFC 791 and is the basic communication protocol in the Internet protocol suite.

IP specifies the exact format of all data as it passes across the Internet. IP software performs the routing function, choosing the path over which data will be sent. IP includes

Wireless Mobile Internet Security, Second Edition. Man Young Rhee.
© 2013 John Wiley & Sons, Ltd. Published 2013 by John Wiley & Sons, Ltd.

a set of rules that embody the idea of unreliable packet delivery. IP is an unreliable and connectionless datagram protocol. The service is called unreliable because delivery is not guaranteed. The service is called connectionless because each packet is treated independently from all others. If reliability is important, IP must be paired with a reliable protocol such as TCP. However, IP does its best to get a transmission through to its destination, but carries no guarantees.

IP transports the datagram in packets, each of which is transported separately. Datagrams can travel along different routes and can arrive out of sequence or be duplicated. IP does not keep track of the routes taken and has no facility for reordering datagrams once they arrive at their destination. In short, the packet may be lost, duplicated, delayed or delivered out of order.

IP is a connectionless protocol designed for a packet switching network which uses the datagram mechanism. This means that each datagram is separated into segments (packets) and is sent independently following a different route to its destination. This implies that if a source sends several datagrams to the same destination, they could arrive out of order. Even though IP provides limited functionality, it should not be considered a weakness. Figure 2.1 shows the format of an IP datagram. Since datagram processing occurs in software, the content of an IP datagram is not constrained by any hardware.

IP Datagrams

Packets in the IP layer are called *datagrams*. Each IP datagram consists of a header (20–60 bytes) and data. The IP datagram header consists of a fixed 20-byte section and a variable options section with a maximum of 40 bytes. The Internet header length (HLEN) is the total length of the header, including any option fields, in 32-bit words. The minimum value for the Internet HLEN is 5 (five 32-bit words or 20 bytes of the IPv4 header).

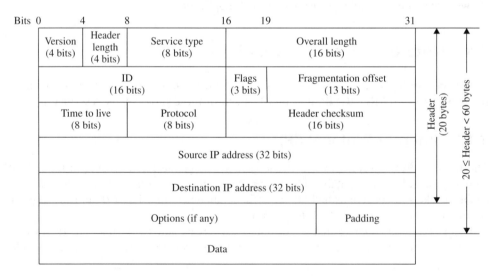

Figure 2.1 IP datagram format.

The maximum permitted length of an IP datagram is 65 536 bytes. However, such large packets would not be practical, particularly on the Internet where they would be heavily fragmented. RFC 791 states that all hosts must accept IP datagrams up to 576 bytes. An IPv4 datagram consists of three primary components. The header is 20 bytes long and contains a number of fields. The option is a variable-length set of fields, which may or may not be present. Data is the encapsulated payload from the higher level, usually a whole TCP segment or UDP datagram. The datagram header contains the source and destination IP addresses, fragmentation control, precedence, a checksum used to detect transmission errors, and IP options to record routing information or gathering timestamps. A brief explanation of each field in an IP datagram is described below:

- *Version* (*VER, 4 bits*). Version 4 of the IP (IPv4) has been in use since 1981, but Version 6 (IPv6 or IPng) will soon replace it. The first 4-bit field in a datagram contains the version of the IP that was used to create the datagram. It is used to verify that the sender, receiver, and any routers in-between them agree on the format of datagram. In fact, this field is an indication to the IP software running in the processing machine that it is required to check the version field before processing a datagram to ensure it matches the format the software expects.

- *HLEN* (*4 bits*). This 4-bit field defines the total length of the IPv4 datagram header measured in 32-bit words. This field is needed because the length of the header varies between 20 and 60 bytes. All fields in the header have fixed lengths except for the IP options and corresponding padding field.

- *Type of service* (*TOS, 8 bits*). This 8-bit field specifies how the datagram should be handled by the routers. This TOS field is divided into two subfields: precedence (3 bits) and TOS (5 bits) as shown in Figure 2.2. *Precedence* is a 3-bit subfield with values ranging from 0 (000 in binary, normal precedence) to 7 (111 in binary, network control), allowing senders to indicate the importance of each datagram. Precedence defines the priority of the datagram in issues such as congestion. If a router is congested and needs to discard some datagrams, those datagrams with lowest precedence are discarded first. A datagram in the Internet used for network management is much more important than a datagram used for sending optional information to a group of users. Many routers use a precedence value of 6 or 7 for routing traffic to make it possible for routers to exchange

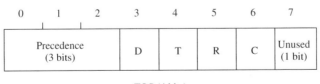

TOS (4 bits)

D : Minimize delay (1000)   R : Maximize reliability (0010)
T : Maximize throughput (0100)   C : Minimize cost (0001)

Figure 2.2   The 8-bit service type field.

routing information even when networks are congested. At present, the precedence subfield is not used in version 4, but it is expected to be functional in future versions.

The TOS field is a 5-bit subfield, each bit having a special meaning. Bits D, T, R, and C specify the type of transport desired for the datagram. When they are set, the D bit requests low delay, the T bit requests high throughput, the R bit requests high reliability, and the C bit requires low cost. Of course, it may not be possible for the Internet to guarantee the type of transport requested. Therefore, the transport request may be thought of as a hint to the routing algorithms, not as a demand. Datagrams carrying keystrokes from a user to a remote computer could set the D bit to request that they be delivered as quickly as possible, while datagrams carrying a bulk file transfer could have the T bit set requesting that they travel across the high-capacity path.

Although a bit in TOS bits can be either 0 or 1, only 1 bit can have the value 1 in each datagram. The bit patterns and their descriptions are given in Table 2.1.

In the late 1990s, the Internet Engineering Task Force (IETF) redefined the meaning of the 8-bit service type field to accommodate a set of differentiated services (DSs). The DS defines that the first 6 bits comprise a codepoint and the last 2 bits are left unused. A codepoint value maps to an underlying service through an array of pointers. Although it is possible to design 64 separate services, designers suggest that a given router will only have a few services, and multiple codepoints will map to each service. When the last 3 bits of the codepoint field contains zero, the precedence bits define eight broad classes of service that adhere to the same guidelines as the original definition. When the last three bits are zero, the router must map a codepoint with precedence 6 or 7 into the higher-priority class and other codepoint values into the lower-priority class.

- *Overall length (16 bits)*. The IPv4 datagram format allots 16 bits to the total length field, limiting the datagram to at most 65 535 bytes. This 16-bit field defines the total length (header plus data) of the IP datagram in bytes. To find the data length coming from the upper layer, subtract the HLEN from the total length. Since the field length is 16 bits, the total length of the IP datagram is limited to $2^{16} - 1 = 65\ 535$ bytes, of which 20–60 bytes are the header and the rest are data from the upper layer. In practice, some physical networks are unable to encapsulate a datagram of 65 535 bytes in the process of fragmentation.

- *Identification (ID, 16 bits)*. This 16-bit field specifies to identify a datagram originating from the source host. The ID field is used to help a destination host to reassemble a fragmented packet. It is set by the sender and uniquely identifies a specific IP datagram

Table 2.1 Type of service (TOS)

TOS bit	Description
0000	Normal (default)
0001	Minimize cost
0010	Maximize reliability
0100	Maximize throughput
1000	Minimize delay

sent by a source host. The combination of the identification and source IP address must uniquely define the same datagram as it leaves the source host. To guarantee uniqueness, the IP uses a counter to label the datagrams. When a datagram is fragmented, the value in the identification field is copied in all fragments. Hence, all fragments have the same identification number, which is the same as in the original datagram. The identification number helps the destination in reassembling the datagram. RFC 791 suggests that the ID number is set by the higher-layer protocol, but in practice it tends to be set by IP.

- *Flags (3 bits).* This 3-bit field is used in fragmentation. The flag field is 3 bits long. Bit 0: Reserved, Bit 1: May fragment or may not fragment, Bit 2: Last fragment or more fragments. The first bit is reserved. The second bit is called the *don't fragment* bit. If its value is 1, don't fragment the datagram. If it cannot pass the datagram through any available physical network, it discards the datagram and sends an ICMP error message to the source host. The third bit is called the *more fragment* bit. If its value is 1, it means the datagram is not the last fragment; there are more fragments to come. If its value is 0, it means that it is the last or only fragment.

- *Fragmentation offset (13 bits).* The small pieces into which a datagram is divided are called *fragments*, and the process of dividing a datagram is known as *fragmentation*. This 13-bit field denotes an offset to a nonfragmented datagram, used to reassemble a datagram that has become fragmented. This field shows the relative position of each fragment with respect to the whole datagram. The offset states where the data in a fragmented datagram should be placed in the datagram being reassembled. The offset value for each fragment of a datagram is measured in units of 8 bytes, starting at offset zero. Since the length of the offset field is only 13 bits, it cannot represent a sequence of bytes greater than $2^{13} - 1 = 8191$.

   Suppose a datagram with a data size of $x < 8191$ bytes is fragmented into i fragments. The bytes in the original datagram are numbered from 0 to $(x - 1)$ bytes. If the first fragment carries bytes from 0 to x_1, then the offset for this fragment is $0/8 = 0$. If the second fragment carries $(x_1 + 1)$ bytes to x_2 bytes, then the offset value for this fragment is $(x_1 + 1)/8$. If the third fragment carries bytes $x_2 + 1$ to x_3, then the offset value for the third fragment is $(x_2 + 1)/8$. Continue this process within the range under 8191 bytes. Thus, the offset value for these fragments is 0, $(x_{i-1} + 1)/8, i = 2, 3, \ldots$. Consider what happens if a fragment itself is fragmented. In this case the value of the offset field is always relative to the original datagram.

   Fragment size is chosen such that each fragment can be sent across the network in a single frame. Since IP represents the offset of the data in multiples of 8 bytes, the fragment size must be chosen to be a multiple of 8. Of course, choosing the multiple of 8 bytes nearest to the network's maximum transfer unit (MTU) does not usually divide the datagram into equal-sized fragments; the last piece or fragment is often shorter than the others. The MTU is the maximum size of a physical packet on the network. If datagram, including the 20-byte IP header, to be transmitted is greater than the MTU, then the datagram is fragmented into several small fragments. To reassemble the datagram, the destination must obtain all fragments starting with the fragment that has offset 0 through the fragment with the highest offset.

- *Time to live* (*TTL, 8 bits*). A datagram should have a limited lifetime in its travel through an Internet. This 8-bit field specifies how long (in seconds) the datagram is allowed to remain in the Internet.

    Routers and hosts that process datagrams must decrement this TTL field as time passes and remove the datagram from the Internet when its time expires. Whenever a host computer sends the datagram to the Internet, it sets a maximum time that the datagram should survive. When a router receives a datagram, it decrements the value of this field by 1. Whenever this value reaches 0 after being decremented, the router discards the datagram and returns an error message to the source.

- *Protocol* (*8 bits*). This 8-bit field defines the higher-level protocol that uses the services of the IP layer. An IP datagram can encapsulate data from several higher-level protocols such as TCP, UDP, ICMP, and IGMP. This field specifies the final destination protocol to which the IP datagram should be delivered. Since the IP multiplexes and demultiplexes data from different higher-level protocols, the value of this field helps the demultiplexing process when the datagram arrives at its final destination.

- *Header checksum* (*16 bits*). The error detection method used by most TCP/IP protocols is called the *checksum*. This 16-bit field ensures the integrity of header values. The checksum (redundant bits added to the packet) protects against errors which may occur during the transmission of a packet.

    At the sender, the checksum is calculated and the result obtained is sent with the packet. The packet is divided into *n*-bit sections. These sections are added together using arithmetic in such a way that the sum also results in *n* bits. The sum is then complemented to produce the checksum.

    At the receiver, the same calculation is repeated on the whole packet including the checksum. The received packet is also divided into *n*-bit sections. The sum is then complemented. The final result will be zero if there are no errors in the data during transmission or processing. If the computed result is satisfactorily met, the packet is accepted; otherwise it is rejected.

    It is important to note that the checksum only applies to values in the IP header, and not in the data. Since the header usually occupies fewer bytes than the data, the computation of header checksums will lead to reduced processing time at routers.

Example 2.1   Consider a checksum calculation for an IP header without options. The header is divided into 16-bit fields. All the fields are added and the sum is complemented to obtain the checksum. The result is inserted in the checksum field.

4	5	0	28	
	1		0	0
4		17	0 (checksum)*	
10.12.14.5				
12.6.7.9				

4, 5, and 0:	01000101	00000000
28:	00000000	00011100
1:	00000000	00000001
0 and 0:	00000000	00000000
4 and 17:	00000100	00010001
0:	00000000	00000000
10.12:	00001010	00001100
14.5:	00001110	00000101
12.6:	00001100	00000110
7.9:	00000111	00001001
Sum:	01110100	01001110
*Checksum:	10001011	10110001

- *Source IP address (32 bits).* This 32-bit field specifies the IP address of the sender of the IP datagram.

- *Destination IP address (32 bits).* This 32-bit field designates the IP address of the host to which this datagram is to be sent. Source and destination IP addresses are discussed in more detail in the section "IP Addressing."

- *Options (variable length).* The IP header option is a variable-length field, consisting of 0, 1, or more individual options. This field specifies a set of fields, which may or may not be present in any given datagram, describing specific processing that takes place on a packet. RFC 791 defines a number of option fields, with additional options defined in RFC 3232. The most common options include:

  - The *security option* tends not to be used in most commercial networks. Refer to RFC 1108 for more details.
  - A *record route option* is used to record the Internet routers that handle the datagram. Each router records its IP address in the option field, which can be useful for tracing routing problems.
  - The *timestamp option* is used to record the time of datagram processing by a router. This option requests each router to record both the router address and the time. This option is useful for debugging router problems.
  - A *source routing option* is used by the source to predetermine a route for the datagram as it travels through the Internet. This option enables a host to define the routers the packet is to be transmitted through. Dictation of a route by the source is useful for several reasons. The sender can choose a route with a specific TOS, such as minimum delay or maximum throughput. It may also choose a route that is safer or more reliable for the sender's purpose. Because the option fields are of variable length, it may be necessary to add additional bytes to the header to make it a whole number of 32-bit words. Since the IP option fields represent a significant overhead, they tend not to be used, especially for IP routers. If required, additional padding bytes are added to the end of any specific options.

IP Addressing

Addresses belonging to three different layers of TCP/IP architecture are shown in Table 2.2.

- *Physical (local or link) address.* At the physical level, the hosts and routers are recognized by their physical addresses. The physical address is the lowest-level address which is specified as the node or local address defined by LAN (Local Area Network) or WAN. This local address is included in the frame used by the network access layer. A local address is called a physical address because it is usually (but not always) implemented in hardware. Ethernet or token ring uses a 6-byte address that is imprinted on the network interface card (NIC) installed in the host or router. The physical address should be unique locally, but not necessary universally. Physical addresses can be either unicast (one single recipient), multicast (a group of recipients), or broadcast (all recipients on the network). The physical addresses will be changed as a packet moves from network to network.

- *IP address.* An IP address is called a *logical address* at the network level because it is usually implemented in software. A logical address identifies a host or router at the network level. TCP/IP calls this logical address an IP address. Internet addresses can be either unicast, multicast, or broadcast. IP addresses are essentially needed for universal communication services that are independent of underlying physical networks. IP addresses are designed for a universal addressing system in which each host can be identified uniquely. An Internet address is currently a 32-bit address which can uniquely define a host connected to the Internet.

- *Port address.* The data sequences need the IP address and the physical address to move data from a source to the destination host. In fact, delivery of a packet to a host or router requires two levels of addresses, logical and physical. Computers are devices that can run multiple processes at the same time. For example, computer A communicates with computer B using TELNET (TErminaL NETwork). At the same time, computer A can communicate with computer C using File Transfer Protocol (FTP). If these processes occur simultaneously, we need a method to label different processes. In TCP/IP architecture, the label assigned to a process is called a *port address*. A port address in TCP/IP is 16 bits long.

  The Internet Assigned Numbers Authority (IANA) manages the well-known port numbers between 1 and 1023 for TCP/IP services. Ports between 256 and 1023 were normally used by UNIX systems for UNIX-specific services, but are probably not found on other operating systems.

Table 2.2 TCP/IP architecture and corresponding addresses

Layer	TCP/IP protocol	Address
Application	HTTP, FTP, SMTP DNS, and other protocols	Port address
Transport	TCP, UDP	–
Internet	IP, ICMP, IGMP	IP address
Network access	Physical network	Physical (link) address

Servers are normally known by their port number. For few examples, every TCP/IP implementation that provides an FTP server provides that service on TCP port 21. TELNET is a TCP/IP standard with a port number of 23 and can be implemented on almost any operating system. Hence, every TELNET server is on TCP port 23. Every implementation of the Trivial File Transfer Protocol (TFTP) is on UDP port 69. The port number for the Domain Name System (DNS) is on TCP port 53.

Addressing schemes

Each IP address is made of two parts in such a way that the *netid* defines a network and the *hostid* identifies a host on that network. An IP address is usually written as four decimal integers separated by decimal points, that is, 239.247.135.93. If this IP address changes from decimal point notation to binary form, it becomes 11101111 11110111 10000111 01011101. Thus, we see that each integer gives the value of 1 octet (byte) of the IP address.

IP addresses are divided into five different classes: A, B, C, D, and E. Classes A, B, and C differ in the number of hosts allowed per network. Class D is used for multicasting, and class E is reserved for future use. Table 2.3 shows the number of networks and hosts in the five different IP address classes. Note that the binary numbers in brackets denote class prefixes.

The relationship between IP address classes and dotted decimal numbers is summarized in Table 2.4, which shows the range of values for each class. The use of leading bits as class prefixes means that the class of a computer's network can be determined by the numerical value of its address.

A number of IP addresses have specific meanings. The address 0.0.0.0 is reserved and 224.0.0.0 is left unused. Addresses in the range 10.0.0.0 through to 10.255.255.255 are available for use in private intranets. Addresses in the range 240.0.0.0 through 255.255.255.255 are class E addresses and are reserved for future use when new protocols are developed. Address 255.255.255.255 is the broadcast address, used to reach all systems on a local link. Although the multicast address of class D may extend from 224.0.0.0 to 239.255.255.255, address 224.0.0.0 is never used and 224.0.0.1 is assigned to the permanent group of all IP hosts, including gateways. A packet addressed to 224.0.0.1 will reach all multicast hosts on the directly connected network. In addition, a hostid of

Table 2.3 Number of networks and hosts in each address class

Address class	Netid	Hostid	Number of networks and hosts	
			Netid	Hostid
A (0)	First octet (8 bits)	Three octets (24 bits)	$2^7 - 2 = 126$	$2^{24} - 2 = 16\,777\,214$
B (10)	Two octets (16 bits)	Two octets (16 bits)	$2^{14} = 16\,384$	$2^{16} - 2 = 65\,534$
C (110)	Three octets (24 bits)	Last octet (8 bits)	$2^{21} = 2\,097\,152$	$2^8 - 2 = 254$
D (1110)	–	–	No netid	No hostid
E (1111)	–	–	No netid	No hostid

D (1110): Multicast address only.
E (1111): Reserved for special use.

Table 2.4 Dotted decimal values corresponding to IP address classes

Class	Prefix	Address range	
		Lowest	Highest
A	0	0.0.0.0	127.255.255.255
B	10	128.0.0.0	191.255.255.255
C	110	192.0.0.0	223.255.255.255
D	1110	224.0.0.0	239.255.255.255
E	1111	240.0.0.0	255.255.255.255

255 specifies all systems within a given subnet, and a subnetid of 255 specifies all subnets within a network.

When an IP address is given, the address class can be determined. Once the address class is determined, it is easy to extract the netid and hostid. Figure 2.3 shows how to extract the netid and hostid by the octets and how to determine the number of networks and hosts.

According to Table 2.3 or Figure 2.3, the two-layer hierarchy established in IP address pairs (netid, hostid) lacks the flexibility needed for any sophisticated size of network. To begin with, a class A network can contain 16 777 214 host identifiers (hostids). These are too many identifiers to configure and manage as an address space. Many of these hosts are likely to reside on various locally administered LANs, with different media and data link protocols, different access needs, and, in all likelihood, different geographical locations. In fact, the IP addressing scheme has no way to reflect these subdivisions within a large organization WAN. In addition, classes A, B, and C network identifiers (netids) are a limited and scarce resource, whose use under the class addressing scheme was often inefficient. In reality, many medium-sized organizations found class C hostids to be too small, containing fewer than 256 hosts. On the other hand, they often requested class B identifiers despite having far fewer than 65 534 hostids. As a result, many of the (netid, hostid) pairs were allocated but unused, being superfluous to the network owner and unusable by other organizations.

Subnetting and supernetting

The increasing number of hosts connected to the Internet and restrictions imposed by the Internet addressing scheme led to the idea of subnetting and supernetting. In subnetting, one large network is divided into several smaller subnetworks, and classes A, B, and C addresses can be subnetted. In supernetting, several networks are combined into one large network, bringing several class C addresses to create a large range of addresses. Classes A, B, and C in IP addressing are designed by two levels of hierarchy such that a portion of the address indicates a netid and a portion of address indicates a hostid on the network.

Consider an organization with two-level hierarchical addressing. With this scheme, the organization has one network with many hosts because all of the hosts are at the same level. Subnetting is accomplished by the further division of a network into smaller subnetworks. When a network is subnetted, it has three portions: netid, subnetid, and

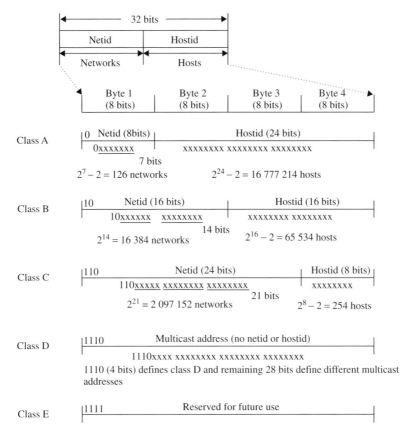

Figure 2.3   The number of networks and hosts corresponding to IP address classes.

hostid. When the datagram arrives at a router, it knows that the first 2 octets (bytes) denote netid and the last 2 octets (bytes) define subnetid and hostid, respectively. For example, for a 32-bit IP address of 141.14.5.23, the router uses the first 2 octets (141.14) as the netid, the third octet (5) as the subnetid, and the fourth octet (23) as the hostid. Thus, the routing of an IP datagram now involves three steps: delivery to the network site, delivery to the subnetwork, and delivery to the host.

Example 2.2   Consider the IP address in decimal point notation (141.14.2.21).

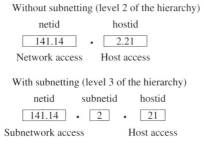

To accommodate the growth of address space, by 1993, the supernetting scheme had begun to take an approach that is complementary to subnet addressing. Supernetting allows addresses to assign a single organization to span multiple-classed prefixes. A class C address cannot accommodate more than 254 hosts and a class B address has suffi-cient bits to make subnetting convenient. Therefore, one solution to this is supernetting. An organization that needs 1000 addresses can be granted four class C addresses. The organization can then use these addresses in one supernetwork. Suppose an organization requests a class B address and intends to subnet using the third octet as a subnet field. Instead of a single class B number, supernetting assigns the organization a block of 256 contiguous class C numbers that the organization can then assign to physical networks.

Mapping by mask

Masking is a process that extracts the physical network address from an IP address. Masking can be accomplished regardless of whether it has subnetting or not. Consider two cases in which a network is either subnetted or is not. With no subnetting, masking extracts the network address from an IP address, while with subnetting, masking also extracts the subnetwork address from an IP address. The masking operation can be done by performing a 32-bit IP address on another 32-bit mask. A masking pattern consists of a contiguous string of 1s and 0s. The contiguous mask means a string of 1s precedes a string of 0s. To get either the network address or the subnet address, the logical AND operation with the bit-by-bit basis must be applied on the IP address and the mask. An example is shown below.

Example 2.3   Suppose a 32-bit IP address is 141.14.5.23 and the mask 255.255.0.0. Find the network address and subnetwork address.

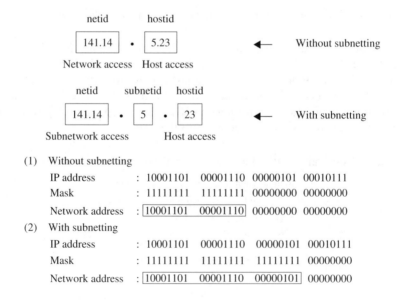

(1)   Without subnetting

IP address	:	10001101 00001110 00000101 00010111
Mask	:	11111111 11111111 00000000 00000000
Network address	:	10001101 00001110 00000000 00000000

(2)   With subnetting

IP address	:	10001101 00001110 00000101 00010111
Mask	:	11111111 11111111 11111111 00000000
Network address	:	10001101 00001110 00000101 00000000

Mapping of a logical address to a physical address can be static or dynamic. Static mapping involves a list of logical and physical address correspondences, but maintenance of the list requires high overhead. ARP is a dynamic mapping method that finds a physical address given a logical address. An ARP request is broadcast to all devices on the network, while an ARP reply is unicast to the host requesting the mapping. RARP is a form of dynamic mapping in which a given physical address is associated with a logical address. ARP and RARP use unicast and broadcast physical addresses. These subjects will be discussed in a later section.

IP Routing

In a connectionless packet delivery system, the basic unit of transfer is the IP datagram. The routing problem is characterized by describing how routers forward IP datagrams and deliver them to their destinations. In a packet switching system, "routing" refers to the process of choosing a path over which to send packets. Unlike routing within a single network, the IP routing must choose the appropriate algorithm for how to send a datagram across multiple physical networks. In fact, routing over the Internet is generally difficult because many computers have multiple physical network connections.

To understand IP routing, a TCP/IP architecture should be reviewed completely. The Internet is composed of multiple physical networks interconnected by routers. Each router has direct connections to two or more networks, while a host usually connects directly to one physical network. However, it is possible to have a multihomed host connected directly to multiple network.

Packet delivery through a network can be managed at any layer in the OSI stack model. The physical layer is governed by the Media Access Control (MAC) address; the data link layer includes the Logical Link Control (LLC); and the network layer is where most routing takes place.

Delivery

The delivery of an IP packet to its final destination is accomplished by means of either direct or indirect delivery. Direct delivery occurs when the source and destination of the packet are located on the same physical network. The sender can easily determine whether the delivery is direct or not by extracting the network (IP) address of the destination packet and comparing this address with the addresses of the networks to which it is connected. If a match is found, the delivery is direct. In direct delivery, the sender uses the senders IP address to find the destination physical address. This mapping process can be done by ARP.

If the destination host is not on the same network as the source host, the packet will be delivered indirectly. In an indirect delivery, the packet goes from router to router through a number of networks until it reaches one that is connected to the same physical network as its final destination. Thus, the last delivery is always a direct delivery, which always occurs after zero or more indirect deliveries. In an indirect delivery, the sender uses the destination IP address and a routing table to find the IP address of the next router to which the packet should be delivered. The sender then uses the ARP to find the physical address of the next router.

2.1.2   Address Resolution Protocol (ARP)

IP (logical) addresses are assigned independently from physical (hardware) addresses. The logical address is called a *32-bit IP address*, and the physical address is a 48-bit MAC address in Ethernet and token ring protocols. The delivery of a packet to a host or a router requires two levels of addressing, such as logical (IP) address and physical (MAC) addresses. When a host or a router has an IP datagram forwarding to another host or router, it must know the logical IP address of the receiver. Since the IP datagram is encapsulated in a form to be passed through the physical network (such as a LAN), the sender needs the physical MAC address of the receiver.

Mapping of an IP address to a physical address can be done by either static or dynamic mapping. Static mapping means creating a table that associates an IP address with a physical address. But static mapping has some limitations because table lookups are inefficient. As a consequence, static mapping creates a huge overhead on the network. Dynamic mapping can employ a protocol to find the other. Two protocols (ARP and RARP) have been designed to perform dynamic mapping. When a host needs to find the physical address of another host or router on its network, it sends an ARP query packet. The intended recipient recognizes its IP address and sends back an ARP response which contains the recipient IP and physical addresses. An ARP request is broadcast to all devices on the network, while an ARP reply is unicast to the host requesting the mapping.

Figure 2.4 shows an example of simplified ARP dynamic mapping. Let a host or router call a machine. A machine uses ARP to find the physical address of another machine by

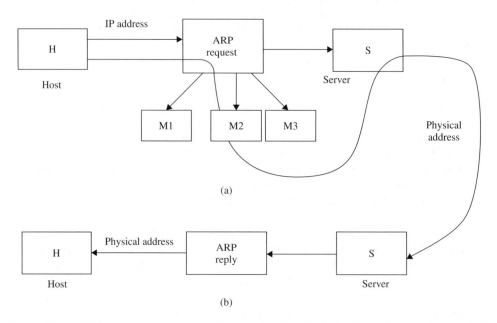

Figure 2.4   ARP dynamic mapping. (a) Request for the physical address by broadcast and (b) reply for the physical address by unicast.

broadcasting an ARP request. The request contains the IP address of the machine for which a physical address is needed. All machines (M1, M2, M3, ...) on the network receive an ARP request. If the request matches the M2 machine's IP address, the machine responds by sending a reply that contains the requested physical address. Note that Ethernet uses the 48-bit address of all 1s (FFFFFFFFFFFF) as the broadcast address.

A proxy ARP is an ARP that acts on behalf of a set of hosts. Proxy ARP can be used to create a subnetting effect. In proxy ARP, a router represents a set of hosts. When an ARP request seeks the physical address of any host in this set, the router sends its own physical address. This creates a subnetting effect. Whenever looking for the IP address of one of these hosts, the router sends an ARP reply announcing its own physical address.

To make address resolution easy, choose both IP and physical addresses of the same length. Address resolution is difficult for Ethernet-like networks because the physical address of the Ethernet interface is 48 bits long and the high-level IP address is 32 bits long. In order for the 48-bit physical address to encode a 32-bit IP address, the next generation of IP is being designed to allow 48-bit physical (hardware) addresses P to be encoded in IP addresses I by the functional relationship of $P = f(I)$. Conceptually, it will be necessary to choose a numbering scheme that makes address resolution efficient by selecting a function f that maps IP addresses to physical addresses.

As shown in Figure 2.5, the ARP software package consists of the following five components:

- The *cache table* has an array of entries used and updated by ARP messages. It is inefficient to use the ARP for each datagram destined for the same host or router. The solution is to use the cache table. The cache table is implemented as an array of entries. When a host or router receives the corresponding physical address for an IP datagram, the address can be saved in the cache table within the next few minutes. However, mapping in the cache should not be retained for an unlimited time, due to the limited cache space.
- A *queue* contains packets going to the same destination. The ARP package maintains a set of queues to hold the IP packets, while ARP tries to resolve the physical address. The output module sends unresolved packets to the corresponding queue. The input module removes a packet from a queue and sends it to the physical access layer for transmission.
- The *output module* takes an IP packet from the IP layer and sends it to a queue as well as the physical access layer. The output module checks the cache table to find an entry corresponding to the destination IP address of this packet. If the entry is found and the state of the entry is resolved, the packet, along with the destination physical address, is passed to the physical access layer (or data link layer) for transmission. If the entry is found and the state of the entry is pending, the packet should wait until the destination physical address is found. If no entry is found, the module creates a queue and enqueues the packet. A new cache entry ("pending") is created for the destination and the attempt field is set to 1. An ARP request is then broadcast.
- The *input module* waits until an ARP request or reply arrives. The input module checks the cache table to find an entry corresponding to this packet (request or reply). If the entry is found and the state of the entry is "pending," the module updates the entry by copying the target physical address in the packet to the physical address field

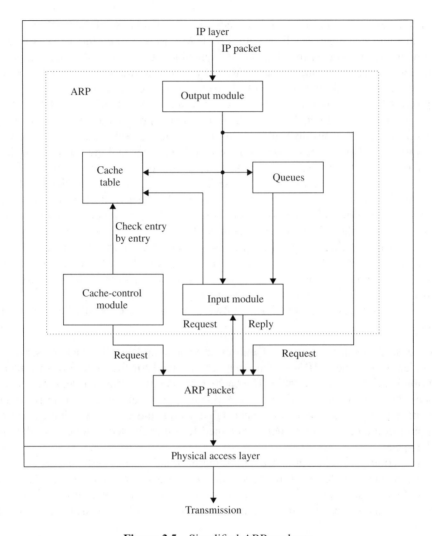

Figure 2.5 Simplified ARP package.

of the entry and changing the state to "resolved." The module also sets the value of the time-out for the entry and then dequeues the packets from the corresponding queue, one by one, and delivers them along with the physical address to the physical access layer for transmission.

If the entry is found and the state is "resolved," the module still updates the entry. This is because the target physical address could have been changed. The value of the time-out field is also reset. If the entry is not found, the module creates a new entry and adds it to the cache table.

Now the module checks to see if the arrived ARP packet is a request. If it is, the input module immediately creates an ARP reply message and sends it to the sender.

The ARP reply packet is created by changing the value of the operation field from request to reply and filling in the target physical address.

- The *cache-control module* is responsible for maintaining the cache table. It checks the cache table periodically, entry by entry. If the entry is free, it continues to the next entry. If the state is "pending," the module increments the value of the attempts field by 1. It then checks the value of the attempts field. If this value is greater than the maximum number of attempts allowed, the state is changed to "free" and the corresponding queue is destroyed. However, if the number of attempts is less than the maximum, the input module creates and sends another ARP request. If the state of the entry is "resolved," the module decrements the value of the "time-out" field by the amount of the time elapsed since the last check. If this value is less than or equal to zero, the state is changed to free and the queue is destroyed.

2.1.3 Reverse Address Resolution Protocol (RARP)

To create an IP datagram, a host or a router needs to know its own IP address, which is independent of the physical address. The RARP is designed to resolve the address mapping of a machine in which its physical address is known, but its logical (IP) address is unknown. The machine can get its physical address, which is unique locally. It can then use the physical address to get the logical IP address using the RARP protocol. In reality, RARP is a protocol of dynamic mapping in which a given physical address is associated with a logical IP address, as shown in Figure 2.6.

To get the IP address, an RARP request is broadcast to all systems on the network. Every host or router on the physical network will receive the RARP request packet, but the RARP server will only answer it as shown in Figure 2.6b. The server sends a RARP reply packet including the IP address of the requestor.

2.1.4 Classless Interdomain Routing (CIDR)

Classless Interdomain Routing (CIDR) is the standard that specifies the details of both classless addressing and an associated routing scheme. Accordingly, the name is slightly inaccurate designation because CIDR specifies addressing as well as routing.

The original IPv4 model built on network classes was a useful mechanism for allocating identifiers (netid and hostid) when the primary users of the Internet were academic and research organizations. But, this mode proved insufficiently flexible and inefficient as the Internet grew rapidly to include gateways into corporate enterprises with complex networks. By September 1993, it was clear that the growth in Internet users would require an interim solution while the details of IP version 6 (IPv6) were being finalized. The resulting proposal was submitted as RFC 1519 titled "Classless Inter-Domain Routing (CIDR): an Address Assignment and Aggregation Strategy." CIDR is classless, representing a move away from the original IPv4 network class model. CIDR is concerned with interdomain routing rather than host identification. CIDR has a strategy for the allocation and use of IPv4 addresses, rather than a new proposal.

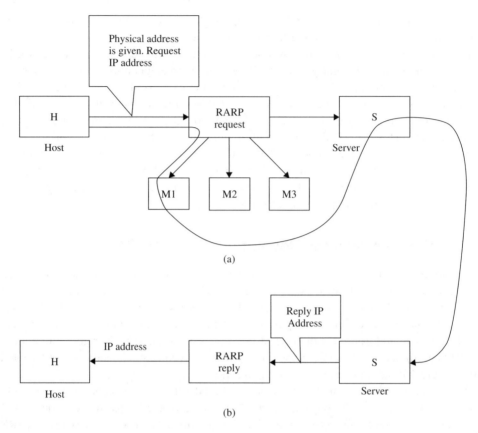

Figure 2.6 RARP dynamic mapping. (a) Request for the physicSal address by broadcast and (b) reply IP address by unicast.

2.1.5   IP Version 6 (IPv6 or IPng)

The evolution of the TCP/IP technology has led on to attempts to solve problems that improve service and extend functionalities. Most researchers seek new ways to develop and extend the improved technology, and millions of users want to solve new networking problems and improve the underlying mechanisms. The motivation behind revising the protocols arises from changes in underlying technology: first, computer and network hardware continues to evolve; second, as programmers invent new ways to use TCP/IP, additional protocol support is needed; and third, the global Internet has experienced huge growth in size and use. This section examines a proposed revision of the Internet protocol, which is one of the most significant engineering efforts so far.

The network layer protocol is currently IPv4. IPv4 provides the basic communication mechanism of the TCP/IP suite. Although IPv4 is well designed, data communication has evolved since the inception of IPv4 in the 1970s. Despite its sound design, IPv4 has some deficiencies that make it unsuitable for the fast-growing Internet. The IETF decided to assign the new version of IP and to name it IPv6 to distinguish it from the current IPv4. The proposed IPv6 retains many of the features that contributed to the success of IPv4. In fact, the designers have characterized IPv6 as being basically the same as IPv4 with a few modifications: IPv6 still supports connectionless delivery, allows the sender to choose the size of a datagram, and requires the sender to specify the maximum number of hops a datagram can make before being terminated. In addition, IPv6 also retains most of IPv4's options, including facilities for fragmentation and source routing.

IPv6, also known as the Internet Protocol next generation (IPng), is the new version of the IP, designed to be a full replacement for IPv4. IPv6 has an 128-bit address space, a revised header format, new options, an allowance for extension, support for resource allocation, and increased security measures. However, due to the huge number of systems on the Internet, the transition from IPv4 to IPv6 cannot occur at once. It will take a considerable amount of time before every system in the Internet can move from IPv4 to IPv6. RFC 2460 defines the new IPv6 protocol. IPv6 differs from IPv4 in a number of significant ways:

- The IP address length in IPv6 is increased from 32 to 128 bits.
- IPv6 can automatically configure local addresses and locate IP routers to reduce configuration and setup problems.
- The IPv6 header format is simplified and some header fields dropped. This new header format improves router performance and makes it easier to add new header types.
- Support for authentication, data integrity, and data confidentiality are part of the IPv6 architecture.
- A new concept of flows has been added to IPv6 to enable the sender to request special handling of datagrams.

IPv4 has a two-level address structure (netid and hostid) categorized into five classes (A, B, C, D, and E). The use of address space is inefficient. For instance, when an organization is granted a class A address, 16 million addresses from the address space are assigned for the organization's exclusive use. On the other hand, if an organization is granted a class C address, only 256 addresses are assigned to this organization, which may not be enough. Soon there will be no addresses left to assign to any new system that wants to be connected to the Internet.

Although the subnetting and supernetting strategies have alleviated some addressing problems, subnetting and supernetting make routing more complicated. The encryption and authentication options in IPv6 provide confidentiality and integrity of the packet. However, no encryption or authentication is provided by IPv4.

IPv6 Addressing

In December 1995, the network working group of IETF proposed a longer-term solution for specifying and allocating IP addresses. RFC 2373 describes the address space associated with the IPv6. The biggest concern with Internet developers will be the migration process from IPv4 to IPv6.

IPv4 addressing has the following shortcoming: IPv4 was defined when the Internet was small and consisted of networks of limited size and complexity. It offered two layers of address hierarchy (netid and hostid) with three address formats (classes A, B, and C) to accommodate varying network sizes. Both the limited address space and the 32-bit address size in IPv4 proved to be inadequate for handling the increase in the size of the routing table caused by the immense numbers of active hosts and servers. IPv6 is designed to improve upon IPv4 in each of these areas. IPv6 allocates 128 bits for addresses. Analysis shows that this address space will suffice to incorporate flexible hierarchies and to distribute the responsibility for allocation and management of the IP address space.

Like IPv4, IPv6 addresses are represented as a string of digits (128 bits or 32 hex digits) which are further broken down into eight 16-bit integers separated by colons (:). The basic representation takes the form of eight sections, each 2 bytes in length.

xx:xx:xx:xx:xx:xx:xx:xx

where each xx represents the hexadecimal form of 16 bits of address. IPv6 uses hexadecimal colon notation with abbreviation methods.

Example 2.4 An IPv6 address consists of 16 bytes (octets), which is 128 bits long. The IPv6 address consists of 32 hexadecimal digits, with every four digits separated by a colon.

IPv6 address:	flea:1075:fffb:110e:0000:0000:7c2d:a65f
	↓
Abbreviated address:	flea:1075:fffb:110e::7c2d:a65f
	↓
Binary address:	1111000111101010 ... 1010011001011111

Many of the digits in IPv6 addresses are zeros. In this case, the abbreviated address can be obtained by omitting the leading zeros of a section (four hex digits between two colons), but not the trailing zeros.

Example 2.5 Assume that the IPv6 address is given as

fedc:ab98:0052:4310:000f:bccf:0000:ff1f (unabbreviated)

Using the abbreviated form, 0052 can be written as 52, 000f as f, and 0000 as 0. But the trailing zeros cannot be dropped, so that 4310 would not be abbreviated. Thus, the given IP address becomes fedc:ab98:52:4310:f:bccf:0:ff1f (abbreviated).

Example 2.6 Consider an abbreviated address with consecutive zeros. When consecutive sections are composed of zeros, further abbreviations are possible. We can remove the zeros altogether and replace them with a double colon.

fedc:0:0:0:0:abf8:0:f75f (abbreviated)

fedc::abf8:0:f75f (more abbreviated)

IPv6 address types

IPv6 has identified three types of addresses:

- *Unicast.* To associate with a specific physical interface to a network. Packets sent to a unicast address are delivered to the interface uniquely specified by the address.
- *Anycast.* To associate with a set of physical interfaces, generally on different modes. Packets sent to an anycast address will be delivered to at least one interface specified by the address.
- *Multicast.* To associate with a set of physical interfaces, generally on multiple hosts (nodes). Packets sent to a multicast address will be delivered to all the interfaces to which the address refers.

Figure 2.7 illustrates three address types.

IPv6 addresses divide the address space into two parts with the type prefix for each type of address, rest of address, and the fraction of each type of address relative to the whole address space. Table 2.5 illustrates the address space assignment for type prefixes.

IPv6 Packet Format

The IPv6 protocol consists of two parts: the basic elements of the IPv6 header and IPv6 extension headers. The IPv6 datagram is composed of a base header (40 bytes) followed by the payload. The payload consists of two parts: optional extension headers and data from the upper layer. The extension headers and data packet from the upper layer usually occupy up to 65 535 bytes of information. Figure 2.8 shows the base header with its eight fields. Each IPv6 datagram begins with a base header. The IPv6 header has a fixed length of 40 octets, consisting of the following fields:

- *Version.* This 4-bit field defines the version number of the IP. For IPv6, the value is 6.
- *Priority.* This 4-bit priority field defines the priority of the packet with respect to traffic congestion. So, this field is a measure of the importance of a datagram. The IPv4 service class field has been renamed the IPv6 traffic class field.
- *Flow label.* This 24-bit field is designed to provide special handling for a particular flow of data. This field contains information that routers use to associate a datagram with a specific flow and priority.
- *Payload length.* This 16-bit payload length field defines the total length of the IP datagram excluding the base header. A payload consists of optional extension headers plus data from the upper layer. It occupies up to $2^{16} - 1 = 65\ 535$ bytes.
- *Next header.* The next header is an 8-bit field defining the header that follows the base header in the datagram. The next header is either one of the optional extension headers used by IP or a header for an upper-layer protocol such as UDP or TCP. Extension headers add functionality to the IPv6 datagram.

Table 2.6 shows the values of next headers (i.e., IPv6 extension headers).

Six types of extension header have been defined: the hop-by-hop option, source routing, fragmentation, authentication, encrypted security payload (ESP), and destination option. These are discussed below.

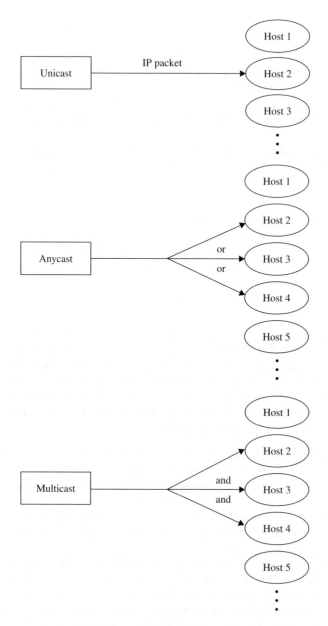

Figure 2.7   IPv6 address types.

The *hop-by-hop option* is used when the source needs to pass information to all routers (in the path) visited by the datagram.

The *source routing* extension header combines the concepts of the strict source route and the loose source route options of IPv4. The source routing extension is used when the source wants to specify the transmission path.

Table 2.5 Type prefixes for IPv6 addresses

Type prefix(binary)	Type of address	Fraction of address space
0000 0000	Reserved	1/256
0000 0001	Reserved	1/256
0000 001	NSAP (Network Service Access Point)	1/128
0000 010	IPX (Novell)	1/128
0000 011	Reserved	1/128
0000 100	Reserved	1/128
0000 101	Reserved	1/128
0000 110	Reserved	1/128
0000 111	Reserved	1/128
0001	Reserved	1/16
001	Reserved	1/8
010	Provider-based unicast addresses	1/8
011	Reserved	1/8
100	Geographic unicast addresses	1/8
101	Reserved	1/8
110	Reserved	1/8
1110	Reserved	1/16
1111 0	Reserved	1/32
1111 10	Reserved	1/64
1111 110	Reserved	1/128
1111 1110 0	Reserved	1/512
1111 1110 10	Link local addresses	1/1024
1111 1110 11	Site local addresses	1/1024
1111 1111	Multicast addresses	1/256

|←— Prefix (variable) —→|←——— Rest of address (variable) ———→|

|←———————————— 128 bits ————————————→|

The source routing header contains a minimum of seven fields which are expressed in a unified form as follows:

- The next header and HLEN are identical to that of hop-by-hop extension header.
- The type field defines loose or strict routing.
- The address left field indicates the number of hops still needed to reach the destination.
- The strict/loose mask field determines the rigidity of routing.
- The destination address in source routing changes from router to router.

The *fragmentation* extension is used if the payload is a fragment of a message. The concept of fragmentation is the same as that in IPv4 except that where fragmentation takes place differs. In IPv4, the source or router is required to fragment if the size of the datagram is larger than the MTU of the network. In IPv6, only the original source can fragment using the Path MTU Discovery technique. If the source does not use this

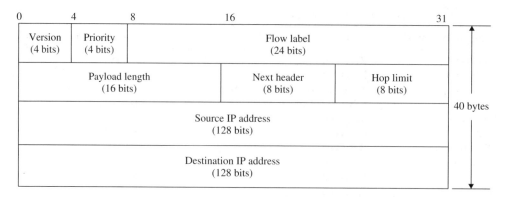

Figure 2.8   IPv6 base header with its eight fields.

Table 2.6   Next header codes

Code	Next header
0	Hop-by-hop option
2	ICMP
6	TCP
17	UDP
43	Source routing
44	Fragmentation
50	Encrypted security payload
51	Authentication
59	Null (no next header)
60	Destination option

technique, it should fragment the datagram to a size of 576 bytes or smaller, which is the minimum size of MTU required for each network connected to the Internet.

The *ESP* is an extension that provides confidentiality between sender and receiver and guards against eavesdropping. The ESP format contains the security parameter index (SPI) field and the encrypted data field. The SPI field is a 32-bit word that defines the type of encryption/decryption used. The encrypted data field contains the data being encrypted along with any extra parameters needed by the algorithm. Encryption can be implemented in two ways: transport mode and tunnel mode, as shown in Figure 2.9. The transport-mode method encrypts a TCP segment or UDP user datagram first and then encapsulated along with its base header, extension headers, and SPI as shown in Figure 2.9a. The tunnel-mode method encrypts the entire IP datagram together with its base header and extension headers and then encapsulates it in a new IP packet as shown in Figure 2.9b.

The *authentication* extension validates the sender of the message and protects the data from hackers. The authentication extension field has a dual purpose: sender identification and data integrity. The sender verification is needed because the receiver can be sure that a message is from the genuine sender and not from an imposter. The data integrity is

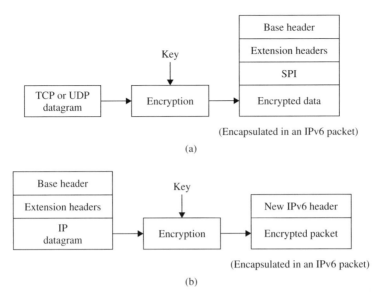

Figure 2.9 Encrypted security payload. (a) Transport-mode encryption and (b) tunnel-mode encryption.

needed to check that the data is not altered in transition by some hackers. The format of authentication extension header consists of the SPI field and the authentication data field. The former defines the algorithm used for authentication, and the latter contains the actual data generated by the algorithm.

The *destination* extension passes information from the source to the destination exclusively. This header contains optional information to be examined by the destination mode. It is worth comparing the options in IPv4 with the extension headers in IPv6.

1. The record route option in IPv4 is not used in IPv6.
2. The timestamp option in IPv4 is not implemented in IPv6.
3. The source router option in IPv4 is called the source route extension header in IPv6.
4. The fragmentation fields in the base header section of IPv4 have moved to the fragmentation extension header in IPv6.
5. The ESP extension header is new in IPv6.

- *Hop limit.* This 8-bit hop limit field decrements by 1 each node that forwards the packet. The packet is discarded if the hop limit is decremented to zero. This field serves the same purpose as the TTL field in IPv4. IPv6 interprets the value as giving a strict bound on the maximum number of hops a datagram can make before being discarded.
- *Source address.* The source address field is a 128-bit originator address that identifies the initial sender of the packet.
- *Destination address.* The destination address field specifies a 128-bit recipient address that usually identifies the final destination of the datagram. However, if source routing is used, this field contains the address of the next router.

To summarize, each IPv6 datagram begins with a 40-octet base header that includes fields for the source and destination addresses, the maximum hop limit, the traffic class (priority), the flow label, and the type of the next header. Thus, an IPv6 datagram should contain at least 40 octets in addition to the data.

Comparison between IPv4 and IPv6 Headers

Despite many conceptual similarities, IPv6 changes most of the protocol scopes. Most important, IPv6 completely revises the datagram format by replacing IPv4's variable-length options field with a series of fixed-format headers. A comparison between IPv4 and IPv6 headers will be examined in the following section.

- The HLEN field is eliminated in IPv6 because the length of the header is fixed in IPv6.
- The service type field is eliminated in IPv6. The priority and flow label fields together take over the function of the service type field in IPv4.
- The total length field is eliminated in IPv6 and replaced by the payload length field.
- The identification, flag, and offset fields in IPv4 are eliminated from the base header in IPv6. They are included in the fragmentation extension header.
- The TTL field in IPv4 is called the hop limit in IPv6.
- The protocol field is replaced by the next header field.
- The header checksum field in IPv4 is eliminated because the checksum is provided by upper-level protocols. It is thereby not needed at this level.
- The option fields in IPv4 are implemented as extension headers in IPv6.

The length of the base header is fixed at 40 bytes. However, to give more functionality to the IP datagram, the base header can be followed by up to six extension headers.

2.1.6   Internet Control Message Protocol (ICMP)

The ICMP is an extension to the IP which is used to communicate between a gateway and a source host, to manage errors and generate control messages.

The IP is not designed to be absolutely reliable. The purpose of control messages (ICMP) is to provide feedback about problems in the communication environment, not to make IP reliable.

There are still no guarantees that a datagram will be delivered or a control message will be returned. Some datagrams may still be undelivered without any report of their loss. The higher-level protocols that use TCP/IP must implement their own reliability procedures if reliable communication is required.

IP is an unreliable protocol that has no mechanisms for error checking or error control. ICMP was designed to compensate for this IP deficiency. However, ICMP does not correct errors, simply reports them. ICMP uses the source IP address to send the error message to the source of the datagram. ICMP messages consist of error-reporting messages and query messages. The error-reporting messages report problems that a router or a destination host may encounter when it processes an IP packet. In addition to error reporting, ICMP can diagnose some network problems through the query messages. The query messages (in pairs) give a host or a network manager specific information from a router or another host.

2.1.7 Internet Group Management Protocol (IGMP)

The IGMP is used to facilitate the simultaneous transmission of a message to a group of recipients. IGMP helps multicast routers to maintain a list of multicast addresses of groups. "Multicasting" means sending of the same message to more than one receiver simultaneously. When the router receives a message with a destination address that matches one on the list, it forwards the message, converting the IP multicast address to a physical multicast address. To participate in the IP on a local network, the host must inform local multicast routers. The local routers contact other multicast routers, passing on the membership information and establishing route.

IGMP has only two types of messages: report and query. The report message is sent from the host to the router. The query message is sent from the router to the host. A router sends in an IGMP query to determine if a host wishes to continue membership in a group. The query message is multicast using the multicast address 244.0.0.1. The report message is multicast using a destination address equal to the multicast address being reported. IP addresses that start with $1110_{(2)}$ are multicast addresses. Multicast addresses are class D addresses.

The IGMP message is encapsulated in an IP datagram with the protocol value of 2. When the message is encapsulated in the IP datagram, the value of TTL must be 1. This is required because the domain of IGMP is the LAN.

The multicast backbone (MBONE) is a set of routers on the Internet that supports multicasting. MBONE is based on the multicasting capability of IP. Today, MBONE uses the services of UDP at the transport layer.

2.2 Transport Layer Protocols

Two protocols exist for the transport layer: TCP and UDP. Both TCP and UDP lie between the application layer and the network layer. As a network layer protocol, IP is responsible for host-to-host communication at the computer level, whereas TCP or UDP is responsible for process-to-process communication at the transport layer.

2.2.1 Transmission Control Protocol (TCP)

This section describes the services provided by TCP for the application layer. TCP provides a connection-oriented byte stream service, which means two end points (normally a client and a server) communicating with each other on a TCP connection. TCP is responsible for flow/error controls and delivering the error-free datagram to the receiving application program.

TCP needs two identifiers, IP address and port number, for a client/server to make a connection offering a full-duplex service. To use the services of TCP, the client socket address and server socket address are needed for the client/server application programs.

The sending TCP accepts a datagram from the sending application program, creates segments (or packets) extracted from the datagram, and sends them across the network. The receiving TCP receives packets, extracts data from them, orders them if they arrived out of order, and delivers them as a byte stream (datagram) to the receiving application program.

TCP Header

TCP data is encapsulated in an IP datagram as shown in Figure 2.10. The TCP packet (or segment) consists of a 20- to 60-byte header, followed by data from the application program. The header is 20 bytes if there is no option and up to 60 bytes if it contains some options. Figure 2.11 illustrates the TCP packet format, whose header is explained in the following.

- *Source and destination port numbers (16 bits each).* Each TCP segment contains a 16-bit field each that defines the source and destination port number to identify the sending and receiving application. These two port numbers, along with the source and destination IP addresses in the IP header, uniquely identify each connection. The combination of an IP address and a port number is sometimes called a *socket*. The socket pair, consisting of the client IP address and port number and the server IP address and port number, specifies two end points that uniquely identify each TCP connection in the Internet.

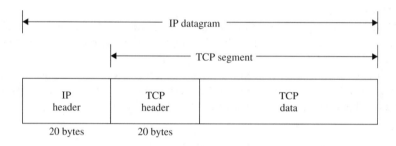

Figure 2.10 Encapsulation of TCP data in an IP datagram.

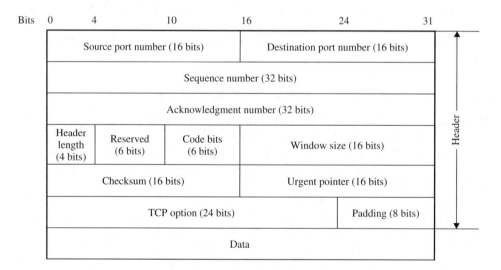

Figure 2.11 TCP packet format.

- *Sequence number* (*32 bits*). This 32-bit sequence field defines the sequence number assigned to the first byte of data stream contained in this segment. To ensure connectivity, each byte to be transmitted is numbered. This sequence number identifies the byte in the data stream from the sending TCP to the receiving TCP. Considering the stream of bytes following in one direction between two applications, TCP will number each byte with a sequence number. During connection establishment, each party uses a random number generator to create an initial sequence number (ISN) that is usually different in each direction. The 32-bit sequence number is an unsigned number that wraps back around to 0 after reaching $2^{32} - 1$.

- *Acknowledgment number* (*32 bits*). This 32-bit field defines the byte number that the sender of the segment is expecting to receive from the receiver. Since TCP provides a full-duplex service to the application layer, data can flow in each direction, independent of the other direction. The sequence number refers to the stream flowing in the same direction as the segment, while the acknowledgment number refers to the stream flowing in the opposite direction from the segment. Therefore, the acknowledgment number is the sequence number plus 1 of the last successfully received byte of data. This field is only valid if the ACK flag is on.

- *HLEN* (*4 bits*). This field indicates the number of 4-byte words in the TCP header. Since the HLEN is between 20 and 60 bytes, an integer value of this field can be between 5 and 15, because $5 \times 4 = 20$ bytesand $15 \times 4 = 60$ bytes.

- *Reserved* (*6 bits*). This is a 6-bit field reserved for future use.

- *Code bits* (*6 bits*). There are six flag bits (or control bits) in the TCP header. One or more can be turned on at the same time. Below is a brief description of each flag to determine the purpose and contents of the segment.

URG	The urgent point field is valid.
ACK	The acknowledgment number is valid.
PSH	This segment requests a push.
RST	Reset the connection.
SYN	Synchronize sequence number to initiate a connection.
FIN	The sender has finished sending data.

- *Window size* (*16 bits*). This 16-bit field defines the size of window in bytes. Since the window size of this field is 16 bits, the maximum size of the window is $2^{16} - 1 = 65\,535$ bytes. TCP's flow control is provided by each end, advertising a window size. This is the number of bytes, starting with the one specified by the acknowledgment number field, that the receiver is willing to accept.

- *Checksum* (*16 bits*). This 16-bit field contains the checksum. The checksum covers the TCP segment, TCP header, and TCP data. This is a mandatory field that must be calculated and stored by the sender, and then verified by the receiver.

- *Urgent pointer* (*16 bits*). This 16-bit field is valid only if the URG flag is set. The urgent point is used when the segment contains urgent data. It defines the number that must be added to the sequence number to obtain the number of the last urgent byte in the data section of the segment.

- *Options* (*24 bits*). The options field (if any) varies in length, depending on which options have been included. The size of the TCP header varies depending on the options selected. The TCP header can have up to 40 bytes of optional information. The options are used to convey additional information to the destination or to align other options. The options are classified into two categories: single-byte options contain end of option and no operation and multiple-byte operations contain maximum segment size, window scale factor, and timestamp.

TCP is a connection-oriented byte stream transport layer protocol in the TCP/IP suite. TCP provides a full-duplex connection between two applications, allowing them to exchange large volumes of data efficiently. Since TCP provides flow control, it allows systems of widely varying speeds to communicate. To accomplish flow control, TCP uses a sliding window protocol so that it can make efficient use of the network. Error detection is handled by the checksum, acknowledgment, and time-out. TCP is used by many popular applications such as HTTP (Hypertext Transfer Protocol, World Wide Web), TELNET, Rlogin (Remote login), FTP, and SMTP (Simple Mail Transfer Protocol) for e-mail.

2.2.2 User Datagram Protocol (UDP)

UDP lies between the application layer and IP layer. Like TCP, UDP serves as the intermediary between the application programs and network operations. UDP uses port numbers to accomplish a process-to-process communication. The UDP provides a flow-and-control mechanism at the transport level. In fact, it performs very limited error checking. UDP can only receive a data unit from the process and deliver it to the receiver unreliably. The data unit must be small enough to fit in a UDP packet. If a process wants to send a small message and does not care much about reliability, it will use UDP. UDP is a connectionless protocol. It is often used for broadcast-type protocols, such as audio or video traffic. It is quicker and uses less bandwidth because a UDP connection is not continuously maintained. This protocol does not guarantee delivery of information, nor does it repeat a corrupted transfer, as does TCP.

UDP Header

UDP receives the data and adds the UDP header. UDP then passes the user datagram to the IP with the socket addresses. IP adds its own header. The IP datagram is then passed to the data link layer. The data link layer receives the IP datagram, adds its own header and a trailer (possibly), and passes it to the physical layer. The physical layer encodes bits into electrical or optical signals and sends it to the remote machine. Figure 2.12 shows the encapsulation of a UDP datagram as an IP datagram. The IP datagram contains its total length in bytes, so the length of the UDP datagram is this total length minus the length of the IP header.

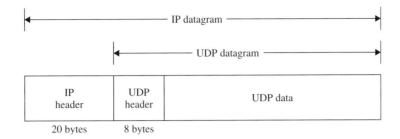

Figure 2.12   UDP encapsulation.

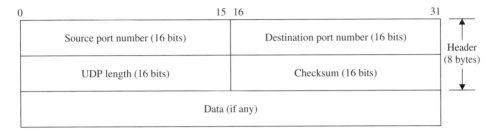

Figure 2.13   UDP header.

The UDP header is shown by the fields illustrated in Figure 2.13.

- *Source port numbers* (*16 bits*). This 16-bit port number identifies the sending process running on the source host. Since the source port number is 16 bits long, it can range from 0 to 65 656 bytes. If the source host is the client, the client program is assigned a random port number called the *ephemeral port number* requested by the process and chosen by the UDP software running on the source host. If the source host is the server, the port number is a universal port number.

- *Destination port numbers* (*16 bits*). This is the 16-bit port number used by the process running on the destination host. If the destination host is the server, the port number is a universal port number, while if the destination host is the client, the port number is an ephemeral port number.

- *Length* (*16 bits*). This is a 16-bit field that contains a count of bytes in the UDP datagram, including the UDP header and the user data. This 16-bit field can define a total length of 0–65 535 bytes. However, the minimum value for length is 8, which indicates a UDP datagram with only header and no data. Therefore, the length of data can be between 0 and 65 507 bytes, subtracting the total length 65 535 bytes from 20 bytes for an IP header and 8 bytes for a UDP header. The length field in a UDP user datagram is redundant. The IP datagram contains its total length in bytes, so the length of the UDP datagram is this total length minus the length of the IP header.

- *Checksum (16 bits).* The UDP checksum is used to detect errors over the entire user datagram covering the UDP header and the UDP data. UDP checksum calculations include a pseudoheader, the UDP header, and the data coming from the application layer. The value of the protocol field for UDP is 17. If this value changes during transmission, the checksum calculation at the receiver will detect it and UDP drops the packet.

The checksum computation at the sender is as follows:

1. Add the pseudoheader to the UDP datagram.
2. Fill the checksum field with zero.
3. Divide the total bits into 16-bit words.
4. If the total number of bytes is not even, add padding of all 0s.
5. Complement the 16-bit result and insert it in the checksum field.
6. Drop the pseudoheader and any added padding.
7. Deliver the UDP datagram to the IP software for encapsulation.

The checksum computation at the receiver is as follows:

1. Add the pseudoheader to the UDP datagram.
2. Add padding if needed.
3. Divide the total bits into 16-bit words.
4. Add all 16-bit sections using arithmetic.
5. Complement the result.
6. If the result is all 0s, drop the pseudoheader and any added padding and accept the user datagram. Otherwise, discard the user datagram.

Multiplexing and Demultiplexing

In a host running a TCP/IP suite, there is only one UDP but there may be several processes that may want to use the services of UDP. To handle this situation, UDP needs multiplexing and demultiplexing.

- *Multiplexing.* At the sender side, it may have several processes that need user datagrams. But there is only one UDP. This is a many-to-one relationship and requires multiplexing. UDP accepts messages from different processes, differentiated by their assigned port numbers. After adding the header, UDP passes the user datagram to IP.

- *Demultiplexing.* At the receiver side, there is only one UDP. However, it may happen to be many processes that can receive user datagrams. This is a one-to-many relationship and requires demultiplexing. UDP receives user datagrams from IP. After error checking and dropping of header, UDP delivers each message to the appropriate process based on the port numbers.

UDP is suitable for a process that requires simple request–response communication with little concern for flow and error control. It is not suitable for a process that needs to send bulk data, like FTP. However, UDP can be used for a process with internal flow and error control mechanisms such as the TFTP process. UDP is also used for management processes such as SNMP (Simple Network Management Protocol).

2.3 World Wide Web

The World Wide Web (WWW) is a repository of information spread all over the world and linked together. The WWW is a distributed client–server service, in which a client using a browser can access a service using a server. The Web consists of Web pages that are accessible over the Internet.

The Web allows users to view documents that contain text and graphics. The Web grew to be the largest source of Internet traffic since 1994 and continues to dominate, with a much higher growth rate than the rest of the Internet. By 1995, Web traffic overtook FTP to become the leader. By 2001, Web traffic completely overshadowed other applications.

2.3.1 Hypertext Transfer Protocol (HTTP)

The protocol used to transfer a Web page between a browser and a Web server is known as *HTTP*. HTTP operates at the application level. HTTP is a protocol used mainly to access data on the World Wide Web. HTTP functions like a combination of FTP and SMTP. It is similar to FTP because it transfers files, while HTTP is like SMTP because the data transferred between the client and the server looks like SMTP messages. However, HTTP differs from SMTP in the way that SMTP messages are stored and forwarded; HTTP messages are delivered immediately.

As a simple example, a browser sends an HTTP GET command to request a Web page from a server. A browser contacts a Web server directly to obtain a page. The browser begins with a URL, extracts the hostname section, uses DNS to map the name into an equivalent IP address, and uses the IP address to form a TCP connection to the server. Once the TCP connection is in place, the browser and Web server use HTTP to communicate. Thus, if the browser sends a request to retrieve a specific page, the server responds by sending a copy of the page.

A browser requests a Web page, and the server transfers a copy to the browser. HTTP also allows transfer from a browser to a server. HTTP allows browsers and servers to negotiate details such as the character set to be used during transfers. To improve response time, a browser caches a copy of each Web page it retrieves. HTTP allows a machine along the path between a browser and a server to act as a proxy server that caches Web pages and answers a browser's request from its cache. Proxy servers are an important part of the Web architecture because they reduce the load on servers.

In summary, a browser and server use HTTP to communicate. HTTP is an application-level protocol with explicit support for negotiation, proxy servers, caching, and persistent connections.

2.3.2 Hypertext Markup Language (HTML)

The browser architecture is composed of the controller and the interpreters to display a Web document on the screen. The controller can be one of the protocols such as HTTP, FTP, Gopher, or TELNET. The interpreter can be HTML or Java, depending on the type of document.

HTML is a language used to create Web pages. A markup language such as HTML is embedded in the file itself, and formatting instructions are stored with the text. Thus, any browser can read the instructions and format the text according to the workstation being used. Suppose a user creates formatted text on a Macintosh computer and stores it in a Web page, another user who is on an IBM computer is not able to receive the Web page because the two computers are using different formatting procedures. Consider a case where different word processors use different techniques or procedures to format text. To overcome these difficulties, HTML uses only ASCII characters for both main text and formatting instructions. Therefore, every computer can receive the whole document as an ASCII document.

Web Page

A Web page consists of two parts: the head and body. The head is the first part of a Web page. The head contains the file of the page and other parameters that the browser will use. The body contains the actual content of a page. The body includes the text and tags (marks). The text is the information contained in a page, whereas the tags define the appearance of the document.

Tags

Tags are marks that are embedded into the text. Every HTML tag is a name followed by an optional list of attributes. An attribute is followed by an equals sign (=) and the value of the attribute. Some tags are used alone; some are used in pairs. The tags used in pairs are called *starting and ending tags*. The starting tag can have attributes and values. The ending tag cannot have attributes or values, but must have a slash before the name. An example of starting and ending tags is shown below:

< TagName Attribute = Value Attribute = Value ... > (Starting tag)

< Tag Name > (Ending tag)

A tag is enclosed in two angled brackets like <A>and usually comes in pairs as <A>and </A>. The starting tag starts with the name of the tag, and the ending tag starts with a backslash followed by the name of the tag. A tag can have a list of attributes, each of which can be followed by an equals sign and a value associated with the attribute.

2.3.3   Common Gateway Interface (CGI)

A dynamic document is created by a Web server whenever a browser requests the document. When a request arrives, the Web server runs an application program that creates the dynamic document. Common Gateway Interface (CGI) is a technology that creates and handles dynamic documents. CGI is a set of standards that defines how a dynamic document should be written, how the input data should be supplied to the program, and how the output result should be used. CGI is not a new language, but it allows programmers to use any of several languages such as C, C++, Bourne Shell, Korn Shell, or Perl. A CGI program in its simplest form is code-written in one of the languages supporting the CGI.

2.3.4   Java

Java is a combination of a high-level programming language, a run-time environment, and a library that allows a programmer to write an active document and a browser to run it. It can also be used as a stand-alone program without using a browser. However, Java is mostly used to create a small application program of an applet.

2.4   File Transfer

The file transfer application allows users to send or receive a copy of a data file. Access to data on remote files takes two forms: whole-file copying and shared online access. FTP is the major file transfer protocol in the TCP/IP suite. TFTP provides a small, simple alternative to FTP for applications that need only file transfer. Network file system (NFS) provides online shared file access.

2.4.1   File Transfer Protocol (FTP)

FTP is the standard mechanism provided by TCP/IP for copying a file from one host to another. The FTP protocol is defined in RFC959. It is further defined in RFC 2227, 2640, 2773 for updated documentation.

In transferring files from one system to another, two systems may have different ways to represent text and data. Two systems may have different directory structures. All of these problems have been solved by FTP in a very simple and elegant way.

FTP differs from other client–server applications in that it establishes two connections between the hosts. One connection is used for data transfer (port 20), the other for control information (port 21). The control connection port remains open during the entire FTP session and is used to send control messages and client commands between the client and server. A data connection is established using an ephemeral port. The data connection is created each time a file is transferred between the client and server. Separation of commands and data transfer makes FTP more efficient. FTP allows the client to specify whether a file contains text (ASCII or EBCDIC character sets) or binary integers. FTP requires clients to authorize themselves by sending a log name and password to the server before requesting file transfers.

Since FTP is used only to send and receive files, it is very difficult for hackers to exploit.

2.4.2   Trivial File Transfer Protocol (TFTP)

TFTP is designed to simply copy a file without the need for all of the functionalities of the FTP protocol. TFTP is a protocol that quickly copies files because it does not require all the sophistication provided in FTP. TFTP can read or write a file for the client. Since TFTP restricts operations to simple file transfer and does not provide authentication, TFTP software is much smaller than FTP.

2.4.3   Network File System (NFS)

The NFS, developed by Sun Microsystems, provides online shared file access that is transparent and integrated. The file access mechanism accepts the request and automatically passes it to either the local file system software or to the NFS client, depending on whether the file is on the local disk or on a remote machine. When it receives a request, the client software uses the NFS protocol to contact the appropriate server on a remote machine and performs the requested operation. When the remote server replies, the client software returns the results to the application program.

Since Sun's Remote Procedure Call (RPC) and eXternal Data Representation (XDR) are defined separately from NFS, programmers can use them to build distributed applications.

2.5   E-Mail

In this section, we consider e-mail service and the protocols that support it. An e-mail facility allows users to send small notes or large voluminous memos across the Internet. E-mail is popular because it offers a fast, convenient method of transferring information and communicating.

2.5.1   Simple Mail Transfer Protocol (SMTP)

The SMTP provides a basic e-mail facility. SMTP is the protocol that transfers e-mail from one server to another. It provides a mechanism for transferring messages among separate servers. Features of SMTP include mailing lists, return receipts, and forwarding. SMTP accepts the incoming message and makes use of TCP to send it to an SMTP module on another server. The target SMTP module will make use of a local e-mail package to store the incoming message in a user's mailbox. Once the SMTP server identifies the IP address for the recipient's e-mail server, it sends the message through standard TCP/IP routing procedures.

Since SMTP is limited in its ability to queue messages at the receiving end, it is usually used with one of two other protocols, POP3 (Post Office Protocol Version 3) or IMAP (Internet Message Access Protocol), that let the user save messages in a server mailbox and download them periodically from the server. In other words, users typically use a program that uses SMTP for sending e-mail and either POP3 or IMAP for receiving messages that have been received for them at their local server. Most mail programs (such as Eudora) let you specify both an SMTP server and a POP server. On UNIX-based systems, sendmail is the most widely used SMTP server for e-mail. Earlier versions of sendmail presented many security risk problems. Through the years, however, sendmail has become much more secure, and can now be used with confidence. A commercial package, sendmail, includes a POP3 server and there is also a version for Windows NT.

Hackers often use different forms of attack with SMTP. A hacker might create a fake e-mail message and send it directly to an SMTP server. Other security risks associated with SMTP servers are denial-of-service attacks. Hackers will often flood an SMTP server with so many e-mails that the server cannot handle legitimate e-mail traffic. This type of flood effectively makes the SMTP server useless, thereby denying service to legitimate e-mail users. Another well-known risk of SMTP is the sending and receiving of viruses and Trojan horses. The information in the header of an e-mail message is easily forged. The body of an e-mail message contains standard text or a real message. Newer e-mail programs can send messages in HTML format. No viruses and Trojans can be contained within the header and body of an e-mail message, but they may be sent as attachments. The best defense against malicious attachments is to purchase an SMTP server that scans all messages for viruses, or to use a proxy server that scans all incoming and outgoing messages.

SMTP is usually implemented to operate over TCP port 25. The details of SMTP are in RFC 2821 of the IETF. An alternative to SMTP that is widely used in Europe is X.400.

2.5.2 Post Office Protocol Version 3 (POP3)

The most popular protocol used to transfer e-mail messages from a permanent mailbox to a local computer is known as the *POP3*. The user invokes a POP3 client, which creates a TCP connection to a POP3 server on the mailbox computer. The user first sends a login and a password to authenticate the session. Once authentication has been accepted, the user client sends commands to retrieve a copy of one or more messages and to delete the message from the permanent mailbox. The messages are stored and transferred as text files in RFC 2822 standard format.

Note that computers with a permanent mailbox must run two servers – an SMTP server accepts mail sent to a user and adds each incoming message to the user's permanent mailbox, and a POP3 server allows a user to extract messages from the mailbox and delete them. To ensure correct operation, the two servers must coordinate with the mailbox so that if a message arrives via SMTP while a user extracts messages via POP3, the mailbox is left in a valid state.

2.5.3 Internet Message Access Protocol (IMAP)

The IMAP is a standard protocol for accessing e-mail from your local server. IMAP4 (the latest version) is a client–server protocol in which e-mail is received and held for you by your Internet server. You (or your e-mail client) can view just the subject and the sender of the e-mail and then decide whether to download the mail. You can also create, manipulate, and delete folders or mailboxes on the server; delete messages; or search for certain e-mails. IMAP requires continual access to the server during the time that you are working with your mail.

A less sophisticated protocol is POP3. With POP3, your mail is saved for you in your mailbox on the server. When you read your mail, it is immediately downloaded to your computer and no longer maintained on the server.

IMAP can be thought of as a remote file server. POP can be thought of as a "store-and-forward" service.

POP and IMAP deal with receiving e-mail from your local server and are not to be confused with SMTP, a protocol for transferring e-mail between points on the Internet. You send e-mail by SMTP and a mail handler receives it on your recipient's behalf. Then the mail is read using POP or IMAP.

2.5.4   Multipurpose Internet Mail Extension (MIME)

The Multipurpose Internet Mail Extension (MIME) is defined to allow transmission of non-ASCII data via e-mail. MIME allows arbitrary data to be encoded in ASCII and then transmitted in a standard e-mail message. SMTP cannot be used for languages that are not supported by 7-bit ASCII characters. It cannot also be used for binary files or to send video or audio data.

MIME is a supplementary protocol that allows non-ASCII data to be sent through SMTP. MIME is a set of software functions that transforms non-ASCII data to ASCII data and vice versa.

2.6   Network Management Service

This section takes a look at a protocol that more directly supports administrative functions. RFC 1157 defines the SNMP.

2.6.1   Simple Network Management Protocol (SNMP)

The SNMP is an application-layer protocol that facilitates the exchange of management information between network devices. It is part of the TCP/IP protocol suite. SNMP enables network administrators to manage network performance, find and solve network problems, and plan for network growth.

There are two versions of SNMP, v1 and v2. Both versions have a number of features in common, but SNMP v2 offers enhancements, such as additional protocol operations.

SNMP version 1 is described in RFC 1157 and functions within the specifications of the Structure of Management Information (SMI). SNMP v1 operates over protocols such as the UDP, IP, OSI Connectionless Network Service (CLNS), Apple-Talk Datagram-Delivery Protocol (DDP), and Novell Internet Packet Exchange (IPX). SNMP v1 is widely used and is the *de facto* network management protocol in the Internet community.

SNMP is a simple request–response protocol. The network management system (NMS) issues a request, and managed devices return responses. This behavior is implemented using one of four protocol operations: Get, GetNext, Set, and Trap. The Get operation is used by the NMS to retrieve the value of one or more object instances from an agent. If the agent responding to the Get operation cannot provide values for all the object instances in a list, it provides no values. The GetNext operation is used by the NMS to retrieve the value of the next object instance in a table or list within an agent. The Set operation is used by the NMS to set the values of object instances within an agent. The Trap operation is used by agents to asynchronously inform the NMS of a significant event.

SNMP v2 is an evolution of the SNMP v1. It was originally published as a set of proposed Internet Standards in 1993. SNMP v2 functions within the specifications of the SMI which defines the rules for describing management information, using Abstract Syntax Notation One (ASN.1). The Get, GetNext, and Set operations used in SNMP v1 are exactly the same as those used in SNMP v2. However, SNMP v2 adds and enhances some protocol operations. SNMP v2 also defines two new protocol operations: GetBulk and Inform. The GetBulk operation is used by the NMS to efficiently retrieve large blocks of data, such as multiple rows in a table. GetBulk fills a response message with as much of the requested data as will fit. The Inform operation allows one NMS to send Trap information to another NMS and receive a response.

SNMP lacks any authentication capabilities, which results in vulnerability to a variety of security threats. These include masquerading, modification of information, message sequence, and timing modifications and disclosure.

2.7 Converting IP Addresses

To identify an entity, TCP/IP protocols use the IP address, which uniquely identifies the connection of a host to the Internet. However, users prefer a system that can map a name to an address or an address to a name. This section considers converting a name to an address and vice versa, mapping between high-level machine names and IP addresses.

2.7.1 Domain Name System (DNS)

The DNS uses a hierarchical naming scheme known as *domain names*. The mechanism that implements a machine name hierarchy for TCP/IP is called *DNS*. DNS has two conceptual aspects: the first specifies the name syntax and rules for delegating authority over names and the second specifies the implementation of a distributed computing system that efficiently maps names to addresses.

DNS is a protocol that can be used in different platforms. In the Internet, the domain name space is divided into three different sections: generic domain, country domain, and inverse domain. A DNS server maintains a list of hostnames and IP addresses, allowing computers that query them to find remote computers by specifying hostnames rather than IP addresses. DNS is a distributed database, and therefore, DNS servers can be configured to use a sequence of name servers, based on the domains in the name being looked for.

2.8   Routing Protocols

The Internet is a combination of networks connected by routers. When a datagram goes from a source to a destination, it will probably pass through many routers until it reaches the router attached to the destination network. A router chooses the route with the shortest metric. The metric assigned to each network depends on the type of protocol. The Routing Information Protocol (RIP) is a simple protocol which treats each network as equal. The Open Shortest Path First (OSPF) protocol is an interior routing protocol that is becoming very popular. Border Gateway Protocol (BGP) is an interautonomous system routing protocol which first appeared in 1989.

2.8.1   Routing Information Protocol (RIP)

The RIP is a protocol used to propagate routing information inside an autonomous system (AS). Today, the Internet is so large that one routing protocol cannot handle the task of updating the routing tables of all routers.

Therefore, the Internet is divided into ASs. An AS is a group of networks and routers under the authority of a single administration. Routing inside an AS is referred to as *interior routing*. RIP and OSPF are popular interior routing protocols used to update routing tables in an AS. Routing between ASs is referred to as *exterior routing*. RIP is a popular protocol which belongs to the interior routing protocol. It is a very simple protocol based on distance-vector routing, which uses the Bellman–Ford algorithm for calculating routing tables. An RIP routing table entry consists of a destination network address, the hop count to that destination, and the IP address of the next router. RIP uses three timers: the periodic timer controls the advertising of the update message, the expiration timer governs the validity of a route, and the garbage collection timer advertises the failure of a route. However, two shortcomings associated with the RIP are slow convergence and instability.

2.8.2   Open Shortest Path First (OSPF)

The OSPF is a new alternative to RIP as an interior routing protocol. It overcomes all the limitations of RIP. Link-state routing is a process by which each router shares its knowledge about its neighborhood with every other router in the area. OSPF uses link-state routing to update the routing tables in an area, as opposed to RIP, which is a distance-vector protocol. The term *distance-vector* means that messages sent by RIP contain a vector of distances (hop counts). In reality, the important difference between two protocols is that a link-state protocol always converges faster than a distance-vector protocol.

OSPF divides an AS in areas, defined as collections of networks, hosts, and routers. At the border of an area, area border routers summarize information about the area and send it to other areas. There is a special area called the *backbone* among the areas inside an AS. All the areas inside an AS must be connected to the backbone whose area identification is zero. OSPF defines four types of links: point-to-point, transient, stub, and virtual. Point-to-point links between routers do not need an IP address at each end. Unnumbered links can save IP addresses. A transient link is a network with several routers attached to it. A stub link is a network that is connected to only one router. When the link between two routers is broken, the administration may create a virtual link between them using a longer path that probably goes through several routers.

A simple authentication scheme can be used in OSPF. OSPF uses multicasting rather than broadcasting in order to reduce the load on systems not participating in OSPF. Distance-vector Multicast Routing Protocol (DVMRP) is used in conjunction with IGMP to handle multicast routing. DVMRP is a simple protocol based on distance-vector routing and the idea of MBONE. Multicast Open Shortest Path First (MOSPF), an extension to the OSPF protocol, adds a new type of packet (called the *group membership packet*) to the list of link-state advertisement packets. MOSPF also uses the configuration of MBONE and islands.

2.8.3   Border Gateway Protocol (BGP)

BGP is an exterior gateway protocol for communication between routers in different ASs. BGP is based on a routing method called *path-vector routing*. Refer to RFC 1772 (1991) which describes the use of BGP in the Internet. BGP version 3 is defined in RFC 1267 (1991) and BGP version 4 in RFC 1467 (1993).

Path-vector routing is different from both distance-vector routing and link-state routing. Path-vector routing does not have the instability or looping problems of distance-vector routing. Each entry in the routing table contains the destination network, the next router, and the path to reach the destination. The path is usually defined as an ordered list of ASs that a packet should travel through to reach the destination.

BGP is different from RIP and OSPF in that BGP uses TCP as its transport protocol. There are four types of BGP messages: open, update, keepalive, and notification. BGP detects the failure of either the link or the host on the other end of the TCP connection by sending a keepalive message to its neighbor on a regular basis.

2.9   Remote System Programs

High-level services allow users and programs to interact with automated services on remote machines and with remote users. This section describes programs that include Rlogin and TELNET.

2.9.1   TELNET

TELNET is a simple remote terminal protocol that allows a user to log on to a computer across an Internet. TELNET establishes a TCP connection, and then passes keystrokes from the user's keyboard directly to the remote computer as if they had been typed on a keyboard attached to the remote machine. TELNET also carries output from the remote machine back to the user's screen. The service is called *transparent* because it looks as if the user's keyboard and display attach directly to the remote machine. TELNET client software allows the user to specify a remote machine either by giving its domain name or IP address.

TELNET offers three basic services. First, it defines a network virtual terminal (NVT) that provides a standard interface to remote systems. Second, TELNET includes a mechanism that allows the client and server to negotiate options. Finally, TELNET treats both ends of the connection symmetrically.

2.9.2   Remote Login (Rlogin)

Rlogin was designed for remote login only between UNIX hosts. This makes it a simpler protocol than TELNET because option negotiation is not required when the operating system on the client and server are known in advance. Over the past few years, Rlogin has also ported to several non-UNIX environments. RFC 1282 specifies the Rlogin protocol.

When a user wants to access an application program or utility located on a remote machine, the user performs Rlogin. The user sends the keystrokes to the terminal driver where the local operating system accepts the characters but does not interpret them. The characters are sent to the TELNET client, which transforms the characters into a universal character set called NVT characters and delivers them to the local TCP/IP stack.

The commands or text (in NVT form) travel through the Internet and arrive at the TCP/IP stack at the remote machine. Here the characters are delivered to the operating system and passed to the TELNET server, which changes the characters to the corresponding characters understandable by the remote computer.

2.10   Social Networking Services

2.10.1   Facebook

Facebook is a social networking service launched by Mark Zuckerberg with his friends in February 2004. After registering, users may post messages on their own Web pages and leave messages and comments on other users' Web pages once accepted as friends. Facebook is not just a social networking service but a service platform on which users can develop commercial services such as online game and other online services based on the social network. As of February 2012, Facebook has more than 845 million active users. Facebook is ranked as the most used social networking service.

2.10.2   Twitter

Twitter is a microblogging service launched by a company, Odeo, in July 2006. It was an internal service for Odeo employees and later introduced in public. Twitter is known as the *SMS of the Internet*. Users post text-based messages up to 140 characters, known as *tweets*. Anyone may read the messages unless they are blocked by the owner. Users also can follow particular users' postings. A user following others is called a *follower*. As of January 2012, Twitter has more than 462 million active users.

2.10.3   Linkedin

Linkedin is a social networking service mainly used for professional networking, launched in December 2003. Users maintain a list of contact details of people with whom they have relationship, called *Connections*. Using the connections, employers can search for potential candidates and job seekers can contact hiring managers and find jobs for them. Plus, people also can find business opportunities recommended by people in the contact network. As of Q3 2011, Linkedin has more than 131 million members.

2.10.4   Groupon

Groupon, derived from "group coupon," is a Web site for discounted gift certificates, launched in November 2008. Groupon offers one coupon called *Groupon* per day. If a certain number of people sign up for the offer, the deal becomes available to all. If the predetermined minimum is not met, no one gets the deal. As of January 2011, Groupon has more than 50 million subscribers and 22 million Groupons have been sold in North America. The users have saved more than $980 million in North America.

2.11   Smart IT Devices

2.11.1   Smartphones

A smartphone is a mobile phone build on a mobile computing platform. This phone have more advanced computing ability and high-speed connectivity to the network via Wi-Fi and mobile broadband. The first smartphone was a feature phone with the functions of a personal digital assistant (PDA) and a mobile phone or camera phone. But today's models have more functions like e-mail, Internet Web browsing, online banking, GPS location tracking service, and low-end compact digital cameras. The most common mobile computing platform (OS) used by modern smartphones include Apple's iOS, Google's Android, Nokia's Symbian, RIM's BlackBerry OS, and embedded Linux distributions such as Maemo and MeeGo.

2.11.2   Smart TV

A smart TV is a television set that offers more advanced computing ability and connectivity to the network. This device often allows the user to install and run more advanced applications. Based on the complete operation system inside this device, this device services contents delivery from other computers or network-attached storage devices like photos, movies, and music. For this, DLNA service program like Windows Media Player on Windows PC or NAS, or Apple's iTunes are used. And the device services access to Internet-based services including Video-On-Demand (VOD), EPG, games, and social networking.

2.11.3   Video Game Console

A video game console is a customized computer system for playing video games on a TV. The video game consoles have a complete operating system. Based on this, the devices serve various functions like those of smart TV. The most common modern game consoles are Microsoft's Xbox360, Sony's Playstation 3, and Nintendo's Wii U.

2.12   Network Security Threats

2.12.1   Worm

A computer worm is a stand-alone malware program which replicates itself and spreads to other devices via network by utilizing vulnerabilities on the target devices. Unlike a computer virus, it does not need to attach itself to an existing program. Worms are usually made to harm the network by consuming bandwidth, whereas viruses usually corrupt or modify files on a targeted device.

2.12.2   Virus

A computer virus is a program that can replicate itself and spread from one computer to another. Viruses increase their chances of spreading to other computers by infecting files on an NFS or a file system that is accessed by other computers. Recently, viruses are distributed mainly to exploit personal computers for distributed denial-of-service (DDoS) attacks.

2.12.3   DDoS

A denial-of-service attack or DDoS attack is an attempt to make a computer or network resource unavailable to its users. Attackers typically target web sites or services such as search engines, banks, credit card payment gateways, and even servers in national security agencies. DDoS attack overloads and saturates the target machine with external communication requests, such that it cannot respond to legitimate traffic, or responds so slowly as to be rendered effectively unavailable.

2.13   Internet Security Threats

2.13.1   Phishing

Phishing is to acquire confidential information such as usernames, passwords, and credit card numbers by masquerading a trustworthy entity in an electronic communication such as e-mail and Web. An example of e-mail phishing is an e-mail which is disguised as an official e-mail from a well-known bank. The sender is trying to trick the recipient by keying in confidential information in order for putting a link in the e-mail leading to his own fake phishing site.

2.13.2 SNS Security Threats

Privacy Infringement

With the advancement of data mining technology, personal information on Social Network Services (SNS) such as Facebook and Twitter are easily obtained by an ordinary person, even a child. The revealed personal information is exploited to embarrass, to blackmail, or even to damage the image of its owner.

Spamming

Malicious SNS users can easily create unsolicited messages and produce overloaded traffic in the social networks, called *Social Network Spam*. The Spam not only overloads SNS servers so as to make it difficult in using the SNS, but also phishes the SNS users and directs them to malicious commercial sites like pornographic Web pages.

Identity Theft

A malicious SNS user can create a fake profile and ID impersonating a famous person or commercial brand. Such a squatting profile usually damages the reputation of the original person or brand, leading to financial and social loss.

2.14 Computer Security Threats

2.14.1 Exploit

An exploit is a piece of software, a chunk of data, or sequence of commands that takes advantage of a bug, glitch, or vulnerability in order to cause unintended or unanticipated behavior to occur on computer software, hardware, or electronic devices. This often refers to things like gaining control of a computer system or allowing privilege escalation or a denial-of-service attack.

Buffer Overflow

Buffer overflow is an anomaly where a program, while writing data to a buffer, overruns the buffer's boundary and overwrites adjacent memory. This is a special case of violation of memory safety. Buffer overflows can be triggered by inputs that are designed to execute code, or alter the way the program operates. This may result in erratic program behavior, including memory access errors, incorrect results, a crash, or a breach of system security. Thus, they are the basis of many software vulnerabilities and can be maliciously exploited.

Cross-Site Scripting

Cross-site scripting (XSS) is a type of computer security vulnerability typically found in Web applications that enables attackers to inject client-side script into Web pages viewed by other users. An XXS vulnerability may be used by attackers to bypass access controls such as the same origin policy.

Cross-Site Request Forgery

Cross-site request forgery, also known as a *one-click attack* or *session riding* and abbreviated as CSRF or XSRF, is a type of malicious exploit of a web site whereby unauthorized commands are transmitted from a user that the web site trusts. Unlike XSS, which exploits the trust a user has for a particular site, CSRF exploits the trust that a site has in a user's browser.

2.14.2 Password Cracking

In cryptanalysis and computer security, password cracking is the process of recovering passwords from data that has been stored in or transmitted by a computer system. A common approach is to repeatedly try guesses for the password. Another common approach is to say that you have "forgotten" the password and then changing it. The purpose of password cracking might be to help a user recover a forgotten password, to gain unauthorized access to a system, or as a preventive measure by system administrators to check for easily crackable passwords.

2.14.3 Rootkit

A rootkit is a malicious software designed to hide the existence of certain processes or programs from normal methods of detection and enables continued privileged access to a computer. Rootkit installation may be either automated or when an attacker installs it once they have obtained root or administrator access. Obtaining this access is a result of direct attack on a system. Once installed it becomes possible to hide the intrusion as well as to maintain privileged access. Like any software they can have a good purpose or a malicious purpose. Rootkit detection is difficult because a rootkit may be able to subvert the software that is intended to find it.

2.14.4 Trojan Horse

A Trojan horse is a stand-alone malicious program that does not attempt to infect other computers in a completely automatic manner without help from outside forces like other programs and human intervention. The term is derived from the Trojan Horse story in Greek mythology. Others rely on drive-by downloads in order to reach target computers.

Trojan may allow a hacker remote access to a target computer system. Once a Trojan has been installed on a target computer system, a hacker may have access to the computer remotely and perform various operations, limited by user privileges on the target computer system and the design of the Trojan. Popular Trojan Horses include Netbus, Back Orifice, Schoolbus, Executor, Silencer, and Striker.

2.14.5 Keylogging

Keylogging is the action of tracking (or logging) the keys struck on a keyboard, typically in a covert manner so that the person using the keyboard is unaware that their actions are being monitored. There are numerous keylogging methods, ranging from hardware- and software-based approaches to electromagnetic and acoustic analysis.

2.14.6 Spoofing Attack

A spoofing attack is a situation in which one person or program successfully masquerades as another by falsifying data and thereby gaining an illegitimate advantage.

ARP Spoofing

ARP spoofing is a computer hacking technique whereby an attacker sends fake ARP messages onto a LAN. ARP spoofing may allow an attacker to intercept data frames on a LAN, modify the traffic, or stop the traffic altogether.

IP Spoofing

IP spoofing refers to the creation of IP packets with a forged-source IP address, called *spoofing*, with the purpose of concealing the identity of the sender or impersonating another computing system.

E-mail Spoofing

E-mail spoofing is e-mail activity in which the sender address and other parts of the e-mail header are altered to appear as though the e-mail originated from a different source. Because core SMTP does not provide any authentication, it is easy to impersonate and forge e-mails.

Web Site Spoofing

Web site spoofing is the act of creating a web site, as a hoax, with the intention of misleading readers that the web site has been created by a different person or organization. Normally, the spoof web site will adopt the design of the target web site and sometimes has a similar URL. Another technique is to use a "cloaked" URL. By using domain forwarding, or inserting control characters, the URL can appear to be genuine while concealing the address of the actual web site.

2.14.7   Packet Sniffer

A packet sniffer is a computer program or a piece of computer hardware that can intercept and log traffic passing over a digital network or part of a network. As data streams flow across the network, the sniffer captures each packet and, if needed, decodes the packet's raw data, showing the values of various fields in the packet, and analyzes its content according to the appropriate RFC or other specifications.

2.14.8   Session Hijacking

Session hijacking is the exploitation of a valid computer session to gain unauthorized access to information or services in a computer system. In particular, it is used to refer to the theft of a magic cookie used to authenticate a user to a remote server. It has particular relevance to Web developers, as the HTTP cookies used to maintain a session on many Web sites can be easily stolen by an attacker using an intermediary computer or with access to the saved cookies on the victim's computer.

3

Global Trend of Mobile Wireless Technology

This chapter presents an outline of the evolution and migration from first-generation (1G) to third-generation (3G) mobile radio technologies, moving rapidly from 1G circuit-switched analog voice-only communications to 2G, 2.5G, and 3G packet-switched voice and data mobile wireless communications.

This chapter covers the technological development of wireless mobile communications in compliance with each iterative generation over the past decade. At present, mobile data service has been rapidly transforming to facilitate and ultimately profit from the increased demand for nonvoice service. Through aggressive 3G deployment plans, the world's major operators boast attractive and homogeneous portal offerings in all of their markets, notably in data, audio, and video multimedia application services. Despite the improbability of any major changes in the next 4–5 years, rapid technological advances have already bolstered talks for 3.5G and even 4G systems. New All-IP Wireless systems to reach a position to compete with current cellular networks; Wi-Fi technology, along with WiMAX (Worldwide Interoperability for Microwave Access) and TDD mode, may make it possible for new entrants to compete with incumbent mobile operators.

The chapter is separated into six parts and progresses in a systematic manner on wireless mobile communications.

3.1   1G Cellular Technology

1G technology refers to the earliest wireless networks. Data rates available for 1G wireless technologies were 9.6 kbps or lower.

Wireless Mobile Internet Security, Second Edition. Man Young Rhee.
© 2013 John Wiley & Sons, Ltd. Published 2013 by John Wiley & Sons, Ltd.

3.1.1 AMPS (Advanced Mobile Phone System)

- AT&T's Bell Labs developed the AMPS (Advanced Mobile Phone System) and deployed throughout North America.

3.1.2 NMT (Nordic Mobile Telephone)

- NMT (Nordic Mobile Telephone) is a 1G wireless technology that is the 1981 Nordic countries standard for analog cellular service.

3.1.3 TACS (Total Access Communications System)

- This is an analog FM communication system used in some parts of Europe and Asia (United Kingdom, Malaysia, China, etc.)

3.2 2G Mobile Radio Technology

There are two major 2G mobile telecommunications standards that have been dominating the global wireless market: GSM (Global System for Mobile Communications), developed at the beginning of the 1990s by the ETSI in Western Europe, and TDMA-136/CDMA IS-95 (Code Division Multiple Access), developed by the TIA in North America. Since the GSM standard was originally designed for voice, GSM was ill-suited to data transmission. Although GSM is the most widely used circuit-switched cellular system for voice communications, GSM networks were not optimized for high-speed data, image, and other multimedia applications and services. The ETSI hence upgraded the GSM standard, albeit still in the circuit-switched mode. The High-Speed Circuit-Switched Data (HSCSD) technology was first deployed to enable higher data rates. The General Packet Radio Service (GPRS) soon followed, introducing packet-switched mode. Finally, EDGE (Enhanced Data Rate for GSM Evolution or Enhanced Data Rate for Global Evolution) was introduced to further increase the data speeds provided by GPRS. At present, most GSM-based networks are expected to evolve 3G Universal Mobile Telecommunication System (UMTS).

Qualcomm developed a mobile communication technology based on the CDMA spectrum-sharing technique. This network technology, which is modulated by codes, is called *cdmaOne IS-95A/B* and was standardized by the TIA. Under the influence of Qualcomm, the IS-95 technology continues to evolve steadily to provide higher data rates, such as 3G CDMA2000 1x networks that were standardized by the 3GPP2 (3rd Generation Partnership Project).

The IS-54, based on the TDMA mode, was the first North American digital telephony standard. This standard was also adapted for use in wideband Personal Communications Systems (PCS) networks under the name TDMA-136. The IS-54 was primarily used by US operators, but limitation in data transfers due to the use of a relatively narrowband stopped its use in December 2001.

In addition, two other proprietary technologies classified as 2G systems include NTT DoCoMo's i-mode in Japan and the Wireless Application Protocol (WAP), the *de facto* standard created by the WAP Forum, which was founded in 1997 through the initiative of Nokia, Motorola, Ericsson, Phone.com, and other such companies. NTT DoCoMo's i-mode is the mobile Internet access system that provides service over the packet-switched network. The i-mode gives users a new range of capabilities, offering voice and data cellular service in one convenient package. The WAP Forum's initial aim was to establish a universal and open standard to provide wireless users access to the Internet. The WAP was designed to deliver Internet content by adapting to the features and constraints of mobile phones. The WAP technology cannot be exactly defined as a 2G system; it was designed to work with all wireless network technologies, beginning with a majority of 2G (GSM, GPRS, PDC (Personal Digital Cellular), IS-95, TDMA-136) and 3G systems. However, with the commercial failure of the launch of WAP at the beginning of 2000, the European industry missed the mobile data service explosion. Only SMS appeared to offer access to the WAP portal via GSM networks. The MMS technology will make it possible to overcome the technological constraints of SMS and to further enhance existing services with SMS. The i-mode service was started in February 1999; WAP 2.0 was released by the WAP Forum in August 2001; and M-service Phase 2 has been started in GSM-A.

2G technologies are digital in nature and provide improved system performance and security. 2G technology data rates vary from 9.6 to 14.4 kbps.

3.2.1 CDPD (Cellular Digital Packet Data), North American Protocol

- Cellular Digital Packet Data (CDPD) is a TCP/IP-based mobile data-only service that runs on AMPS networks.
- The first service was launched in 1999.
- Since CDPD runs on analog cellular networks, it requires a modem to convert the TCP/IP-based data into analog signals when sending and receiving.

3.2.2 GSM (Global System for Mobile Communications)

Since the GSM standard was originally designed for voice, GSM is ill-switched to data transmission. GSM networks were not optimized for high-speed data, image, and other multimedia applications because GSM is the most widely used circuit-switched cellular system for voice communications. The ETSI hence upgraded the GSM standard, albeit still in circuit-switched mode. Since 2G systems transmit voice traffic over wireless and limit their data handling capacity, evolution is therefore needed for providing high data-rate services.

- First digital cellular system (DCS) (first standard in Europe).
- Most widely deployed digital network in the world to date.
- Not optimized for high-speed data, image, and other multimedia services.

GSM can operate four distinct frequency bands.

- *GSM 450.* GSM 450 supports very large cells in the 450-MHz band. It was designed for countries with a low user density, such as in Africa. It may also replace the original 1981 NMT 450 analog networks used in the 450-MHz band. NMT is a 1G wireless technology.

- *GSM 900*. The original GSM system was called *GSM 900* because the original frequency band was represented by 900 MHz. To provide additional capacity and to enable higher subscriber densities, two other systems were added afterward, namely, GSM 1800 and GSM 1900.

- *GSM 1800*. GSM 1800 (or DCS 1800) is an adapted version of GSM 900, operating in the 1800-MHz frequency range. Any GSM system operating in a higher frequency band requires a large number of base stations than that required for an original GSM system. The availability of a wider band of spectrum and a reduction in cell size will enable GSM 1800 to handle more subscribers than GSM 900. The smaller cells, in fact, give improved indoor coverage and low power requirements.

- *GSM 1900*. GSM 1900 (or PCS 1900) is a GSM 1800 variation designed for use in North America, which uses the 1900-MHz band. Since 1993, phase 2 of the specifications has included both the GSM 900 and DCS 1800 in common documents. The GSM 1900 system has been added to the IS-136 D-AMPS (Digital Advanced Mobile Phone System) and IS-95 CDMA system, both operated at the 1900-MHz band.

The ITU (International Telecommunication Union) has allocated the GSM radio spectrum with the following bands.

- GSM 900. Uplink: 890–915 MHz
             Downlink: 935–960 MHz
- GSM 1800. Uplink: 1710–1785 MHz
             Downlink: 1805–1880 MHz
- GSM 1900. Uplink: 1850–1910 MHz
             Downlink: 1930–1990 MHz

In this list, uplink designates connection from the mobile station to the base station and downlink denotes connection from the base station to the mobile station.

3.2.3   TDMA-136 or IS-54

- It is the first US digital standard developed by the TIA in 1992.
- It was used as the basis for GSM and implemented in North America in some PCS for Personal Communication Services.
- TDMA divides the frequency range into a series of channels that are divided into time slots. Each slot can carry one voice or data transmission. Owing to the limitation in data transfers, its use was stopped in December 2001.

3.2.4   iDEN (Integrated Digital Enhanced Network)

- The Integrated Digital Enhanced Network (iDEN) digital technology developed by Motorola integrates four network communication services into one device: dispatch radio, full-duplex telephone interconnect, 140-character SMS, and data transmission.
- iDEN uses TDMA and is considered a 2G technology.

3.2.5 cdmaOne IS-95A

- Qualcomm developed the CDMA IS-95 standard in 1993, based on the CDMA spectrum-sharing technique. This network technology is called *IS-95*.
- IS-95 was followed by IS-95A revision in May 1995 and was marketed under the name.
- cdmaOne IS-95A transmits signal over 1.25-MHz channels that are modulated by codes.
- CDMA consistently provides better capacity for voice and data communications.
- It reached 100 million subscribers within only 6 years of commercial deployment.

3.2.6 PDC (Personal Digital Cellular)

- PDC (a Japanese proprietary 2G technology) was developed on a national basis by two of the three mobile carriers.
- PDC is a TDMA-based Japanese standard for digital cellular service.

3.2.7 i-mode

- NTT DoCoMo's i-mode is a mobile Internet access system.
- A significant part of the Japanese market experienced a dynamic growth preferring proprietary systems, but it may have a rather negative effect of absence of roaming between Japan and other countries.
- KDDI is developing the CDMA 1x technology (12 million CDMA 2000 1x subscribers) to provide service over the packet-switched network, thus offering a new range of capabilities for voice and data cellular service.

3.2.8 WAP (Wireless Application Protocol)

- The goal of WAP technology is to find a universal and open standard to provide wireless access to the Internet.
- WAP technology is not exactly classified as a 2G system, but rather designed for working with all wireless network technologies starting from 2G to 3G system.
- With the commercial failure of the launch of WAP at the beginning of 2000, the European industry missed the mobile data service explosion.
- Only SMS appeared to offer access to the WAP portal via GSM networks.

3.3 2.5G Mobile Radio Technology

2G mobile radio technologies enable voice traffic and limited data traffic, such as SMS, to transmit over wireless. Improvements must be made in order to facilitate high data-rate services that ultimately allow transmitting and receiving high-quality data and video to and from the Internet. However, the data handling capabilities of 2G mobile systems are limited.

For 2.5G systems, HSCSD in circuit-switched mode was the first step in GSM evolution in increased data transmission rates, reaching maximum speeds of about 43 kbps (three simultaneous GSM circuits running at 14.4 kbps). The drawback of HSCSD, when compared to GPRS, is the several time slots used in circuit mode, whereas GPRS uses several time slots in packet mode. HSCSD is considered an interim technology to GPRS, which offers instant connectivity at higher speeds.

GPRS is the evolution of GSM for higher data rates within the GSM carrier spacing. GPRS introduces packet transmission for data services, replacing GSM's circuit-switched mode. EDGE (an upgraded version of GPRS) was designed for a network to evolve its current 2G GSM system to support faster throughput and to give operators the opportunity to understand the new technology before the complete 3G rollout. EDGE is a higher-bandwidth version of GPRS, with transmission rates up to 384 kbps. Such high speeds can aptly support wireless multimedia applications.

ITU defined the IMT 2000 program for the 3GPP as well as 3GPP2 as a main part of the 3G technical framework. The primary objective of the standardization activities for IMT 2000 is to develop a globally unified standard for worldwide roaming and mobile multimedia services. In order to achieve these goals, the ITU has strived to create harmonized recommendations supported by technical forums such as the 3GPP and 3GPP2. The ITU-R and ITU-T are the main bodies that produce recommendations for IMT 2000. The 3GPP, created in late 1998, is the group responsible for standardizing UMTS with the Wideband Code Division Multiple Access (WCDMA) technology.

The 3GPP has so far released: Release 99, Release 4, Release 5, and Release 6. Release 99 includes the basic capabilities and functionalities of UMTS. Release 4 was contributed by the CWTS and incorporated as WCDMA/TDD, but it was frozen in 2003. As Release 5 was successfully completed in March 2002, 3GPP is moving toward the next release, Release 6, to further improve performance and to enhance capabilities.

The 3GPP2 specifies an air interface based on cdmaOne technology and the cdma2000 interface to increase capability and to enable faster data communication. The technical area of 3GPP2 is similar to that of the 3GPP.

cdmaOne IS-95B is enhanced through the migration from cdmaOne IS-95A. In the non-GSM regions (notably the United States and South Korea), network operators are preparing next-generation wireless systems based on cdmaOne IS-95A/B. The first phase with IS-95A was fully covered in TIA/EIA/IS-95 + TSB74. In late 1997, the second phase with IS-95B brought about improvements in terms of capacity, allowing data transmission at 64 kbps. The 2.5G equivalent for CDMA operators is a technology called *CDMA2000 1x*. The IS-95 technology continues to improve to provide higher data rates toward the natural evolution of CDMA2000 1x networks for 3G, that is, CDMA2000 1xEV-DO (1x Evolution Data Only) and 1xEV-DV (1x Evolution Data and Voice), both standardized by 3GPP2, which are covered in Section 3.4.

2G mobile radio systems are originally designed for transmission of voice traffic over wireless, but soon after found some limitation for their data handling capacity. Henceforth, evolution is needed to provide their high rate services.

2.5G systems are aimed to transmit and receive both high-quality images and data to and from the Internet.

3.3.1   ECSD (Enhanced Circuit-Switched Data)

- Enhanced Circuit-Switched Data (ECSD) is implemented over GSM-type systems.

3.3.2   HSCSD (High-Speed Circuit-Switched Data)

- This is a Circuit-Switched protocol based on GSM. Its transmission rates are up to 38.4 kbps, higher than the usual 9.6 kbps of GSM. Even though some of the European operators have already started to offer HSCSD services, the technology still lacks widespread support. It may be considered as an interim technology to GPRS.

3.3.3   GPRS (General Packet Radio Service)

- GPRS is the packet-mode extension of GSM.
- GPRS is an IP-based packet-switched wireless protocol that allows for burst transmission speeds between 30 and 50 kbps, compared to the 10 kbps of GSM.
- The first commercial launch was in 2001.
- Several problems were encountered in services until the end of 2002.
- The number of GPRS users in Western Europe was estimated at 28 million at the end of 2003.

3.3.4   EDGE (Enhanced Data rate for GSM Evolution)

- EDGE is the first stage in the evolution of GSM technology.
- It is the upgraded version of the GPRS System.
- It has a transmission rate of up to 150 kbps, allowing wireless multimedia services.
- It uses 8PSK modulator, compared to GSM's GMSK (Gaussian Minimum-Shift Keying) modulation.
- TDMA-136 and GSM/GPRS operators plan to use in North American deployments.
- A few European operators indicated their interest in EDGE as a complement to UMTS.
- The first EDGE service was marketed at the end of 2004.

3.3.5   cdmaOne IS-95B

- 2.5G IS-95B was followed by 2G IS-95A in late 1997.
- It brought about improvements in terms of capacity and allows data transmission at 64 kbps.
- IS-95B forms compatibility standard for 800-MHz cellular mobile system and 1.8- to 2.0-GHz CDMA PCS.
- They ensure that a mobile station can obtain service in a cellular system or PCS manufactured according to this standard.

3.4 3G Mobile Radio Technology (Situation and Status of 3G)

3G mobile technologies referred to cellular radio systems for mobile technology. ITU defined the 3G technical framework as a part of the IMT 2000 program. The 3GPPis responsible for standardizing UMTS at a global level. It is composed of several international standardization bodies involved with defining 3G technologies. TDMA-136 and GSM/GPRS operators plan to use UMTS (3G), which is an advanced version of EDGE (2.5G). GSM/GPRS operators plan to deploy UMTS with WCDMA technology. Unlike CDMA2000, WCDMA will be deployed in the frequency bandwidths identified for 3G, leading some American operators to adopt EDGE. TDMA networks are steadily being replaced with GSM-evolved technology, that is, from GPRS, to EDGE, and finally to the 3G WCDMA (UMTS) standard.

The 3GPP2 is the international organization in charge of the standardization of CDMA2000, which in turn is the 3G evolution of the IS-95A/B standards. CDMA2000 represents a family of ITU-approved IMT 2000 (3G) standards including, CDMA2000 1x networks, CDMA2000 1xEV-DO, and 1xEV-DV technologies. The first CDMA2000 1x networks were launched in Korea in October 2000 by SK Telecom and LG Telecom. CDMA2000 1xEV-DO was recognized as an IMT 2000 technology with data rates of 2.4 Mbps on 1.25-MHz CDMA carrier. As 1xEV-DO makes use of the existing suite of IP, operating systems, and software applications, it builds on the architecture of CDMA2000 1x network, while preserving seamless backward compatibility with IS-95A/B and CDMA2000 1x. CDMA2000 1xEV-DV provides integrated voice with simultaneous high-speed packet data services at a speed of up to 3.09 Mbps. But 1xEV-DV is still in the developmental stage. The CDMA2000 family of air interfaces operates with an IS-4 network, and an IP network, or a GSM-WAP network. This provides operators with tremendous flexibility with the network and assures backward compatibility with deployed terminal base.

In June 2000, the Ministry of Postal and Transportation of Japan awarded 3G licenses to three mobile operators, namely, NTT DoCoMo, KDDI, and Vodafone KK (J-Phone), via a comparative bidding process. NTT DoCoMo's i-mode (2G) is the first mobile Internet service in the world with 42 million subscribers at the end of March 2004. NTT DoCoMo commercially launched a WCDMA network based on the UMTS standard in Tokyo under the name FOMA. Since the launch of the 3G FOMA service in October 2001, many new and exclusive services have been made accessible to FOMA subscribers, including video telephony and i-motion's video clip distribution service, as well as i-motion's mail messaging service.

KDDI, unlike its two competitors (NTT DoCoMo and Vodafone KK), opted for 3G CDMA2000 technology. In April 2002, KDDI opened its CDMA2000 1x network using 3G bandwidths. KDDI began deployment of the 3G version of 1x. KDDI launched its commercial CDMA2000 1xEV-DO services nationwide in the first quarter of 2004. KDDI has offered WIN services based on CDMA 1xEV-DO technology from November 2003. Since March 2004, KDDI has offered a BREW (a new application platform) terminal with Bluetooth technology.

The three major South Korean mobile operators, namely, SK Telecom, KTF, and LG Telecom, provide 2G and 2.5G mobile services using the Qualcomm-developed CDMA IS-95 system and its successor CDMA2000 1x, thus enabling a maximum bandwidth capacity of 144 kbps. In fact, SK Telecom was the first operator in the world to launch a CDMA2000 1x service in October 2000. As for 3G technologies, the Ministry of Information and Communication of the Korean government decided to grant licenses according to the type of technology, either WCDMA or CDMA2000. The Korean government decided to grant WCDMA licenses to SK Telecom and KTF, and a CDMA2000 license to LG Telecom. Thus, SK Telecom and KTF, holders of 3G WCDMA license, are deploying networks based on the CDMA2000 1xEV-DO standard in their existing frequency bandwidths. This resulting system is capable of providing 3G services with a maximum bandwidth of 2 Mbps.

SK Telecom was the first in the world to launch a CDMA2000 1xEV-DO network in January 2002, followed by KTF in May 2002. It was one of the very first operators in the world, including KDDI (2003), to launch 3G service. It launched the WCDMA service at the end of 2003.

KTF is the mobile subsidiary of KT Corporation. KTF launched services based on CDMA2000 1x technology in June 2001 and on CDMA2000 1xEV-DO technology in May 2002. Deployment uses Qualcomm's BREW platform, but the Java WIPI platform was used by all operators toward the end of 2003. Launch of the WCDMA service started at the end of 2003.

LG Telecom is the subsidiary of LG Corporation. Launch of services using CDMA2000 1x technology was in August 2001. LG Telecom is the holder of a 3G license based on the CDMA2000 standard.

Seven operators control the market for mobile telephony in the United States. Instead of facing open competition between three or even four major mobile operators, the North American market is now structured around two main operators, namely, AT&T Wireless + Cingular wireless announced in February 2004 and the less important merger between Verizon Wireless and Qwest Wireless. In response to this situation, other operators may have to join forces in terms of operations and capital.

Verizon Wireless is a leader in the United States in terms of number of subscribers; it covers 40 of the 50 key markets in the United States. Launched in January 2002, Verizon Wireless was the first major US operator to commercially provide a CDMA2000 1x network. Verizon Wireless launched its DMA2000 1xEV-DO broadband access in Washington, DC and San Diego in 2003, and plans to continue deploying its market on a national level.

Cingular Wireless is the subsidiary of the regional operators SBC and BellSouth. It originally was the operator of a TDMA-136 network, but Cingular Wireless decided to migrate to GPRS, with the launch in March 2001. Cingular Wireless is also the first US operator to launch EDGE in June 2003.

AT&T Wireless is the North American operator that was acquired by Cingular Wireless in February 2004. AT&T Wireless launched GPRS in mid-2001 and coverage of all markets was done by the end of 2002. In April 2002, AT&T Wireless launched an i-mode type service called *mMode* on the GPRS network. The EDGE deployment plan was established in mid-2002 and nationally launched in November 2003. AT&T Wireless announced its UMTS deployment plan at the beginning of 2003, and on 26 December 2003, the company announced the four markets (San Francisco, San Diego, Seattle, and Dallas) in which the first WCDMA networks were to be deployed by the end of 2004. In partnership with NTT DoCoMo, the first UMTS call between New York and Tokyo was carried out on 12 November 2002.

The Wireless industry worldwide will put their continuous efforts to derive the technology evolution to support even greater data throughput and better network capacity than those offered by 3G. In a 4G environment, an aggressive and iterative generation of all wireless mobile communications (a combination of 2G, 2.5G, and 3G) as well as Bluetooth and IEEE 802.11 could all coexist for attaining faster data throughput and greater network capacity.

As IMT 2000 has just been commercialized, new standardization work should commence for the systems beyond IMT 2000. Those new systems will be expected to provide more sophisticated services to meet the further demands of the wireless community. The overall objectives of the future development of IMT 2000 and of systems beyond IMT 2000 include new radio access capabilities and a new IP-based core network for resulting in another phase of harmonization.

As 3GPP's Release 5 of UMTS was almost completed in March 2002, 3GPP is moving toward Release 6, which aims to further improve performance and to enhance capabilities. The interworking between WLAN and UMTS has been proved to be one of the keys for providing both flexibility when accessing multiple radio resources and mobility between WLAN and the 3G system in various mobile environments.

As one of the major applications, MBMS (Multimedia Broadcast Multicast Service) may pioneer a new service which allows broadcast of multimedia messaging and video/music streaming capabilities. HSDPA (High-Speed Data Packet Access) will represent a change in WCDMA systems and could be compatible with existing networks. HSDPA may enable packet transmission to provide speeds of 8–10 Mbps for the downlink in UMTS channels of 5 MHz. With the MIMO (Multiple Input Multiple Output) function, a speed of 20 Mbps could even be reached for providing the throughput. HSDPA systems will be in a position to compete with Wi-Fi services at certain mobile markets: NTT DoCoMo in Japan, several Western European operators, and Cingular Wireless and Verizon Wireless in the United States.

OFDM (Orthogonal Frequency Division Multiple Access) is being studied as a radio access technology that may drastically increase data rates using a large number of orthogonal frequencies. It is also foreseen that OFDM could be a promising candidate for what is called the *4G mobile system*.

The 3GPP2 has been working on an evolution of CDMA technology to enhance new features. CDMA2000, backed by the United States (primarily by Qualcomm), is the direct successor to cdmaOne IS-95A/B networks. There are two phases to deploy CDMA2000, that is, CDMA2000 1x and CDMA2000 1xEV. CDMA2000 1xEV is the final stage in the evolution of cdmaOne network to 3G. The transition from 1x to 1xEV takes place in two phases: CDMA2000 1xEV-DO and 1xEV-DV.

CDMA2000 1xEV-DO is the first phase, which uses a separate carrier for traffic and data. 1xEV-DO may function on a bi-mode operation (1x for voice and EV-DO for data only). By the end of 2000, a specification for High Rate Packet Data (HRPD) was issued to enhance downlink data transmission. HRPD, sometimes called the *1xEV-DO*, allows mobile terminals to easily access the IP network through a high-speed data communication link. As the 1xEV-DO was primarily devised for data communication only, another radio channel was required for speed communication, which led to the development of the 1xEV-DV. CDMA2000 1xEV-DV builds on the architecture of CDMA2000 1x while preserving seamless backward compatibility with cdmaOne IS-95A/B and CDMA2000 1x. CDMA2000 1xEV-DV (was approved by the 3GPP2 in June 2000 and was submitted to ITU for approval in July 2002) provides integrated voice with simultaneous high-speed packet data services such as video, videoconferencing, and other multimedia services at speeds of up to 3.09 Mbps. In order to support multimedia services, it is necessary to provide simultaneous speech and data communication using the same carrier frequency.

The systems beyond IMT 2000 include new radio access capabilities and a new IP-based core network to be realized in the future, around 2010. Owing to the tireless efforts of ITU and 3GPPs, a global consensus has been recognized to further develop a worldwide harmonized standard that will make it easier to improve mobile services and stimulate the mobile market.

3G mobile technologies are referred to cellular radio systems for mobile technology. The ITU defined the 3G technical framework as a part of IMT 2000 program.

3.4.1   UMTS (Universal Mobile Telecommunication System)

- UMTS (the name given by ETSI) is the first European implementation of IMT 2000 standard.
- UMTS has a proposed data rate of 2 Mbps using a combination of TDMA and WCDMA operations at 2 GHz.
- UMTS standardization is based on the WCDMA radio interface and a GSM/GPRS core network, allowing easy migration from 2G to 3G for GSM operators.

3.4.2   HSDPA (High-Speed Downlink Packet Access)

HSDPA system of UMTS Release 5 was introduced to provide high-speed data rates in downlink. UMTS (WCDMA) uses fixed modulation and coding scheme, but HSDPA, using HS-DSCH (High-Speed Downlink Shared Channel), helps obtain flexible high-data-rate service by Adaptive Modulation and Coding (AMC) technique. The HS-DSCH operates in an environment where certain cells are updated with HSDPA functionality. HSDPA operational features are based on AMC, HARQ, FCS (Fast Cell Selection), Scheduling at Node B, and MIMO antenna technique. These operational techniques are primarily aimed for increasing throughput, reducing delay, and achieving high peak rates. These functionalities should rely on a new type of HS-DSCH transport channel to which Node B is terminated. HS-DSCH is only applicable to packet-switched-domain Radio Access Bearers (RABs). HSDPA representing a change in WCDMA systems could be compatible with existing networks (Wi-Fi/Bluetooth service markets). HSDPA may enable packet transmission providing speeds of 8–10 Mbps for the downlink in UMTS channels of 5 MHz. With the MIMO function, a speed of 20 Mbps could even be reached for providing faster throughput. New technologies such as Release 6 and Release 7 have been proposed for further enhancement of the HSDPA system.

3.4.3 CDMA2000 1x

- CDMA2000 1x evolved from the IS-95 network in late 1999, and the ITU designated 1x as a 3G technology.
- It provides double voice and an increased data speed of 60–100 kbps, but it can theoretically provide a data speed of up to 144 kbps.
- The natural evolution of the CDMA network for 3G is toward the deployment of CDMA 1xEV-DO and EV-DV, both standardized by 3GPP2.
- CDMA2000 1x technology supports voice and data on a single CDMA channel with a bandwidth of 1.25 MHz and constitutes the migration from CDMA to 3G networks.

3.4.4 CDMA2000 1xEV (1x Evolution)

- CDMA2000 1xEV standardization within 3GPP2 is mostly driven by Qualcomm. 1xEV can be deployed both in the frequency bands used by 2G systems and in IMT 2000 frequency bands.

3.4.5 CDMA2000 1xEV-DO (1x Evolution Data Only)

- CDMA2000 1xEV-DO uses a separate carrier for traffic and data. It offers data speeds of up to 2.4 Mbps on one CDMA 1.25-MHz carrier.
- The 1xEV-DO provides bi-mode functions (1x for voice and 1xEV-DO for data).
- The first CDMA2000 1xEV-DO networks were launched by SK Telecom in January 2002 and by KTF in May 2002 in South Korea.

3.4.6 CDMA2000 1xEV-DV (1x Evolution Data Voice)

- CDMA 1xEV-DV builds on the architecture of CDMA2000 1x while preserving seamless backward compatibility with IS-95A/B and CDMA2000 1x.
- The migration requires only simple upgrades to the BTS (base transceiver station), BSC (base station controller), PNSN, and AAA (authentication, authorization, and accounting).
- CDMA2000 1xEV-DV provides integrated voice with simultaneous high-speed packet data services such as video, videoconferencing, and other multimedia services at a speed of up to 3.09 Mbps.
- The standard was approved by 3GPP2 in June 2002 and was submitted to the ITU for approval in July 2002.

3.5   3G UMTS Security-Related Encryption Algorithm

In spring 2002, the SAGE (Security Algorithms Group of Experts) initiated the task of designing a new encryption algorithm for GSM, ECSD, GPRS, and UMTS encryptions. These new algorithms were intended to implement dual-mode handsets for operating with both GSM and UMTS modes.

The 3GPP Task Force specified three encryption algorithms.

- A5/3 algorithms for GSM and ECSD;
- GEA3 algorithms for GPRS and EDGE;
- *f*8 algorithm for UMTS.

The common aspect of all these encryption algorithms is given by the name *KGCORE function* that is based on the KASUMI block cipher.

3.5.1   KASUMI Encryption Function

KASUMI algorithm is based on a form of output-feedback mode (OFM) as a keystream generator illustrated in Figure 3.1.

The three ciphering algorithms are all very similar and they use KASUMI as a keystream generator. KASUMI, as a keystream generator, is a block cipher that produces a 64-bit output from a 64-bit input under the control of a 128-bit key, as shown in Figure 3.2.

The 64-bit input is divided into 32-bit block L_0 and R_0. The outputs of each round are

$$R_i = L_{i-1}$$
$$L_i = R_{i-1} + f_i(L_{i-1}, RK_i) \quad \text{for } i = 1, 2, \ldots, 8,$$

where f_i denotes the round function with L_{i-1} and RK_i denotes the round key. The subscript i denotes the number of cipher rounds.

$$RK_i = (KL_i, KO_i, KI_i)$$

At the end of each round, the left 32-bit block and right 32-bit block should be swapped.

The produced ciphertext at the final round 8 should be the 64-bit string from $L_8 || R_8$.

Round function f_i is constructed from two associated subfunctions FL_i and FO_i with their respective subkeys KL_i for FL_i and (KO_i, KI_i) for FO_i, followed by a bitwise *XOR* operation with the previous branch stream (R_i or L_i).

In the odd round, the *FL* subfunction is performed first, followed by the *FO* subfunction.

In the even round, the *FO* subfunction is performed first, followed by the *FL* subfunction.

- The *FO* function consists of a three-round network with a 16-bit nonlinear *FI* function.

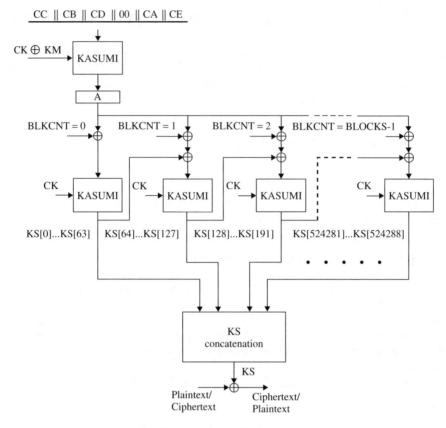

$$CC \parallel CB \parallel CD \parallel 00 \parallel CA \parallel CE$$

CC: Frame-dependent 32-bit input
CB: 5-bit bearer input
CD: 1-bit input indicating the direction of transmission
CA: 8-bit input initialized as 0000 0000
CE: 16-bit input initialized as 0000 0000 0000 0000

Figure 3.1   The *f8* algorithm architecture.

- The *FI* nonlinear function consists of four-round operations with two S-boxes of S9 and S7. These two S-boxes perform in the binary extension field $GF(2^m)$. Specifically, the S9 performs the x^5 in $GF(2^9)$ operation and S7 performs the x^{81} in $GF(2^7)$ operation.

- The *FL* function transforms the 32-bit data with two 16-bit subkeys KL_{i1} and KL_{i2} by means of *AND*, *OR*, and 1-bit cyclic left shift ($<<<1$) operations.

- The *FO* function divides the 32-bit input data into two 16-bit blocks. The left block is *XOR*ed with the 16-bit subkey KO_{ij}, transferred by *FI* function with a 16-bit subkey KI_{ij} and *XOR*ed with the right block R_0. This routine is iterated three times with swaps of the left and right blocks. A 16-bit data block entering the *FI* function is

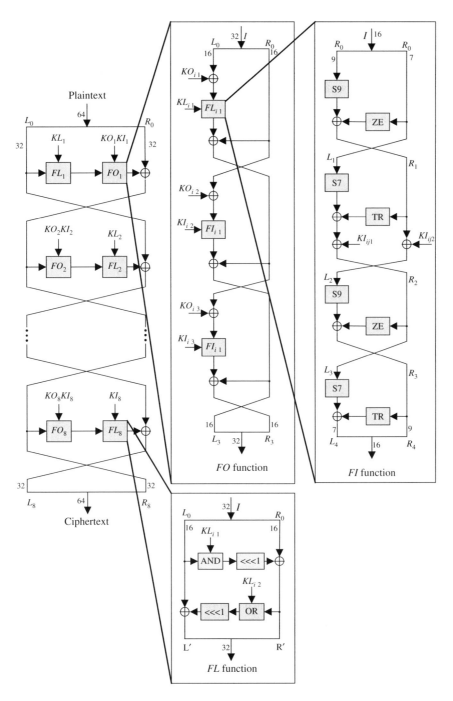

Figure 3.2 The KASUMI encryption data path.

divided into two smaller blocks for S-box transformation. The leftmost 9 bits become one block, and the rightmost 7 bits become another block. Thus, they are transformed twice using the 9-bit S-box S9 and the 7-bit S-box S7, respectively. As the bit length is different, Zero-Extension (ZE) should be done to the 7-bit blocks by adding two 0s, and the two most significant bits of the 9-bit blocks are truncated (TR). In the middle of the four-round network, XOR operation will be done with the 16-bit subkey KI_{ij}, where KI_{ij1} is 7 bits and KI_{ij2} is 9 bits.

- The 32-bit FL function transforms the 32-bit data, which is divided into two 16-bit blocks L_0 and R_0 each. These L_0 and R_0 are processed with two 16-bit subkeys KL_{i1} and KL_{i2}, respectively, by using AND, OR, XOR, and 1-bit left shift ($<<<1$), and finally, the two 16-bit blocks L' and R' are derived from L_0 and R_0.

FL Function

The FL_i function comprises a 32-bit data input E and a 32-bit subkey KL_i, which is also split into two 16-bit subkeys, KL_{i1} and KL_{i2}, so that $KL_i = KL_{i1}||KL_{i2}$.

The input data I is split into two 16-bit data units L_0 and R_0 so that $I = L_0||R_0$. If the 32-bit output of the FL function is defined as L' and R', it can then be expressed as

$$L' = L_0 \oplus ((R' \odot KL_{i2}) <<< 1)$$

$$R' = R_0 \oplus ((L_0 \boxplus KL_{i1}) <<< 1)$$

The FL function is a linear function, which is expressed using the following logical definitions:

$X \odot Y$ (or XY): The bitwise AND of X and Y (logical multiplication of two 16-bit X and Y).
$X \boxplus Y$ (or $X + Y$): The bitwise OR of X and Y (logical addition of two 16-bit X and Y).
$\oplus$: The bitwise XORing.
$X <<< Y$: Rotation X to the left by Y, that is, $<<<1$ denotes shifting 1 bit to the left.

The FL function is a linear function whose main purpose is to make individual bits harder to track through the rounds by executing scrambling.

FO Function

The FO function constitutes the nonlinear part of the KASUMI round function. The FO is a permutation of 32-bit blocks, but due to its three-round structure, it can be distinguished from a randomly chosen permutation when used for given plaintexts. Consideration was given to improving the diffusion properties of a three-round FO that can be used in an attack on the full eight-round KASUMI.

The input to the function FO_i comprises a 32-bit data input I and two sets of subkeys, that is, a 48-bit KO_i and a 48-bit KI_i. The 32-bit data input is split into two halves,

L_0 and R_0, so that $I = L_0 || R_0$, whereas the 48-bit subkeys are subdivided into three 16-bit subkeys.

$$KO_i = KO_{i1} || KO_{i2} || KO_{i3}$$

$$KI_i = KI_{i1} || KI_{i2} || KI_{i3}$$

For each integer j with $1 \leqslant j \leqslant 3$, the operation of the jth round of the function FO_i is defined as

$$R_1 = FI_{i1}(L_0 \oplus KO_{i1}, KI_{i1}) \oplus R_0$$

$$L_1 = R_0$$

Generally,

$$R_j = FI_{ij}(L_{j-1} \oplus KO_{ij}, KI_{ij}) \oplus R_{j-1}$$

$$L_j = R_{j-1}$$

The output from the FO_i function is defined as the 32-bit data block $L_3 || R_3$.

FI Function

The FI function is depicted in the rightmost block of Figure 3.2.

The FI function FI_{ij} takes a 16-bit data input I and a 16-bit subkey KI_{ij}. The input I is split into two unequal components, a 9-bit left half L_0 and a 7-bit right half R_0, so that $I = L_0 || R_0 = 16$ bits. Similarly, the subkey KI_{ij} is split into a 7-bit component KI_{ij1} and 9-bit component KI_{ij2}, so that $KI_{ij} = KI_{ij1} || KI_{ij2}$.

Each FI function FI_{ij} uses two S-boxes, S7 and S9. S7 maps a 7-bit input R_0 to a 7-bit output, while S9 maps a 9-bit input to a 9-bit output. The FI function also uses two additional functions, which are designated by ZE and TR. These functions are defined as follows:

- ZE(d) takes a 7-bit data string d and converts it to a 9-bit data string by appending two 0 bits to the most significant end of d.
- TR(d) takes a 9-bit data string d and converts it to a 7-bit value by discarding the two most significant bits of d.

The functions FI_{ij} is defined by the following series of operations:

$$L_1 = R_0$$

$$R_1 = S9[L_0] \oplus ZE(R_0)$$

$$L_2 = R_1 \oplus KI_{ij2}$$

$$R_2 = S7[L_1] \oplus TR(R_1) \oplus KI_{ij1}$$

$$L_3 = R_2$$

$$R_3 = S9[L_2] \oplus ZE(R_2)$$

$$L_4 = S7[L_3] \oplus TR(R_3)$$

$$R_4 = R_3$$

Finally, the output of the FI_{ij} function is the 16-bit data block $L_4 \| R_4$. The FI function is the basic randomizing function of KASUMI with 16-bit input and 16-bit output. It is composed of a four-round structure using two nonlinear substitution boxes S7 and S9. The S-boxes S7 and S9 have been designed to avoid linear structures in the FI functions. The interested reader is recommended to study the $f8$ algorithm presented in the author's book "Mobile Communication Systems and Security."

4

Symmetric Block Ciphers

This chapter deals with some important block ciphers that have been developed in the past. They are IDEA (International Data Encryption Algorithm; 1992), RC5 (1995), RC6 (1996), DES (Data Encryption Standard; 1977), and AES (Advanced Encryption Standard; 2001). The AES specifies an FIPS-approved symmetric block cipher, which will soon come to be used in lieu of Triple DES or RC6.

4.1   Data Encryption Standard (DES)

In the late 1960s, IBM initiated a Lucifer research project, led by Horst Feistel, for computer cryptography. This project ended in 1971 and LUCIFER was first known as a *block cipher* that operated on blocks of 64 bits, using a key size of 128 bits. Soon after this IBM embarked on another effort to develop a commercial encryption scheme, which was later called *DES*. This research effort was led by Walter Tuchman. The outcome of this effort was a refined version of Lucifer that was more resistant to cryptanalysis.

   In 1973, the National Bureau of Standards (NBS), now the National Institute of Standards and Technology (NIST), issued a public request for proposals for a national cipher standard. IBM submitted the research results of the DES project as a possible candidate. The NBS requested the National Security Agency (NSA) to evaluate the algorithm's security and to determine its suitability as a federal standard. In November 1976, the DES was adopted as a federal standard and authorized for use on all unclassified US government communications. The official description of the standard, FIPS PUB 46, *Data Encryption Standard* was published on 15 January 1977. The DES algorithm was the best one proposed and was adopted in 1977 as the DES even though there was much criticism of its key length (which had changed from Lucifer's original 128 bits to 64 bits) and the design criteria for the internal structure of DES, that is, S-box. Nevertheless, DES has survived remarkably well over 20 years of intense cryptanalysis and has been a worldwide standard for over 18 years. The recent work on differential cryptanalysis seems to indicate that DES has a very strong internal structure.

Wireless Mobile Internet Security, Second Edition. Man Young Rhee.
© 2013 John Wiley & Sons, Ltd. Published 2013 by John Wiley & Sons, Ltd.

Since the terms of the standard stipulate that it be reviewed every 5 years, on 6 March 1987, the NBS published in the *Federal Register* a request for comments on the second 5-year review. The comment period closed on 10 December 1992. After much debate, DES was reaffirmed as a US government standard until 1992 because there was still no alternative for DES. The NIST again solicited a review to assess the continued adequacy of DES to protect computer data. In 1993, NIST formally solicited comments on the recertification of DES. After reviewing many comments and technical inputs, NIST recommended that the useful lifetime of DES would end in the late 1990s. In 2001, the AES, known as the *Rijndael algorithm*, became an FIPS-approved advanced symmetric cipher algorithm. AES will be a strong advanced algorithm in lieu of DES.

The DES is now a basic security device employed by worldwide organizations. Therefore, it is likely that DES will continue to provide network communications, stored data, passwords, and access control systems.

4.1.1 Description of the Algorithm

DES is the most notable example of a conventional cryptosystem. Since it has been well documented for over 20 years, it will not be discussed in detail here.

DES is a symmetric block cipher, operating on 64-bit blocks using a 56-bit key. DES encrypts data in blocks of 64 bits. The input to the algorithm is a 64-bit block of plaintext and the output from the algorithm is a 64-bit block of ciphertext after 16 rounds of identical operations. The key length is 56 bits by stripping off the 8 parity bits, ignoring every eighth bit from the given 64-bit key.

As with any block encryption scheme, there are two inputs to the encryption function: the 64-bit plaintext to be encrypted and the 56-bit key. The basic building block of DES is a suitable combination of permutation and substitution on the plaintext block (16 times). Substitution is accomplished via table lookups in S-boxes. Both encryption and decryption use the same algorithm except for processing the key schedule in the reverse order.

The plaintext block X is first transposed under the initial permutation IP, giving $X_0 = \text{IP}(X) = (L_0, R_0)$. After passing through 16 rounds of permutation, XORs, and substitutions, it is transposed under the inverse permutation IP^{-1} to generate the ciphertext block Y. If $X_i = (L_i, R_i)$ denotes the result of the ith round encryption, then we have

$$L_i = R_{i-1}$$
$$R_i = L_{i-1} \oplus \text{f}(R_{i-1}, K_i)$$

The ith round encryption of DES algorithm is shown in Figure 4.1. The block diagram for computing the f(R, K) function is shown in Figure 4.2. The decryption process can be derived from the encryption terms as follows:

$$R_{i-1} = L_i$$
$$L_{i-1} = R_i \oplus \text{f}(R_{i-1}, K_i) = R_i \oplus \text{f}(L_i, K_i)$$

If the output of the ith round encryption is $L_i \| R_i$, then the corresponding input to the $(16-i)$th round decryption is $R_i \| L_i$. The input to the first-round decryption is equal to

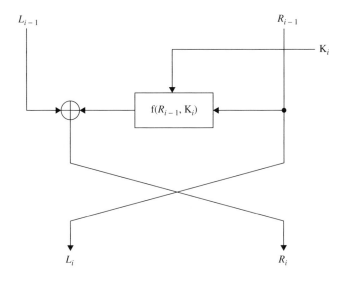

Figure 4.1   The ith round of DES algorithm.

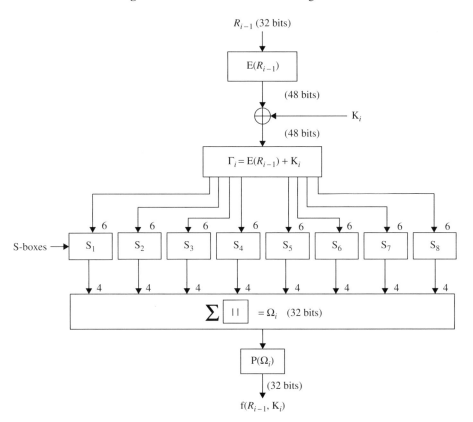

Figure 4.2   Computation of the f function.

the 32-bit swap of the output of the sixteenth round encryption process. The output of the first-round decryption is $L_{15}\|R_{15}$, which is the 32-bit swap of the input to the sixteenth round of encryption.

4.1.2 Key Schedule

The 64-bit input key is initially reduced to a 56-bit key by ignoring every eighth bit. This is described in Table 4.1. These ignored 8 bits, $k_8, k_{16}, k_{24}, k_{32}, k_{40}, k_{48}, k_{56}, k_{64}$, are used as a parity check to ensure that each byte is of old parity and no errors have entered the key.

After the 56-bit key is extracted, they are divided into two 28-bit halves and loaded into two working registers. The halves in registers are shifted left either one or two positions, depending on the round. The number of bits shifted is given in Table 4.2.

After being shifted, the halves of 56 bits (C_i, D_i), $1 \leqslant i \leqslant 16$, are used as the key input to the next iteration. These halves are concatenated in the ordered set and serve as input to the permuted choice 2 (Table 4.3), which produces a 48-bit key output. Thus, a different 48-bit key is generated for each round of DES. These 48-bit keys, $K_1, K_2, \ldots, K_{16}$, are used for encryption at each round in the order from K_1 through K_{16}. The key schedule for DES is illustrated in Figure 4.3.

With a key length of 56 bits, these are $2^{56} = 7.2 \times 10^{16}$ possible keys. Assuming that, on average, half the key space has to be searched, a single machine performing one DES encryption per microsecond would take more than 1000 years to break the cipher. Therefore, a brute-force attack on DES appears to be impractical.

Table 4.1 Permuted choice 1 (PC-1)

57	49	41	33	25	17	9	1	58	50	42	34	26	18
10	2	59	51	43	35	27	19	11	3	60	52	44	36
63	55	47	39	31	23	15	7	62	54	46	38	30	22
14	6	61	53	45	37	29	21	13	5	28	20	12	4

Table 4.2 Schedule for key shifts

Round number	1	2	3	4	5	6	7	8	9	10	11	12	13	14	15	16
Number of left shifts	1	1	2	2	2	2	2	2	1	2	2	2	2	2	2	1

Table 4.3 Permuted choice 2 (PC-2)

14	17	11	24	1	5	3	28	15	6	21	10
23	19	12	4	26	8	16	7	27	20	13	2
41	52	31	37	47	55	30	40	51	45	33	48
44	49	39	56	34	53	46	42	50	36	29	32

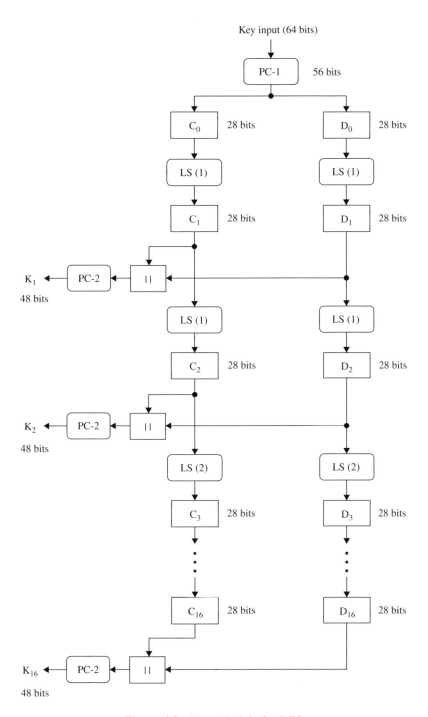

Figure 4.3 Key schedule for DES.

Example 4.1   Assume that a 64-bit key input is $K = 581fbc94d3a452ea$, including 8 parity bits. Find the first three round keys only: K_1, K_2, and K_3.

The register contents C_0 (left) and D_0 (right) are computed using Table 4.1:

$$C_0 = bcd1a45$$

$$D_0 = d22e87f$$

Using Table 4.2, the blocks C_1 and D_1 are obtained from the block C_0 and D_0, respectively, by shifting 1 bit to the left as follows:

$$C_1 = 79a348b$$

$$D_1 = a45d0ff$$

The 48-bit key K_1 is derived using Table 4.3 (PC-2) by inputting the concatenated block $(C_1 \| D_1)$ such that $K_1 = 27a169e58dda$.

The concatenated block $(C_2 \| D_2)$ is computed from $(C_1 \| D_1)$ by shifting 1 bit to the left as shown below:

$$(C_2 \| D_2) = f346916 \quad 48ba1ff$$

Using Table 4.3 (PC-2), the 48-bit key K_2 at round 2 is computed as $K_2 = da91ddd7b748$. Similarly, $(C_3 \| D_3)$ is generated from shifting $(C_2 \| D_2)$ by 2 bits to the left as follows:

$$(C_3 \| D_3) = cd1a456 \quad 22e87fd$$

Using Table 4.3, we have

$$K_3 = 1dc24bf89768$$

In a similar fashion, all the other 16-round keys can be computed and the set of entire DES keys is listed as follows:

$K_1 = 27a169e58dda$	$K_2 = da91ddd7b748$
$K_3 = 1dc24bf89768$	$K_4 = 2359ae58fe2e$
$K_5 = b829c57c7cb8$	$K_6 = 116e39a9787b$
$K_7 = c535b4a7fa32$	$K_8 = d68ec5b50f76$
$K_9 = e80d33d75314$	$K_{10} = e5aa2dd123ec$
$K_{11} = 83b69cf0ba8d$	$K_{12} = 7c1ef27236bf$
$K_{13} = f6f0483f39ab$	$K_{14} = 0ac756267973$
$K_{15} = 6c591f67a976$	$K_{16} = 4f57a0c6c35b$

4.1.3 DES Encryption

DES operates on a 64-bit block of plaintext. After initial permutation, the block is split into two blocks L_i (left) and R_i (right), each 32 bits in length. This permuted plaintext

Table 4.4 Initial permutation (IP)

	58	50	42	34	26	18	10	2
L_i	60	52	44	36	28	20	12	4
	62	54	46	38	30	22	14	6
	64	56	48	40	32	24	16	8
	57	49	41	33	25	17	9	1
R_i	59	51	43	35	27	19	11	3
	61	53	45	37	29	21	13	5
	63	55	47	39	31	23	15	7

Table 4.5 E bit-selection table

32	1	2	3	4	5
4	5	6	7	8	9
8	9	10	11	12	13
12	13	14	15	16	17
16	17	18	19	20	21
20	21	22	23	24	25
24	25	26	27	28	29
28	29	30	31	32	1

(Table 4.4) has bit 58 of the input as its first bit, bit 50 as its second bit, and so on down to bit 7 as the last bit. The right half of the data, R_i, is expanded to 48 bits according to Table 4.5 of an expansion permutation.

The expansion symbol E of $E(R_i)$ denotes a function which takes the 32-bit R_i as input and produces the 48-bit $E(R_i)$ as output. The purpose of this operation is twofold – to make the output the same size as the key for the XOR and to provide a longer result that is compressed during the S-box substitution operation.

After the compressed key K_i is XORed with the expanded block $E(R_{i-1})$ such that $\Gamma_i = E(R_{i-1}) \oplus K_i$ for $1 \leqslant i \leqslant 15$, this 48-bit Γ_i moves to substitution operations that are performed by eight S_i-boxes. The 48-bit Γ_i is divided into eight 6-bit blocks. Each 6-bit block is operated on by a separate S_i-box, as shown in Figure 4.2. Each S_i-box is a table of 4 rows and 16 columns as shown in Table 4.6. This 48-bit input Γ_i to the S-boxes are passed through a nonlinear S-box transformation to produce the 32-bit output.

If each S_i denotes a matrix box defined in Table 4.6 and A denotes an input block of 6 bits, then $S_i(A)$ is defined as follows: the first and last bits of A represent the row number of the matrix S_i, while the middle 4 bits of A represent a column number of S_i in the range from 0 to 15.

For example, for the input (101110) to S_5-box, denoted as S_5^{10} (0111), the first and last bits combine to form 10, which corresponds to the row 2 (actually third row) of S_5. The middle 4 bits combine to form 0111, which corresponds to the column 7 (actually

Table 4.6 S-boxes

		0	1	2	3	4	5	6	7	8	9	10	11	12	13	14	15
	0	14	4	13	1	2	15	11	8	3	10	6	12	5	9	0	7
S_1	1	0	15	7	4	14	2	13	1	10	6	12	11	9	5	3	8
	2	4	1	14	8	13	6	2	11	15	12	9	7	3	10	5	0
	3	15	12	8	2	4	9	1	7	5	11	3	14	10	0	6	13
	0	15	1	8	14	6	11	3	4	9	7	2	13	12	0	5	10
S_2	1	3	13	4	7	15	2	8	14	12	0	1	10	6	9	11	5
	2	0	14	7	11	10	4	13	1	5	8	12	6	9	3	2	15
	3	13	8	10	1	3	15	4	2	11	6	7	12	0	5	14	9
	0	10	0	9	14	6	3	15	5	1	13	12	7	11	4	2	8
S_3	1	13	7	0	9	3	4	6	10	2	8	5	14	12	11	15	1
	2	13	6	4	9	8	15	3	0	11	1	2	12	5	10	14	7
	3	1	10	13	0	6	9	8	7	4	15	14	3	11	5	2	12
	0	7	13	14	3	0	6	9	10	1	2	8	5	11	12	4	15
S_4	1	13	8	11	5	6	15	0	3	4	7	2	12	1	10	14	9
	2	10	6	9	0	12	11	7	13	15	1	3	14	5	2	8	4
	3	3	15	0	6	10	1	13	8	9	4	5	11	12	7	2	14
	0	2	12	4	1	7	10	11	6	8	5	3	15	13	0	14	9
S_5	1	14	11	2	12	4	7	13	1	5	0	15	10	3	9	8	6
	2	4	2	1	11	10	13	7	8	15	9	12	5	6	3	0	14
	3	11	8	12	7	1	14	2	13	6	15	0	9	10	4	5	3
	0	12	1	10	15	9	2	6	8	0	13	3	4	14	7	5	11
S_6	1	10	15	4	2	7	12	9	5	6	1	13	14	0	11	3	8
	2	9	14	15	5	2	8	12	3	7	0	4	10	1	13	11	6
	3	4	3	2	12	9	5	15	10	11	14	1	7	6	0	8	13
	0	4	11	2	14	15	0	8	13	3	12	9	7	5	10	6	1
S_7	1	13	0	11	7	4	9	1	10	14	3	5	12	2	15	8	6
	2	1	4	11	13	12	3	7	14	10	15	6	8	0	5	9	2
	3	6	11	13	8	1	4	10	7	9	5	0	15	14	2	3	12
	0	13	2	8	4	6	15	11	1	10	9	3	14	5	0	12	7
S_8	1	1	15	13	8	10	3	7	4	12	5	6	11	0	14	9	2
	2	7	11	4	1	9	12	14	2	0	6	10	13	15	3	5	8
	3	2	1	14	7	4	10	8	13	15	12	9	0	3	5	6	11

the eighth column) of the same S_5-box. Thus, the entry under row 2, column 7 of S_5-box is computed as:

$$S_5^{10}(0111) = S_5^2(7) = 8 \text{ (hexadecimal)} = 1000 \text{ (binary)}$$

Thus, the value of 1000 is substituted for 101110; that is, the 4-bit output 1000 from S_5 is substituted for the 6-bit input 101110 to S_5. Eight 4-bit blocks are the S-box output resulting from the substitution phase, which recombine into a single 32-bit block Ω_i by concatenation. This 32-bit output Ω_i of the S-box substitution is permuted according to Table 4.7. This permutation maps each input bit of Ω_i to an output position of $P(\Omega_i)$. The output $P(\Omega_i)$ are obtained from the input Ω_i by taking the sixteenth bit of Ω_i as the first bit of $P(\Omega_i)$, the seventh bit as the second bit of $P(\Omega_i)$, and so on until the twenty-fifth

Table 4.7 Permutation function P

16	7	20	21
29	12	28	17
1	15	23	26
5	18	31	10
2	8	24	14
32	27	3	9
19	13	30	6
22	11	4	25

Table 4.8 Inverse of initial permutation, IP^{-1}

40	8	48	16	56	24	64	32
39	7	47	15	55	23	63	31
38	6	46	14	54	22	62	30
37	5	45	13	53	21	61	29
36	4	44	12	52	20	60	28
35	3	43	11	51	19	59	27
34	2	42	10	50	18	58	26
33	1	41	9	49	17	57	25

bit of Ω_i is taken as the thirty-second bit of $P(\Omega_i)$. Finally, the permuted result is XORed with the left half L_i of the initial permuted 64-bit block. Then the left and right halves are swapped and another round begins. The final permutation is the inverse of the initial permutation, and is described in Table 4.8 (IP^{-1}). Note here that the left and right halves are not swapped after the last round of DES. Instead, the concatenated block $R_{16}\|L_{16}$ is used as the input to the final permutation of Table 4.8 (IP^{-1}). Thus, the overall structure for DES algorithm is shown in Figure 4.4.

Example 4.2 Suppose the 64-bit plaintext is $X = 3570e2f1ba4682c7$, and the same key as used in Example 4.1, $K = 581fbc94d3a452ea$ is assumed again. The first two-round keys are, respectively, $K_1 = 27a169e58dda$ and $K_2 = da91ddd76748$.

For the purpose of demonstration, the DES encryption aims to limit the first two rounds only. The plaintext X splits into two blocks (L_0, R_0) using Table 4.4 (IP), such that $L_0 = ae1ba189$ and $R_0 = dc1f10f4$.

The 32-bit R_0 is expanded to the 48-bit $E(R_0)$ such that $E(R_0) = 6f80fe8a17a9$.

The key-dependent function Γ_i is computed by XORing $E(R_0)$ with the first-round key K_1, such that

$$\Gamma_1 = E(R_0) \oplus K_1$$

$$= 4821976f9a73$$

This 48-bit Γ_1 is first divided into eight 6-bit blocks and then fed into eight S_i-boxes. The output Ω_1 resulting from the S-box substitution phase is computed as $\Omega_1 = a1ec961c$.

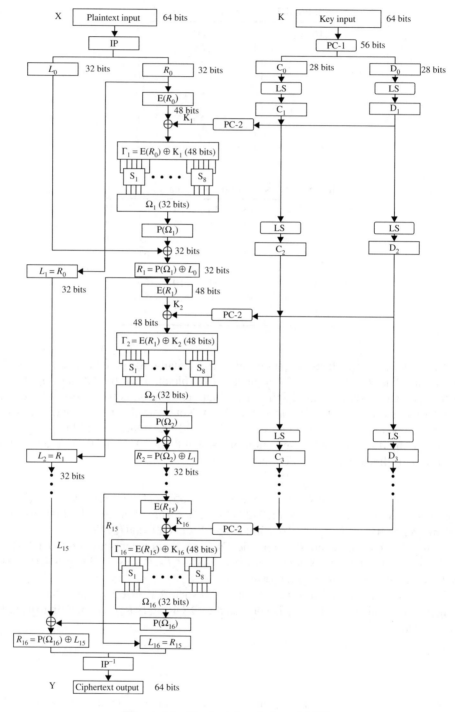

Figure 4.4 Block cipher design of DES.

Using Table 4.7, the permuted values of Ω_1 are $P(\Omega_1) = 2ba1536c$. Modulo-2 addition of $P(\Omega_1)$ with L_0 becomes

$$R_1 = P(\Omega_1) \oplus L_0$$
$$= 85baf2e5$$

Since $L_1 = R_0$, this gives $L_1 = dc1f10f4$.

Consider next the second-round encryption. Expanding R_1 with the aid of Table 4.5 yields $E(R_1) = c0bdf57a570b$. XORing $E(R_1)$ with K_2 produces

$$\Gamma_2 = E(R_1) \oplus K_2$$
$$= 1a2c28ade043$$

The substitution operations with S-boxes yields the 32-bit output Ω_2, such that $\Omega_2 = 1ebcebdf$. Using Table 4.7, the permutation $P(\Omega_2)$ becomes $P(\Omega_2) = 5f3e39f7$. Thus, the right-half output R_2 after round 2 is computed as

$$R_2 = P(\Omega_2) \oplus L_1$$
$$= 83212903$$

The left-half output L_2 after round 2 is immediately obtained as

$$L_2 = R_1 = 85baf2e5$$

Concatenation of R_2 with L_2 is called the *preoutput block* in our 2-round cipher system. The preoutput is then subjected to the inverse permutation of Table 4.8. Thus, the output of the DES algorithm at the end of the second round becomes the ciphertext Y:

$$Y = IP^{-1}(R_2 \| L_2)$$
$$= d7698224283e0aea$$

4.1.4 DES Decryption

The decryption algorithm is exactly identical to the encryption algorithm except that the round keys are used in the reverse order. Since the encryption keys for each round are $K_1, K_2, \ldots, K_{16}$, the decryption keys for each round are $K_{16}, K_{15}, \ldots, K_1$. Therefore, the same algorithm works for both encryption and decryption. The DES decryption process will be explained in the following example.

Example 4.3 Recover the plaintext X from the ciphertext $Y = d7698224283e0aea$ (computed in Example 4.2). Using Table 4.4 in the first place, divide the ciphertext Y into the two blocks:

$$R_2 = 83212903$$

$$L_2 = 85baf2e5$$

Applying Table 4.5 to L_2 yields $E(L_2) = c0bdf57a570b$.

$E(L_2)$ is XORed with K_2 such that

$$\Gamma_2 = E(L_2) \oplus K_2$$
$$= 1a2c28ade043$$

This is the 48-bit input to the S-boxes.

After the substitution phase of S-boxes, the 32-bit output Ω_2 from the S-boxes is computed as $\Omega_2 = 1ebcebdf$. From Table 4.7, the permuted values of Ω_2 are $P(\Omega_2) = 5f3e39f7$.

Moving up to the first round, we have $L_1 = P(\Omega_2) \oplus R_2 = dc1f10f4$.
Applying Table 4.5 for L_1 yields $E(L_1) = 6f80fe8a17a9$.

XORing $E(L_1)$ with K_1, we obtain the 48-bit input to the S-boxes.

$$\Gamma_1 = E(L_1) \oplus K_1$$
$$= 4821976f9a73$$

The 32-bit output from the S-boxes is computed as:

$$\Omega_1 = a1ec961c$$

Using Table 4.7 for permutation, we have

$$P(\Omega_1) = 2ba1536c$$

The preoutput block can be computed as follows:

$$L_0 = P(\Omega_1) \oplus R_1 = ae1ba189$$
$$R_0 = L_1 = dc1f10f4$$
$$L_0 \| R_0 = ae1ba189dc1f10f4 (\text{preoutput block})$$

Applying Table 4.8 (IP^{-1}) to the preoutput block, the plaintext X is restored as follows:

$$X = IP^{-1}(L_0 \| R_0)$$
$$= 3570e2f1ba4682c7$$

Example 4.4   Consider the encryption problem of plaintext

$X = 785ac3a4bd0fe12d$ with the original input key

$K = 38a84ff898b90b8f$.

The 48-bit round keys from K_1 through K_{16} are computed from the 56-bit key blocks through a series of permutations and left shifts, as shown below:

Compressed round keys

$K_1 = 034b8fccfd2e$	$K_2 = 6e26890ddd29$
$K_3 = 5b9c0cca7c70$	$K_4 = 48a8dae9cb3c$
$K_5 = 34ec2e915e9a$	$K_6 = e22d02dd1235$
$K_7 = 68ae35936aec$	$K_8 = c5b41a30bb95$
$K_9 = c043eebe209d$	$K_{10} = b0d331a373c7$
$K_{11} = 851b6336a3a3$	$K_{12} = a372d5f60d47$
$K_{13} = 1d57c04ea3da$	$K_{14} = 5251f975f549$
$K_{15} = 9dc1456a946a$	$K_{16} = 9f2d1a5ad5fa$

The 64-bit plaintext X is split into two blocks (L_0, R_0), according to Table 4.4 (IP), such that

$L_0 = 4713b8f4$

$R_0 = 5cd9b326$

The 32-bit R_0 is spread out and scrambled in 48 bits, using Table 4.5, such that $E(R_0) = 2f96f3da690c$.

The 48-bit input to the S-box, Γ_1, is computed as:

$\Gamma_1 = E(R_0) \oplus K_1$

$\quad = 2cdd7c169422$

The 32-bit output from the S-box is $\Omega_1 = 28e8293b$.

Using Table 4.7, $P(\Omega_1)$ becomes

$P(\Omega_1) = 1a0b2fc4$

XORing $P(\Omega_1)$ with L_0 yields

$R_1 = P(\Omega_1) \oplus L_0$

$\quad = 5d189730$

which is the right-half output after round 1.

Since $L_1 = R_0$, the left-half output L_1 after round 1 is $L_1 = 5cd9b326$. The first round of encryption has been completed.

In a similar fashion, the 16-round output block (L_i, R_i), $2 \leqslant i \leqslant 16$, can be computed as follows:

Table for encryption blocks (L_i, R_i), $1 \leqslant i \leqslant 16$

i	L_i	R_i
1	5cd9b326	5d189730
2	5d189730	e0e7a039
3	e0e7a039	61123d5d

i	L_i	R_i
4	61123d5d	a6f29581
5	a6f29581	c1fe0f05
6	c1fe0f05	8e6f6798
7	8e6f6798	6bc34455
8	6bc34455	ec6d1ab8
9	ec6d1ab8	d0d10423
10	d0d10423	56a0e201
11	56a0e201	b6c73726
12	b6c73726	6ff2ef60
13	6ff2ef60	f04bf1ad
14	f04bf1ad	f0d35530
15	f0d35530	07b5cf74
16	07b5cf74	09ef5b69

The preoutput block (R_{16}, L_{16}) is the concatenation of R_{16} with L_{16}. Using Table 4.8 (IP^{-1}), the ciphertext Y, which is the output of the DES, can be computed as:

$Y = $ fd9cba5d26331f38

Example 4.5 Consider the decryption process of the ciphertext
$Y = $ fd9cba5d26331f38, which was obtained in Example 4.4. Applying Table 4.4 (IP) to the 64-bit ciphertext Y, the two blocks (R_{16}, L_{16}) after swap yields

$R_{16} = $ 07b5cf74,

$L_{16} = $ 09ef5b69

Expansion of R_{16}: $E(R_{16}) = $ 00fdabe5eba8

S-box input: $\Gamma_{16} = E(R_{16}) \oplus K_{16}$

$$= 9fd0b1bf3e52$$

S-box output: $\Omega_{16} = $ 2e09ee9
Permutation of Ω_{16}: $P(\Omega_{16}) = $ f93c0e59
The left-half output R_{15} after round 16:

$R_{15} = P(\Omega_{16}) \oplus L_{16}$

$$= \text{f0d35530}$$

Since $L_{15} = R_{16}$, the right-half output L_{15} is $L_{15} = $ 07b5cf74.

Thus, the sixteenth round decryption process is accomplished counting from the bottom up. In a similar fashion, the rest of the decryption processes are summarized in the following table.

Table for decryption blocks (R_i, L_i), $15 \leqslant i \leqslant 0$

i	R_i	L_i
15	f0d35530	07b5cf74
14	f04bf1ad	f0d35530
13	6ff2ef60	f04bf1ad
12	b6c73726	6ff2ef60
11	56a0e201	b6c73726
10	d0b10423	56a0e201
9	ec6d1ab8	d0b10423
8	6bc34499	ec6d1ab8
7	8e6f6798	6bc34499
6	c1fe0f05	8e6f6798
5	a6f29581	c1fe0f05
4	61125d5d	a6f29581
3	e0e7a039	61125d5d
2	5d189730	e0e7a039
1	5cd9b326	5d189730
0	4713b8f4	5cd9b326

The preoutput block is $(R_0 \| L_0) = 4713b8f45cd9b326$.
Using Table 4.8 (IP^{-1}), the plaintext is recovered as X = 785ac3a4bd0fe12d.

4.1.5 Triple DES

Triple DES is popular in Internet-based applications, including PGP and S/MIME. The possible vulnerability of DES to a brute-force attack brings us to find an alternative algorithm. Triple DES is a widely accepted approach which uses multiple encryption with DES and multiple keys, as shown in Figure 4.5. The three-key triple DES is the preferred alternative, whose effective key length is 168 bits.

Triple DES with two keys (K$_1$ = K$_3$, K$_2$) is a relatively popular alternative to DES. But triple DES with three keys (K$_1$, K$_2$, K$_3$) is preferred, as it results in a great increase in cryptographic strength. However, this alternative raises the cost of the known-plaintext attack to 2^{168}, which is beyond what is practical.

Referring to Figure 4.5, the ciphertext C is produced as

$$C = E_{K_3}[D_{K_2}[E_{K_1}(P)]]$$

The sender encrypts with the first key K$_1$, then decrypts with the second key K$_2$, and finally encrypts with the third key K$_3$. Decryption requires that the keys are applied in reverse order:

$$P = D_{K_1}[E_{K_2}[D_{K_3}(C)]]$$

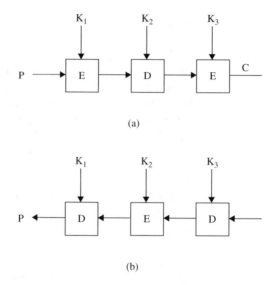

Figure 4.5   Triple DES (a) encryption and (b) decryption.

The receiver decrypts with the third key K_3, then encrypts with the second key K_2, and finally decrypts with the first key K_1. This process is sometimes known as *Encrypt-Decrypt-Encrypt (EDE) mode*.

Example 4.6   Using Figure 4.5, the triple DES computation is considered here. Given three keys:

$K_1 = 0x260b152f31b51c68$

$K_2 = 0x321f0d61a773b558$

$K_3 = 0x519b7331bf104ce3$

and the plaintext P = 0x403da8a295d3fed9.

The 16-round keys corresponding to each given key K_1, K_2, and K_3 are computed as shown below.

Round	K_1	K_2	K_3
1	000ced9158c9	5a1ec4b60e98	03e4ee7c63c8
2	588490792e94	710c318334c6	8486dd46ac65
3	54882eb9409b	c5a8b4ec83a5	575a226a8ddc
4	a2a006077207	96a696124ecf	aab9e009d59b
5	280e26b621e4	7e16225e9191	98664f4f5421
6	e03038a08bc7	ea906c836569	615718ca496c
7	84867056a693	88c25e6abb00	4499e580db9c

Round	K_1	K_2	K_3
8	c65a127f0549	245b3af0453e	93e853d116b1
9	2443236696a6	76d38087dd44	cc4a1fa9f254
10	a311155c0deb	1a915708a7f0	27b30c31c6a6
11	0d02d10ed859	2d405ff9cc05	0a1ce39c0c87
12	1750b843f570	2741ac4a469a	f968788e62d5
13	9e01c0a98d28	9a09b19d710d	84e78833e3c1
14	1a4a0dc85e16	9d2a39a252e0	521f17b28503
15	09310c5d42bc	87368cd0ab27	6db841ce2706
16	53248c80ee34	30258f25c11d	c9313c0591e3

Encryption: Compute the ciphertext C through the EDE mode operation of P. Each stage in the triple DES-EDE sequence is computed as:

First stage: $E_{K_1}(P) = 0x7a39786f7ba32349$
Second stage: $D_{K_2} = 0x9c60f85369113aea$
Third stage: $E_{K_3} = 0xe22ae33494beb930 = C$ (ciphertext)

Decryption: Using the ciphertext C obtained above, the plaintext P is recovered as:

Fourth stage: $D_{K_3}(C) = 0x9c60f85369113aea$
Fifth stage: $E_{K_2} = 0x7a39786f7ba32349$
Final stage: $D_{K_1} = 0x403da8a295d3fed9 = P$ (plaintext)

4.1.6 DES-CBC Cipher Algorithm with IV

This section describes the use of the DES cipher algorithm in Cipher Block Chaining (CBC) mode as a confidentiality mechanism within the context of the Encapsulating Security Payload (ESP). ESP provides confidentiality for IP datagrams by encrypting the payload data to be protected (Chapter 8).

DES-CBC requires an explicit Initialization Vector (IV) of 64 bits that is the same size as the block size. The IV must be a random value which prevents the generation of identical ciphertext. IV implementations for inner CBC must not use a low Hamming distance between successive IVs. The IV is XORed with the first plaintext block before it is encrypted. For successive blocks, the previous ciphertext block is XORed with the current plaintext before it is encrypted.

DES-CBC is a symmetric secret key algorithm. The key size is 64 bits, but it is commonly known as a *56-bit key*. The key has 56 significant bits; the least significant bit in every byte is the parity bit.

There are several ways to specify triple DES encryption, depending on the decision which affects both security and efficiency. For using triple encryption with three different keys, there are two possible triple-encryption modes (i.e., three DES-EDE modes): inner CBC and outer CBC, as shown in Figure 4.6.

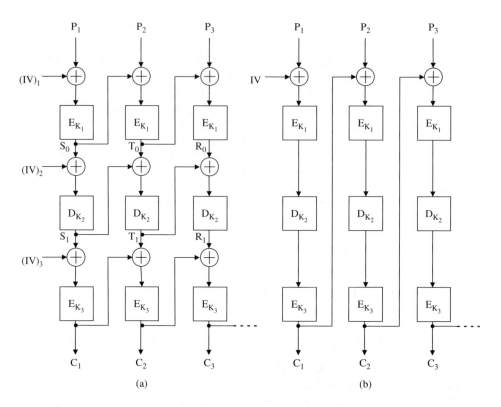

Figure 4.6 Triple DES-EDE in CBC mode. (a) Inner CBC and (b) outer CBC.

Inner CBC

This mode requires three different IVs.

$$S_0 = E_{K_1}(P_1 \oplus (IV)_1), T_0 = E_{K_1}(P_2 \oplus S_0), R_0 = E_{K_1}(P_3 \oplus T_0)$$
$$S_1 = D_{K_2}(S_0 \oplus (IV)_2), T_1 = D_{K_2}(T_0 \oplus S_1), R_1 = D_{K_2}(R_0 \oplus T_1)$$
$$C_1 = E_{K_3}(S_1 \oplus (IV)_3), C_2 = E_{K_3}(T_1 \oplus C_1), C_3 = E_{K_3}(R_1 \oplus C_2)$$

Outer CBC

This mode requires one IV.

$$C_1 = E_{K_3}(D_{K_2}(E_{K_1}(P_1 \oplus IV)))$$
$$C_2 = E_{K_3}(D_{K_2}(E_{K_1}(P_2 \oplus C_1)))$$
$$C_3 = E_{K_3}(D_{K_2}(E_{K_1}(P_3 \oplus C_2)))$$

Example 4.7 Consider the triple DES-EDE operation in outer CBC mode as shown in Figure 4.6b.

Suppose three plaintext blocks P_1, P_2, and P_3 and IV are given as:

$P_1 = 0x317f2147a6d50c38$

$P_2 = 0xc6115733248f702e$

$P_3 = 0x1370f341da552d79$

$IV = 0x714289e53306f2e1$

Assume that three keys K_1, K_2, and K_3 used in this example are exactly the same keys as those given in Example 4.6. The computation of ciphertext blocks (C_1, C_2, C_3) at each EDE stage is shown as follows:

1. C_1 computation with first EDE operation

   $P_1 \oplus IV = 0x403da8a295d3fed9$

   $E_{K_1}(P_1 \oplus IV) = 0x7a39786f7ba32349$

   $D_{K_2}(E_{K_1}(P_1 \oplus IV)) = 0x9c60f85369113aea$

   $C_1 = E_{K_3}(D_{K_2}(E_{K_1}(P_1 \oplus IV))) = 0xe22ae33494beb930$

2. C_2 computation with second EDE operation

   $P_2 \oplus C_1 = 0x243bb407b031c91e$

   $E_{K_1}(P_2 \oplus C_1) = 0xfeb7c33e747abf74$

   $D_{K_2}(E_{K_1}(P_2 \oplus C_1)) = 0x497f548f78af6e6f$

   $C_2 = E_{K_3}(D_{K_2}(E_{K_1}(P_2 \oplus C_1))) = 0xe4976149de15ca176$

3. C_3 computation with third EDE operation

   $P_3 \oplus C_2 = 0x5a06e7dc3b098c0f$

   $E_{K_1}(P_3 \oplus C_2) = 0x0eb878e2680e7f78$

   $D_{K_2}(E_{K_1}(P_3 \oplus C_2)) = 0xc6c8441ee3b5dd1c$

   $C_3 = E_{K_3}(D_{K_2}(E_{K_1}(P_3 \oplus C_2))) = 0xf980690fc2db462d$

Thus, all three ciphertext blocks (C_1, C_2, C_3) are obtained using the outer CBC mechanism.

4.2 International Data Encryption Algorithm (IDEA)

In 1990, Xuejia Lai and James Massey of the Swiss Federal Institute of Technology devised a new block cipher. The original version of this block-oriented encryption algorithm was called the *Proposed Encryption Standard* (*PES*). Since then, PES has been strengthened against differential cryptographic attacks. In 1992, the revised version of PES appeared to be strong and was renamed as the *International Data Encryption Algorithm* (*IDEA*). IDEA is a block cipher that uses a 128-bit key to encrypt 64-bit data blocks.

Pretty Good Privacy (PGP) provides a privacy and authentication service that can be used for e-mail and file storage applications. PGP uses IDEA for conventional block encryption, along with RSA for public-key encryption and MD5 for hash coding. The 128-bit key length seems to be long enough to effectively prevent exhaustive key searches. The 64-bit input block size is generally recognized as sufficiently strong enough to deter statistical analysis, as experienced with DES. The ciphertext depends on the plaintext and key, which are largely involved in a complicated manner. IDEA achieves this goal by mixing three different operations. Each operation is performed on two 16-bit inputs to produce a single 16-bit output. IDEA has a structure that can be used for both encryption and decryption, like DES.

4.2.1   Subkey Generation and Assignment

The 52 subkeys are all generated from the 128-bit original key. IDEA uses fifty-two 16-bit key sub-blocks, that is, six subkeys for each of the first eight rounds and four more for the ninth round of output transformation.

The 128-bit encryption key is divided into eight 16-bit subkeys. The first eight subkeys, labeled Z_1, Z_2, ... , Z_8 are taken directly from the key, with Z_1 being equal to the first 16 bits, Z_2 to the next 16 bits, and so on. The first eight subkeys for the algorithm are assigned such that the six subkeys are for the first round, and the first two for the second round. After that, the key is circularly shifted 25 bits to the left and again divided into eight subkeys. This procedure is repeated until all 52 subkeys have been generated. Since each round uses the 96-bit subkey (16 bits × 6) and the 128-bit subkey (16 bits × 8) is extracted with each 25-bit rotation of the key, there is no way to expect a simple shift relationship between the subkeys of one round and that of another. Thus, this key schedule provides an effective technique for varying the key bits used for subkeys in the eight rounds. Figure 4.7 illustrates the subkey generation scheme for making use of IDEA encryption/decryption.

If the original 128-bit key is labeled as $Z(1, 2, ..., 128)$, then the entire subkey blocks of the eight rounds have the following bit assignments (Table 4.9).

Only six 16-bit subkeys are needed in each round, whereas the final transformation uses four 16-bit subkeys. But eight subkeys are extracted from the 128-bit key with the left shift of 25 bits. That is why the first subkey of each round is not in order, as shown in Table 4.9.

Example 4.8   Suppose the 128-bit original key Z is given as

$$Z = (5a14 \quad fb3e \quad 021c \quad 79e0 \quad 6081 \quad 46a0 \quad 117b \quad ff03)$$

The 52 16-bit subkey blocks are computed from the given key Z as follows: for the first round, the first eight subkeys are taken directly from Z. After that, the key Z is circularly shifted 25 bits to the left and again divided into eight 16-bit subkeys. This shift-divide

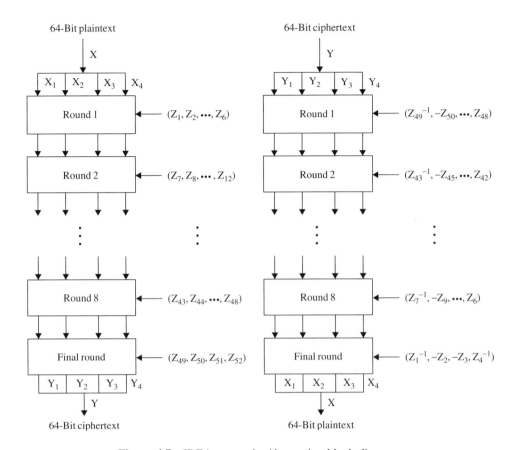

Figure 4.7   IDEA encryption/decryption block diagrams.

procedure is repeated until all 52 subkeys are generated, as shown in Table 4.9. The IDEA encryption key is computed as shown in Table 4.10.

4.2.2   IDEA Encryption

The overall scheme for IDEA encryption is illustrated in Figure 4.8. As with all block ciphers, there are two inputs to the encryption function, that is, the plaintext block and encryption key. IDEA is unlike DES (which relies mainly on the XOR and on nonlinear S-boxes). In IDEA, the plaintext is 64 bits in length and the key size is 128 bits long. The design methodology behind IDEA is based on mixing three different operations:

$\oplus$ Bit-by-bit XOR of 16-bit sub-blocks
$\boxplus$ Addition of 16-bit integers modulo 2^{16}
$\odot$ Multiplication of 16-bit integers modulo $2^{16} + 1$

IDEA utilizes both confusion and diffusion by using these three different operations. For example, for the additive inverse modulo 2^{16}, $-Z_i \boxplus Z_i = 0$, where the notation $-Z_i$

Table 4.9   Generation of IDEA 16-bit subkeys

Round 1

$Z_1 = Z(1, 2, \ldots, 16)$ $Z_4 = Z(49, 50, \ldots, 64)$
$Z_2 = Z(17, 18, \ldots, 32)$ $Z_5 = Z(65, 66, \ldots, 80)$
$Z_3 = Z(33, 34, \ldots, 48)$ $Z_6 = Z(81, 82, \ldots, 96)$

Round 2

$Z_7 = Z(97, 98, \ldots, 112)$ $Z_{10} = Z(42, 43, \ldots, 57)$
$Z_8 = Z(113, 114, \ldots, 128)$ $Z_{11} = Z(58, 59, \ldots, 73)$
$Z_9 = Z(26, 27, \ldots, 41)$ $Z_{12} = Z(74, 75, \ldots, 89)$

Round 3

$Z_{13} = Z(90, 91, \ldots, 105)$ $Z_{16} = Z(10, 11, \ldots, 25)$
$Z_{14} = Z(106, 107, \ldots, 121)$ $Z_{17} = Z(51, 52, \ldots, 66)$
$Z_{15} = Z(122, 123, \ldots, 128, 1, 2, \ldots, 9)$ $Z_{18} = Z(67, 68, \ldots, 82)$

Round 4

$Z_{19} = Z(83, 84, \ldots, 98)$ $Z_{22} = Z(3, 4, \ldots, 18)$
$Z_{20} = Z(99, 100, \ldots, 114)$ $Z_{23} = Z(19, 20, \ldots, 34)$
$Z_{21} = Z(115, 116, \ldots, 128, 1, 2)$ $Z_{24} = Z(35, 36, \ldots, 50)$

Round 5

$Z_{25} = Z(76, 77, \ldots, 91)$ $Z_{28} = Z(124, 125, \ldots, 128, 1, 2, \ldots, 11)$
$Z_{26} = Z(92, 93, \ldots, 107)$ $Z_{29} = Z(12, 13, \ldots, 27)$
$Z_{27} = Z(108, 109, \ldots, 123)$ $Z_{30} = Z(28, 29, \ldots, 43)$

Round 6

$Z_{31} = Z(44, 45, \ldots, 59)$ $Z_{34} = Z(117, 118, \ldots, 128, 1, 2, 3, 4)$
$Z_{32} = Z(60, 61, \ldots, 75)$ $Z_{35} = Z(5, 6, \ldots, 20)$
$Z_{33} = Z(101, 102, \ldots, 115)$ $Z_{36} = Z(21, 22, \ldots, 36)$

Round 7

$Z_{37} = Z(37, 38, \ldots, 52)$ $Z_{40} = Z(85, 86, \ldots, 100)$
$Z_{38} = Z(53, 54, \ldots, 68)$ $Z_{41} = Z(126, 127, 128, \ldots, 1, 2, \ldots, 13)$
$Z_{39} = Z(69, 70, \ldots, 84)$ $Z_{42} = Z(14, 15, \ldots, 29)$

Round 8

$Z_{43} = Z(30, 31, \ldots, 45)$ $Z_{46} = Z(78, 79, \ldots, 93)$
$Z_{44} = Z(46, 47, \ldots, 61)$ $Z_{47} = Z(94, 95, \ldots, 109)$
$Z_{45} = Z(62, 63, \ldots, 77)$ $Z_{48} = Z(110, 111, \ldots, 125)$

Round 9 (final transformation stage)

$Z_{49} = Z(23, 24, \ldots, 38)$ $Z_{51} = Z(55, 56, \ldots, 70)$
$Z_{50} = Z(39, 40, \ldots, 54)$ $Z_{52} = Z(71, 72, \ldots, 86)$

denotes the additive inverse, and for the multiplicative inverse modulo $2^{16} + 1, Z_i \odot Z_i^{-1} = 1$, where the notation Z_i^{-1} denotes the multiplicative inverse.

In Figure 4.8, IDEA consists of eight rounds followed by a final output transformation. The 64-bit input block is divided into four 16-bit sub-blocks, labeled X_1, X_2, X_3, and X_4. These four sub-blocks become the input to the first round of IDEA. The subkey generator generates a total of 52 subkey blocks that are all generated from the original 128-bit encryption key. Each subkey block consists of 16 bits. The first round makes use of six 16-bit subkeys ($Z_1, Z_2, \ldots, Z_6$), whereas the final output transformation uses

Table 4.10   Subkeys for encryption

$Z_1 = 5a14$	$Z_{27} = dff8$
$Z_2 = fb3e$	$Z_{28} = 1ad0$
$Z_3 = 021c$	$Z_{29} = a7d9$
$Z_4 = 79e0$	$Z_{30} = f010$
$Z_5 = 6081$	$Z_{31} = e3cf$
$Z_6 = 46a0$	$Z_{32} = 0304$
$Z_7 = 117b$	$Z_{33} = 17bf$
$Z_8 = ff03$	$Z_{34} = f035$
$Z_9 = 7c04$	$Z_{35} = a14f$
$Z_{10} = 38f3$	$Z_{36} = b3e0$
$Z_{11} = c0c1$	$Z_{37} = 21c7$
$Z_{12} = 028d$	$Z_{38} = 9e06$
$Z_{13} = 4022$	$Z_{39} = 0814$
$Z_{14} = f7fe$	$Z_{40} = 6a01$
$Z_{15} = 06b4$	$Z_{41} = 6b42$
$Z_{16} = 29f6$	$Z_{42} = 9f67$
$Z_{17} = e781$	$Z_{43} = c043$
$Z_{18} = 8205$	$Z_{44} = 8f3c$
$Z_{19} = 1a80$	$Z_{45} = 0c10$
$Z_{20} = 45ef$	$Z_{46} = 28d4$
$Z_{21} = fc0d$	$Z_{47} = 022f$
$Z_{22} = 6853$	$Z_{48} = 7fe0$
$Z_{23} = ecf8$	$Z_{49} = cf80$
$Z_{24} = 0871$	$Z_{50} = 871e$
$Z_{25} = 0a35$	$Z_{51} = 7818$
$Z_{26} = 008b$	$Z_{52} = 2051$

four 16-bit subkeys ($Z_{49}, Z_{50}, Z_{51}, Z_{52}$). The final transformation stage also produces four 16-bit blocks, which are concatenated to form the 64-bit ciphertext. In each round of Figure 4.8, the four 16-bit sub-blocks are XORed, added, and multiplied with one another and with six 16-bit key sub-blocks. Between each round, the second and third sub-blocks are interchanged. This swapping operation increases the mixing of the bits being processed and makes the IDEA more resistant to differential cryptanalysis.

In each round, the sequential operations will be taken into the following steps:

1. $X_1 \odot Z_1$
2. $X_2 \boxplus Z_2$
3. $X_3 \boxplus Z_3$
4. $X_4 \odot Z_4$
5. $(X_1 \odot Z_1) \oplus (X_3 \boxplus Z_3) = (1) \oplus (3)$
6. $(X_2 \boxplus Z_2) \oplus (X_4 \odot Z_4) = (2) \oplus (4)$
7. $((X_1 \odot Z_1) \oplus (X_3 \boxplus Z_3)) \odot Z_5 = ((1) \oplus (3)) \odot Z_5$
8. $(((X_2 \boxplus Z_2) \oplus (X_4 \odot Z_4)) \boxplus (((X_1 \odot Z_1) \oplus (X_3 \boxplus Z_3)) \odot Z_5)) = ((2) \oplus (4)) \boxplus (((1) \oplus (3)) \odot Z_5)$

Plaintext X = (X$_1$, X$_2$, X$_3$, X$_4$)

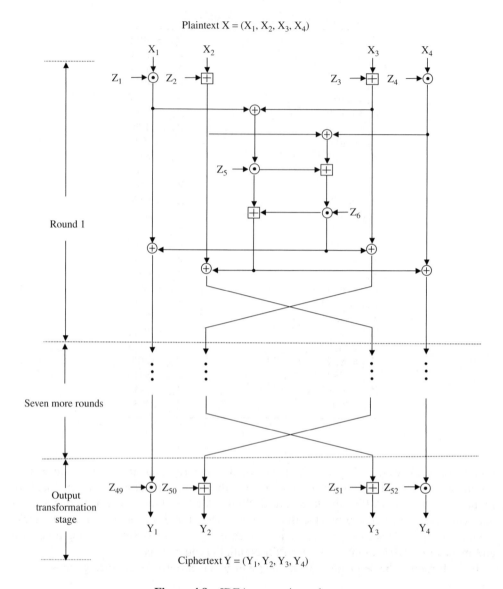

Figure 4.8   IDEA encryption scheme.

9. $(8) \odot Z_6$
10. $(7) \boxplus (9) = (((1) \oplus (3)) \odot Z_5) \boxplus ((8) \odot Z_6)$
11. $(X_1 \odot Z_1) \oplus ((8) \odot Z_6) = (1) \oplus (9)$
12. $(X_3 \boxplus Z_3) \oplus (9) = (3) \oplus (9)$
13. $(X_2 \boxplus Z_2) \oplus (10) = (2) \oplus (10)$
14. $(X_4 \odot Z_4) \oplus (10) = (4) \oplus (10)$

The output of each round is the four sub-blocks that result from steps 11–14. The two inner blocks (12) and (13) are interchanged before being applied to the next-round input. The final output transformation after the eighth round will involve the following steps:

1. $\overline{X}_1 \odot Z_{49}$
2. $\overline{X}_2 \boxplus Z_{50}$
3. $\overline{X}_3 \boxplus Z_{51}$
4. $\overline{X}_4 \odot Z_{52}$

where $\overline{X}_i, 1 \leqslant i \leqslant 4$, represents the output of the eighth round. As you see, the final ninth stage requires only four subkeys, Z_{49}, Z_{50}, Z_{51}, and Z_{52}, compared to six subkeys for each of the first eight rounds. Note also that no swap is required for the two inner blocks at the output transformation stage.

Example 4.9 Assume that the 64-bit plaintext X is given as

$$X = (X_1, X_2, X_3, X_4) = (7fa9\ 1c37\ ffb3\ df05) \qquad (4.1)$$

In the IDEA encryption, the plaintext is 64 bits in length and the encryption key consists of the 52 subkeys as computed in Example 4.8.

As shown in Figure 4.8, the four 16-bit input sub-blocks, X_1, X_2, X_3, and X_4, are XORed, added, and multiplied with one another and with six 16-bit subkeys. Following the sequential operation starting from the first round to the final transformation stage, the ciphertext $Y = (Y_1, Y_2, Y_3, Y_4)$ is computed as shown in Table 4.11.

The ciphertext Y represents the output of the final transformation stage:

$$Y = (Y_1, Y_2, Y_3, Y_4) = (106b\ dbfd\ f323\ 0876)$$

Table 4.11 Ciphertext computation through IDEA encryption rounds

Plaintext input X

7fa9	1c37	ffb3	df05
(X_1)	(X_2)	(X_3)	(X_4)

Round	Round output			
1	C579	F2ff	0fbd	0ffc
2	D7a2	80cb	9a61	27c5
3	ab6c	e2f9	f3be	36bd
4	ef5b	9cd2	6808	3019
5	7e09	2445	d223	d639
6	4a6e	d7ac	ac8c	8b09
7	244d	6f5c	4459	3a9c
8	0f86	7b0b	54df	759f
9 (final	106b	dbfd	f323	0876	← Ciphertext Y
transformation)	(Y_1)	(Y_2)	(Y_3)	(Y_4)	

4.2.3 IDEA Decryption

IDEA decryption is exactly the same as the encryption process, except that the key sub-blocks are reversed and a different selection of subkeys is used. The decryption subkeys are either the additive or multiplicative inverse of the encryption subkeys. The decryption key sub-blocks are derived from the encryption key sub-blocks shown in Table 4.12.

Looking at the decryption key sub-blocks in Table 4.12, we see that the first four decryption subkeys at round i are derived from the first four subkeys at encryption round $(10 - i)$, where the output transformation stage is counted as round 9. For example, the first four decryption subkeys at round 2 are derived from the first four encryption subkeys of round 8, as shown in Table 4.13.

Note that the first and fourth decryption subkeys are equal to the multiplicative inverse modulo $(2^{16} + 1)$ of the corresponding first and fourth encryption subkeys. For rounds 2–8, the second and third decryption subkeys are equal to the additive inverse modulo 2^{16} of the corresponding subkeys' third and second encryption subkeys. For rounds 1 and 9, the second and third decryption subkeys are equal to the additive inverse modulo 2^{16} of the corresponding second and third encryption subkeys. Note also that, for the first

Table 4.12 IDEA encryption and decryption subkeys

Round	Encryption subkeys	Decryption subkeys
1	$Z_1\ Z_2\ Z_3\ Z_4\ Z_5\ Z_6$	$Z_{49}^{-1}\ -Z_{50}\ -Z_{51}Z_{52}^{-1}Z_{47}Z_{48}$
2	$Z_7\ Z_8\ Z_9\ Z_{10}\ Z_{11}\ Z_{12}$	$Z_{43}^{-1}\ -Z_{45}\ -Z_{44}Z_{46}^{-1}Z_{41}Z_{42}$
3	$Z_{13}\ Z_{14}\ Z_{15}\ Z_{16}\ Z_{17}\ Z_{18}$	$Z_{37}^{-1}\ -Z_{39}\ -Z_{38}Z_{40}^{-1}Z_{35}Z_{36}$
4	$Z_{19}\ Z_{20}\ Z_{21}\ Z_{22}\ Z_{23}\ Z_{24}$	$Z_{31}^{-1}\ -Z_{33}\ -Z_{32}Z_{34}^{-1}Z_{29}Z_{30}$
5	$Z_{25}\ Z_{26}\ Z_{27}\ Z_{28}\ Z_{29}\ Z_{30}$	$Z_{25}^{-1}\ -Z_{27}\ -Z_{26}Z_{28}^{-1}Z_{23}Z_{24}$
6	$Z_{31}\ Z_{32}\ Z_{33}\ Z_{34}\ Z_{35}\ Z_{36}$	$Z_{19}^{-1}\ -Z_{21}\ -Z_{20}Z_{22}^{-1}Z_{17}Z_{18}$
7	$Z_{37}\ Z_{38}\ Z_{39}\ Z_{40}\ Z_{41}\ Z_{42}$	$Z_{13}^{-1}\ -Z_{15}\ -Z_{14}Z_{16}^{-1}Z_{11}Z_{12}$
8	$Z_{43}\ Z_{44}\ Z_{45}\ Z_{46}\ Z_{47}\ Z_{48}$	$Z_7^{-1}\ -Z_9\ -Z_8Z_{10}^{-1}Z_5Z_6$
9	$Z_{49}\ Z_{50}\ Z_{51}\ Z_{52}$	$Z_1^{-1}\ -Z_2\ -Z_3Z_4^{-1}$

Table 4.13 Decryption subkeys derived from encryption subkeys

Round i	First four decryption subkeys at i	Round $(10 - i)$	First four encryption subkeys at $(10 - i)$
1	$Z_{49}^{-1}\ -Z_{50}\ -Z_{51}Z_{52}^{-1}$	9	$Z_{49}\ Z_{50}\ Z_{51}\ Z_{52}$
2	$Z_{43}^{-1}\ -Z_{45}\ -Z_{44}Z_{46}^{-1}$	8	$Z_{43}\ Z_{44}\ Z_{45}\ Z_{46}$
.		.	
.		.	
.		.	
8	$Z_7^{-1}\ -Z_9\ -Z_8Z_{10}^{-1}$	2	$Z_7\ Z_8\ Z_9\ Z_{10}$
9	$Z_1^{-1}\ -Z_2\ -Z_3Z_4^{-1}$	1	$Z_1\ Z_2\ Z_3\ Z_4$

eight rounds, the last two subkeys of decryption round i are equal to the last two subkeys of encryption round $(9 - i)$ (Table 4.12).

Example 4.10   Using Table 4.12, compute the decryption subkeys corresponding to the encryption key sub-blocks obtained in Table 4.10. The IDEA decryption key is computed as shown in Table 4.14.

IDEA decryption is exactly the same as the encryption process, but the decryption subkeys are composed of either the additive or multiplicative inverse of the encryption subkeys, as indicated in Table 4.12.

The IDEA decryption scheme for recovering plaintext is shown in Figure 4.9.

Example 4.11   Restore the plaintext $X = (7fa9 \quad 1c37 \quad ffb3 \quad df05)$ using the ciphertext $Y = (106b \; dbfd \; f323 \; 0876)$ that was computed in Example 4.9.

The recovering steps are shown by the round-after-round process as indicated in Table 4.15.

Thus, the recovered plaintext is $X = (X_1, X_2, X_3, X_4) = (7fa9 \quad 1c37 \quad ffb3 \quad df05)$.

Table 4.14   Subkey blocks for decryption

$Z_{49}^{-1} = 9194$	$-Z_{26} = ff75$
$-Z_{50} = 78e2$	$Z_{28}^{-1} = 24f6$
$-Z_{51} = 87e8$	$Z_{23} = ecf8$
$Z_{52}^{-1} = 712a$	$Z_{24} = 0871$
$Z_{47} = 022f$	$Z_{19}^{-1} = 4396$
$Z_{48} = 7fe0$	$-Z_{21} = 03f3$
$Z_{43}^{-1} = a24c$	$-Z_{20} = ba11$
$-Z_{45} = f3f0$	$Z_{22}^{-1} = dfa7$
$-Z_{44} = 70c4$	$Z_{17} = e781$
$Z_{46}^{-1} = 3305$	$Z_{18} = 8205$
$Z_{41} = 6b42$	$Z_{13}^{-1} = 18a7$
$Z_{42} = 9f67$	$-Z_{15} = f94c$
$Z_{37}^{-1} = c579$	$-Z_{14} = 0802$
$-Z_{39} = f7ec$	$Z_{16}^{-1} = 9a13$
$-Z_{38} = 61fa$	$Z_{11} = c0c1$
$Z_{40}^{-1} = bf28$	$Z_{12} = 028d$
$Z_{35} = a14f$	$Z_{7}^{-1} = 55ed$
$Z_{36} = b3e0$	$-Z_9 = 83fc$
$Z_{31}^{-1} = c53c$	$-Z_8 = 00fd$
$-Z_{33} = e841$	$Z_{10}^{-1} = 2cd9$
$-Z_{32} = fcfc$	$Z_5 = 6081$
$Z_{34}^{-1} = 3703$	$Z_6 = 46a0$
$Z_{29} = a7d9$	$Z_1^{-1} = 0dd8$
$Z_{30} = f010$	$-Z_2 = 04c2$
$Z_{25}^{-1} = cc14$	$-Z_3 = fde4$
$-Z_{27} = 2008$	$Z_4^{-1} = 4fd0$

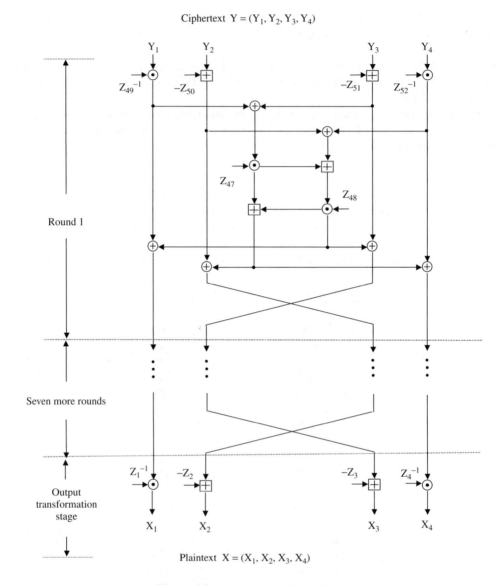

Figure 4.9 IDEA decryption scheme.

4.3 RC5 Algorithm

The RC5 encryption algorithm was designed by Ronald Rivest of the Massachusetts Institute of Technology (MIT) and it first appeared in December 1994. RSA Data Security, Inc. estimates that RC5 and its successor, RC6, are strong candidates for potential successors to DES. RC5 analysis (RSA Laboratories) is still in progress and is periodically updated to reflect any additional findings.

Table 4.15 Plaintext computation through IDEA decryption steps

Ciphertext input Y

↓

106b	dbfd	f323	0876
(Y_1)	(Y_2)	(Y_3)	(Y_4)

Round		Round output			
1	24e4	5069	fe98	dfd8	
2	ffb1	b4a0	75b2	0b77	
3	7420	e9e2	2749	00cc	
4	124c	4800	9d5d	9947	
5	9c42	efcb	28e8	70f9	
6	ed80	a415	78c9	bdca	
7	dca8	8bc1	f202	48a6	
8	3649	01cf	1775	1734	
9 (final	7fa9	1c37	ffb3	df05	→ Recovered
transformation)	(X_1)	(X_2)	(X_3)	(X_4)	plaintext X

4.3.1 Description of RC5

RC5 is a symmetric block cipher designed to be suitable for both software and hardware implementation. It is a parameterized algorithm, with a variable block size, a variable number of rounds, and a variable-length key. This provides the opportunity for great flexibility in both performance characteristics and the level of security.

A particular RC5 algorithm is designated as RC5-$w/r/b$. The number of bits in a word, w, is a parameter of RC5. Different choices of this parameter result in different RC5 algorithms. RC5 is iterative in structure, with a variable number of rounds. The number of rounds, r, is the second parameter of RC5. RC5 uses a variable-length secret key. The key length b (in bytes) is the third parameter of RC5. These parameters are summarized as follows:

w: The word size, in bits. The standard value is 32 bits; allowable values are 16, 32, and 64. RC5 encrypts two-word blocks so that the plaintext and ciphertext blocks are each $2w$ bits long.

r: The number of rounds. Allowable values of r are 0, 1, ... , 255. Also, the expanded key table S contains $t = 2(r + 1)$ words.

b: The number of bytes in the secret key K. Allowable values of b are 0, 1, ... , 255.

K: The b-byte secret key; $K[0], K[1], ... , K[b - 1]$

RC5 consists of three components: a key-expansion algorithm, an encryption algorithm, and a decryption algorithm. These algorithms use the following three primitive operations:

1. ⊞ Addition of words modulo 2^w
2. ⊕ Bitwise XOR of words
3. <<< Rotation symbol: the rotation of x to the left by y bits is denoted by $x <<< y$.

One design feature of RC5 is its simplicity, which makes RC5 easy to implement. Another feature of RC5 is its heavy use of data-dependent rotations in encryption; this feature is very useful in preventing both differential and linear cryptanalysis.

Example 4.12   Given RC5-32/16/10. This particular RC5 algorithm has 32-bit words, 16 rounds, a 10-byte (80-bit) secret key variable, and an expanded key table S of $t = 2(r + 1) = 2(16 + 1) = 34$ words. Rivest proposed RC5-32/12/16 as a block cipher providing a normal choice of parameters, that is, 32-bit words, 12 rounds, 16-byte (128-bit) secret key variable, and an expanded key table of 26 words.

4.3.2   Key Expansion

The key-expansion algorithm expands the user's key K to fill the expanded key table S, so that S resembles an array of $t = 2(r + 1)$ random binary words determined by K. It uses two word-size magic constants P_w and Q_w defined for arbitrary w as shown below:

$$P_w = \text{Odd} ((e - 2)2^w)$$

$$Q_w = \text{Odd} ((\phi - 1)2^w)$$

where

$e = 2.71828 \ldots$ (base of natural logarithms)
$\phi = (1 + \sqrt{5})/2 = 1.61803 \ldots$ (golden ratio)
$\text{Odd}(x) =$ the odd integer nearest to x.

First algorithmic step of key expansion. This step is to copy the secret key $K[0, 1, \ldots, b - 1]$ into an array $L[0, 1, \ldots, c - 1]$ of $c = \lceil b/u \rceil$ words, where $u = w/8$ is the number of bytes/word.

   This first step will be achieved by the following pseudocode operation: for $i = b - 1$ down to 0 do $L[i/u] = (L[i/u] <<< 8) + K[i]$, where all bytes are unsigned and the array L is initially zeroes.

Second algorithmic step of key expansion. This step is to initialize array S to a particular fixed pseudorandom bit pattern, using an arithmetic progression modulo 2^w determined by two constants P_w and Q_w.

$S[0] = P_w$:
for $i = 1$ to $t - 1$ do $S[i] = S[i - 1] + Q_w$

Third algorithmic step of key expansion. This step is to mix in the user's secret key in three passes over the arrays S and L. More precisely, due to the potentially different sizes of S and L, the larger array is processed three times and the other array will be handled more after.

$i = j = 0$
$A = B = 0$

do 3^* max (t, c) times:

$A = S[i] = (S[i] + A + B) <<< 3$

$B = L[j] = (L[j] + A + B) <<< (A + B)$

$i = (i + 1) \pmod{t}$

$j = (j + 1) \pmod{c}$

Note that with the key-expansion function it is not so easy to determine K from S due to the one-wayness.

Example 4.13   Consider RC5-32/12/16. Since $w = 32, r = 12$, and $b = 16$, we have

$u = w/8 = 32/8 = 4$ bytes/word

$c = \lceil b/u \rceil = \lceil 16/4 \rceil = 4$ words

$t = 2(r + 1) = 2(12 + 1) = 26$ words

The plaintext and the user's secret key are given as follows:

  Plaintext = eedba521   6d8f4b15

  Key = 915f 46 19 be 41 b2 51 63 55 a5 01 10 a9 ce 91

1. Key expansion
Two magic constants

$P_{32} = 3084996963 = 0xb7e15163$

$Q_{32} = 2654435769 = 0x9e3779b9$

Step 1
    For $i = b - 1$ down to 0 do $L[i/u] = (L[i/u] <<< 8) + K[i]$ where $b = 16, u = 4$, and L is initially 0.

$L[i/4] = L[3]$ for $i = 15, 14, 13$, and 12

  $L[3] = (L[3] <<< 8) + K[15] = 00 + 91 = 0091$

  $L[3] = (L[3] <<< 8) + K[14] = 9100 + ce = 91ce$

  $L[3] = (L[3] <<< 8) + K[13] = 91ce00 + a9 = 91cea9$

  $*L[3] = (L[3] <<< 8) + K[12] = 91cea900 + 10 = 91cea910$

$L[i/4] = L[2]$ for $i = 11, 10, 9$, and 8

  $L[2] = (L[2] <<< 8) + K[11] = 00 + 01 = 0001$

  $L[2] = (L[2] <<< 8) + K[10] = 0100 + a5 = 01a5$

  $L[2] = (L[2] <<< 8) + K[9] = 01a500 + 55 = 01a555$

$*L[2] = (L[2] <<< 8) + K[8] = 01a55500 + 63 = 01a55563$

$L[i/4] = L[1]$ for $i = 7, 6, 5,$ and 4

$\quad L[1] = (L[1] <<< 8) + K[7] = 00 + 51 = 0051$

$\quad L[1] = (L[1] <<< 8) + K[6] = 5100 + b2 = 51b2$

$\quad L[1] = (L[1] <<< 8) + K[5] = 51b200 + 41 = 51b241$

$*L[1] = (L[1] <<< 8) + K[4] = 51b24100 + be = 51b241be$

$L[i/4] = L[0]$ for $i = 3, 2, 1,$ and 0

$\quad L[0] = (L[0] <<< 8) + K[3] = 00 + 19 = 0019$

$\quad L[0] = (L[0] <<< 8) + K[2] = 1900 + 46 = 1946$

$\quad L[0] = (L[0] <<< 8) + K[1] = 194600 + 5f = 19465f$

$*L[0] = (L[0] <<< 8) + K[0] = 19465f00 + 91 = 19465f91$

Thus, converting the secret key from bytes to words (*) yields:

$L[0] = 19465f91$

$L[1] = 51b241be$

$L[2] = 01a55563$

$L[3] = 91cea910$

Step 2

$S[0] = P_{32}$. For $i = 1$ to 25, do $S[i] = S[i-1] + Q_{32}$:

$S[0] = b7e15163$

$S[1] = S[0] + Q_{32} = b7e15163 + 9e3779b9 = 5618cb1c$

$S[2] = S[1] + Q_{32} = 5618cb1c + 9e3779b9 = f45044d5$

$S[3] = S[2] + Q_{32} = f45044d5 + 9e3779b9 = 9287be8e$

$$\vdots \qquad\qquad\qquad\qquad\qquad \vdots$$

$S[25] = S[24] + Q_{32} = 8f14babb + 9e3779b9 = 2b4c3474$

When the iterative processes continue up to $t - 1 = 2(r + 1) - 1 = 25$, we can obtain the expanded key table S as shown below:

$S[0] = b7e15163$	$S[09] = 47d498e4$	$S[18] = d7c7e065$
$S[1] = 5618cb1c$	$S[10] = e60c129d$	$S[19] = 75ff5a1e$
$S[2] = f45044d5$	$S[11] = 84438c56$	$S[20] = 1436d3d7$
$S[3] = 9287be8e$	$S[12] = 227b060f$	$S[21] = b26e4d90$

$S[4] = 30\text{bf}3847$ $S[13] = \text{c0b}27\text{fc8}$ $S[22] = 50\text{a5c}749$

$S[5] = \text{cef6b}200$ $S[14] = 5\text{ee9f}981$ $S[23] = \text{eedd}4102$

$S[6] = 6\text{d2e}2\text{bb}9$ $S[15] = \text{fd}21733\text{a}$ $S[24] = 8\text{d}14\text{babb}$

$S[7] = 0\text{b}65\text{a}572$ $S[16] = 9\text{b}58\text{ecf}3$ $S[25] = 2\text{b4c}3474$

$S[8] = \text{a}99\text{d}1\text{f2b}$ $S[17] = 399066\text{ac}$

Step 3

$i = j = 0; A = B = 0$

$3 \times \max(t, c) = 3 \times 26 = 78$ times

$A = S[i] = (S[i] + A + B) <<< 3$

$B = L[j] = (L[j] + A + B) <<< (A + B)$

$i = i + 1 \pmod{26}$

$j = j + 1 \pmod{4}$

$A = S[0] = (\text{b7e}15163 + 0 + 0) <<< 3$
$$= \text{b7e}15163 <<< 3 = \text{bf0a8b1d}$$

$B = L[0] = (19465\text{f}91 + \text{bf0a8b1d}) <<< (A + B)$
$$= \text{d}850\text{eaae} <<< \text{bf0a8b1d} = \text{db0a1d}55$$

$A = S[1] = (5618\text{cb1c} + \text{bf0a8b1d} + \text{db0a1d}55) <<< 3$
$$= \text{f02d}738\text{e} <<< 3 = 816\text{b9c}77$$

$B = L[1] = (51\text{b}241\text{be} + 816\text{b9c}77 + \text{db0a1d}55) <<< (A + B)$
$$= \text{ae}27\text{fb8a} <<< 5\text{c}75\text{b9cc} = 7\text{fb8aae}2$$

$A = S[2] = (\text{f}45044\text{d}5 + 816\text{b9c}77 + 7\text{fb8aae}2) <<< 3$
$$= \text{f}5748\text{c2e} <<< 3 = \text{aba}46177$$

$B = L[2] = (01\text{a}55563 + \text{aba}46177 + 7\text{fb8aae}2) <<< (A + B)$
$$= 2\text{d}0261\text{bc} <<< 2\text{b}5\text{d}0\text{c}59 = 785\text{a}04\text{c}3$$

$A = S[3] = (9287\text{be8e} + \text{aba}46177 + 785\text{a}04\text{c}3) <<< 3$
$$= \text{b}68624\text{c}8 <<< 3 = \text{b}4312645$$

$B = L[3] = (91\text{cea}910 + \text{b}4312645 + 785\text{a}04\text{c}3) <<< (A + B)$
$$= \text{be}59\text{d}418 <<< 2\text{c8b2b}08 = 59\text{d}418\text{be}$$

$\ldots$

$$A = S[25] = (4\text{e}0\text{d}4\text{c}36 + \text{f}66\text{a}1\text{aaf} + 6\text{d}7\text{f}672\text{f}) <<< 3$$

$$= \text{b}1\text{f}6\text{ce}14, <<< 3 = 8\text{fb}670\text{a}5,$$

$$B = L[1] = (\text{cdfc}2657 + 8\text{fb}670\text{a}5 + 6\text{d}7\text{f}672\text{f}) <<< (A + B)$$

$$= \text{cb}31\text{fe}2\text{b} <<< \text{fd}35\text{d}7\text{d}4 = \text{e}2\text{bcb}31\text{f}$$

This is the first group expanded key table S resulted from iterative process over $0 \leqslant r \leqslant 25$.

The second group expanded key table S over $26 \leqslant r \leqslant 53$ is S[0] = bfcda6f9, $S[1] =$ 4dd05d18, ... , $S[25] = 255565\text{cd}$.

The third group expanded key table S resulted from iterative operation over $54 \leqslant r \leqslant 77$ is $S[0] = 6\text{d}835\text{afc}$, $S[1] = 7\text{d}15\text{cd}97$, $S[2] = 32\text{f}9\text{c}923$, ... , $S[25] = 30726\text{d}5\text{a}$.

The continuous processing of repeated computations is shown in Table 4.16.

4.3.3 Encryption

The input block to RC5 consists of two w-bit words given in two registers, A and B. The output is also placed in the registers A and B. Recall that RC5 uses an expanded key table, $S[0, 1, \dots, t - 1]$, consisting of $t = 2(r + 1)$ words. The key-expansion algorithm initializes S from the user's given secret key parameter K. However, the S table in RC5 encryption is not like an S-box used by DES. The encryption algorithm is given in the pseudocode as shown below:

$$A = A + S[0]$$

$$B = B + S[1]$$

for $i = 1$ to r do

$$A = ((A \oplus B) <<< B) + S[2i]$$

$$B = ((B \oplus A) <<< A) + S[2i + 1]$$

The output is in the registers A and B.

Example 4.14 Consider again RC 5 - 32/12/16. Suppose the plaintext X is given as:

$$X = \text{eedba}521 \ 6\text{d}8\text{f}4\text{b}15 = A \| B$$

To encrypt the 64-bit input block X, the following steps will be taken:

- Store the plaintext in two 32-bit registers, A and B.
- Choose the third group expanded key table S[0], S[1], ..., $S[25]$ over $54 \leqslant r \leqslant 77$ as shown in Table 4.16.
- Compute the ciphertext using the RC5 encryption algorithm according to Figure 4.10.

Table 4.16 Expanded key table S resulted from groupwise iterative operation over $0 \leqslant r \leqslant 77$

Round	Value	Round	Value
1	A = S[0] = bf0a8b1d,B = L[0] = db0a1d55	40	A = S[13] = 60e93e12, B = L[3] = 160c2277
2	A = S[1] = 816b9c77,B = L[1] = 7fb8aae2	41	A = S[14] = 8595c842, B = L[0] = c517db63
3	A = S[2] = aba46177,B = L[2] = 785a04c3	42	A = S[15] = 26d9d406, B = L[1] = 3cc0d68d
4	A = S[3] = b4312645,B = L[3] = 59d418be	43	A = S[16] = 5d4e600c, B = L[2] = 1d9e8680
5	A = S[4] = f623ba51,B = L[0] = 8321580	44	A = S[17] = 9a469d73, B = L[3] = 3356f8a
6	A = S[5] = ea64de8d,B = L[1] = d9ddec49	45	A = S[18] = 1ee6853d, B = L[0] = aa681507
7	A = S[6] = 8b813479,B = L[2] = 76e49617	46	A = S[19] = 98464d27, B = L[1] = ce2edfdb
8	A = S[7] = 6e5b8010,B = L[3] = 8a17729f	47	A = S[20] = 1309c416, B = L[2] = 54e3fdae
9	A = S[8] = 10808ed5,B = L[0] = 6f492ca1	48	A = S[21] = 652071c0, B = L[3] = b7be3b56
10	A = S[9] = 3cf2a2d6,B = L[1] = e0430cdd	49	A = S[22] = 1eafced6, B = L[0] = 61f3380d
11	A = S[10] = 1a0e1280,B = L[2] = 8e26b6ae	50	A = S[23] = a8500d9, B = L[1] = 29c63076
12	A = S[11] = 63c2ac21,B = L[3] = 6ab73e00	51	A = S[24] = 704825b0, B = L[2] = bc94f53b
13	A = S[12] = 87a78187,B = L[0] = d3f61430	52	A = S[25] = 255565cd, B = L[3] = a8965e99
14	A = S[13] = e280abf8,B = L[1] = b9cd0596	53	A = S[0] = 6d835afc, B = L[0] = 344f019e
15	A = S[14] = d9bd587f,B = L[2] = 98643622	54	A = S[1] = 7d15cd97, B = L[1] = f57b655f
16	A = S[15] = 7a180edb,B = L[3] = afa6705f	55	A = S[2] = 0942b409, B = L[2] = 530ea3bb
17	A = S[16] = 28bb61ce,B = L[0] = fcbfb58a	56	A = S[3] = 32f9c923, B = L[3] = cba7b2dd
18	A = S[17] = f85bed22,B = L[1] = 8a842aee	57	A = S[4] = a811fb02, B = L[0] = d40457be
19	A = S[18] = d53fc3aa,B = L[2] = baf82824	58	A = S[5] = 64f121e8, B = L[1] = 9c37c14b
20	A = S[19] = 31ba2f60,B = L[3] = c58c7e39	59	A = S[6] = d1cc8b4e, B = L[2] = a98225e0
21	A = S[20] = 5bec0b80,B = L[0] = 863c707e	60	A = S[7] = e887c6f, B = L[3] = 8b962ed8
22	A = S[21] = a4b64c74,B = L[1] = 9f82d5db	61	A = S[8] = 61399bbb, B = L[0] = 128e06a1
23	A = S[22] = a6f74cc4,B = L[2] = 80b92561	62	A = S[9] = f1b91926, B = L[1] = 3f708950
24	A = S[23] = b46d9938,B = L[3] = a5f56679	63	A = S[10] = ac661520, B = L[2] = c4509558
25	A = S[24] = 3bbdd367,B = L[0] = 67efaa5e	64	A = S[11] = a21a31c9, B = L[3] = e401ebf3
26	A = S[25] = 77cd91ce,B = L[1] = 01207ff4	65	A = S[12] = d424808d, B = L[0] = cab47321
27	A = S[0] = bfc4a6f9,B = L[2] = c889c833	66	A = S[13] = fe118e07, B = L[1] = 368a7808
28	A = S[1] = 4dd05d18,B = L[3] = 7c5e25e2	67	A = S[14] = d18e728d, B = L[2] = fdb98d2f
29	A = S[2] = ae97238b,B = L[0] = 9e79725c	68	A = S[15] = abac9e17, B = L[3] = 5a05e63
30	A = S[3] = 0a0de160,B = L[1] = 0a9a7cbb	69	A = S[16] = 18066433, B = L[0] = 6dcf3029
31	A = S[4] = 5660c360,B = L[2] = 714c2842	70	A = S[17] = 00e18e79, B = L[1] = 94ecdaaa
32	A = S[5] = 9087d17d,B = L[3] = bf190fd0	71	A = S[18] = 65a77305, B = L[2] = ed6f7c26
33	A = S[6] = d910ae36,B = L[0] = a8cc188d	72	A = S[19] = 5ae9e297, B = L[3] = 144be5a4
34	A = S[7] = 81c2369f,B = L[1] = 8cbe7352	73	A = S[20] = 11fc628c, B = L[0] = 78599417
35	A = S[8] = f809c630,B = L[2] = d8518713	74	A = S[21] = 7bb3431f, B = L[1] = 78223e6c
36	A = S[9] = 6a6f80c8,B = L[3] = 580ed0bd	75	A = S[22] = 942a8308, B = L[2] = d9af9bc3
37	A = S[10] = e463202e,B = L[0] = f04bc729	76	A = S[23] = b2f8fd20, B = L[3] = 07a3f43d
38	A = S[11] = c38c9bc1,B = L[1] = 5b58f102	77	A = S[24] = 57288869, B = L[0] = c902f75
39	A = S[12] = 34687255,B = L[2] = 35340975	78	A = S[25] = 3072d5a, B = L[1] = 6d9db912

Encryption Process

Round 0:

$A + S[0] = \text{eedba521} + \text{6d835afc} = \text{5c5f001d}$

$B + S[1] = \text{6d8f4b15} + \text{7d15cd97} = \text{eaa518ac}$

Round 1:

$A \oplus B = \text{5c5f001d} \oplus \text{eaa518ac}$

$\qquad = \text{b6fa18b1}$

$A = \text{b6fa18b1} <<< \text{eaa518ac}$

Convert 0xeaa518ac to decimal number as follows:

$\text{eaa518a} = 12 \times 16^0 + 10 \times 16^1 + 8 \times 16^2 + 1 \times 16^3 + 5 \times 16^4 + 10 \times 16^5$

$\qquad\qquad + 10 \times 16^6 + 14 \times 16^7$

$\qquad\quad = 3936688300 (\text{mod } 32)$

$\qquad\quad = 12$

Thus, $A = \text{b6fa18b1} <<< 12$

$\qquad\quad = \text{0xa18b1b6f}$

$A = \text{a18b1b6f} + S[2]$

$\quad = \text{a18b1b6f} + \text{0942b409}$

$\quad = \text{aacdcf78}$

On the other hand, B-routine computation becomes:

$B = (B + S[1]) \oplus A$

$\quad = (\text{eaa518ac} \oplus \text{aacdcf78}) <<< \text{aacdcf78}$

Since $\text{0xaacdcf78} = 2865614712_{(10)} (\text{mod } 32) = 24$,

$B = \text{4068d7d4} <<< 24$

$\quad = \text{0xd44068d}$

Finally,

$B = B + S[3]$

$\quad = \text{d44068d7} + \text{32f9c923}$

$\quad = \text{073a31fa}$

We have computed the first two rounds 0 and 1. Repeated computation up to round 12 will be resulted in the following encryption process table.

Encryption process

Round	A	B
0	5c5f001d	eaa518ac
1	aacdcf78	073A31fa
2	b2c9dafc	d0506098
3	362f2508	67cccf55
4	ace3d838	5f84483d
5	6ad30720	d77180e6
6	3cc6723c	accd0d34
7	c2177344	9954851d
8	436ee2fe	f7702871
9	fac6db42	91c5af63
10	6a180397	f63131f5
11	e07e082e	816fc2b3
12	ac13c0f7	52892b5b

Ciphertext = ac13c0f7 52892b5b.

4.3.4 Decryption

RC5 decryption is given in the pseudocode as shown below.

For $i = r$ down to 1 do

$$B = ((B - S[2i + 1]) >>> A) \oplus A$$

$$A = ((A - S[2i]) >>> B) \oplus B$$

$$B = B - S[1]$$

$$A = A - S[0]$$

The decryption routine is easily derived from the encryption routine. The RC5 encryption/decryption algorithms are illustrated as shown in Figures 4.10 and 4.11, respectively.

Example 4.15 Consider the decryption problem of RC5-32/12/16. To decrypt the ciphertext obtained in Example 4.14, the output of round 11 is inputted into two 32-bit registers, A and B in Figure 4.11, and the following steps are taken according to the RC5 decryption algorithm.

Decryption Process

The cyphertext Y is

$$Y = A \| B = (ac13c0f7) \| (52892b5b)$$

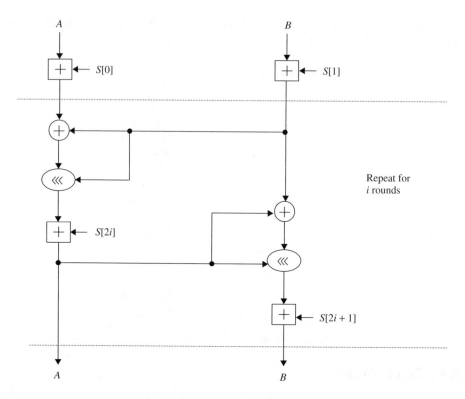

Figure 4.10　RC5 encryption algorithm.

Solution for Round 12

- B-routine computation:

For $i = 12, S[2i + 1] = S[25] = 30726d5a$

$$= 0011\ 0000\ 0111\ 0010\ 0110\ 1101\ 0101\ 1010$$

(Refer to third group expended key in Table 4.16)
Taking 2's compliment of $S[25]$, we can compute additive inverse $-S[25]$ as:

$-S[25] = 1100\ 1111\ 1000\ 1101\ 1001\ 0010\ 1010\ 0110$

$\quad = cf8d92a6$

Convert $A = 0xac13c0f7$ to decimal number such that

$0xac13c0f7 = 28866975735_{(10)} (\text{mod } 32) = 23$

Hence, $B + \{-S[25]\} = 0xac13c0f7 >>> 23$

$\quad = 2d7c0244$

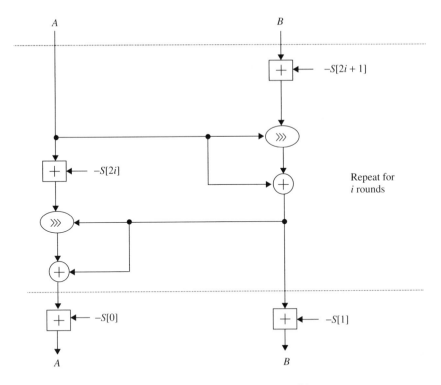

Figure 4.11 RC5 decryption algorithm.

Next step : $B + \{-S[25]\} \oplus A$

$$= \text{2d7c0244} \oplus \text{ac13c0f7}$$

$$= \text{0x816fc2b3}$$

Thus, the B-routine computation for round 12 is completed.

- A-routine computation

For $i = 12, S[2i] = S[24] = \text{0x5728b869}$ (from Table 4.16)

Convert the hexadecimal $S[24]$ to its binary sequence and then take the 2's complement of its binary sequence to obtain the additive inverse $-S[24]$ as follows.

$2's$ complement $=$ 1010 1000 1101 0111 0100 0111 1001 0111

$-S[24] = \text{a8d74797}$

Next step : $A + (-S[24])$

$$= \text{ac13c0f7} + \text{a8d74797}$$

$$= \text{0x54eb088e}$$

$A + (-S[24]) >>> B$

$\quad$ 0x54eb088e $>>>$ 816fc2b3

Convert Hexadecimal to Decimal:

0x816fc2b3 $= 2171585203 \pmod{32} = 19$

Thus, $A = $ 0x54eb088e $>>> 19$

$\qquad = 0110\ 0001\ 0001\ 0001\ 1100\ 1010\ 1001\ 1101$

$\qquad = $ 0x6111ca9d

Finally, $A \oplus B = $ (6111ca9d)(816fc2b3)

$\qquad\qquad A = $ e07e082e

This result shows the end of A-routine computation for Round 12.

Solution for Round 11

Input to Round 11 is

$A = $ e07e082e, $B = $ 816fc2b3

B-routine computation for round 11:
For $i = 11$, $S[2i + 1] = S[23] = $ b2f8fd20 (from Table 4.16)
Since $-S[23] = $ 4d0702e0,

$B + (-S[23]) = $ 816fc2b3 $+$ 4d0702e0

$\qquad\qquad = $ ce76c593
$B >>> A :$ ce76c593 $>>>$ e07e082e,

where 0xe07e082e $= 14$ (decimal)

ce76c593 $>>> 14 = $ 0x164f39db

Finally, B-routine computation yields

$B = B + A = $ 164f39db $+$ e07e082e

$\quad = $ (1111 0110 0011 0001 0011 0001 0001 1111 0101)

$\quad = $ 0xf63131f5

This is the result of B-routine computation for Round 11.

A-routine computation for Round 11:
For $i = 11$, $S[2i] = S[22] = 942a8308$ (see Table 4.16)

$$S[22] = 1001\ 0100\ 0010\ 1010\ 1000\ 0011\ 0000\ 1000$$

$\text{2's complement} = 0110\ 1011\ 1101\ 0101\ 0111\ 1100\ 1111\ 0111$

$$-S[22] = 0110\ 1011\ 1101\ 0101\ 0111\ 1100\ 1111\ 1000$$

$$= 6bd57cf8$$

$$A + (-S[22]) = e07e082e + 6bd57cf8$$

$$= 4c538526$$

$$A + (-S[22]) >>> f63131f5 = 4130419189 \pmod{23} = 21$$

$$= 4c538526 >>> 21$$

Finally, $A = (A + (-S[22]) >>> 21) \oplus B$

$$= 9c293262 \oplus f63131f5$$

$$= 6a180397$$

This is the result of A-routine computation for Round 11.

We have computed Round 12 and 11 only as shown above. The rest of Round 10~1 is listed in the following decryption process table.

Decryption process

Round	A	B
12	e07e082e	816fc2b3
11	6a180397	f63131f5
10	fac6db42	91c5af63
9	436ee2fe	f7702871
8	c2177344	9954851d
7	3cc6723c	accd0d34
6	6ad30720	d77180e6
5	ace3d838	5f84483d
4	362f2508	67cccf55
3	b2c9dafc	d0506098
2	aacdcf78	073a31fa
1	5c5f001d	eaa518ac

Deciphered plaintext = eedba521 6d8f4b15.

Example 4.16 Consider RC5-32/16/10. Since $w = 32$-bit words, $r = 16$ rounds, and $b = 10$-byte key, the parameters to compute are $u = w/8 = 4$ bytes/word, $c = \lceil b/u \rceil = 3$ words in key, and $t = 2(r + 1) = 34$ words in S.

Key mixing

$S[0] = $ ce9e9457	$S[1] = $ 9b2aa851	$S[2] = $ 37cde42b	$S[3] = $ c74caeb7
$S[4] = $ 12f39eef	$S[5] = $ 66ba64e2	$S[6] = $ aec49188	$S[7] = $ 4699fa2b
$S[8] = $ 0f1e2ae7	$S[9] = $ ae384da7	$S[10] = $ 9ad0a8ed	$S[11] = $ 31200c4f
$S[12] = $ f67fd8f0	$S[13] = $ 8ddf1681	$S[14] = $ 3a7c135e	$S[15] = $ 22d6c9ed
$S[16] = $ 4516534e	$S[17] = $ 82472626	$S[18] = $ 383c9ba7	$S[19] = $ 1c2074e9
$S[20] = $ 3e10bde0	$S[21] = $ 4215fa75	$S[22] = $ f8dfa01c	$S[23] = $ cda35bac
$S[24] = $ a1d40dae	$S[25] = $ 8ef11ef1	$S[26] = $ d4409560	$S[27] = $ 043199d0
$S[28] = $ e820a877	$S[29] = $ 1899687c	$S[30] = $ 011db658	$S[31] = $ 72062f23
$S[32] = $ 7f05f007	$S[33] = $ eef913ed		

Encryption

Round	A	B
0	bd7a3978	08b9f366
1	a8c06bd8	85ed284f
2	b4bf3585	90fe1e28
3	eff03eac	28a2421b
4	cd58becc	5e05cc06
5	722d5b91	604e64a0
6	08e31821	5f3a0f83
7	f944d070	02ca706b
8	ba17322a	f7542d09
9	be78e241	ae7a1379
10	ae30c3c2	43413d61
11	d3c39d63	51b85bc0
12	244fd451	ae140ae0
13	5e9c7411	02157ae0
14	44a9b768	d566f0c2
15	485ad502	e6f6c625
16	548854fc	8a20fd1a

Ciphertext = 548854fc 8a20fd1a.

Decryption

Round	A	B
16	485ad502	e6f6c625
15	44a9b768	d566f0c2
14	5e9c7411	02157ae0
13	244fd451	ae140ae0
12	d3c39d63	51b85bc0
11	ae30c3c2	43413d61
10	be78e241	ae7a1379
9	ba17332a	f7542d09
8	f944d070	02ca706b
7	08e31821	5f3a0f83
6	722d5b91	604e64a0
5	cd58becc	5e05cc06
4	eff03eac	28a2421b
3	b4bf3585	90fe1e28
2	a8c06bd8	85ed284f
1	bd7a3978	08b9f366
0	eedba521	6d8f4615

Plaintext (deciphered text) = eedba52 6d8f4b15.

4.4   RC6 Algorithm

RC6 is an improvement to RC5, designed to meet the requirements of increased security and better performance. Like RC5, which was proposed in 1995, RC6 makes use of data-dependent rotations. One new feature of RC6 is the use of four working registers instead of two. While RC5 is a fast block cipher, extending it to act on 128-bit blocks using two 64-bit working registers, RC6 is modified its design to use four 32-bit registers rather than two 64-bit registers. This has the advantage that it can be done two rotations per round rather than the one found in a half-round of RC5.

4.4.1   Description of RC6

Like RC5, RC6 is a fully parameterized family of encryption algorithms. A version of RC6 is also specified as RC6-$w/r/b$ where the word size is w bits, encryption consists of a number of rounds r, and b denotes the encryption key length in bytes.

RC6 was submitted to NIST for consideration as the new AES. Since the AES submission is targeted at $w = 32$ and $r = 20$, the parameter values specified as RC6-w/r are used as shorthand to refer to such versions. For all variants, RC6-$w/r/b$ operates on four w-bit words using the following six basic operations:

$a + b$: Integer addition modulo 2^w
$a - b$: Integer subtraction modulo 2^w
$a \oplus b$: Bitwise XOR of w-bit words
$a \times b$: Integer multiplication modulo 2^w
$a <<< b$: Rotate the w-bit word a to the left by the amount given by the least significant lg w bits of b
$a >>> b$: Rotate the w-bit word a to the right by the amount given by the least significant lg w bits of b (where lg w denotes the base-two logarithm of w).

RC6 exploits data-dependent operations such that 32-bit integer multiplication is efficiently implemented on most processors. Integer multiplication is a very effective diffusion and is used in RC6 to compute rotation amounts so that these amounts are dependent on all of the bits of another register. As a result, RC6 has much faster diffusion than RC5.

4.4.2 Key Schedule

The key schedule of RC6-$w/r/b$ is practically identical to that of RC5-$w/r/b$. In fact, the only difference is that in RC6-$w/r/b$, more words are derived from the user-supplied key for use during encryption and decryption.

The user supplies a key of b bytes, where $0 \leqslant b \leqslant 255$. Sufficient zero bytes are appended to give a key length equal to a nonzero integral number of words; these key bytes are then loaded into an array of c w-bit words $L[0], L[1], \ldots, L[c - 1]$. The number of w-bit words generated for additive round keys is $2r + 4$, and these are stored in the array $S[0, 1, \ldots, 2r + 3]$.

The key schedule algorithm is as shown below.

Key Schedule for RC6-$w/r/b$
Input: User-supplied b byte key preloaded into the c-word array $L[0, 1, \ldots, c - 1]$
Number of rounds, r
Output: w-bit round keys $S[0, 1, \ldots, 2r + 3]$
Key expansion:

Definition of the magic constants

$$P_w = \text{Odd}((e - 2)2^w)$$
$$Q_w = \text{Odd}((\phi - 1)2^w)$$

where

$e = 2.71828182 \ldots$ (base of natural logarithms)
$\phi = 1.618033988 \ldots$ (golden ratio)

Converting the secret key from bytes to words
for $i = b - 1$ down to 0 do

$$L[i/u] = (L[i/u] <<< 8 + K[i]$$

Initializing the array S
$S[0] = P_w$

for $i = 1$ to $2r + 3$ do

$S[i] = S[i - 1] + Q_w$

Mixing in the secret key S
$A = B = i = j = 0$

$v = 3 \times \max\{c, 2r + 4\}$

for $s = 1$ to v do

{

$A = S[i] = (S[i] + A + B) <<< 3$

$B = L[j] = (L[j] + A + B) <<< (A + B)$

$i = (i + 1) \bmod (2r + 4)$

$j = (j + 1) \bmod c$

}

4.4.3 Encryption

RC6 encryption works with four w-bit registers A, B, C, and D which contain the initial input plaintext. The first byte of plaintext is placed in the least significant byte of A. The last byte of plaintext is placed into the most significant byte of D. The arrangement of $(A, B, C, D) = (B, C, D, A)$ is like that of the paralleled assignment of values (bytes) on the right to the registers on the left, as shown in Figure 4.12.

The RC6 encryption algorithm is shown below:

Encryption with RC6-w/r/b
   Input: Plaintext stored in four w-bit input registers A, B, C, D
   Number of rounds, r
   w-bit round keys $S[0, 1, \ldots, 2r + 3]$
   Output: Ciphertext stored in A, B, C, D

Procedure: $B = B + S[0]$

   $D = D + S[1]$

   for $i = 1$ to r do

   {

   $t = (B \times (2B + 1)) <<< \lg w$

   $u = (D \times (2D + 1)) <<< \lg w$

   $A = ((A \oplus t) <<< u) + S[2i]$

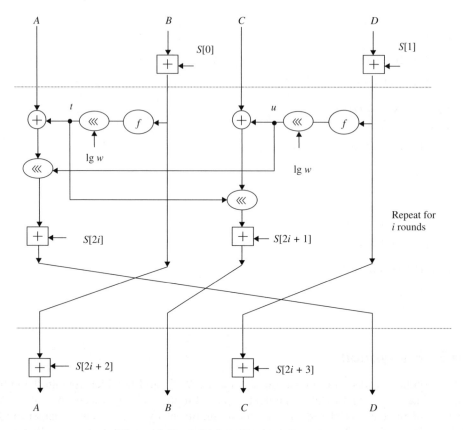

Figure 4.12 RC6-$w/r/b$ encryption scheme.

$$C = ((C \oplus u) <<< t) + S[2i + 1]$$
$$(A, B, C, D) = (B, C, D, A)$$
$$\}$$
$$A = A + S[2r + 2]$$
$$C = C + S[2r + 3]$$

Example 4.17   Consider RC6-$w/r/b$ where $w = 32, r = 20$, and $b = 16$. Suppose the plaintext and user key are given as follows.

Plaintext: 02 13   24   35   46   57   68   79   8a   9b   ac   bd   ce   df   e0   f1

Key: 01 23   45   67   89   ab   cd   ef   01   12   23   34   45   56   67   78

Key expansion
  Parameters:

$c = 4$ (number of words in key)

$t = 44$ (number of words in S)

$u = 4$ (number of bytes in word)

Magic constants
$P_w = \text{b7e15163}$

$Q_w = \text{9e3779b9}$

Converting the secret key from bytes to words:
$L[0] = 67452301 \quad L[1] = \text{efcdab89}$

$L[2] = 34231201 \quad L[3] = 78675645$

Mixing in the secret key S

$S[0] = 05479d38$	$S[1] = \text{e4a3e582}$	$S[2] = \text{fbcc7a4b}$	$S[3] = \text{e878faa4}$
$S[4] = 8\text{ed}14980$	$S[5] = 5\text{f}5873\text{fd}$	$S[6] = \text{aec05ae6}$	$S[7] = \text{aafffe1d}$
$S[8] = 6\text{bf}8\text{b7e3}$	$S[9] = 64\text{e}27682$	$S[10] = 23\text{c4d46f}$	$S[11] = \text{da521c4b}$
$S[12] = 662\text{b}9392$	$S[13] = \text{c51ae971}$	$S[14] = \text{be84587a}$	$S[15] = 473\text{c}1481$
$S[16] = \text{ab246684}$	$S[17] = \text{b9770047}$	$S[18] = 98327\text{b6a}$	$S[19] = 529\text{be}229$
$S[20] = \text{b992809a}$	$S[21] = 79\text{c1fa56}$	$S[22] = 617\text{cd18d}$	$S[23] = 1\text{bcb9a08}$
$S[24] = 8\text{babbbb3}$	$S[25] = 0\text{dd061bd}$	$S[26] = 8\text{c1ec8a2}$	$S[27] = 20\text{f286d0}$
$S[28] = \text{faf8eff4}$	$S[29] = 46\text{b87c92}$	$S[30] = \text{c5096b01}$	$S[31] = \text{dbdcc9b0}$
$S[32] = \text{d1b212b4}$	$S[33] = \text{dd0f3d38}$	$S[34] = 27\text{c02df3}$	$S[35] = 0\text{fb21526}$
$S[36] = 46\text{e0faa6}$	$S[37] = \text{e9d9748f}$	$S[38] = \text{e274fdcc}$	$S[39] = 09\text{ae3f8e}$
$S[40] = 95\text{f85e40}$	$S[41] = \text{a9f90a40}$	$S[42] = \text{f0e51469}$	$S[43] = 45\text{f060d1}$

Encryption
  Using Figure 4.12, compute the ciphertext of RC6-32/20/16.
  Initial value in each register:

$A = 35241302 \quad B = 7\text{eaff47e}$

$C = \text{bdac9b8a} \quad D = \text{d684c550}$

Encryption process

Round	A	B	C	D
1	7eaff47e	a17a48d4	d684c550	fdbc336a
2	a17a48d4	Fd35085f	fdbc336a	8d81f7b9

Round	A	B	C	D
3	fd35085f	9300620e	8d81f7b9	2d144999
4	9300620e	5013ef46	2d144999	53caa736
5	5013ef46	8c83dd52	53caa736	ef7cbe5d
6	8c83dd52	f8754ace	ef7cbe5d	8cc61508
7	f8754ace	49dd0a20	8cc61508	0035d1db
8	49dd0a20	662fc8cb	0035d1db	7e9553f1
9	662fc8cb	8fde9634	7e9553f1	84ceecec
10	8fde9634	Ce5ac268	84ceecec	42aa5994
11	ce5ac268	4a1d83c3	42aa5994	31cdfe66
12	4a1d83c3	113537e5	31cdfe66	5db94923
13	113537e5	4b1b6674	5db94923	e3632504
14	4b1b6674	f60dd47f	e3632504	0750ccfe
15	f60dd47f	95a4e7a0	0750ccfe	b1e27064
16	95a4e7a0	442babe9	b1e27064	f229c1dc
17	442babe9	cb3a05f9	f229c1dc	fadd06ef
18	cb3a05f9	4ce5dc7b	fadd06ef	a76a5ba6
19	4ce5dc7b	3e3439e9	a76a5ba6	f105f04e
20	3e3439e9	23c61547	f105f04e	183fa47e

Final value in each register:

$A = 2f194e52 \quad B = 23c61547$

$C = 36f6511f \quad D = 183fa47e$

Thus, the ciphertext is computed as:

52 4e 19 2f 47 15 c6 23 1f 51 f6 36 7e a4 3f 18

4.4.4 Decryption

RC6 decryption works with four w-bit registers A, B, C, D which contain the initial output ciphertext at the end of encryption. The first byte of ciphertext is placed into the least significant byte of A. The last byte of ciphertext is placed into the most significant byte of D.

The RC6 decryption algorithm is illustrated as shown below:

Decryption with RC6-$w/r/b$

Input: Ciphertext stored in four w-bit input registers A, B, C, D

Number of rounds, r

w-bit round keys $S[0, 1, \ldots, 2r + 3]$

Output: Plaintext stored in A, B, C, D

Procedure: $C = C - S[2r + 3]$

$\quad A = A - S[2r + 2]$

$\quad$ for $i = r$ down to 1 do

$\quad$ {

$\quad (A, B, C, D) = (D, A, B, C)$

$\quad u = (D \times (2D + 1)) <<< \lg w$

$\quad t = (B \times (2B + 1)) <<< \lg w$

$\quad C = ((C - S[2i + 1] >>> t) \oplus u$

$\quad A = ((A - S[2i]) >>> u) \oplus t$

$\quad$ }

$\quad D = D - S[1]$

$\quad B = B - S[0]$

The decryption of RC6 is depicted as shown in Figure 4.13.

Example 4.18 Consider again RC6-32/20/16. Utilizing Figure 4.13 for RC6 decryption, the input is the ciphertext stored in four 32-bit input registers A, B, C, and D.

Initial value in each register:

$A = 3e3439e9 \quad B = 23c61547$

$C = f105f04e \quad D = 183fa47e$

Decryption process

Round	A	B	C	D
1	4ce5dc7b	3e3439e9	a76a5ba6	f105f04e
2	cb3a05f9	4ce5dc7b	fadd06ef	a76a5ba6
3	442babe9	cb3a05f9	f229c1dc	fadd06ef
4	95a4e7a0	442babe9	b1e27064	f229c1dc
5	f60dd47f	95a4e7a0	0750ccfe	b1e27064
6	4b1b6674	f60dd47f	e3632504	0750ccfe
7	113537e5	4b1b6674	5db94923	e3632504
8	4a1d83c3	113537e5	31cdfe66	5db94923
9	ce5ac268	4a1d83c3	42aa5994	31cdfe66
10	8fde9634	ce5ac268	84ceecec	42aa5994
11	662fc8cb	8fde9634	7e9553f1	84ceecec
12	49dd0a20	662fc8cb	0035d1db	7e9553f1

Round	A	B	C	D
13	f8754ace	49dd0a20	8cc61508	0035d1db
14	8c83dd52	f8754ace	ef7cbe5d	8cc61508
15	5013ef46	8c83dd52	53caa736	ef7cbe5d
16	9300620e	5013ef46	2d144999	53caa736
17	fd35085f	9300620e	8d81f7b9	2d144999
18	a17a48d4	fd35085f	fdbc336a	8d81f7b9
19	7eaff47e	a17a48d4	d684c550	fdbc336a
20	35241302	7eaff47e	bdac9b8a	d684c550

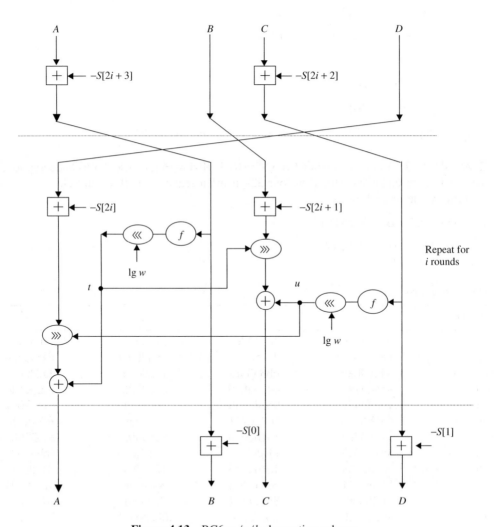

Figure 4.13 RC6-w/r/b decryption scheme.

Final value in each register:

$A = 35241302 \quad B = 79685746$

$C = \text{bdac9b8a} \quad D = \text{f1e0dfce}$

Thus, the recovered plaintext is computed as:

02 13 24 35 46 57 68 79 8a 9b ac bd ce df e0 f1

Example 4.19 Consider RC6-32/20/16. Assume that the plaintext and user key are given as follows.

Plaintext: b267af31   6d8259e7   b16ac385   f2a072be

User key: de 37 a1 fd 84 92 d8 ef e7 14 f1 b7 cc 78 3a ad

Converting the secret key from bytes to words:

$L[0] = \text{f2baabd4} \quad L[1] = \text{73e727d4}$

$L[2] = \text{edc4db16} \quad L[3] = \text{45c0de8b}$

Mixing in the secret key S:

$S[0] = \text{62e429de}$	$S[1] = \text{3bdc27f1}$	$S[2] = \text{daf4e1c8}$	$S[3] = \text{16c26209}$
$S[4] = \text{b22edecc}$	$S[5] = \text{509c1331}$	$S[6] = \text{3487c3db}$	$S[7] = \text{2b8adb1e}$
$S[8] = \text{5e4c1907}$	$S[9] = \text{14458ba5}$	$S[10] = \text{18da3591}$	$S[11] = \text{8fcdd4b5}$
$S[12] = \text{5a76c846}$	$S[13] = \text{2085c465}$	$S[14] = \text{78c44f1a}$	$S[15] = \text{344b8269}$
$S[16] = \text{cd810b25}$	$S[17] = \text{a4c787e8}$	$S[18] = \text{4fcc683d}$	$S[19] = \text{f0d0d987}$
$S[20] = \text{1d1a587a}$	$S[21] = \text{b55757dc}$	$S[22] = \text{c3d68827}$	$S[23] = \text{bfcc8533}$
$S[24] = \text{094a038c}$	$S[25] = \text{5c4b0c8e}$	$S[26] = \text{4aa837e7}$	$S[27] = \text{ae2430af}$
$S[28] = \text{2e5e3577}$	$S[29] = \text{305afc61}$	$S[30] = \text{3e3b932a}$	$S[31] = \text{3db9bd11}$
$S[32] = \text{9a891917}$	$S[33] = \text{1982ee95}$	$S[34] = \text{eabbfb7a}$	$S[35] = \text{4da6c90}$
$S[36] = \text{0b0945ad}$	$S[37] = \text{16059bf7}$	$S[38] = \text{a4fcfe21}$	$S[39] = \text{aa2c586f}$
$S[40] = \text{3e05d045}$	$S[41] = \text{5fbe7c05}$	$S[42] = \text{974646ea}$	$S[43] = \text{d4af0053}$

Encryption:
Compute the ciphertext of RC6-32/20/16.
Initial values in registers:

$A = \text{b267af31} \quad B = \text{6d8a59e7}$

$C = \text{b16ac385} \quad D = \text{f2a072be}$

Encryption process

Round	A	B	C	D
1	d06e83c5	0fbe58ad	2e7c9aaf	82122047
2	0fbe58ad	aebf8fe0	82122047	4c5209fc
3	aebf8fe0	1eea2af6	4c5209fc	671ab020
4	1eea2af6	4c0793b9	671ab020	7dcf4468
5	4c0793b9	d02f880f	7dcf4468	e1f57f20
6	d02f880f	76e50556	e1f57f20	040efeb0
7	76e50556	9226cc1b	040efeb0	6bc6f374
8	9226cc1b	06a119a3	6bc6f374	97683738
9	06a119a3	85830598	97683738	250fbfe5
10	85830598	1c28dc0a	250fbfe5	c89c019f
11	1c28dc0a	adb7d6c6	c89c019f	c28f0f4b
12	adb7d6c6	1911f356	c28f0f4b	d547cb27
13	1911f356	8a0b16e8	d547cb27	2c1d3ae4
14	8a0b16e8	08ddf156	2c1d3ae4	bed49d1e
15	08ddf156	c77d14d5	bed49d1e	4fc7085f
16	c77d14d5	474b1fd6	4fc7085f	67ffbcff
17	474b1fd6	327894f2	67ffbcff	99d3105c
18	327894f2	438277f7	99d3105c	7351c0e7
19	438277f7	ff8422c8	7351c0e7	3e0b9530
20	ff8422c8	ce15ebd7	3e0b9530	f3ca4bd4

Final value in each register:

$A = 96ca69b2$ $B = ce15ebd7$

$C = 12ba9583$ $D = f3ca4bd4$

Thus, the ciphertext is computed as: $A\|B\|C\|D =$

96ca69b2   ce15ebd7   12ba9583   f3ca4bd4

Decryption:
The initial values in registers A, B, C, and D are output at round 19 at the end of encryption.

Decryption process

Round	A	B	C	D
1	438277f7	ff8422c8	7351c0e7	3e0b9530
2	327894f2	438277f7	99d3105c	7351c0e7
3	474b1fd6	327894f2	67ffbcff	99d3105c

Round	A	B	C	D
4	c77d14d5	474b1fd6	4fc7085f	67ffbcff
5	08ddf156	c77d14d5	bed49d1e	4fc7085f
6	8a0b16e8	08ddf156	2c1d3ae4	bed49d1e
7	1911f356	8a0b16e8	d547cb27	2c1d3ae4
8	adb7d6c6	1911f356	c28f0f4b	d547cb27
9	1c28dc0a	adb7d6c6	c89c019f	c28f0f4b
10	85830598	1c28dc0a	250fbfe5	c89c019f
11	06a119a3	85830598	97683738	250fbfe5
12	9226c1b	06a119a3	6bc6f374	97683738
13	76e50556	9226cc1b	040efeb0	6bc6f374
14	d02f880f	76e50556	e1f57f20	040efeb0
15	4c0793b9	d02f880f	7dcf4468	e1f57f20
16	1eea2af6	4c0793b9	671ab020	7dcf4468
17	aebf8fe0	1eea2af6	4c5209fc	671ab020
18	0fbe58ad	aebf8fe0	82122047	4c5209fc
19	d06e83c5	0fbe58ad	2e7c9aaf	82122047
20	b267af31	d06e83c5	b16ac385	2e7c9aaf

The final decrypted plaintext is:

b267af31   6d8a59e7   b16ac385   f2a072be

Example 4.20   Consider RC6-32/20/24. Suppose the plaintext and user key are given as follows:

Plaintext: 35241302   79685746   bdac9b8a   f1e0dfce

User key: 01   23   45   67   89   ab   cd   ef
             01   12   23   34   45   56   67   78
             89   9a   ab   bc   cd   de   ef   f0

The user supplies a key of $b = 24$ bytes, where $0 \leqslant b \leqslant 255$. From this key, $2r + 4 = 44$ words are derived and stored in the array $S[0, 1, \ldots, 2r + 3]$. This array is used in both encryption and decryption.

Key array

$S[0] = $ 4d80ade	$S[1] = $ c85296a3	$S[2] = $ c7ca853c	$S[3] = $ d665bea0
$S[4] = $ 4d34492f	$S[5] = $ e110bf65	$S[6] = $ 9f4acf83	$S[7] = $ ed85cb10
$S[8] = $ f9f0f8eb	$S[9] = $ 2275ea3f	$S[10] = $ e5dc8714	$S[11] = $ a1b4b8b4
$S[12] = $ 1a28cd0a	$S[13] = $ 618fbe87	$S[14] = $ 6fc1ede0	$S[15] = $ 8eaf634d
$S[16] = $ 7d213901	$S[17] = $ bed7ab73	$S[18] = $ 79ba092e	$S[19] = $ 6179bc8a
$S[20] = $ aa35b6f6	$S[21] = $ 0091b3ca	$S[22] = $ 65f970e9	$S[23] = $ 687e9e94

$S[24] = $ f17e5188	$S[25] = $ 7ec55cf7	$S[26] = $ fe2c8e93	$S[27] = $ 2e7b3dae
$S[28] = $ 56093cb8	$S[29] = $ ed28fa03	$S[30] = $ ab2eaaec	$S[31] = $ d049366f
$S[32] = $ fcd4cbd3	$S[33] = $ 84b3906f	$S[34] = $ 8eced9f1	$S[35] = $ e02a2453
$S[36] = $ 123b6e03	$S[37] = $ a6192a81	$S[38] = $ 8648252c	$S[39] = $ b29fbd04
$S[40] = $ 735d2dc1	$S[41] = $ 97447b58	$S[42] = $ 362b46b2	$S[43] = $ 7c310342

RC6 works with four 32-bit registers A, B, C, and D which contain the initial input plaintext as well as the output ciphertext at the end of encryption. Both encryption and decryption using RC6-32/20/24 are processed as shown below.

Initial values in registers:

$A = 35241302 \quad B = 79685746$

$C = $ bdac9b8a $\quad D = $ f1e0dfce

Encryption with RC6-32/20/24

Round	A	B	C	D
0	35241302	7e406224	bdac9b8a	ba337671
1	7e406224	bf73145b	ba337671	ae7fec22
2	bf73145b	8223f9cc	ae7fec22	d96ddcb2
3	8223f9cc	823d1be2	d96ddcb2	8ad786e7
4	823d1be2	30fa9e1e	8ad786e7	3439983d
5	30fa9e1e	69de30e7	3439983d	41340557
6	69de30e7	1e5076a4	41340557	5cbef6d9
7	1e5076a4	a3202136	5cbef6d9	90578218
8	a3202136	48cd17be	90578218	36536a30
9	48cd17be	89b9dc8a	36536a30	6f54b847
10	89b9dc8a	e21b47ad	6f54b847	4927a4a1
11	e21b47ad	51ea2335	4927a4a1	21e33ea6
12	51ea2335	f6288913	21e33ea6	8dfa1819
13	f6288913	74cc2d40	8dfa1819	23c3a852
14	74cc2d40	3cfc9386	23c3a852	99050d00
15	3cfc9386	f0cd5501	99050d00	4f93af72
16	f0cd5501	f3d82818	4f93af72	096f38cb
17	f3d82818	1e600aa7	096f38cb	13e79bec
18	1e600aa7	f3af0e5c	13e79bec	38d4defa
19	f3af0e5c	99fe3cb6	38d4defa	aeb84edc
20	99fe3cb6	0405e519	aeb84edc	d49152f9

Thus, the output ciphertext at the end of encryption is:

d0298368   0405e519   2ae9521e   d49152f9.

Decryption with RC6-32/20/24

Round	A	B	C	D
1	f3af0e5c	99fe3cb6	38d4defa	aeb84edc
2	1e600aa7	f3af0e5c	13e79bec	38d4defa
3	f3d82818	1e600aa7	096f38cb	13e79bec
4	f0cd5501	f3d82818	4f93af72	096f38cb
5	3cfc9386	f0cd5501	99050d00	4f93af72
6	74cc2d40	3cfc9386	23c3a852	99050d00
7	f6288913	74cc2d40	8dfa1819	23c3a852
8	51ea2335	f6288913	21e33ea6	8dfa1819
9	e21b47ad	51ea2335	4927a4a1	21e33ea6
10	89b9dc8a	e21b47ad	6f54b847	4927a4a1
11	48cd17be	89b9dc8a	36536a30	6f54b847
12	a3202136	48cd17be	90578218	36536a30
13	1e5076a4	a3202136	5cbef6d9	90578218
14	69de30e7	1e5076a4	41340557	5cbef6d9
15	30fa9e1e	69de30e7	3439983d	41340557
16	823d1be2	30fa9e1e	8ad786e7	3439983d
17	8223f9cc	823d1be2	d96ddcb2	8ad786e7
18	bf73145b	8223f9cc	ae7fec22	d96ddcb2
19	7e406224	bf73145b	ba337671	ae7fec22
20	35241302	7e406224	bdac9b8a	ba337671

Thus, the final decrypted plaintext is:

$$35241302 \quad 79685746 \quad bdac9b8a \quad f1e0dfce.$$

4.5 AES (Rijndael) Algorithm

The AES specified the Rijndael algorithm which is an FIFS-approved cryptographic algorithm developed by Daemen and Rijmen as an AES candidate algorithm in 1999. The Rijndael algorithm is a symmetric block cipher that can process data blocks of 128 bits using cryptographic keys of 128, 192, and 256 bits.

In this section, we will cover the algorithm specification such as the key-expansion routine, encryption by cipher, and decryption by inverse cipher.

4.5.1 Notational Conventions

- The cipher key for the Rijndael algorithm is a sequence of 128, 196, or 256 bits such that the index attached to a bit falls in between the range $0 \leqslant i \leqslant 128, 0 \leqslant i \leqslant 192$, or $0 \leqslant i \leqslant 256$, respectively.

- All byte values of the AES Rijndael algorithm are presented by a vector notation $(b_7, b_6, b_5, b_4, b_3, b_2, b_1, b_0)$ which corresponds to a polynomial representation as:

$$b_7 x^7 + b_6 x^6 + b_5 x^5 + b_4 x^4 + b_3 x^3 + b_2 x^2 + b_1 x + b_0 = \sum_{i=0}^{7} b_i x^i$$

For example, $(01001011) \rightarrow x^6 + x^3 + x + 1$.
- If there is an additional bit b_8 to the left of an 8-bit byte, it will appear immediately to the left of the left bracket such as $1(00101110) = 1(2e)$.
- Arrays of bytes, $a_0, a_1, a_2, \ldots, a_{15}$, are defined from the 128-bit input sequence, $ip_0, ip_1, ip_2, \ldots, ip_{126}, ip_{127}$, as follows:

$a_0 = (ip_0, ip_1, \ldots, ip_7)$

$a_1 = (ip_8, ip_9, \ldots, ip_{15})$

$\vdots$

$a_{15} = (ip_{120}, ip_{121}, \ldots, ip_{127})$

where ip_k denotes $input_k$ for $k = 0, 1, 2, \ldots, 127$.

In general, the pattern extended to longer sequence like 192- and 256-bit keys is expressed as:

$a_n = (ip_{8n}, ip_{8n+1}, \ldots, ip_{8n+7}), \quad n \leqslant 16.$

- The AES algorithm's operations are internally performed on a two-dimensional array of bytes called the *state*. The state consists of four rows of bytes. The state array $S_{r,c}$ has a row number $r, 0 \leqslant r < 4$, and a column number $c, 0 \leqslant c < Nb$, where Nb bytes are the block length divided by 32.
The input-byte array $(in_0, in_1, \ldots, in_{15})$ at the cipher is copied into the state array according to the following scheme:

$S_{r,c} = in(r + 4c)$ for $0 \leqslant r < 4$ and $0 \leqslant c < Nb$

and at the Inverse Cipher, the state is copied into the output array as follows:

$out(r + 4c) = S_{r,c}$ for $0 \leqslant r < 4$ and $0 \leqslant c < Nb.$

An individual byte of the state is referred to as either $S_{r,c}$ or $S(r, c)$. The cipher and Inverse Cipher operations are conducted on the state array as illustrated in Figure 4.14. For example, if $r = 0$ and $c = 3$, then $in(0 + 12) = in(12) = S_{0,3}$; if $r = 3$ and $c = 2$, then $in(3 + 8) = in(11) = S_{3,2}$.
The 4 bytes in each column of the state form a 32-bit word, where the row number r provides an index for the 4 bytes within each word, and the column number c provides an index representing the column in this array.

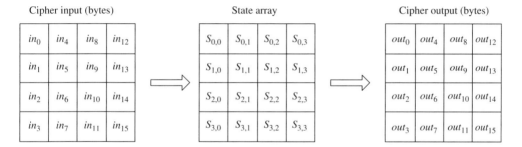

Figure 4.14 State array input and output.

4.5.2 Mathematical Operations

Finite field elements (all bytes in the AES algorithm) can be added and multiplied. The basic mathematical operations will be introduced in the following.

Addition

The addition of two elements in a finite field is achieved by XORing the coefficients for the corresponding powers in the polynomials for two elements. For example,

$(x^5 + x^3 + x^2 + 1) + (x^7 + x^5 + x + 1) = x^7 + x^3 + x^2 + x$ (polynomial)

$(00101101) \oplus (10100011) = (10001110)$ (binary)

$(2d) \oplus (a3) = (8e)$ (hexadecimal)

Multiplication

The polynomial multiplication in $GF(2^8)$ corresponds to the multiplication of polynomial modulo $m(x)$ that an irreducible (or primitive) polynomial of degree 8 for the AES algorithm:

$m(x) = x^8 + x^4 + x^3 + x + 1$

Example 4.21 Prove $(73) \cdot (a5) = (e3)$

$(01110011) \cdot (10100101)$

$(x^6 + x^5 + x^4 + x + 1) \cdot (x^7 + x^5 + x^2 + 1)$

$= x^{13} + x^{12} + x^{10} + x^9 + x^6 + x^4 + x^3 + x^2 + x + 1$

The modular reduction by $m(x)$ results in

$$x^{13} + x^{12} + x^{10} + x^9 + x^6 + x^4 + x^3 + x^2 + x + 1 \bmod (x^8 + x^4 + x^3 + x + 1)$$
$$= x^7 + x^6 + x^5 + x + 1$$
$$= (11100011) = (e3)$$

Since the multiplication is associative, it holds that

$$a(x)(b(x) + c(x)) = a(x)b(x) + a(x)c(x)$$

The element $(01) = (00000001)$ is called the *multiplicative identity*. For any polynomial $b(x)$ of degree less than 8, the multiplicative inverse of $b(x)$, denoted by $b^{-1}(x)$, can be found by using the extended euclidean algorithm such that

$$b(x)a(x) + m(x)c(x) = 1$$

from which $b(x)a(x) \bmod m(x) \equiv 1$. Thus, the multiplicative inverse of $b(x)$ becomes

$$b^{-1}(x) = a(x) \bmod m(x)$$

The set of 256 possible byte values has the structure of the finite field $GF(2^8)$ by means of XOR used as both addition and multiplication.

Multiplication by x
Let the binary polynomial be $b(x) = \sum_{i=0}^{7} b_i x^i$. Multiplying $b(x)$ by x results in $xb(x) = \sum_{i=0}^{7} b_i x^{i+1}$, but it can be reduced by modulo $m(x)$.

   If $b_7 = 1$, the reduction is achieved by XORing $m(x)$. It follows that implication by x (i.e., $(00000010)_{(2)} = (02)_{(16)}$) can be implemented at the byte level with a left shift and bitwise XOR with (1b). This operation on bytes is denoted by *xtime()*. Multiplication by higher powers of x can be implemented by repeated application of *xtime()*.

Example 4.22   Compute $(57) \cdot (13) = (fe)$

$$(57) = (01010111)$$

$$(57) \cdot (02) = xtime(57) = (10101110) = (ae)$$

$$(57) \cdot (04) = xtime(ae) = (01011100) \oplus (00011011)$$

$$= (01000111) = (47)$$

$$(57) \cdot (08) = xtime(47) = (10001110) = (8e)$$

$$(57) \cdot (10) = xtime(8e) = (00011100) \oplus (00011011) = (07)$$

Thus, it follows that

$$(57) \cdot (13) = (57) \cdot \{(01) \oplus (02) \oplus (10)\}$$

$$= (57) \oplus (57) \cdot (02) \oplus (57) \cdot (10)$$

$$= (57) \oplus (ae) \oplus (07)$$

$$= (01010111) \oplus (10101110) \oplus (00000111)$$

$$= (11111110) = (fe)$$

Polynomials with Finite Field Elements in GF(2^8)

A polynomial $a(x)$ with byte-coefficient in GF(2^8) can be expressed in word form as:

$$a(x) = a_3x^3 + a_2x^2 + a_1x + a_0 \Leftrightarrow a = (a_0, a_1, a_2, a_3)$$

To illustrate the addition and multiplication operations, let

$$b(x) = b_3x^3 + b_2x^2 + b_1x + b_0 \Leftrightarrow b = (b_0, b_1, b_2, b_3)$$

be a second polynomial.

Addition is performed by adding the finite field coefficients of like powers of x such that

$$a(x) + b(x) = (a_3 \oplus b_3)x^3 + (a_2 \oplus b_2)x^2 + (a_1 \oplus b_1)x + (a_0 \oplus b_0)$$

This addition corresponds to an XOR between the corresponding bytes in each of the words. Multiplication is achieved as shown below:

The polynomial product $c(x) = a(x) \cdot b(x)$ is expanded and like powers are collected to give

$$c(x) = a(x) \cdot b(x) = c_6x^6 + c_5x^5 + c_4x^4 + c_3x^3 + c_2x^2 + c_1x + c_0$$

$$= (c_6, c_5, c_4, c_3, c_2, c_1, c_0)$$

where

$$c_0 = a_0b_0 \qquad\qquad\qquad c_4 = a_3b_1 \oplus a_2b_2 \oplus a_1b_3$$

$$c_1 = a_1b_0 \oplus a_0b_1 \qquad\qquad c_5 = a_3b_2 \oplus a_2b_3$$

$$c_2 = a_2b_0 \oplus a_1b_1 \oplus a_0b_2 \qquad\qquad c_6 = a_3b_3$$

$$c_3 = a_3b_0 \oplus a_2b_1 \oplus a_1b_2 \oplus a_0b_3$$

The next step is to reduce $c(x)$ mod $(x^4 + 1)$ for the AES algorithm, so that x^i mod $(x^4 + 1) = x^{i \bmod 4}$.

The modular product, $a(x) \otimes b(x)$, of two four-term polynomials $a(x)$ and $b(x)$, is given by

$$d(x) = a(x) \otimes b(x) = d_3x^3 + d_2x^2 + d_1x + d_0$$

where

$$d_0 = a_0b_0 \oplus a_3b_1 \oplus a_2b_2 \oplus a_1b_3$$
$$d_1 = a_1b_0 \oplus a_0b_1 \oplus a_3b_2 \oplus a_2b_3$$
$$d_2 = a_2b_0 \oplus a_1b_1 \oplus a_0b_2 \oplus a_3b_3$$
$$d_3 = a_3b_0 \oplus a_2b_1 \oplus a_1b_2 \oplus a_0b_3.$$

Thus, $d(x)$ in matrix form is written as:

$$\begin{bmatrix} d_0 \\ d_1 \\ d_2 \\ d_3 \end{bmatrix} = \begin{bmatrix} a_0a_3a_2a_1 \\ a_1a_0a_3a_2 \\ a_2a_1a_0a_3 \\ a_3a_2a_1a_0 \end{bmatrix} \begin{bmatrix} b_0 \\ b_1 \\ b_2 \\ b_3 \end{bmatrix}$$

The AES algorithm also defines the inverse polynomials as:

$$a(x) = (03)x^3 + (01)x^2 + (01)x^1 + (02)$$
$$a^{-1}(x) = (0b)x^3 + (0d)x^2 + (09)x^1 + (0e)$$

4.5.3 AES Algorithm Specification

For the AES algorithm, Nb denotes the number of 32-bit words with respect to the 128-bit block of the input, output, or state ($128 = Nb \times 32$, from which $Nb = 4$).

Nk represents the number of 32-bit words with respect to the cipher-key length of 128, 192, or 256 bits:

$$128 = Nk \times 32, \quad Nk = 4$$
$$196 = Nk \times 32, \quad Nk = 6$$
$$256 = Nk \times 32, \quad Nk = 8$$

The number of rounds are 10, 12, and 14, respectively.

Key Expansion

The AES algorithm takes the cipher key K and performs a key-expansion routine to generate a key schedule. The key expansion generates a total of $Nb(Nr + 1)$ words: an initial set of Nb words for $Nr = 0$ and $2Nb$ for $Nr = 1$, $3Nb$ for $Nr = 2$, ..., $11Nb$ for $Nr = 10$. Thus, the resulting key schedule consists of a linear array of 4-byte words $[w_i]$, $0 \leqslant i < Nb(Nr + 1)$.

RotWord() takes a 4-byte input word $[a_0, a_1, a_2, a_3]$ and performs a cyclic permutation such as $[a_1, a_2, a_3, a_0]$.

		y														
		0	1	2	3	4	5	6	7	8	9	a	b	c	d	e	f
	0	63	7c	77	7b	f2	6b	6f	c5	30	01	67	2b	fe	d7	ab	76
	1	ca	82	c9	7d	fa	59	47	f0	ad	d4	a2	af	9c	a4	72	c0
	2	b7	fd	93	26	36	3f	f7	cc	34	a5	e5	f1	71	d8	31	15
	3	04	c7	23	c3	18	96	05	9a	07	12	80	e2	eb	27	b2	75
	4	09	83	2c	1a	1b	6e	5a	a0	52	3b	d6	b3	29	e3	2f	84
	5	53	d1	00	ed	20	fc	b1	5b	6a	cb	be	39	4a	4c	58	cf
	6	d0	ef	aa	fb	43	4d	33	85	45	f9	02	7f	50	3c	9f	a8
x	7	51	a3	40	8f	92	9d	38	f5	bc	b6	da	21	10	ff	f3	d2
	8	cd	0c	13	ec	5f	97	44	17	c4	a7	7e	3d	64	5d	19	73
	9	60	81	4f	dc	22	2a	90	88	46	ee	b8	14	de	5e	0b	db
	a	e0	32	3a	0a	49	06	24	5c	c2	d3	ac	62	91	95	e4	79
	b	e7	c8	37	6d	8d	d5	4e	a9	6c	56	f4	ea	65	7a	ae	08
	c	ba	78	25	2e	1c	a6	b4	c6	e8	dd	74	1f	4b	db	8b	8a
	d	70	3e	b5	66	48	03	f6	0e	61	35	57	b9	86	c1	1d	9e
	e	e1	f8	98	11	69	d9	8e	94	9b	1e	87	e9	ce	55	28	df
	f	8c	a1	89	0d	bf	e6	42	68	41	99	2d	0f	b0	54	bb	16

Figure 4.15 AES S-box (FIPS Publication, 2001).

SubWord() takes a 4-byte input word and applies the S-box (Figure 4.15) to each of the 4 bytes to produce an output word.

Rcon[i] represents the round constant word array and contains the values given by $[x^{i-1}, \{00\}, \{00\}, \{00\}]$ with x^{i-1} starting i at 1.

Example 4.23 Compute the round constant words Rcon [i]:

$$\text{Rcon}[i] = [x^{i-1}, \{00\}, \{00\}, \{00\}]$$

$$\text{Rcon}[1] = [x^0, \{00\}, \{00\}, \{00\}] = [\{01\}, \{00\}, \{00\}, \{00\}] = 01000000$$

$$\text{Rcon}[2] = [x^1, \{00\}, \{00\}, \{00\}] = 02000000$$

$$\text{Rcon}[3] = [x^2, \{00\}, \{00\}, \{00\}] = 04000000$$

$$\text{Rcon}[4] = [x^3, \{00\}, \{00\}, \{00\}] = 08000000$$

$$\text{Rcon}[5] = [x^4, \{00\}, \{00\}, \{00\}] = 10000000$$

$$\text{Rcon}[6] = [x^5, \{00\}, \{00\}, \{00\}] = 20000000$$

$$\text{Rcon}[7] = [x^6, \{00\}, \{00\}, \{00\}] = 40000000$$

$$\text{Rcon}[8] = [x^7, \{00\}, \{00\}, \{00\}] = 80000000$$

$$\text{Rcon}[9] = [x^8, \{00\}, \{00\}, \{00\}] = [x^7 \cdot x, \{00\}, \{00\}, \{00\}] = 1b000000$$

$$x^7 \cdot x = xtime(x^7) = xtime(80) = \{leftshift(80)\} \oplus \{1b\} = 1b$$

$$\text{Rcon}[10] = [x^9, \{00\}, \{00\}, \{00\}] = [x^8 \cdot x, \{00\}, \{00\}, \{00\}] = 36000000$$

$$\text{Rcon}[11] = [x^{10}, \{00\}, \{00\}, \{00\}] = [x^9 \cdot x, \{00\}, \{00\}, \{00\}] = 6c000000$$

$$\text{Rcon}[12] = [x^{11}, \{00\}, \{00\}, \{00\}] = [x^{10} \cdot x, \{00\}, \{00\}, \{00\}] = d8000000$$

$$\text{Rcon}[13] = [x^{12}, \{00\}, \{00\}, \{00\}] = [x^{11} \cdot x, \{00\}, \{00\}, \{00\}] = ab000000$$

$$x^{11} \cdot x = xtime(x^{11}) = xtime(d8) = \{leftshift(d8)\} \oplus \{1b\} = ab$$

Rcon[i] is a useful component for the round constant ward array in order to compute the key-expansion routine.

The input key expansion into the key schedule proceeds as shown in Figure 4.16.

Example 4.24   Suppose the cipher key K is given as

$K = 36$ 8a c0 f4 ed cf 76 a6 08 a3 b6 78 31 31 27 6e

```
Key Expansion(byte key[4*Nk], word w[Nb*(Nr+1)],Nk)
begin
    i= 0
    while(i< Nk)
        w[i]=word[key[4*i],key[4*i+1],key[4*i+2],key[4*i+3]]
        i=i+1
    end while

    i= Nk
    while (i< Nb*(Nr+1))
        word temp=w[i-1]
        if(i mod Nk =0)
            temp=SubWord(RotWord(temp)) xor Rcon[i/Nk]
        else if (Nk= 8 and i mod Nk= 4)
            temp=SubWord(temp)
        endif
    w[i]=w[i-Nk] xor temp
    i= i+1
    end while
end
```

Figure 4.16   Pseudocode for key expansion (FIPS Publication, 2001).

The first four words of K for $Nk = 4$ results in $w[0] = 368ac0f4$, $w[1] = edcf76a6$, $w[2] = 08a3b678$, and $w[3] = 3131276e$.

Computation of $w[4]$ for $i = 4$ is as follows:

Temp $= w[3] = 3131276e$
A cyclic permutation of $w[3]$ by 1 byte produces
RotWord (w [3]) $= 31276e31$

Taking each byte of RotWord$(w[3])$ at a time and applying to the S-box yields

SubWord$(31276e31) = c7cc9fc7$

Compute a round constant Rcon$[i/Nk]$:

Rcon$[4/4]$ = Rcon$[1]$ = 01000000

XORing SubWord() with Rcon[1] yields
SubWord() $\oplus$ Rcon[1] $= c6cc9fc7$

$w[i - Nk] = w[0] = 368ac0f4$

Finally, $w[4]$ is computed as:

$w[4] = c6cc9fc7 \oplus 368ac0f4 = f0465f33$.

Continuing in this fashion, the remaining $w[i]$, $4 \leqslant i \leqslant 43$, can be computed as shown in Table 4.17.

Cipher

The 128-bit cipher input is fed in a column-by-column manner, comprising each column with a 4-byte word. In other words, the input is copied to the state array as shown in Table 4.18.

The cipher is described in the pseudocode in Figure 4.17.

Individual transformations for the pseudocode computation consist of SubBytes(), ShiftRows(), MixColumns(), and AddRoundKey(). These transformations play a role in processing the state and are briefly described below.

SubBytes() Transformation
The SubBytes() transformation is a nonlinear byte substitution that operates independently on each byte of the state using an S-box (Figure 4.18).

For example, if $s_{2,1} = \{8f\}$, then the substitution value is determined by the intersection of the row with index 8 and the column with index f in Figure 4.15. The resulting $s'_{2,1}$ would be a value of $\{73\}$.

Table 4.17   AES key expansion

i	Temp.	After RotWord	After SubWord	Rcon[i/Nk]	After XOR with Rcon	$w[i] = $ temp $\oplus w[i - Nk]$
4	3131276e	31276e31	c7cc9fc7	01000000	c6cc9fc7	f0465f33
5	f0465f33					1d892995
6	1d892995					152a9fed
7	152a9fed					241bb883
8	241bb883	1bb88324	af6cec36	02000000	ad6cec36	5d2ab305
9	5d2ab305					40a39a90
10	40a39a90					5589057d
11	5589057d					7192bdfe
12	7192bdfe	92bdfe71	4f7abba3	04000000	4b7abba3	165008a6
13	165008a6					56f39236
14	56f39236					037a974b
15	037a974b					72e82ab5
16	72e82ab5	e82ab572	9be5d540	08000000	93e5d540	85b5dde6
17	85b5dde6					d3464fd0
18	d3464fd0					d03cd89b
19	d03cd89b					a2d4f22e
20	a2d4f22e	d4f22ea2	4889313a	10000000	5889313a	dd3cecdc
21	dd3cecdc					0e7aa30c
22	0e7aa30c					de467b97
23	de467b97					7c9289b9
24	7c9289b9	9289b97c	4fa75610	20000000	6fa75610	b29bbacc
25	b29bbacc					bce119c0
26	bce119c0					62a76257
27	62a76257					1e35ebee
28	1e35ebee	35ebee1e	96e92872	40000000	d6e92872	647292be
29	647292be					d8938b7e
30	d8938b7e					ba34e929
31	ba34e929					a40102c7
32	a40102c7	0102c7a4	7c77c649	80000000	fc77c649	980554f7
33	980554f7					4096df89
34	4096df89					faa236a0
35	faa236a0					5ea33467
36	5ea33467	a334675e	0a188558	1b000000	11188558	891dd1af
37	891dd1af					c98b0e26
38	c98b0e26					33293886
39	33293886					6d8a0ce1
40	6d8a0ce1	8a0ce16d	7efef83c	36000000	48fef83c	c1e32993
41	c1e32993					086827b5
42	086827b5					3b411f33
43	3b411f33					56cb13d2

Table 4.18 A 16-byte cipher input array

Row No.	Mapping of input block into column-by-column array			
0	a_0	a_4	a_8	a_{12}
1	a_1	a_5	a_9	a_{13}
2	a_2	a_6	a_{10}	a_{14}
3	a_3	a_7	a_{11}	a_{15}

```
Cipher(byte in [4*Nb], byte out[4*Nb], word w[Nb*(Nr+1)])
begin
    byte    state[4, Nb]
    state=in
    AddRopundKey(state, w)
    for round=1 step 1 to Nr-1
        SubBytes(state)
        ShiftRows(state)
        MixColumns(state)
        AddRoundKey(state, w+round*Nb)
    end for
    SubBytes(state)
    ShiftRows(state)
    AddRoundKey(state, w+Nr*Nb)
    out=state
end
```

Figure 4.17 Pseudocode for the cipher (FIPS Publication, 2001).

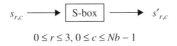

$$0 \le r \le 3, 0 \le c \le Nb - 1$$

Figure 4.18 SubBytes() transformation by the S-box.

ShiftRows() Transformation

In the ShiftRows(), the first row (row 0) is not shifted and the remaining rows proceed as follows:

$$s'_{r,c} = s'_{r,(c+\text{shift}(r,Nb))} \mod Nb, \text{ for } 0 < r < 4 \text{ and } 0 \leqslant c < Nb$$

where the shift value $shift(r, Nb) = shift(r, 4)$ depends on the row number r as follows:

$$shift(1,4) = 1; shift(2,4) = 2; shift(3,4) = 3;$$

This has the effect of shifting the leftmost bytes around into the rightmost positions over different numbers of bytes in a given row.

MixColumns() Transformation
The MixColumns() transformation operates on the state column-by-column, treating each column as a four-term polynomial over GF(2^8) and multiplied modulo $x^4 + 1$ with a fixed polynomial $a(x)$ as:

$$s'(x) = a(x) \otimes s(x)$$

where $a(x) = \{03\}x^3 + \{01\}x^2 + \{01\}x + \{02\}$, $s(x)$ is the input polynomial, and $s'(x)$ is the corresponding polynomial after the MixColumns() transformation.

The matrix multiplication of $s'(x)$ is

$$\begin{bmatrix} s'_{0,c} \\ s'_{1,c} \\ s'_{2,c} \\ s'_{3,c} \end{bmatrix} = \begin{bmatrix} 02 & 03 & 01 & 01 \\ 01 & 02 & 03 & 01 \\ 01 & 01 & 02 & 03 \\ 03 & 01 & 01 & 02 \end{bmatrix} \begin{bmatrix} s_{0,c} \\ s_{1,c} \\ s_{2,c} \\ s_{3,c} \end{bmatrix} \quad \text{for } 0 \leqslant c < Nb$$

The 4 bytes in a column after the matrix multiplication are

$$s'_{0,c} = (\{02\} \cdot s_{0,c}) \oplus (\{03\} \cdot s_{1,c}) \oplus s_{2,c} \oplus s_{3,c}$$
$$s'_{1,c} = s_{0,c} \oplus (\{02\} \cdot s_{1,c}) \oplus (\{03\} \cdot s_{2,c}) \oplus s_{3,c}$$
$$s'_{2,c} = s_{0,c} \oplus s_{1,c} \oplus (\{02\} \cdot s_{2,c}) \oplus (\{03\} \cdot s_{3,c})$$
$$s'_{3,c} = (\{03\} \cdot s_{0,c}) \oplus s_{1,c} \oplus s_{2,c} \oplus (\{02\} \cdot s_{3,c})$$

AddRoundKey() Transformation
In AddRoundKey() transformation, a round key is added to the state by a simple bitwise XOR. Each round key consists of Nb words from the key schedule. These Nb words are added into the columns of the state such that

$$[s'_{0,c}, s'_{1,c}, s'_{2,c}, s'_{3,c}] = [s_{0,c}, s_{1,c}, s_{2,c}, s_{3,c}] \oplus [w_{round*Nb+c}] \text{ for } 0 \leqslant c < Nb$$

where $[w_i]$ is the key schedule words and *round* is a value in the range $0 \leqslant round \leqslant Nr$. The initial round key addition occurs when $round = 0$, prior to the first application of the round function. The application of the AddRoundKey() transformation to the Nr rounds of the cipher occurs when $1 \leqslant round \leqslant Nr$.

Example 4.25　Assume that the input block and a cipher key whose length of 16 bytes each are given as:

```
Plaintext   = a3  c5  08  08  78  a4  ff  d3  00  ff  36  36  28  5f  01  02
Cipher key  = 36  8a  c0  f4  ed  cf  76  a6  08  a3  b6  78  31  31  27  6e
```

Using the algorithm for the pseudocode computation described in Figure 4.17, compute the intermediate values in the state array:

The round key values w[i] are taken from Example 4.24.

r : Round, w[i] : ith Round key values, $N_b = 4$, $N_r = 10$, $w[N_b \times (N_r + 1)] = w[4 \times 11] = w[44]$

$r = 0$: No shift
Input state array $\oplus$ Round key value $w[0]$.

Plaintext block		a3 c5 08 08	→ 1010 0011 1100 0101 0000 1000 0000 1000
Cipher key	⊕	36 8a c0 f4	→ 0011 0110 1000 1010 1100 0000 1111 0100
		95 4f c8 fc	← 1001 0101 0100 1111 1100 1000 1111 1100

Plaintext block		78 a4 ff d3
Cipher key	⊕	ed cf 76 a6
		95 6b 89 75

Plaintext block		00 ff 36 36
Cipher key	⊕	08 a3 b6 78
		08 5c 80 4e

Plaintext block		28 5f 01 02
Cipher key	⊕	31 31 27 6e
		19 6e 26 6c

$r = 1$:
SubBytes()
Compute SubByte() transformation using S-box

\<first column\>	95 → 2a	4f → 84	c8 → e8	fc → b0
\<second column\>	95 → 2a	6b → 7f	89 → a7	75 → 9d
\<third column\>	08 → 30	5c → 4a	80 → cd	4e → 28
\<fourth column\>	19 → d4	6e → 9f	26 → f7	6c → 50

ShiftRows()

\<first row\>	2a 2a 30 d4 → 2a 2a 30 d4
\<second row\>	84 7f 4a 9f → 7f 4a 9f 84
\<third row\>	e8 a7 cd f7 → cd f7 e8 a7
\<fourth row\>	b0 9d 28 50 → 50 b0 9d 28

MixColumns()
The 4 bytes in a column after matrix multiplication denote:

$$s'_{0,0} = (\{02\} \cdot s_{0,0}) \oplus (\{03\} \cdot s_{1,0}) \oplus s_{2,0} \oplus s_{3,0}$$

$$= (\{02\} \cdot 2a) \oplus (\{03\} \cdot 7f) \oplus cd \oplus 50$$

$$= (\{0000 \quad 0010\} \cdot \{00101010\}) \oplus (\{0000 \quad 0011\} \cdot \{0111 \quad 1111\}) \oplus cd \oplus 50$$

$$= \{x(x^5 + x^3 + x)\} \oplus \{(x + 1)(x^6 + x^5 + x^4 + x^3 + x^2 + x + 1)\} \oplus cd \oplus 50$$

$$= \{x^6 + x^4 + x^2\} \oplus \{x^7 + 1\} \oplus cd \oplus 50$$

$$= (0101 \quad 0100) \oplus (1000 \quad 0001) \oplus cd \oplus 50$$

$$= 54 \oplus 81 \oplus cd \oplus 50$$

$$= 48$$

$$s'_{1,0} = s_{0,0} \oplus (\{02\} \cdot s_{1,0}) \oplus (\{03\} \cdot s_{2,0}) \oplus s_{3,0}$$

$$= 2a \oplus (\{02\} \cdot 7f) \oplus (\{03\} \cdot cd) \oplus 50$$

$$= 2a \oplus \{x(x^6 + x^5 + x^4 + x^3 + x^2 + x + 1)\}$$

$$\quad \oplus \{(x + 1)(x^7 + x^6 + x^3 + x^2 + 1)\}cd \oplus 50$$

$$= 2a \oplus \{x^7 + x^6 + x^5 + x^4 + x^3 + x^2 + x\} \oplus \{x^8 + x^6 + x^4 + x^2 + x + 1\} \oplus 50 \quad :$$

$$\quad \text{using } x^8 + x^4 + x^3 + x + 1 = 0$$

$$= 2a \oplus \{x^7 + x^6 + x^5 + x^4 + x^3 + x^2 + x\} \oplus \{(x^6 + x^4 + x^2 + x + 1)$$

$$\quad + (x^4 + x^3 + x + 1)\} \oplus 50$$

$$= 2a \oplus \{x^7 + x^6 + x^5 + x^4 + x^3 + x^2 + x\} \oplus \{(x^6 + x^3 + x^2)\} \oplus 50$$

$$= 2a \oplus fe \oplus 4c \oplus 50$$

$$= c8$$

$$s'_{2,0} = s_{0,0} \oplus s_{1,0} \oplus (\{02\} \cdot s_{2,0}) \oplus (\{03\} \cdot s_{3,0})$$

$$= 2a \oplus 7f \oplus (\{02\} \cdot cd) \oplus (\{03\} \cdot 50)$$

$$= 24$$

$$s'_{3,0} = (\{03\} \cdot s_{0,0}) \oplus s_{1,0} \oplus s_{2,0} \oplus (\{02\} \cdot s_{3,0})$$

$$= (\{03\} \cdot 2a) \oplus 7f \oplus cd \oplus (\{02\} \cdot 50)$$

$$= 6c$$

$$s'_{0,1} = (\{02\} \cdot s_{0,1}) \oplus (\{03\} \cdot s_{1,1}) \oplus s_{2,1} \oplus s_{3,1}$$

$$= (\{02\} \cdot 2a) \oplus (\{03\} \cdot 4a) \oplus f7 \oplus b0$$

$$= cd$$

$$s'_{1,1} = s_{0,1} \oplus (\{02\} \cdot s_{1,1}) \oplus (\{03\} \cdot s_{2,1}) \oplus s_{3,1}$$

$$= 2a \oplus (\{02\} \cdot 4a) \oplus (\{03\} \cdot f7) \oplus b0$$

$$= 0c$$

$$s'_{2,1} = s_{0,1} \oplus s_{1,1} \oplus (\{02\} \cdot s_{2,1}) \oplus (\{03\} \cdot s_{3,1})$$
$$= 2a \oplus 4a \oplus (\{02\} \cdot f7) \oplus (\{03\} \cdot b0)$$
$$= 5e$$

$$s'_{3,1} = (\{03\} \cdot s_{0,1}) \oplus s_{1,1} \oplus s_{2,1} \oplus (\{02\} \cdot s_{3,1})$$
$$= (\{03\} \cdot 2a) \oplus 4a \oplus f7 \oplus (\{02\} \cdot b0)$$
$$= b8$$

$$s'_{0,2} = (\{02\} \cdot s_{0,2}) \oplus (\{03\} \cdot s_{1,2}) \oplus s_{2,2} \oplus s_{3,2}$$
$$= (\{02\} \cdot 30) \oplus (\{03\} \cdot 9f) \oplus e8 \oplus 9d$$
$$= cd$$

$$s'_{1,2} = s_{0,2} \oplus (\{02\} \cdot s_{1,2}) \oplus (\{03\} \cdot s_{2,2}) \oplus s_{3,2}$$
$$= 30 \oplus (\{02\} \cdot 9f) \oplus (\{03\} \cdot e8) \oplus 9d$$
$$= 0c$$

$$s'_{2,2} = s_{0,2} \oplus s_{1,2} \oplus (\{02\} \cdot s_{2,2}) \oplus (\{03\} \cdot s_{3,2})$$
$$= 30 \oplus 9f \oplus (\{02\} \cdot e8) \oplus (\{03\} \cdot 9d)$$
$$= 5e$$

$$s'_{3,2} = (\{03\} \cdot s_{0,2}) \oplus s_{1,2} \oplus s_{2,2} \oplus (\{02\} \cdot s_{3,2})$$
$$= (\{03\} \cdot 30) \oplus 9f \oplus e8 \oplus (\{02\} \cdot 9d)$$
$$= b8$$

$$s'_{0,3} = (\{02\} \cdot s_{0,3}) \oplus (\{03\} \cdot s_{1,3}) \oplus s_{2,3} \oplus s_{3,3}$$
$$= (\{02\} \cdot d4) \oplus (\{03\} \cdot 84) \oplus a7 \oplus ef$$
$$= ac$$

$$s'_{1,3} = s_{0,3} \oplus (\{02\} \cdot s_{1,3}) \oplus (\{03\} \cdot s_{2,3}) \oplus s_{3,3}$$
$$= d4 \oplus (\{02\} \cdot 84) \oplus (\{03\} \cdot a7) \oplus ef$$
$$= 1a$$

$$s'_{2,3} = s_{0,3} \oplus s_{1,3} \oplus (\{02\} \cdot s_{2,3}) \oplus (\{03\} \cdot s_{3,3})$$
$$= d4 \oplus 84 \oplus (\{02\} \cdot a7) \oplus (\{03\} \cdot ef)$$
$$= 74$$

$$s'_{3,3} = (\{03\} \cdot s_{0,3}) \oplus s_{1,3} \oplus s_{2,3} \oplus (\{02\} \cdot s_{3,3})$$

$$= (\{03\} \cdot d4) \oplus 84 \oplus a7 \oplus (\{02\} \cdot ef)$$

$$= 1a$$

XOR with $w[4] = f0\ 46\ 5f\ 33$, $w[5] = 1d\ 89\ 29\ 95$, $w[6] = 15\ 2a\ 9f\ ed$, $w[7] = 24\ 1b$ b8 83

Plaintext block		48 c8 24 6c
Cipher key	$\oplus$	f0 46 5f 33
		b8 8e 7b 5f

Plaintext block		cd 0c 5e b8
Cipher key	$\oplus$	1d 89 29 95
		d0 85 77 2d

Plaintext block		af ab d8 06
Cipher key	$\oplus$	15 2a 9f ed
		ba 81 47 eb

Plaintext block		ac 1a 74 1a
Cipher key	$\oplus$	24 1b b8 83
		88 01 cc 99

Inverse Cipher

The Cipher transformation can be implemented in reverse order to produce an Inverse Cipher for the AES algorithm. The individual transformations used in the Inverse Cipher are InvShiftRows(), InvSubBytes(), InvMixColumns(), and AddRoundKey(). These inverse transformations process the state as described in the following.

InvShiftRows() Transformation
InvShiftRows() is the inverse of the ShiftRows() transformation. The first row (Row 0) is not shifted. The bytes in the last three rows (Row 1, Row 2, Row 3) are cyclically shifted over different numbers of bytes as follows:

$shift(r, Nb)$: shift values, where r is a row number and $Nb = 4$.

$$shift(1,4) = 1, shift(2,4) = 2, shift(3,4) = 3, \text{respectively.}$$

Specifically, the InvShiftRows() transformation proceeds as:

$$s'_{r,(c+shift(r,Nb))\bmod Nb} = s_{r,c}, \text{ for } 0 < r < 4 \text{ and } 0 \leqslant c < Nb$$

InvSubBytes() Transformation
InvSubBytes() is the inverse of the byte substitution transformation, in which the inverse S-box is applied to each byte of the state. The inverse S-box used in the InvSubBytes() transformation is presented in Figure 4.19.

									y							
		0	1	2	3	4	5	6	7	8	9	a	b	c	d	e	f
	0	52	09	6a	d5	30	36	a5	38	bf	40	a3	9e	81	f3	d7	fb
	1	7c	e3	39	82	9d	2f	ff	87	34	8e	43	44	c4	de	e9	cb
	2	54	7b	94	32	a6	c2	23	3d	ee	4c	95	0b	42	fa	c3	4e
	3	08	2e	a1	66	28	d9	24	b2	76	5b	a2	49	6d	8b	d1	25
	4	72	f8	f6	64	86	68	98	16	d4	a4	5c	cc	5d	65	b6	92
	5	6c	70	48	50	fd	ed	b9	da	5e	15	46	57	a7	8d	9d	84
	6	90	d8	ab	00	8c	bc	d3	0a	f7	e4	58	05	b8	b3	45	06
x	7	d0	2c	1e	8f	ca	3f	0f	02	c1	af	bd	03	01	13	8a	6b
	8	3a	91	11	41	4f	67	dc	ea	97	f2	cf	ce	f0	b4	e6	73
	9	96	ac	74	22	e7	ad	35	85	e2	f9	37	e8	1c	75	df	6e
	a	47	f1	1a	71	1d	29	c5	89	6f	b7	62	0e	aa	18	be	1b
	b	fc	56	3e	4b	c6	d2	79	20	9a	db	c0	fe	78	cd	5a	f4
	c	1f	dd	a8	33	88	07	c7	31	b1	12	10	59	27	80	ec	5f
	d	60	51	7f	a9	19	b5	4a	0d	2d	e5	7a	9f	93	c9	9c	ef
	e	a0	e0	3b	4d	ae	2a	f5	b0	c8	eb	bb	3c	83	53	99	61
	f	17	2b	04	7e	ba	77	d6	26	e1	69	14	63	55	21	0c	7d

Figure 4.19   AES algorithm Inverse S-box (FIPS Publication, 2001).

InvMixColumns() Transformation

InvMixColumns() is the inverse of the MixColumns() transformation. This transformation operates column-by-column on the state, treating each column as a four-term polynomial. The columns are considered as polynomials over $GF(2^8)$ and multiplied modulo $x^4 + 1$ with a fixed polynomial $a^{-1}(x)$.

If the inverse state $s'(x)$ is written as a matrix multiplication, then it follows:

$$s'(x) = a^{-1}(x) \otimes s(x)$$

where $a^{-1}(x) = \{0b\}x^3 + \{0d\}x^2 + \{09\}x + \{0e\}$.

The matrix multiplication can be expressed as

$$
\begin{bmatrix} s'_{0,c} \\ s'_{1,c} \\ s'_{2,c} \\ s'_{3,c} \end{bmatrix}
=
\begin{bmatrix}
0e & 0b & 0d & 09 \\
09 & 0e & 0b & 0d \\
0d & 09 & 0e & 0b \\
0b & 0d & 09 & 0e
\end{bmatrix}
\begin{bmatrix} s_{0,c} \\ s_{1,c} \\ s_{2,c} \\ s_{3,c} \end{bmatrix}
\quad \text{for } 0 \leqslant c < Nb
$$

This multiplication will result in 4 bytes in a column as follows:

$$s'_{0,c} = (\{0e\} \bullet s_{0,c}) \oplus (\{0b\} \bullet s_{1,c}) \oplus (\{0d\} \bullet s_{2,c}) \oplus (\{09\} \bullet s_{3,c})$$

$$s'_{1,c} = (\{09\} \bullet s_{0,c}) \oplus (\{0e\} \bullet s_{1,c}) \oplus (\{0b\} \bullet s_{2,c}) \oplus (\{0d\} \bullet s_{3,c})$$

$$s'_{2,c} = (\{0d\} \bullet s_{0,c}) \oplus (\{09\} \bullet s_{1,c}) \oplus (\{0e\} \bullet s_{2,c}) \oplus (\{0b\} \bullet s_{3,c})$$

$$s'_{3,c} = (\{0b\} \bullet s_{0,c}) \oplus (\{0d\} \bullet s_{1,c}) \oplus (\{09\} \bullet s_{2,c}) \oplus (\{0e\} \bullet s_{3,c})$$

```
EqInvCipher (byte in[ 4Nb] , byte out [ 4*Nb] , word dw[ Nb* (Nr + 1) ] ) begin

    byte in state [4, Nb]
    state = in
    AddRoundKey (state, dw + Nr * Nb)
    for round = Nr-1 step -1 to 1
            InvShiftRows (state)
            InvSubBytes (state)
            AddRoundKey (state, dw+round*Nb)
            InvMixColumns (state)
    end for
    InvShiftRows (state)
    InvSubBytes (state)
    AddRoundKey (state, dw)
    out = (state)
end
```

Figure 4.20   Pseudocode for the Inverse Cipher (FIPS Publication, 2001).

Inverse of AddRoundKey() Transformation
AddRoundKey() is its own inverse because it only involves application of the XOR.
    For decrypting ciphertext, the Inverse Cipher is described in the pseudocode shown in Figure 4.20.

Example 4.26   The input to the Inverse Cipher is the cipher encryption values obtained from Example 4.25

Plaintext   = a3 c5 08 78 a4 ff d3 00 36 36 28 5f 01 02

Cipher Key = 36 8a c0 f4 ed cf 76 a6 08 a3 b6 78 31 31 27 6e

 Ciphertext =   a6 24 62 48 34 dd a8 b9 1a f1 73 5d 00 0e cf 61

The round key values are the same as those used in Example 4.25

Round 0

• AddRoundKey()

The round counter value is decremented from 10 for Inverse Cipher and Nb is 4 for the 128-bit length of cipher key. So indexes of the key schedule words w are used from 40 to 43.

$$
\begin{array}{ll}
w_{40} = & \boxed{c1e32993} \\
w_{41} = & \boxed{086827b5} \\
w_{42} = & \boxed{3b411f33} \\
w_{43} = & \boxed{56cb13d2}
\end{array}
\rightarrow
\begin{array}{cccc}
w_{40} & w_{41} & w_{42} & w_{43} \\
\hline
c1 & 08 & 3b & 93 \\
e3 & 68 & 41 & cb \\
29 & 41 & 1f & 13 \\
93 & cb & 33 & d2
\end{array}
$$

$$s'_{0,c}, s'_{1,c}, s'_{2,c}, s'_{3,c}] = [s_{0,c}, s_{1,c}, s_{2,c}, s_{3,c}] \oplus w_{\text{round} \bullet Nb+c}, \text{ for } 0 <= c < Nb$$

$$s'_{0,0} = s_{0,0} \oplus w_{40,0}$$

$$= \text{a6} \oplus \text{c1}$$

$$= 10100110b$$

$$\oplus \quad 11000001b$$

$$= 01100111b$$

$$= 67$$

$$s'_{1,0} = s_{1,0} \oplus w_{40,1} = 24 \oplus \text{e3} = \text{c7}$$

$$s'_{2,0} = s_{2,0} \oplus w_{40,2} = 62 \oplus 29 = \text{4b}$$

$$s'_{3,0} = s_{3,0} \oplus w_{40,3} = 48 \oplus 93 = \text{db}$$

$$s'_{0,1} = s_{0,1} \oplus w_{41,0} = 34 \oplus 08 = \text{3c}$$

$$s'_{1,1} = s_{1,1} \oplus w_{41,1} = \text{dd} \oplus 68 = \text{b5}$$

$$s'_{2,1} = s_{2,1} \oplus w_{41,2} = \text{a8} \oplus 27 = \text{8f}$$

$$s'_{3,1} = s_{3,1} \oplus w_{41,3} = \text{b9} \oplus \text{b5} = \text{0c}$$

$$s'_{0,2} = s_{0,2} \oplus w_{42,0} = \text{1a} \oplus \text{3b} = 21$$

$$s'_{1,2} = s_{1,2} \oplus w_{42,1} = \text{f1} \oplus 41 = \text{b0}$$

$$s'_{2,2} = s_{2,2} \oplus w_{42,2} = 73 \oplus \text{1f} = \text{6c}$$

$$s'_{3,2} = s_{3,2} \oplus w_{42,3} = \text{5d} \oplus 33 = \text{6e}$$

$$s'_{0,3} = s_{0,3} \oplus w_{43,0} = 00 \oplus 56 = 56$$

$$s'_{1,3} = s_{1,3} \oplus w_{43,1} = \text{0e} \oplus \text{cb} = \text{c5}$$

$$s'_{2,3} = s_{2,3} \oplus w_{43,2} = \text{cf} \oplus 13 = \text{dc}$$

$$s'_{3,3} = s_{3,3} \oplus w_{43,3} = 61 \oplus \text{d2} = \text{b3}$$

r	Start of round	After InvShiftRows	After InvSubBytes	After $\oplus$ with $w[]$	After InvMixColumns
0	a6 34 1a 00				67 3c 21 56
	24 dd f1 0e				c7 b5 b0 c5
	62 a8 73 cf				4b 8f 6c dc
	48 b9 5d 61				db 0c 6e b3

Round 1

- InvShiftRows()

For row 0, skip the shift operation	:	67	3c	21	56 $\rightarrow$ 67	3c	21	56
For row 1, shift 1 bytes	:	c7	b5	b0	c5 $\rightarrow$ c5	c7	b5	b0
For row 2, shift 2 bytes	:	4b	8f	6c	dc $\rightarrow$ 6c	dc	4b	8f
For row 3, shift 3 bytes	:	db	0c	6e	b3 $\rightarrow$ 0c	6e	b3	db

The result of InvShiftRows() is

67 3c 21 56		67 3c 21 56
c7 b5 b0 c5	$\rightarrow$	c5 c7 b5 b0
4b 8f 6c dc		6c dc 4b 8f
db 0c 6e b3		0c 6e b3 db

- InvSubBytes()

Compute InvSubByte() transformation using Inverse S-box

$$s_{r,c} \longrightarrow \boxed{\text{Inverse S-Box}} \longrightarrow s'_{r,c}$$

$s_{0,0} = \{67\}$, the substitution value is determined by the intersection of the row with index 6 and the column with index with 7 in Figure 4.19.
Like this, after substituting all values, the result obtained is

67 3c 21 56		0a 6d 7b b9
c5 c7 b5 b0	$\rightarrow$	07 31 d2 fc
6c dc 4b 8f		b8 93 cc 73
0c 6e b3 db		81 45 4b 9f

- AddRoundKey()

$$[s'_{0,c}, s'_{1,c}, s'_{2,c}, s'_{3,c}] = [s_{0,c}, s_{1,c}, s_{2,c}, s_{3,c}] \oplus w_{\text{round} \bullet Nb+c}, \text{ for } 0 <= c < Nb$$

	w_{36}	w_{37}	w_{38}	w_{39}
$w_{36} = \boxed{891dd1af}$	89	c9	33	6d
$w_{37} = \boxed{c98b0e26}$	1d	82	29	8a
$w_{38} = \boxed{33293886}$	d1	0e	38	0c
$w_{39} = \boxed{6d8a0ce1}$	af	26	86	e1

$s'_{0,0} = s_{0,0} \oplus w_{36,0} = 0a \oplus 89 = 83$

$s'_{1,0} = s_{1,0} \oplus w_{36,1} = 07 \oplus 1d = 1a$

$s'_{2,0} = s_{2,0} \oplus w_{36,2} = b8 \oplus d1 = 69$

$s'_{3,0} = s_{3,0} \oplus w_{36,3} = 81 \oplus af = 2e$

$s'_{0,1} = s_{0,1} \oplus w_{37,0} = 6d \oplus c9 = a4$

$s'_{1,1} = s_{1,1} \oplus w_{37,1} = 31 \oplus 8b = ba$

$s'_{2,1} = s_{2,1} \oplus w_{37,2} = 93 \oplus 0e = 9d$

$s'_{3,1} = s_{3,1} \oplus w_{37,3} = 45 \oplus 26 = 63$

$s'_{0,2} = s_{0,2} \oplus w_{38,0} = 7b \oplus 33 = 48$

$s'_{1,2} = s_{1,2} \oplus w_{38,1} = d2 \oplus 29 = fb$

$s'_{2,2} = s_{2,2} \oplus w_{38,2} = cc \oplus 38 = f4$

$s'_{3,2} = s_{3,2} \oplus w_{38,3} = 4b \oplus 86 = cd$

$s'_{0,3} = s_{0,3} \oplus w_{39,0} = b9 \oplus 6d = d4$

$s'_{1,3} = s_{1,3} \oplus w_{39,1} = fc \oplus 8a = 76$

$s'_{2,3} = s_{2,3} \oplus w_{39,2} = 73 \oplus 0c = 7f$

$s'_{3,3} = s_{3,3} \oplus w_{39,3} = 9f \oplus e1 = 7e$

- InvMixColumns()

$$s'_{0,} = (\{0e\} \bullet s_{0,}) \oplus (\{0b\} \bullet s_{1,0}) \oplus (\{0d\} \bullet s_{2,0}) \oplus (\{09\} \bullet s_{3,0})$$

$$= (\{0e\} \bullet 83) \oplus (\{0b\} \bullet 1a) \oplus (\{0d\} \bullet 69) \oplus (\{09\} \bullet 2e)$$

$$= \left(\frac{00001110b \bullet 10000011b}{①} \right) \oplus \left(\frac{00001011b \bullet 00011010b}{②} \right) \oplus$$

$$\left(\frac{00001101b \bullet 01101001b}{③} \right) \oplus \left(\frac{00001001b \bullet 00101110b}{④} \right)$$

For ①

```
  00001110b•10000011b
= (x³+x²+x)•(x⁷+x+1)
= x¹⁰ +                           x⁴ + x̸³
       x⁹ +                          + x̸³ + x̸² +
             x⁸ +                          x̸² + x
_____
= x¹⁰ + x⁹ + x⁸ +                x⁴ +            x
= 11100010010b
```

The result polynomial is considered over $GF(x^8)$.

11100010010b modulo (100011011b)

$= 11100010010$

$\underline{100011011}$

$ 1101111110$

$ \underline{100011011}$

$ 101001000$

$ \underline{100011011}$

$ 01010011b = 53$

$\therefore 00001110b \cdot 10000011b = 53$

Like this, calculate ②, ③, and ④ and the results are

For ②: $00001011b \cdot 00011010b = fe$

For ③: $00001101b \cdot 01101001b = b3$

For ④: $00001001b \cdot 00101110b = 45$

$s'_{0,} = 53 \oplus fe \oplus b3 \oplus 45$

$\phantom{s'_{0,}} = 5b$

The remaining values can be obtained using the same way and the results are shown in the following:

$s'_{1,0} = (\{09\} \cdot s_{0,0}) \oplus (\{0e\} \cdot s_{1,0}) \oplus (\{0b\} \cdot s_{2,0}) \oplus (\{0d\} \cdot s_{3,0})$

$\phantom{s'_{1,0}} = (\{09\} \cdot 83) \oplus (\{0e\} \cdot 1a) \oplus (\{0b\} \cdot 69) \oplus (\{0d\} \cdot 2e)$

$\phantom{s'_{1,0}} = f7 \oplus 8c \oplus de \oplus fd$

$\phantom{s'_{1,0}} = 58$

$s'_{2,0} = (\{0d\} \cdot s_{0,0}) \oplus (\{09\} \cdot s_{1,0}) \oplus (\{0e\} \cdot s_{2,0}) \oplus (\{0b\} \cdot s_{3,0})$

$\phantom{s'_{2,0}} = (\{0d\} \cdot 83) \oplus (\{09\} \cdot 1a) \oplus (\{0e\} \cdot 69) \oplus (\{0b\} \cdot 2e)$

$\phantom{s'_{2,0}} = cd \oplus ca \oplus 08 \oplus 19$

$\phantom{s'_{2,0}} = 16$

$s'_{3,0} = (\{0b\} \cdot s_{0,0}) \oplus (\{0d\} \cdot s_{1,0}) \oplus (\{09\} \cdot s_{2,0}) \oplus (\{0e\} \cdot s_{3,0})$

$\phantom{s'_{3,0}} = (\{0b\} \cdot 83) \oplus (\{0d\} \cdot 1a) \oplus (\{09\} \cdot 69) \oplus (\{0e\} \cdot 2e)$

$\phantom{s'_{3,0}} = ea \oplus a2 \oplus 0c \oplus 8f$

$\phantom{s'_{3,0}} = cb$

$$s'_{0,1} = (\{0e\} \cdot s_{0,1}) \oplus (\{0b\} \cdot s_{1,1}) \oplus (\{0d\} \cdot s_{2,1}) \oplus (\{09\} \cdot s_{3,1})$$
$$= (\{0e\} \cdot a4) \oplus (\{0b\} \cdot ba) \oplus (\{0d\} \cdot 9d) \oplus (\{09\} \cdot 63)$$
$$= a2 \oplus 72 \oplus 5b \oplus 56$$
$$= dd$$

$$s'_{1,1} = (\{09\} \cdot s_{0,1}) \oplus (\{0e\} \cdot s_{1,1}) \oplus (\{0b\} \cdot s_{2,1}) \oplus (\{0d\} \cdot s_{3,1})$$
$$= (\{09\} \cdot a4) \oplus (\{0e\} \cdot ba) \oplus (\{0b\} \cdot 9d) \oplus (\{0d\} \cdot 63)$$
$$= f3 \oplus 16 \oplus 38 \oplus c1$$
$$= 1c$$

$$s'_{2,1} = (\{0d\} \cdot s_{0,1}) \oplus (\{09\} \cdot s_{1,1}) \oplus (\{0e\} \cdot s_{2,1}) \oplus (\{0b\} \cdot s_{3,1})$$
$$= (\{0d\} \cdot a4) \oplus (\{09\} \cdot ba) \oplus (\{0e\} \cdot 9d) \oplus (\{0b\} \cdot 63)$$
$$= 55 \oplus 1d \oplus e7 \oplus 90$$
$$= 3f$$

$$s'_{3,1} = (\{0b\} \cdot s_{0,1}) \oplus (\{0d\} \cdot s_{1,1}) \oplus (\{09\} \cdot s_{2,1}) \oplus (\{0e\} \cdot s_{3,1})$$
$$= (\{0b\} \cdot a4) \oplus (\{0d\} \cdot ba) \oplus (\{09\} \cdot 9d) \oplus (\{0e\} \cdot 63)$$
$$= a0 \oplus c3 \oplus 19 \oplus 64$$
$$= 1e$$

$$s'_{0,2} = (\{0e\} \cdot s_{0,2}) \oplus (\{0b\} \cdot s_{1,2}) \oplus (\{0d\} \cdot s_{2,2}) \oplus (\{09\} \cdot s_{3,2})$$
$$= (\{0e\} \cdot 48) \oplus (\{0b\} \cdot fb) \oplus (\{0d\} \cdot f4) \oplus (\{09\} \cdot cd)$$
$$= 45$$

$$s'_{1,2} = (\{09\} \cdot s_{0,2}) \oplus (\{0e\} \cdot s_{1,2}) \oplus (\{0b\} \cdot s_{2,2}) \oplus (\{0d\} \cdot s_{3,2})$$
$$= (\{09\} \cdot 48) \oplus (\{0e\} \cdot fb) \oplus (\{0b\} \cdot f4) \oplus (\{0d\} \cdot cd)$$
$$= 3e \oplus b5 \oplus e6 \oplus e6$$
$$= 8b$$

$$s'_{2,2} = (\{0d\} \cdot s_{0,2}) \oplus (\{09\} \cdot s_{1,2}) \oplus (\{0e\} \cdot s_{2,2}) \oplus (\{0b\} \cdot s_{3,2})$$
$$= (\{0d\} \cdot 48) \oplus (\{09\} \cdot fb) \oplus (\{0e\} \cdot f4) \oplus (\{0b\} \cdot cd)$$
$$= 05 \oplus 62 \oplus ef \oplus 7e$$
$$= f6$$

$$s'_{3,2} = (\{0b\} \cdot s_{0,2}) \oplus (\{0d\} \cdot s_{1,2}) \oplus (\{09\} \cdot s_{2,2}) \oplus (\{0e\} \cdot s_{3,2})$$
$$= (\{0b\} \cdot 48) \oplus (\{0d\} \cdot fb) \oplus (\{09\} \cdot f4) \oplus (\{0e\} \cdot cd)$$
$$= ae \oplus a3 \oplus 15 \oplus aa$$
$$= b2$$

$$s'_{0,3} = (\{0e\} \cdot s_{0,3}) \oplus (\{0b\} \cdot s_{1,3}) \oplus (\{0d\} \cdot s_{2,3}) \oplus (\{09\} \cdot s_{3,3})$$

$$= (\{0e\} \cdot d4) \oplus (\{0b\} \cdot 76) \oplus (\{0d\} \cdot 7f) \oplus (\{09\} \cdot 7e)$$

$$= 34 \oplus 07 \oplus 4d \oplus a3$$

$$= dd$$

$$s'_{1,3} = (\{09\} \cdot s_{0,3}) \oplus (\{0e\} \cdot s_{1,3}) \oplus (\{0b\} \cdot s_{2,3}) \oplus (\{0d\} \cdot s_{3,3})$$

$$= (\{09\} \cdot d4) \oplus (\{0e\} \cdot 76) \oplus (\{0b\} \cdot 7f) \oplus (\{0d\} \cdot 7e)$$

$$= 2e \oplus b2 \oplus 54 \oplus 40$$

$$= 88$$

$$s'_{2,3} = (\{0d\} \cdot s_{0,3}) \oplus (\{09\} \cdot s_{1,3}) \oplus (\{0e\} \cdot s_{2,3}) \oplus (\{0b\} \cdot s_{3,3})$$

$$= (\{0d\} \cdot d4) \oplus (\{09\} \cdot 76) \oplus (\{0e\} \cdot 7f) \oplus (\{0b\} \cdot 7e)$$

$$= 53 \oplus eb \oplus cc \oplus 5f$$

$$= 2b$$

$$s'_{3,3} = (\{0b\} \cdot s_{0,3}) \oplus (\{0d\} \cdot s_{1,3}) \oplus (\{09\} \cdot s_{2,3}) \oplus (\{0e\} \cdot s_{3,3})$$

$$= (\{0b\} \cdot d4) \oplus (\{0d\} \cdot 76) \oplus (\{09\} \cdot 7f) \oplus (\{0e\} \cdot 7e)$$

$$= 9d \oplus 28 \oplus aa \oplus c2$$

$$= dd$$

r	Start of round	After InvShiftRows	After InvSubBytes	After $\oplus$ with $w[]$	After InvMixColumns
1	67 3c 21 56	67 3c 21 56	0a 6d 7b b9	83 a4 48 d4	5b dd 45 dd
	c7 b5 b0 c5	c5 c7 b5 b0	07 31 d2 fc	1a ba fb 76	58 1c 8b 88
	4b 8f 6c dc	6c dc 4b 8f	b8 93 cc 73	69 9d f4 7f	16 3f f6 2b
	db 0c 6e b3	0c 6e b3 db	81 45 4b 9f	2e 63 cd 7e	cb 1e b2 dd

Continuing in this fashion, the remaining rounds $2 \sim 10$ can be computed.

The following table shows the values in the state array as the Inverse Cipher progresses.

Inverse Cipher (Decrypt)

r	Start of round	After InvShiftRows	After InvSubBytes	After $\oplus$ with $w[]$	After InvMixColumns
0	a6 34 1a 00				7 3c 21 56
	24 dd f1 0e				7 b5 b0 c5
	62 a8 73 cf				b 8f 6c dc
	48 b9 5d 61				b 0c 6e b3

r	Start of round	After InvShiftRows	After InvSubBytes	After $\oplus$ with $w[]$	After InvMixColumns
1	67 3c 21 56 c7 b5 b0 c5 4b 8f 6c dc db 0c 6e b3	67 3c 21 56 c5 c7 b5 b0 6c dc 4b 8f 0c 6e b3 db	0a 6d 7b b9 07 31 d2 fc b8 93 cc 73 81 45 4b 9f	83 a4 48 d4 1a ba fb 76 69 9d f4 7f 2e 63 cd 7e	5b dd 45 dd 58 1c 8b 88 16 3f f6 2b cb 1e b2 dd
2	5b dd 45 dd 58 1c 8b 88 16 3f f6 2b cb 1e b2 dd	5b dd 45 dd 88 58 1c 8b f6 2b 16 3f 1e b2 dd cb	57 c9 68 c9 97 5e c4 ce d6 0b ff 25 e9 3e c9 59	cf 89 92 97 92 c8 66 6d 82 d4 c9 11 1e b7 69 3e	c9 4d d4 71 27 3e dd e5 1d 54 27 46 32 05 7a 07
3	c9 4d d4 71 27 3e dd e5 1d 54 27 46 32 05 7a 07	c9 4d d4 71 e5 27 3e dd 27 46 1d 54 05 7a 07 32	12 65 19 2c 2a 3d d1 c9 3d 98 de fd 36 bd 38 a1	76 bd a3 88 58 ae e5 c8 af 13 37 ff 88 c3 11 66	22 94 04 84 81 e1 ea 63 e9 a7 50 5a 43 11 de 64
4	22 94 04 84 81 e1 ea 63 e9 a7 50 5a 43 11 de 64	22 94 04 84 63 81 e1 ea 50 5a e9 a7 11 de 64 43	94 e7 30 4f 00 91 e0 bb 6c 46 eb 89 e3 9c 8c 64	26 5b 52 51 9b 70 47 8e d6 5f 89 62 2f 5c db 8a	f8 6e 70 ac b5 ed cf c9 80 3a 25 f1 89 91 dd a3
5	f8 6e 70 ac b5 ed cf c9 80 3a 25 f1 89 91 dd a3	f8 6e 70 ac c9 b5 ed cf 25 f1 80 3a 91 dd a3 89	e1 45 d0 aa 12 d2 53 5f c2 2b 3a a2 ac c9 71 f2	3c 4b 0e d6 2e a8 15 cd 2e 88 41 2b 70 c5 e6 4b	4a 1e 34 b7 57 ee 77 ba c0 9a 18 bc 91 c4 e7 ca
6	4a 1e 34 b7 57 ee 77 ba c0 9a 18 bc 91 c4 e7 ca	4a 1e 34 b7 ba 57 ee 77 18 bc c0 9a c4 e7 ca 91	5c e9 28 20 c0 da 99 02 34 78 1f 37 88 b0 10 ac	d9 3a f8 82 75 9c a5 d6 e9 37 c7 c5 6e 60 8b 82	32 61 27 a6 52 34 40 30 54 95 3a c7 1f 31 4c 42
7	32 61 27 a6 52 34 40 30 54 95 3a c7 1f 31 4c 42	32 61 27 a6 30 52 34 40 3a c7 54 95 31 4c 42 1f	a1 d8 3d c5 08 48 28 72 a2 31 fd ad 2e 5d f6 cb	b7 8e 3e b7 58 bb 52 9a aa a3 6a 87 88 6b bd 7e	f9 04 b9 03 35 b2 a7 33 65 c8 80 80 64 83 25 64
8	f9 04 b9 03 35 b2 a7 33 65 c8 80 80 64 83 25 64	f9 04 b9 03 33 35 b2 a7 80 80 65 c8 83 25 64 64	69 30 db d5 66 d9 3e 89 3a 3a bc b1 41 c2 8c 8c	34 70 8e a4 4c 7a b7 1b 89 a0 b9 0c 44 52 f1 72	6c 70 f4 c4 97 0c 7c 19 a0 4b 21 f5 ee cf d8 e9
9	6c 70 f4 c4 97 0c 7c 19 a0 4b 21 f5 ee cf d8 e9	6c 70 f4 c4 19 97 0c 7c 21 f5 a0 4b cf d8 e9 ee	b8 d0 ba 88 8e 85 81 01 7b 77 47 cc 5f 2d eb 99	48 cd af ac c8 0c ab 1a 24 5e d8 74 6c b8 06 1a	2a 2a 30 d4 7f 4a 9f 84 cd f7 e8 a7 50 b0 9d 2f

r	Start of round	After InvShiftRows	After InvSubBytes	After $\oplus$ with $w[]$	After InvMixColumns
10	2a 2a 30 d4	2a 2a 30 d4	95 95 08 19	**a3 78 00 28**	
	7f 4a 9f 84	84 7f 4a 9f	4f 6b 5c 6e	**c5 a4 ff 5f**	
	cd f7 e8 a7	e8 a7 cd f7	c8 89 80 26	**08 ff 36 01**	
	50 b0 9d 2f	b0 9d 2f 50	fc 75 4e 6c	**08 d3 36 02**	

Inverse Cipher Text = a3 c5 08 08 78 a4 ff d3 00 ff 36 36 28 5f 01 02
  Original Plaintext = a3 c5 08 08 78 a4 ff d3 00 ff 36 36 28 5f 01 02.

5

Hash Function, Message Digest, and Message Authentication Code

As digital signature technology becomes more widely understood and utilized, many countries worldwide are competitively developing their own signature standards for their use and applications.

Some electronic applications utilizing digital signatures in electronic commerce (e-commerce) include e-mail and financial transactions. E-mail may need to be digitally signed, where sensitive information is being transmitted and security services such as sender authentication, message integrity, and nonrepudiation are desired. Financial transactions, in which money is being transferred directly or in exchange for services and goods, could also benefit from the use of digital signatures. Signing the message digest rather than the message often improves the efficiency of the process because the message digest is usually much smaller than the message.

In e-commerce, it is often necessary for communication parties to verify each other's identity. One practical way to do this is with the use of cryptographic authentication protocols employing a one-way hash function. Division into fixed-bit blocks can be accomplished by mapping the variable-length message onto the suitable-bit value by padding with all zeros, including 1-bit flag and the original message length in hex. Appropriate padding is needed to force the message to divide conveniently into certain fixed lengths. Several algorithms are introduced in order to compute message digests by employing several hash functions. The hash functions dealt with in this chapter are DMDC (DES-like Message Digest Computation) (1994), MD5 (1992), and SHA-1 (1995).

5.1 DMDC Algorithm

DMDC uses a DES (Data Encryption Standard) variant as a one-way hash function. In 1994, this scheme was introduced to compute the 18-bit authentication data with CDMA

Wireless Mobile Internet Security, Second Edition. Man Young Rhee.
© 2013 John Wiley & Sons, Ltd. Published 2013 by John Wiley & Sons, Ltd.

cellular mobile communications system. DMDC divides messages into blocks of 64 bits. The DMDC hash function generates message digests with variable sizes – 18, 32, 64, or 128 bits. This scheme is appropriate for the use of digital signatures and hence it can be employed to increase Internet security.

The message to be signed is first divided into a sequence of 64-bit blocks:

$$M_1, M_2, \ldots, M_t$$

Appropriate padding rules need to be devised for messages that do not divide conveniently. The adjacent message blocks are hashed together with a self-generated key. A better approach is to use one block (64 bits) of the correct message length as the key.

Figure 5.1 shows a typical scheme for hash code computation for $M = 192$ bits using DMDC.

5.1.1   Key Schedule

One authentication problem in the CDMA mobile system is how to confirm the identity of the mobile station by exchanging information between a mobile station and base station. When the authentication field of the access parameters message is set to "01", the mobile station attempts to register by sending a registration request message on the access channel and the authentication procedure will be performed. Computing the authentication data of mobile station registrations, it is necessary to have a 152-bit message value which complies with RAND (32 bits), ESN (32 bits), MIN (24 bits), and SSD-A (64 bits):

RAND: Authentication random challenge value.
ESN: Electronic serial number.
MIN: Mobile station identification number.
SSD-A: Shared secret data to support the authentication procedure.

The 192-bit value is composed of 152-bit message length and 40-bit padding. Suppose M_1, M_2, and M_3 are decompositions of a 192-bit padded message. $M_1 = 64$ bits will be used as input to the key generation scheme in Figure 5.1. The Permuted Choice 2 operation will produce the 48-bit key that is arranged into a 6×8 array as shown below:

Input (column by column)

$$\Downarrow$$

1	7	13	19	25	31	37	43
2	8	14	20	26	32	38	44
3	9	15	21	27	33	39	45
4	10	16	22	28	34	40	46
5	11	17	23	29	35	41	47
6	12	18	24	30	36	42	48

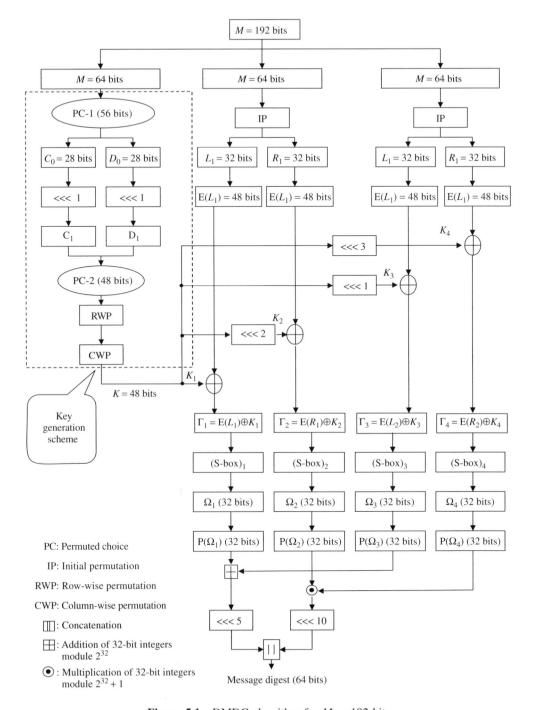

Figure 5.1   DMDC algorithm for $M = 192$ bits.

Row-wise permutation

5	11	17	23	29	35	41	47
1	7	13	19	25	31	37	43
3	9	15	21	27	33	39	45
6	12	18	24	30	36	42	48
2	8	14	20	26	32	38	44
4	10	16	22	28	34	40	46

Column-wise permutation

11	35	5	47	17	41	29	23
7	31	1	43	13	37	25	19
9	33	3	45	16	39	27	21
12	36	6	48	18	42	30	24
8	32	2	44	14	38	26	20
10	34	4	46	16	40	28	22

$\rightarrow$ Output (row by row)

Thus, a 48-bit key generation from M_1 is computed as shown in Table 5.1.

Example 5.1 Assume that division of the 192-bit padded message into 64 bits consists of:

$M_1 = 7a138b2524af17c3$

$M_2 = 17b439a12f51c5a8$

$M_3 = 51cb360000000000$

Note that no 1-bit flag and no message length in hex are inserted in this example. The 48-bit key generation using row/column permutations is given below. Assume that the first data block M_1 is used as the key input. Using Table 3.1 (PC-1), M_1 can be split into two blocks:

$C_0 = a481394, \quad D_0 = e778253$

As shown in Table 3.2, C_1 and D_1 are obtained from C_0 and D_0 by shifting 1 bit to the left, respectively.

$C_1 = 4902729, \quad D_1 = cef04a7$

Table 5.1 A 48-bit key generation by row/column permutations

11	35	5	47	17	41	29	23	7	31	1	43	13	37	25	19
9	33	3	45	15	39	27	21	12	36	6	48	18	42	30	24
8	32	2	44	14	3/8	26	20	10	34	4	46	16	40	28	22

Using Table 3.3 (PC-2), the 48-bit compressed key is computed as:

$K_0 = 058c4517a7a2$.

Finally, using Table 5.1, the 48-bit key with the row/column permutations is computed as:

$K = 5458c42bcc07$

This is the key block to be provided for M_2 and M_3, as shown in Example 5.2.

Example 5.2   Referring to Figure 5.1, $M_2 = 17b439a12f51c5a8$ and $M_3 = 51cb360000000000$ are processed as follows:
    Using Table 3.4, M_2 and M_3 are divided into

$$L_1 = 6027537d, \qquad R_1 = ca9e9411$$

$$\text{and} \quad L_2 = 03050403 \qquad R_2 = 02040206$$

Expansion of these four data blocks using Table 3.5 yields

$$E(L_1) = b0010eaa6bfa \qquad E(R_1) = e554fd4a80a3$$

$$\text{and} \quad E(L_2) = 80680a808006 \qquad E(R_2) = 00400800400c$$

The 48-bit key, $K = 5458c42bcc07$, obtained through row/column permutations, should be shifted 0, 2, 1, and 3 bits to the left such that

$K_1 = 5458c42bcc07$ (zero shift)

$K_2 = a8b18857970e$ (two shifts)

$K_3 = 516310af301d$ (one shift)

$K_4 = a2c6215e603a$ (three shifts)

These four keys are used for XORing (eXclusive ORing) with expanded blocks such that

$\Gamma_1 = E(L_1) \oplus K_1 = e459ca81a7fd$

$\Gamma_2 = E(R_1) \oplus K_2 = b437ede5b0be$

$\Gamma_3 = E(L_2) \oplus K_3 = 28d982d71808$

$\Gamma_4 = E(R_2) \oplus K_4 = a286295e2036$

These four $\Gamma_i, 1 \leqslant i \leqslant 4$'s, are inputs to the (S-box)$_i$, respectively.
    Using Table 3.6, the output Ω_i of S-boxes are computed as:

$\Omega_1 = a4064766$

$\Omega_2 = 1d1dabb8$

$\Omega_3 = f89d0b16$

$\Omega_4 = dabaae4d$

Applying the operation of Table 3.7 to each Ω_i yields:

$P(\Omega_1) = 00f63638$

$P(\Omega_2) = 9f2874d3$

$P(\Omega_3) = 96aab362$

$P(\Omega_4) = 5df889ee$

These four data blocks resulting from Table 3.7 are used for the computation of message digests (or hash codes), as shown in Example 5.3.

5.1.2   Computation of Message Digests

Example 5.3   Compute the hash codes as follows:
32-bit hash code computation
Figure 5.2 shows the processing scheme for the computation of a 32-bit hash code. In this figure, the following symbols are used.

$\odot$: Multiplication of 16-bit integers modulo $2^{16} + 1 = 65537$
$\boxplus$: Addition of 16-bit integers modulo $2^{16} = 65536$
$\oplus$: Bit-by-bit XORing of 16-bit sub-blocks
$\square$: Concatenation

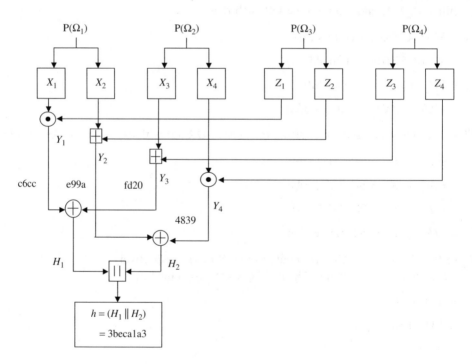

Figure 5.2   A 32-bit hash code computation scheme.

Since we have already calculated $P(\Omega_i)$ in Example 5.2, the message digest of 32 bits is ready to be computed from Figure 5.2:

$Y_1 = $ c6cc

$Y_2 = $ e99a

$Y_3 = $ fd20

$Y_4 = $ 4839

$H_1 = $ 3bec

$H_2 = $ a1a3

Concatenation of H_1 with H_2 results in the 32-bit hash code h such that

$h = (H_1 \| H_2) = $ 3beca1a3

64-bit hash code computation
Referring to Figure 5.3, the 64-bit message digest is computed as follows:

$Y_1 = $ 97a0e99a

$Y_2 = $ 371d4fc8

$H_1 = $ f41d3352

$H_2 = $ 753f20dc

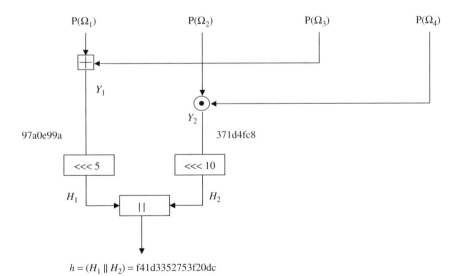

$h = (H_1 \| H_2) = $ f41d3352753f20dc

Figure 5.3   A 64-bit hash code computation scheme.

The 64-bit hash code is thus computed as:

$h = (\text{H}_1 \| \text{H}_2) = \text{f41d3352753f20dc}$

Note that

   $\odot$: Multiplication of 32-bit blocks modulo $2^{32} + 1 = 4294967297$
   $\boxplus$: Addition of 32-bit blocks modulo $2^{32} = 4294967296$
$<<< m$: Shifting m bits to the left

18-bit hash code computation
Utilizing the 64-bit message digest h obtained above, the 18-bit hash code can be computed from the decimation process as shown in Figure 5.4.

$h = \text{f41d3352753f20dc}$ (64 bits)

Discard 6 bits from both ends of the 64-bit message digest h and then pick 1 bit every 3 bits by the rule of decimation such that

$h = 001110011101110001$ (18 bits)

128-Bit hash code computation (using left shift)
Referring to Figure 5.5, each $P(\Omega_i)$ is shifted m bits to the left. Then concatenating them will produce the 128-bit message digest:

$H_1 = \text{7b1b1c00}$

$H_2 = \text{a1d34e7c}$

$H_3 = \text{59b14b55}$

$H_4 = \text{bf113dcb}$

Thus, the 128-bit hash code will be

$h = (H_1 \| H_2 \| H_3 \| H_4)$

 $= \text{7b1b1c00a1d34e7c59b14b55bf113dcb}$

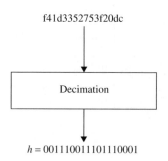

f41d3352753f20dc

Decimation

$h = 001110011101110001$

Figure 5.4 A 18-bit hash code computation scheme.

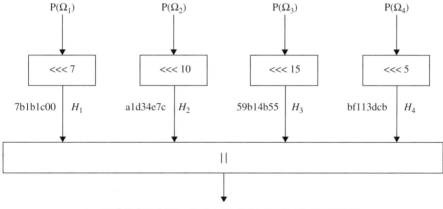

$$h = (H_1 \| H_2 \| H_3 \| H_4) = \text{7b1b1c00 a1d34e7c 59b14b55 bf113dcb}$$

Figure 5.5   A 128-bit hash code computation using a shift left.

128-Bit hash code computation (using inverse)
Based on Figure 5.6, another 128-bit message digest can be computed as follows:

$$
\begin{array}{llll}
X_1 = \text{00f6} & X_2 = \text{3638} & X_3 = \text{9f28} & X_4 = \text{74d3} \\
X_1^{-1} = \text{9b24} & -X_2 = \text{c9c8} & -X_3 = \text{60d8} & X_4^{-1} = \text{8e12} \\
Z_1 = \text{96aa} & Z_2 = \text{b362} & Z_3 = \text{5df8} & Z_4 = \text{89ee} \\
Z_1^{-1} = \text{bf34} & -Z_2 = \text{4c9e} & -Z_3 = \text{a208} & Z_4^{-1} = \text{b652}
\end{array}
$$

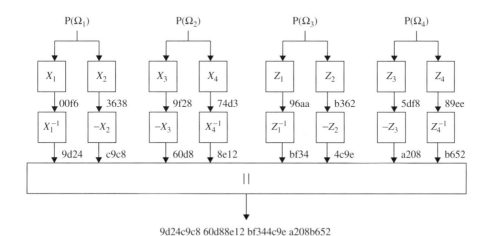

9d24c9c8 60d88e12 bf344c9e a208b652

Figure 5.6   A 128-bit hash code computation using inverse operation.

Thus, the 128-bit hash code is computed from the concatenation of inverse values:

$$h = (X_1^{-1}\| - X_2\| - X_3\|X_4^{-1}\|Z_1^{-1}\| - Z_2\| - Z_3\|Z_4^{-1})$$

$$= 9d24c9c860d88e12bf344c9ea208b652$$

128-Bit hash code computation (using addition and multiplication)
Taking a look at Figure 5.7, computation for the 128-bit message digest proceeds as follows:

$$P(\Omega_1) \boxplus P(\Omega_3) = 97a0e99a <<< 5 = f41d3352$$

$$P(\Omega_2) \odot P(\Omega_4) = 371d4fc8 <<< 10 = 753f20dc$$

$$P(\Omega_1) \odot P(\Omega_3) = 56c9017f <<< 10 = 2405fd5b$$

$$P(\Omega_2) \boxplus P(\Omega_4) = fd20fec1 <<< 5 = a41fd83f$$

$$h = (P(\Omega_1) \boxplus P(\Omega_3)) <<< 5\|(P(\Omega_2) \odot P(\Omega_4)) <<< 10\|$$

$$(P(\Omega_2) \boxplus P(\Omega_4)) <<< 5\|(P(\Omega_2) \odot P(\Omega_3)) <<< 10$$

$$= f41d3352 \quad 753f20dc \quad a41fd83f \quad 2405fd5b \text{ (128 bits)}$$

This is the 128-bit hash code found. So far, we have discussed computation forthe DMDC without appending a 1-bit flag and the message length in hex digits.

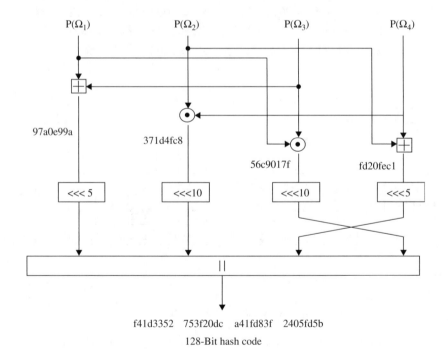

Figure 5.7 A 128-bit hash code computation using addition and multiplication.

5.2   Advanced DMDC Algorithm

This section presents the secure DMDC algorithm for providing an acceptable level of security.

5.2.1   Key Schedule

Figure 5.9 shows the newly devised key generation scheme. The 64-bit input key reshapes to the 56-bit key sequence through Table 3.1 (PC-1). The 56-bit keys are loaded into two 28-bit registers (C_0, D_0). The contents of these two registers are shifted by the S_i^L and S_i^R positions to the left. S_r^L and S_r^R are generated by the state transition function $F(r)$ shown in Figures 5.8 and 5.9. In Figure 5.9, the 64-bit input key is separated into two 32 bits. Each becomes the input S_{in} to $F(r)$. S_r^L and S_r^R are computed from S_{out} (mod 23). LFSR in Figure 5.10 is the device for the generation of a pseudorandom binary sequence (PRBS), whose characteristic function is:

$$f(x) = x^{32} + x^7 + x^5 + x^3 + x^2 + x + 1 \text{ of a period } 2^{32} - 1$$

The 64-bit input key is assumed to be 7a138b2524af17c5. Using Figure 5.11, entire round keys are computed, as shown in Table 5.2.

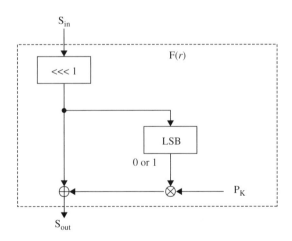

LSB : Least significant bit of input value

$\oplus$: Exclusive OR

$\otimes$: multiplication

P_K : 32-bit constant (e.g., 0x000000AE)

Figure 5.8   State transition function $F(r)$ for PRBS generation.

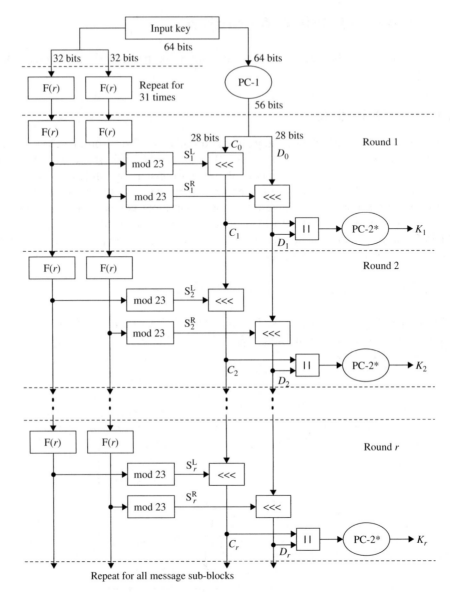

F(r): PRBS state change function
mod 23: modulo 23
PC-1: Permuted choice 1
PC-2*: Permuted choice 2 and row/column wise permutation
<<<: Circular left shift
‖: Concatenation

Figure 5.9   The newly devised DMDC key generation scheme.

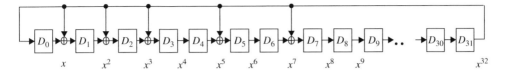

Figure 5.10 LFSR with the primitive polynomial $f(x) = 1 + x + x^2 + x^3 + x^5 + x^7 + x^{32}$ for PRBS generation.

Table 5.2 Round key generation corresponding to (S_r^L, S_r^R)

rth round	(S_r^L, S_r^R)	K_r (rth round key)
1	(2, 21)	36320340397a
2	(14, 19)	9394d0aac24c
3	(0, 15)	91c2c6fcd01e
4	(7, 7)	fcf6701c06a4
5	(21, 13)	c38496e8c45e
6	(1, 20)	12f64d47235d
7	(7, 17)	174a16a3c335
⋮	⋮	⋮
332	(21, 2)	17320b413872
333	(19, 17)	9ad8226cd646
334	(1, 11)	961203c1315b
335	(2, 18)	125ec46f8a55
336	(2, 13)	cd8d4610f0c4
337	(19, 9)	5e40db051358
338	(15, 8)	0414fc86b547

5.2.2 Computation of Message Digests

After the input message M of arbitrary length appends padding, divide the padded message into the integer multiple of 128 bits such that $M_1, M_2, \ldots, M_L$. Each M_i again positions to four 32-bit words as:

$$M_{10}, M_{11}, M_{12}, M_{13}, M_{20}, M_{21}, M_{22}, M_{23}, \ldots, M_{L0}, M_{L1}, M_{L2}, M_{L3}$$

where $M_r = (M_{r0}, M_{r1}, M_{r2}, M_{r3})$ represents the rth round of 128-bit message unit as shown in Figure 5.11. A, B, C, and D denote the four 32-bit buffers in which the data computed at the $(r-1)$th round is to be stored. Thus, $M_{r0} \oplus A$, $M_{r1} \oplus B$, $M_{r2} \oplus C$, and $M_{r3} \oplus D$ will become the rth round input data. Notice that the output at each round is swapped such that the data diffusion becomes very effective.

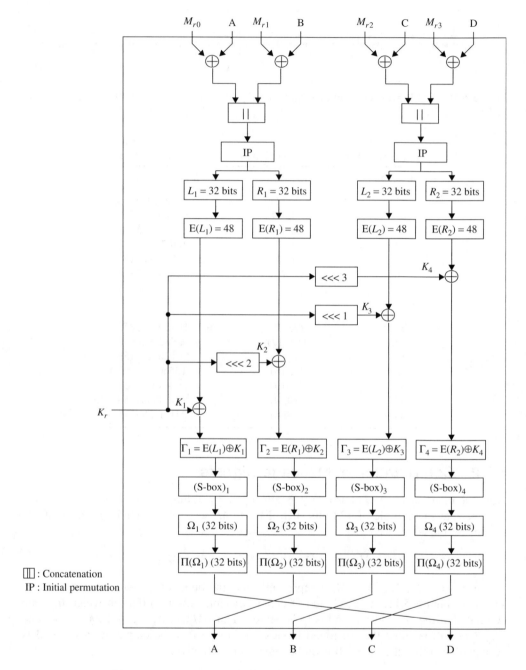

Figure 5.11   The new DMDC algorithm for message digest.

The following example demonstrates motivation, so that the reader can understand the whole process at each round (Figure 5.11). The ASCII file structure for the input message is assumed to be as shown below.

001: 12345678901234567890
002: 23456789012345678901
003: 34567890123456789012

⋮

198: 89012345678901234567
199: 90123456789012345678
200: 01234567890123456789

After receiving this ASCII file as the input, the 128-bit divided blocks are expressed in hexadecimal notation as follows:

3030313a	20313233	34353637	38393031
32333435	36373839	300d0a30	30323a20
32333435	36373839	30313233	34353637
. . . . .			
3a203031	32333435	36373839	30313233
34353637	38398000	00000000	0000a8b0

In the last block, the last three words contain padding and message length. The message length is 0xa8b0 (43184 in decimal).

The swapped outputs A, B, C, and D at each round are computed as shown in Table 5.3.

Table 5.3 The swapped output A, B, C, and D at each round

Round	A	B	C	D
1	3b1b9ba3	d126ddbe	bd3a26d1	67cfb0f3
2	f51e7b49	867a615d	b2990b90	d49538dd
3	06b402c3	a6fd207f	256bdeb5	efdd2572
4	c549ff13b	bceaa5a7	0d1cee9e	a335cf90
5	68433a67	94f78e05	7c72e14f	a32eae10
6	9e53f8b6	5d6b7335	4574651e	9b1b6489
⋮	⋮	⋮	⋮	⋮
333	0b4cbc7b	5abebd16	ccae2d5b	b50606d1
334	36ae1c4b	03b94506	89304464	28457cce
335	c530fa5f	f48260b2	1f8e5c7f	814a2152
336	487df0b3	e046c2c9	999e1066	f27ba5d3
337	58804c4c	223ee9ae	fd265d3a	7894aa4c
338	ee0fd67d	fda0da6a	df5c7095	94287b6c

Table 5.4 Hash code values based on the new DMDC scheme

Hash code length		Hash value
	32 Bits	5f79ee7e
	64 Bits	ad88e2594fe4287a
	18 Bits	32064
	Using left shift	07eb3ef78369abf6384aefae850f6d92
	Using inverse	ad88e2594fe4287a392abad213122695
128 Bits		
	Using addition and multiplication	10c62983026032634cdc8f6b6bd84085

Thus, the hash code computations applied to the new DMDC algorithm are listed in Table 5.4.

The DMDC algorithm is a secure, compact, and simple hash function. The security of DMDC has never been mathematically proven, but it depends on the problem of $F(r)$ generating the PRBS that makes each 28-bit key (left and right) shift to the left. The secure DMDC processes data sequentially, block-by-block of a 128-bit unit when computing the message digest. The computation uses four working registers labeled A, B, C, and D. These registered contents are the swapped outputs at the end of each round. The four 32-bit input units are XORed with the register contents. This process offers good performance and considerable flexibility.

5.3   MD5 Message-Digest Algorithm

The MD5 message-digest algorithm was developed by Ronald Rivest at the MIT in 1992. This algorithm takes an input message of arbitrary length and produces a 128-bit hash value of the message. The input message is processed in 512-bit blocks which can be divided into sixteen 32-bit sub-blocks. The message digest is a set of four 32-bit blocks, which concatenate to form a single 128-bit hash code. MD5 (1992) is an improved version of MD4, but is slightly slower than MD4 (1990).

The following steps are carried out to compute the message digest of the input message.

5.3.1   Append Padding Bits

The message is padded so that its length (in bits) is congruent to 448 modulo 512; that is, the padded message is just 64 bits short of being a multiple of 512. This padding is formed by appending a single "1" bit to the end of the message, and then "0" bits are appended as needed such that the length (in bits) of the padded message becomes congruent to 448 ($= 512 - 64$), modulo 512.

5.3.2   Append Length

A 64-bit representation of the original message length is appended to the result of the previous step. If the original length is greater than 2^{64}, then only the low-order of 64 bits is used for appending two 32-bit words.

The length of the resulting message is an exact multiple of 512 bits. Equivalently, this message has a length that is an exact multiple of sixteen 32-bit words. Let M[0 ... N − 1] denote the word of the resulting message, with N an integer multiple of 16.

5.3.3   Initialize MD Buffer

A four-word buffer represents four 32-bit registers (A, B, C, and D). This 128-bit buffer is used to compute the message digest. These registers are initialized to the following values in hexadecimal (low-order bytes first).

A = 01 23 45 67

B = 89 ab cd ef

C = fe dc ba 98

D = 76 54 32 10

These four variables are then copied into different variables: A as AA, B as BB, C as CC, and D as DD.

5.3.4   Define Four Auxiliary Functions (F, G, H, I)

F, G, H, and I are four basic MD5 functions. Each of these four nonlinear functions takes three 32-bit words as input and produces one 32-bit word as output. They are, one for each round, expressed as:

$F(X, Y, Z) = (X \cdot Y) + (\overline{X} \cdot Z)$

$G(X, Y, Z) = (X \cdot Z) + (Y \cdot \overline{Z})$

$H(X, Y, Z) = X \oplus Y \oplus Z$

$I(X, Y, Z) = Y \oplus (X + \overline{Z})$

where $X \cdot Y$ denotes the bitwise AND of X and Y; $X + Y$ denotes the bitwise OR of X and Y; $\overline{X}$ denotes the bitwise complement of X, that is, NOT(X); and $X \oplus Y$ denotes the bitwise XOR of X and Y.

These four auxiliary functions are designed in such a way that if the bits of X, Y, and Z are independent and unbiased, then at each bit position the function F acts as a conditional: if X then Y else Z. The functions G, H, and I are similar to the function F

Table 5.5 Truth table of four nonlinear functions

XYZ	FGHI
000	0001
001	1010
010	0110
011	1001
100	0011
101	0101
110	1100
111	1110

in that they act in "bitwise parallel" to their product from the bits of X, Y, and Z. Notice that the function H is the bitwise XOR function of its inputs.

The truth table for the computation of the four nonlinear functions (F, G, H, I) is given in Table 5.5.

5.3.5 FF, GG, HH, and II Transformations for Rounds 1, 2, 3, and 4

If M[k], $0 \leqslant k \leqslant 15$, denotes the kth sub-block of the message, and $<<< s$ represents a left shift by s bits, the four operations are defined as follows:

FF(a, b, c, d, M[k], s, i): $a = b + ((a + F(b, c, d) + M[k] + T[i] <<< s)$

GG(a, b, c, d, M[k], s, i): $a = b + ((a + G(b, c, d) + M[k] + T[i] <<< s)$

HH(a, b, c, d, M[k], s, i): $a = b + ((a + H(b, c, d) + M[k] + T[i] <<< s)$

II(a, b, c, d, M[k], s, i): $a = b + ((a + I(b, c, d) + M[k] + T[i] <<< s)$

Computation uses a 64-element table T[i], $i = 1, 2, \ldots, 64$, which is constructed from the sine function. T[i] denotes the ith element of the table, which is equal to the integer part of 4294967296 times abs(sin(i)), where i is in radians.

$$T[i] = \text{integer part of}[2^{32} * |\sin(i)|]$$

where $0 \leqslant |\sin(i)| \leqslant 1$ and $0 \leqslant 2^{32} * |\sin(i)| \leqslant 2^{32}$.

Computation of T[i] for $1 \leqslant i \leqslant 64$ is shown in Table 5.6.

5.3.6 Computation of Four Rounds (64 Steps)

Each round consists of 16 operations. Each operation performs a nonlinear function on three of A, B, C, and D. Let us show FF, GG, HH, and II transformations for rounds 1, 2, 3, and 4 in what follows.

Table 5.6 Computation of T[i] For $1 \leqslant i \leqslant 64$

T[1] = d76aa478	T[17] = f61e2562	T[33] = fffa3942	T[49] = f4292244
T[2] = e8c7b756	T[18] = c050b340	T[34] = 8771f681	T[50] = 432aff97
T[3] = 242070db	T[19] = 265e5a51	T[35] = 69d96122	T[51] = ab9423a7
T[4] = c1bdceee	T[20] = e9b6c7aa	T[36] = fde5380c	T[52] = fc93a039
T[5] = f57c0faf	T[21] = d62f105d	T[37] = a4beea44	T[53] = 655b59c3
T[6] = 4787c62a	T[22] = 02441453	T[38] = 4bdecfa9	T[54] = 8f0ccc92
T[7] = a8304613	T[23] = d8a1e681	T[39] = f6bb4b60	T[55] = ffeff47d
T[8] = fd469501	T[24] = e7d3fbc8	T[40] = bebfbc70	T[56] = 85845dd1
T[9] = 698098d8	T[25] = 21e1cde6	T[41] = 289b7ec6	T[57] = 6fa87e4f
T[10] = 8b44f7af	T[26] = c33707d6	T[42] = eaa127fa	T[58] = fe2ce6e0
T[11] = ffff5bb1	T[27] = f4d50d87	T[43] = d4ef3085	T[59] = a3014314
T[12] = 895cd7be	T[28] = 455a14ed	T[44] = 04881d05	T[60] = 4e0811a1
T[13] = 6b901122	T[29] = a9e3e905	T[45] = d9d4d039	T[61] = f7537e82
T[14] = fd987193	T[30] = fcefa3f8	T[46] = e6db99e5	T[62] = bd3af235
T[15] = a679438e	T[31] = 676f02d9	T[47] = 1fa27cf8	T[63] = 2ad7d2bb
T[16] = 49b40821	T[32] = 8d2a4c8a	T[48] = c4ac5665	T[64] = eb86d391

Round 1

Let FF[a, b, c, d, M[k], s, i] denote the operation

$$a = b + ((a + F(b, c, d) + M[k] + T[i]) <<< s)$$

Then the following 16 operations are computed.

FF[a, b, c, d, M[0], 7, 1], FF[d, a, b, c, M[1], 12, 2], FF[c, d, a, b, M[2], 17, 3],
FF[b, c, d, a, M[3], 22, 4], FF[a, b, c, d, M[4], 7, 5], FF[d, a, b, c, M[5], 12, 6],
FF[c, d, a, b, M[6], 17, 7], FF[b, c, d, a, M[7], 22, 8], FF[a, b, c, d, M[8], 7, 9],
FF[d, a, b, c, M[9], 12, 10], FF[c, d, a, b, M[10], 17, 11], FF[b, c, d, a, M[11], 22, 12],
FF[a, b, c, d, M[12], 7, 13], FF[d, a, b, c, M[13], 12, 14], FF[c, d, a, b, M[14], 17, 15],
FF[b, c, d, a, M[15], 22, 16]

The basic MD5 operation for FF transformations of round 1 is plotted as shown in Figure 5.12. GG, HH, and II transformations for rounds 2, 3, and 4 are similarly sketched.

Round 2

Let GG[a, b, c, d, M[k], s, i] denote the operation

$$a = b + ((a + G(b, c, d) + M[k] + T[i]) <<< s)$$

Then the following 16 operations are computed.

GG[a, b, c, d, M[1], 5, 17], GG[d, a, b, c, M[6], 9, 18], GG[c, d, a, b, M[11], 14, 19],
GG[b, c, d, a, M[0], 20, 20], GG[a, b, c, d, M[5], 5, 21], GG[d, a, b, c, M[10], 9, 22],

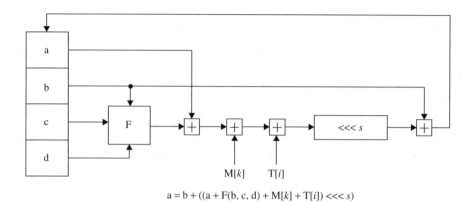

$$a = b + ((a + F(b, c, d) + M[k] + T[i]) <<< s)$$

Figure 5.12   Basic MD5 operation.

GG[c, d, a, b, M[15], 14, 23], GG[b, c, d, a, M[4], 20, 24], GG[a, b, c, d, M[9], 5, 25], GG[d, a, b, c, M[14], 9, 26], GG[c, d, a, b, M[3], 14, 27], GG[b, c, d, a, M[8], 20, 28], GG[a, b, c, d, M[13], 5, 29], GG[d, a, b, c, M[2], 9, 30], GG[c, d, a, b, M[7], 14, 31], GG[b, c, d, a, M[12], 20, 32],

Round 3

Let HH[a, b, c, d, M[k], s, i] denote the operation

$$a = b + ((a + H(b, c, d) + M[k] + T[i]) <<< s)$$

Then the following 16 operations are computed.

HH[a, b, c, d, M[5], 4, 33], HH[d, a, b, c, M[8], 11, 34], HH[c, d, a, b, M[11], 16, 35], HH[b, c, d, a, M[14], 23, 36], HH[a, b, c, d, M[1], 4, 37], HH[d, a, b, c, M[4], 11, 38], HH[c, d, a, b, M[7], 16, 39], HH[b, c, d, a, M[10], 23, 40], HH[a, b, c, d, M[13], 4, 41], HH[d, a, b, c, M[0], 11, 42], HH[c, d, a, b, M[3], 16, 43], HH[b, c, d, a, M[6], 23, 44], HH[a, b, c, d, M[9], 4, 45], HH[d, a, b, c, M[12], 11, 46], HH[c, d, a, b, M[15], 16, 47], HH[b, c, d, a, M[2], 23, 48],

Round 4

Let II[a, b, c, d, M[k], s, i] denote the operation

$$a = b + ((a + I(b, c, d) + M[k] + T[i]) <<< s)$$

Then the following 16 operations are computed.

II[a, b, c, d, M[0], 6, 49], II[d, a, b, c, M[7], 10, 50], II[c, d, a, b, M[14], 15, 51],

II[b, c, d, a, M[5], 21, 52], II[a, b, c, d, M[12], 6, 53], II[d, a, b, c, M[3], 10, 54],
II[c, d, a, b, M[10], 15, 55], II[b, c, d, a, M[1], 21, 56], II[a, b, c, d, M[8], 6, 57],
II[d, a, b, c, M[15], 10, 58], II[c, d, a, b, M[6], 15, 59], II[b, c, d, a, M[13], 21, 60],
II[a, b, c, d, M[4], 6, 61], II[d, a, b, c, M[11], 10, 62], II[c, d, a, b, M[2], 15, 63],
II[b, c, d, a, M[9], 21, 64],

After performing all of the above steps, A, B, C, and D are added to their respective
increments AA, BB, CC, and DD, as

$$A = A + AA, B = B + BB$$

$$C = C + CC, D = D + DD$$

and the algorithm continues with the resulting block of data. The final output is the
concatenation of A, B, C, and D.

Example 5.4   The message-digest problem related to the CDMA cellular system will
be discussed in this example.
   Set the initial buffer contents as follows:

$$A = 67452301 \quad B = efcdab89$$

$$C = 98badcfe \quad D = 10325476$$

The 512-bit padded message is produced from the 152-bit CDMA message by appending
the 360-bit padding as shown below.
   Padded message (512 bits) = Original message (152 bits) + Padding (360 bits):

7a138b25	24af17c3	17b439a1	2f51c5a8
8051cb36	00000000	00000000	00000000
00000000	00000000	00000000	00000000
00000000	00000000	00000098	00000000

I. Round 1 Computation for FF[a, b, c, d, M[k], s, i] $a = b + ((a + F(b, c, d) + M[k] + T[i]) <<< s) = b + U <<< s, 0 \leqslant k \leqslant 15, 1 \leqslant i \leqslant 16$, where $U <<< s$ denotes the 32-bit value obtained by circularly shifting U left by s bit positions.
(1) First-word block process (M[0], T[1], $s = 7$)
   Using Table 5.5, F(b, c, d) is computed as shown below.

b:	1110	1111	1100	1101	1010	1011	1000	1001
c:	1001	1000	1011	1010	1101	1100	1111	1110
d:	0001	0000	0011	0010	0101	0100	0111	0110
F(b, c, d):	1001	1000	1011	1010	1101	1100	1111	1110
	9	8	b	a	d	c	f	e

Compute $U = (a + F(b, c, d) + M[0] + T[1]) <<< s, s = 7$

$$a: 67452301$$
$$F(b, c, d): 98badcfe$$
$$M[0]: 7a138b25$$
$$T[1]: d76aa478$$
$$\overline{}$$
$$U: 517e2f9c$$

$U' = 517e2f9c <<< 7$

$\quad = (0101 \quad 0001 \quad 0111 \quad 1110 \quad 0010 \quad 1111 \quad 1001 \quad 1100) <<< 7$

Since $U <<< 7$ denotes the circular shift of U to the left by 7 bits, the shifted U value yields:

U': 1011	1111	0001	0111	1100	1110	0010	1000
b	f	1	7	c	e	2	8

From $a = b + U'$, we have

$$b: efcdab89$$
$$U': bf17ce28$$
$$\overline{}$$
$$a: aee579b1$$

Hence, FF[a, b, c, d, M[0], 7, 1] of operation (1) can be computed as aee579b1, efcdab89, 98badcfe, 10325476.

(2) Second-word block process (M[1], T[2], $s = 12$)

Using the outcome from operation (1), the second-word block is processed as follows:

$$d: 10325476$$
$$F[a, b, c]: bedfadcf$$
$$M[1]: 24af17c3$$
$$T[2]: e8c7b756$$
$$\overline{}$$
$$U: dc88d15e$$

$U' = U <<< 12 : 8d15edc8$

From $d = a + U'$, we have

$$a: aee57961$$
$$U': 8d15edc8$$
$$\overline{}$$
$$d: 3bfb6779$$

Hence, the result of operation (2) for the second-word block becomes FF[d, a, b, c, M[1], 12, 2] = (aee57961, efcdab89, 98badcfe, 3bfb6779).

All FF transformations for round 1 are similarly computed and consist of the following results from the 16 operations.

[1]	aee57961	efcdab89	98badcfe	10325476
[2]	aee57961	efcdab89	98badcfe	3bfb6779
[3]	aee57961	efcdab89	1e52ee63	3bfb6779
[4]	aee57961	2279e391	1e52ee63	3bfb6779
[5]	65976331	2279e391	1e52ee63	3bfb6779
[6]	65976331	2279e391	1e52ee63	b766cf0e
[7]	65976331	2279e391	e776a653	b766cf0e
[8]	65976331	d4a89062	e776a653	b766cf0e
[9]	140e3c3d	d4a89062	e776a653	b766cf0e
[10]	140e3c3d	d4a89062	e776a653	59a02fdf
[11]	140e3c3d	d4a89062	d62326dc	59a02fdf
[12]	140e3c3d	9d8eb345	d62326dc	59a02fdf
[13]	7dccd1ee	9d8eb345	d62326dc	59a02fdf
[14]	7dccd1ee	9d8eb345	d62326dc	0359415c
[15]	7dccd1ee	9d8eb345	bff77632	0359415c
[16]	7dccd1ee	10821d51	bff77632	0359415c

II. Round 2 Computation for GG
(1) First-word block operation:

$$a = b + ((a + G(b, c, d) + M[1] + T[17]) <<< s)$$

Let $V = a + G(b, c, d) + M[1] + T[17]$ where $a = 7dccd1ee$, $b = 10821d51$, $c = bff77632$, $d = 0359415c$, $M[1] = 24af17c3$, and $T[17] = f61e2562$.

Using Table 5.5, G(b, c, d) is computed as follows:

b:	0001	0000	1000	0010	0001	1101	0101	0001
c:	1011	1111	1111	0111	0111	0110	0011	0010
d:	0000	0011	0101	1001	0100	0001	0101	1100
G(b, c, d):	1011	1100	1010	0110	0011	0111	0111	0010
	b	c	a	6	3	7	7	2

Compute $V = a + G(b, c, d) + M[1] + T[17]$

a:	7dccd1ee
G(b, c, d):	bca63772
M[1]:	24af17c3
T[17]:	f61e2562
V:	55404685

V: 0101 0101 0100 0000 0100 0110 1000 0101

Since $V' = V <<< 5$, V' becomes

$$V' = \begin{array}{cccccccc} 1010 & 1000 & 0000 & 1000 & 1101 & 0000 & 1010 & 1010 \\ a & 8 & 0 & 8 & d & 0 & a & a \end{array}$$

From $a = b + V'$, we have

$$\begin{array}{l} b: 10821d51 \\ V': a808d0aa \\ \hline a: b88aedfb \end{array}$$

Thus, GG[a, b, c, d, M[1], T[17], 5] of operation (1) is computed as:

b88aedfb, 10821d51 bff77632, 0359415c

Through the 16 operations, GG transformation for round 2 can be accomplished as shown below.

[1]	b88aedfb	10821d51	bff77632	0359415c
[2]	b88aedfb	10821d51	bff77632	f14f0cf3
[3]	b88aedfb	10821d51	20aeb48b	f14f0cf3
[4]	b88aedfb	6b6c164c	20aeb48b	f14f0cf3
[5]	80426a6a	6b6c164c	20aeb48b	f14f0cf3
[6]	80426a6a	6b6c164c	20aeb48b	2ac992e7
[7]	80426a6a	6b6c164c	f0263bcd	2ac992e7
[8]	80426a6a	719e1da6	f0263bcd	2ac992e7
[9]	cbec5d78	719e1da6	f0263bcd	2ac992e7
[10]	cbec5d78	719e1da6	f0263bcd	455ddcd7
[11]	cbec5d78	719e1da6	a05494c9	455ddcd7
[12]	cbec5d78	167849a5	a05494c9	455ddcd7
[13]	5b8a2ae8	167849a5	a05494c9	455ddcd7
[14]	5b8a2ae8	167849a5	a05494c9	af92e3c8
[15]	5b8a2ae8	167849a5	2e6d799d	af92e3c8
[16]	5b8a2ae8	29e29554	2e6d799d	af92e3c8

III. Round 3 Computation for HH
(1) First-word block operation:

$a = b + ((a + H(b, c, d) + M[5] + T[33]) <<< 4)$

where $a = 5b8a2ae8$, $b = 29e29554$, $c = 2e6d799d$, $d = af92e3c8$, $M[5] = 00000000$, $T[33] = fffa3942$, and $s = 4$.

Using Table 5.5, H(b, c, d) is computed as follows:

b:	0010	1001	1110	0010	1001	0101	0101	0100
c:	0010	1110	0110	1101	0111	1001	1001	1101
d:	1010	1111	1001	0010	1110	0011	1100	1000

H(b, c, d):	1010	1000	0001	1101	0000	1111	0000	0001
	a	8	1	d	0	f	0	1

Compute W = a + H(b, c, d) + M[5] + T[33]

$$
\begin{array}{rl}
\text{a:} & 5b8a2ae8 \\
\text{H(b, c, d):} & a81d0f01 \\
\text{M[5]:} & 00000000 \\
\text{T[33]:} & fffa3942 \\
\hline
\text{W:} & 03a1732b
\end{array}
$$

W = 0000 0011 1010 0001 0111 0011 0010 1011
Since W' = W <<< 4, we have

W' =	0011	1010	0001	0111	0011	0010	1011	0000
	3	a	1	7	3	2	b	0

From a = b + W', a can be computed as

$$
\begin{array}{l}
\text{b: } 29e29554 \\
\text{W': } 3a1732b0 \\
\hline
\text{a: } 63f9c804
\end{array}
$$

Thus, HH[a, b, c, d, M[5], T[33], 4] of operation (1) is obtained as 63f9c804 29e29554 2e6d799d af92e3c8. Through 16 operations, HH transformation for round 3 can be computed as shown below.

[1]	63f9c804	29e29554	2e6d799d	af92e3c8
[2]	63f9c804	29e29554	2e6d799d	3bf27cdf
[3]	63f9c804	29e29554	38408ad2	3bf27cdf
[4]	63f9c804	39049458	38408ad2	3bf27cdf
[5]	bae75a5e	39049458	38408ad2	3bf27cdf
[6]	bae75a5e	39049458	38408ad2	edcbf07c
[7]	bae75a5e	39049458	02788da0	edcbf07c
[8]	bae75a5e	279f19dc	02788da0	edcbf07c
[9]	e292ec26	279f19dc	02788da0	edcbf07c

[10]	e292ec26	279f19dc	02788da0	937294f5
[11]	e292ec26	279f19dc	784ef22d	937294f5
[12]	e292ec26	67e9dd0d	784ef22d	937294f5
[13]	fbc16051	67e9dd0d	784ef22d	937294f5
[14]	fbc16051	67e9dd0d	784ef22d	9fb3bb46
[15]	fbc16051	67e9dd0d	14f356d2	9fb3bb46
[16]	fbc16051	814dbccf	14f356d2	9fb3bb46

IV. Round 4 Computation for II
(1) First-word block operation:

$$a = b + ((a + I(b, c, d) + M[0] + T[49]) <<< 6)$$

where $a = \text{fbc16051}$, $b = \text{814dbccf}$, $c = \text{14f356d2}$, $d = \text{9fb3bb46}$, $M[0] = \text{7a138b25}$, $T[49] = \text{f4292244}$, and $s = 6$.

Using Table 5.5, $I(b, c, d)$ can be computed as follows:

b:	1000	0001	0100	1101	1011	1100	1100	1111
c:	0001	0100	1111	0011	0101	0110	1101	0010
d:	1001	1111	1011	0011	1011	1011	0100	0110

I(b, c, d):	1111	0101	1011	1110	1010	1010	0010	1101
	f	5	*b*	*e*	*a*	*a*	2	*d*

Compute $Z = a + I(b, c, d) + M[0] + T[49]$

$$
\begin{array}{r}
a: \text{fbc16051} \\
I(b, c, d): \text{f5beaa2d} \\
M[0]: \text{7a138b25} \\
T[49]: \text{f4292244} \\
\hline
Z : \text{5fbcb7e7}
\end{array}
$$

$Z = 0101 \quad 1111 \quad 1011 \quad 1100 \quad 1011 \quad 0111 \quad 1110 \quad 0111$
Since $Z' = Z <<< 6$, we have

$Z' = 1110 \quad 1111 \quad 0010 \quad 1101 \quad 1111 \quad 1001 \quad 1101 \quad 0111$
$\quad\quad e \quad\quad\quad f \quad\quad\; 2 \quad\quad\; d \quad\quad\; f \quad\quad\; 9 \quad\quad\; d \quad\quad\; 7$

From $a = b + Z'$, a is computed as:

$$
\begin{array}{r}
b: \text{814dbccf} \\
Z': \text{ef2df9d7} \\
\hline
a: \text{707bb6a6}
\end{array}
$$

Thus, operation (1) of II[a, b, c, d, M[0], T[49], 6] is obtained as:

707bb6a6 814dbccf 14f356d2 9fb3bb46

The results from the 16 operations are listed in the following.

[1]	707bb6a6	814dbccf	14f356d2	9fb3bb46
[2]	707bb6a6	814dbccf	14f356d2	b374ac1a
[3]	707bb6a6	814dbccf	1dcb5424	b374ac1a
[4]	707bb6a6	ebc0a7cd	1dcb5424	b374ac1a
[5]	e1adb47e	ebc0a7cd	1dcb5424	b374ac1a
[6]	e1adb47e	ebc0a7cd	1dcb5424	2307ce67
[7]	e1adb47e	ebc0a7cd	fc5d488d	2307ce67
[8]	e1adb47e	65cbb221	fc5d488d	2307ce67
[9]	25173275	65cbb221	fc5d488d	2307ce67
[10]	25173275	65cbb221	fc5d488d	e801a803
[11]	25173275	65cbb221	9da76743	e801a803
[12]	25173275	0f04df84	9da76743	e801a803
[13]	d4921a8b	0f04df84	9da76743	e801a803
[14]	d4921a8b	0f04df84	9da76743	400fe907
[15]	d4921a8b	0f04df84	f3d96b57	400fe907
[16]	d4921a8b	24903b0e	f3d96b57	400fe907

A buffer containing four 32-bit registers A, B, C, and D is used to compute the 128-bit message digest. These registers are initialized to the following values.

aa = 67452301   bb = efcdab89

cc = 98badcfe   dd = 10325476

The last operation of this transformation is:

a = d4921a8b   b = 24903b0e

c = f3d96b57   d = 400fe907

After this, the following additions are finally performed to produce the message digest.

A = a + aa

B = b + bb

C = c + cc

D = d + dd

The message digest produced as an output of A, B, C, and D is the concatenation of A, B, C, and D.

a:	d4921a8b	b:	24903b0e
aa:	67452301	bb:	efcdab89

A:	3bd73d8c	B:	145de697

c:	f3d96b57	d:	400fe907
cc:	98badcfe	dd:	10325476

C:	8c944855	D:	50423d7d

The concatenation of the four outputs of A, B, C, and D is the 128-bit message digest such that A$\|$ B$\|$C$\|$D = 3bd73d8c 145de697 8c944855 50423d7d.

In CDMA cellular mobile communications, a shared secret data (SSD) is a 128-bit pattern stored in semipermanent memory in the mobile station. SSD is partitioned into two 64-bit distinct subsets, SSD-A and SSD-B. SSD-A is used to support the authentication process, while SSD-B is used to support voice privacy and message confidentiality.

SSD subsets are generated from the message digest as follows:

SSD-A: 3bd73d8c145de697
SSD-B: 8c94485550423d7d

5.4   Secure Hash Algorithm (SHA-1)

The Secure Hash Algorithm (SHA) was developed by the National Institute of Standards and Technology (NIST) for use with the Digital Signature Algorithm (DSA) and published as a Federal Information Processing Standards (FIPS PUB 180) in 1993. The Secure Hash Standard (SHS) specifies an SHA-1 for computing the hash value of a message or a data file. When a message of any length of less than 2^{64} bits is input, the SHA-1 produces a 160-bit output called a *message digest* (or a *hash code*). The message digest can then be input to the DSA, which generates or verifies the signature for the message. Signing the message digest rather than the message often improves the efficiency of the process because the message digest is usually much smaller than the message.

The SHA-1 (FIPS 180-1, 1995) is a technical revision of SHA (FIPS 180, 1993). The SHA-1 is secure because it is computationally impossible to find a message which corresponds to a given message digest, or to find two different messages which produce the same message digest. Any change to a message in transit will result in a different message digest, and the signature will fail to verify. The SHA-1 is based on the MD4 message-digest algorithm and its design is closely modeled on that algorithm.

5.4.1   Message Padding

The message padding is provided to make a final padded message a multiple of 512 bits. The SHA-1 sequentially processes blocks of 512 bits when computing the hash value (or message digest) of a message or data file that is provided as input. Padding is exactly the same as in MD5. The following specifies how this padding is performed. As a summary, first append a "1" followed by as many "0s" as necessary to make it 64 bits short of a multiple of 512 bits, and finally, a 64-bit integer is appended to the end of the 0-appended message to produce a final padded message of length $n \times 512$ bits. The 64-bit integer "I" represents the length of the original message. Now, the padded message is then processed by the SHA-1 as $n \times 512$-bit blocks.

Example 5.5   Suppose the original message is the bit string.

$$01100001 \ 01100010 \ 01100011$$

This message has length $I = 24$. After "1" is appended, we have 01100001 01100010 011000111. The number of bits of this bit string is 25 because $I = 24$. Therefore, we should append 423 "0s" and the two-word representation of 24, that is, 00000000 00000018 (in hex), for forming the final padded message as follows:

61626380	00000000	00000000	00000000
00000000	00000000	00000000	00000000
00000000	00000000	00000000	00000000
00000000	00000000	00000000	00000018

This final padded message consisting of one block contains 16 words $= 16 \times 8 \times 4 = 512$ bits for $n = 1$ in this case.

5.4.2   Initialize 160-bit Buffer

The 160-bit buffer consists of five 32-bit registers (A, B, C, D, and E). Before processing any blocks, these registers are initialized to the following hexadecimal values.

$$H_0 = 67 \ 45 \ 23 \ 01$$
$$H_1 = ef \ cd \ ab \ 89$$
$$H_2 = 98 \ ba \ dc \ fe$$
$$H_3 = 10 \ 32 \ 54 \ 76$$
$$H_4 = c3 \ d2 \ e1 \ f0$$

Note that the first four values are the same as those used in MD5. The only difference is the use of a different rule for expressing the values, that is, high-order octets first for SHA and low-order octets first for MD5.

5.4.3   Functions Used

A sequence of logical functions $f_0, f_1, \ldots, f_{79}$ is used in SHA-1. Each function $f_t, 0 \leqslant t \leqslant 79$, operates on three 32-bit words B, C, and D and produces a 32-bit word as the output. Each operation performs a nonlinear operation of three of A, B, C, and D and then does shifting and adding as in MD5. The set of SHA primitive functions, f_t (B, C, D), is defined as follows:

$$f_t(B, C, D) = (B \cdot C) + (\overline{B} \cdot D), 0 \leqslant t \leqslant 19$$

$$f_t(B, C, D) = B \oplus C \oplus D, 20 \leqslant t \leqslant 39$$

$$f_t(B, C, D) = (B \cdot C) + (B \cdot D) + (C \cdot D), 40 \leqslant t \leqslant 59$$

$$f_t(B, C, D) = B \oplus C \oplus D, 60 \leqslant t \leqslant 79$$

where

$\quad B \cdot C = $ bitwise logical 'AND' of B and C;

$\quad B \oplus C = $ bitwise logical XOR of B and C;

$\quad \overline{B} = $ bitwise logical 'complement' of B;

$\quad \boxplus = $ addition modulo2^{32}.

As can be seen, only three different functions are used. For $0 \leqslant t \leqslant 19$, the function f_t acts as a conditional: if B then C else D. For $20 \leqslant t \leqslant 39$ and $60 \leqslant t \leqslant 79$, the function f_t is true if two or three of the arguments are true. Table 5.7 is a truth table of these functions.

5.4.4 Constants Used

Four distinct constants are used in SHA-1. In hexadecimal, these values are given by

$$K_t = 5a827999, \quad 0 \leqslant t \leqslant 19$$

$$K_t = 6ed9eba1, \quad 20 \leqslant t \leqslant 39$$

$$K_t = 8f1bbcdc, \quad 40 \leqslant t \leqslant 59$$

$$K_t = ca62c1d6, \quad 60 \leqslant t \leqslant 79$$

Table 5.7 Truth table of four nonlinear functions for SHA-1

B	C	D	$f_{0,1,\ldots,19}$	$f_{20,21,\ldots,39}$	$f_{40,41,\ldots,59}$	$f_{60,61,\ldots,79}$
0	0	0	0	0	0	0
0	0	1	1	1	0	1
0	1	0	0	1	0	1
0	1	1	1	0	1	0
1	0	0	0	1	0	1
1	0	1	0	0	1	0
1	1	0	1	0	1	0
1	1	1	1	1	1	1

5.4.5 Computing the Message Digest

The message digest is computed using the final padded message. To generate the message digest, the 16-word blocks (M_0 to M_{15}) are processed in order. The processing of each M_i involves 80 steps; that is, the message block is transformed from sixteen 32-bit words (M_0 to M_{15}) to eighty 32-bit words (W_0 to W_{79}) using the following algorithm.

Divide M_i into 16 words W_0, W_1, ... , W_{15}, where W_0 is the leftmost word. For $t = 0\text{–}15$, $W_t = M_t$. For $t = 16\text{–}79$, $W_t = S^1(W_{t-16} \oplus W_{t-14} \oplus W_{t-8} \oplus W_{t-3})$.

Let $A = H_0$, $B = H_1$, $C = H_2$, $D = H_3$, $E = H_4$. For $t = 0\text{–}79$ do

$$\text{TEMP} = S^5(A) + F_t(B, C, D) + E + W_t + K_t;$$

$$E = D; \ D = C; \ C = S^{30}(B); \ B = A; \ A = \text{TEMP}$$

where

A, B, C, D, E = five words of the buffer;
t: round number, $0 \leqslant t \leqslant 79$;
S^i = Circular left shift by i bits;
W_t = a 32-bit word derived from the current 512-bit input block;
K_t = an additive constant;
$\boxplus$ = addition modulo 2^{32}.

After all N 512-bit blocks have been processed, the output from the Nth stage is the 160-bit message digest, represented by the five words H_0, H_1, H_2, H_3, and H_4.

The SHA-1 operation looking at the logic in each of the 80 rounds of one 512-bit block is shown in Figure 5.13.

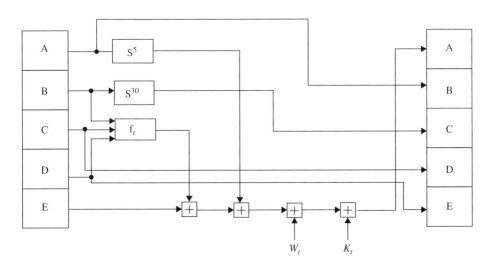

Figure 5.13 SHA-1 operation.

Example 5.6 Show how to derive the 32-bit words $W_t, 0 \leqslant t \leqslant 79$, from the 512-bit message.

t	W_t
0	$W_0 = M_0$
1	$W_1 = M_1$
. . . .	. . . .

t	W_t
15	$W_{15} = M_{15}$
16	$W_{16} = S^1 (W_0 \oplus W_2 \oplus W_8 \oplus W_{13})$
17	$W_{17} = S^1 (W_1 \oplus W_3 \oplus W_9 \oplus W_{14})$
.	. . . .
30	$W_{30} = S^1 (W_{14} \oplus W_{16} \oplus W_{22} \oplus W_{27})$
31	$W_{31} = S^1 (W_{15} \oplus W_{17} \oplus W_{23} \oplus W_{28})$
.	. . . .
59	$W_{59} = S^1 (W_{43} \oplus W_{45} \oplus W_{51} \oplus W_{56})$
60	$W_{60} = S^1 (W_{44} \oplus W_{46} \oplus W_{52} \oplus W_{57})$
.	. . . .
78	$W_{78} = S^1 (W_{62} \oplus W_{64} \oplus W_{70} \oplus W_{75})$
79	$W_{79} = S^1 (W_{63} \oplus W_{65} \oplus W_{71} \oplus W_{76})$

Example 5.7 Let the original message be 1a7fd53b4c. Then, the final padded message consists of the following 16 words.

1a7fd53b	4c800000	00000000	00000000
00000000	00000000	00000000	00000000
00000000	00000000	00000000	00000000
00000000	00000000	00000000	00000028

The initial hex values of $\{H_i\}$ are

$H_0 = 67452301$

$H_1 = \text{efcdab89}$

$H_2 = 98\text{badcfe}$

$H_3 = 10325476$

$H_4 = \text{c3d2e1f0}$

The hex values of A, B, C, D, and E after pass $t(0 \leqslant t \leqslant 79)$ are computed as follows:

t	A	B	C	D	E
			Register output		
0	ba346dee	67452301	7bf36ae2	98badcfe	10325476
1	f9be8ae4	ba346dee	59d148c0	7bf36ae2	98badcfe
2	84e1fdf6	f9be8ae4	ae8d1b7b	59d148c0	7bf36ae2
3	1b82edab	84e1fdf6	3e6fa2b9	ae8d1b7b	59d148c0
4	531f1a75	1b82edab	a1387f7d	3e6fa2b9	ae8d1b7b
5	926052f7	531f1a75	c6e0bb6a	a1387f7d	3e6fa2b9
6	c71cfaac	926052f7	54c7c69d	c6e0bb6a	a1387f7d
7	341b3a4b	c71cfaac	e49814bd	54c7c69d	c6e0bb6a
8	79a59326	341b3a4b	31c73eab	e49814bd	54c7c69d
9	d47fe3c4	79a59326	cd06ce92	31c73eab	e49814bd
10	185db57b	d47fe3c4	9e6964c9	cd06ce92	31c73eab
11	3569d479	185db57b	351ff8f1	9e6964c9	cd06ce92
12	6b01c842	3569d479	c6176d5e	351ff8f1	9e6964c9
13	5d3c5387	6b01c842	4d5a751e	c6176d5e	351ff8f1
14	04434893	5d3c5387	9ac07210	4d5a751e	c6176d5e
15	c1456f97	04434893	d74f14e1	9ac07210	4d5a751e
16	a44dbea6	c1456f97	c110d224	d74f14e1	9ac07210
17	ef0512e1	a44dbea6	f0515be5	c110d224	d74f14e1
18	f3c545ab	ef0512e1	a9136fa9	f0515be5	c110d224
19	b78ca1cc	f3c545ab	7bc144b8	a9136fa9	f0515be5
20	a3d6efd7	b78ca1cc	fcf1516a	7bc144b8	a9136fa9
21	c3880afc	a3d6efd7	2de32873	fcf1516a	7bc144b8
22	a25fd097	c3880afc	e8f5bbf5	2de32873	fcf1516a
23	2263e9cb	a25fd097	30e202bf	e8f5bbf5	2de32873
24	cd820d01	2263e9cb	e897f425	30e202bf	e8f5bbf5
25	9824bad0	cd820d01	c898fa72	e897f425	30e202bf
26	59e04bcd	9824bad0	73608340	c898fa72	e897f425
27	b7581fd3	59e04bcd	26092eb4	73608340	c898fa72
28	7efb6e25	b7581fd3	567812f3	26092eb4	73608340
29	18d1583d	7efb6e25	edd607f4	567812f3	26092eb4
30	42659f77	18d1583d	5fbedb89	edd607f4	567812f3
31	22b4bfef	42659f77	4634560f	5fbedb89	edd607f4
32	a9390191	22b4bfef	d09967dd	4634560f	5fbedb89
33	ffd2919f	a9390191	c8ad2ffb	d09967dd	4634560f
34	a0585c33	ffd2919f	6a4e4064	c8ad2ffb	d09967dd
35	8fae2fc9	a0585c33	fff4a467	6a4e4064	c8ad2ffb
36	5337d670	8fae2fc9	e816170c	fff4a467	6a4e4064
37	7044d0fe	5337d670	63eb8bf2	e816170c	fff4a467

			Register output		
t	A	B	C	D	E
38	78304e61	7044d0fe	14cdf59c	63eb8bf2	e816170c
39	2c5ca6b0	78304e61	9c11343f	14cdf59c	63eb8bf2
40	f304b895	2c5ca6b0	5e0c1398	9c11343f	14cdf59c
41	e89d0d8b	f304b895	b1729ac	5e0c1398	9c11343f
42	79f30210	e89d0d8b	7cc12e25	b1729ac	5e0c1398
43	f37223c6	79f30210	fa274362	7cc12e25	0b1729ac
44	f53bdd27	f37223c6	1e7cc084	fa274362	7cc12e25
45	b1cf753c	f53bdd27	bcdc88f1	1e7cc084	fa274362
46	d9030e9b	b1cf753c	fd4ef749	bcdc88f1	1e7cc084
47	9bf173ff	d9030e9b	2c73dd4f	fd4ef749	bcdc88f1
48	bae46f3c	9bf173ff	f640c3a6	2c73dd4f	fd4ef749
49	e8be1481	bae46f3c	e6fc5cff	f640c3a6	2c73dd4f
50	4a0bb5b8	e8be1481	2eb91bcf	e6fc5cff	f640c3a6
51	6d99dcd5	4a0bb5b8	7a2f8520	2eb91bcf	e6fc5cff
52	5e0e5623	6d99dcd5	1282ed6e	7a2f8520	2eb91bcf
53	422c7e52	5e0e5623	5b667735	1282ed6e	7a2f8520
54	e6ca43ae	422c7e52	d7839588	5b667735	1282ed6e
55	835bd439	e6ca43ae	908b1f94	d7839588	5b667735
56	32a7862d	835bd439	b9b290eb	908b1f94	d7839588
57	250ada00	32a7862d	60d6f50e	b9b290eb	908b1f94
58	a46d627b	250ada00	4ca9e18b	60d6f50e	b9b290eb
59	0588823a	a46d627b	942b680	4ca9e18b	60d6f50e
60	2d9bba2e	588823a	e91b589e	0942b680	4ca9e18b
61	8d8fb303	2d9bba2e	8162208e	e91b589e	0942b680
62	860d6a4f	8d8fb303	8b66ee8b	8162208e	e91b589e
63	14b64733	860d6a4f	e363ecc0	8b66ee8b	8162208e
64	7f486fbe	14b34733	e1835a93	e363ecc0	8b66ee8b
65	7d3d3745	7f486fbe	c52cd1cc	e1835a93	e363ecc0
66	d17b4506	7d3d3745	9fd21bef	c52cd1cc	e1835a93
67	2e4967ee	d17b4506	5f4f4dd1	9fd21bef	c52cd1cc
68	cc1e45de	2e4967ee	b45ed141	5f4f4dd1	9fd21bef
69	b3f80c20	cc1e45de	8b9259fb	b45ed141	5f4f4dd1
70	f124837a	b3f80c20	b3079177	8b9259fb	b45ed141
71	56ed70b1	f124837a	2cfe0308	b3079177	8b9259fb
72	d8b0d990	56ed70b1	bc4920de	2cfe0308	b3079177
73	1d849b17	d8b0d990	55bb5c2c	bc4920de	2cfe0308
74	84257988	1d849b17	362c3664	55bb5c2c	bc4920de
75	9eec3055	84257988	c76126c5	362c3664	55bb5c2c
76	6240e72c	9eec3055	21095e62	c76126c5	362c3664
77	8243ecda	6240e72c	67bb0c15	21095e62	c76126c5
78	a8342af0	8243ecda	189039cb	67bb0c15	21095e62
79	e1426096	a8342af0	a090fb36	189039cb	67bb0c15

After all 512-bit blocks have been processed, the output represented by the five words, H_0, H_1, H_2, H_3, and H_4, is the 160-bit message digest as shown below.

H_0 : 48878397

H_1 : 9801d679

H_2 : 394bd834

H_3 : 28c28e41

H_4 : 2b8dee05

The 160-bit message digest is then the data concatenation of $\{H_i\}$:

$$H_0\|H_1\|H_2\|H_3\|H_4 = 488783979801d679394bd83428c28e412b8dee05$$

As discussed previously, the digitized document or message of any length can create a 160-bit message digest which is produced using the SHA-1 algorithm.

Any change to a digitized message in transit results in a different message digest. In fact, changing a single bit of the data modifies at least half of the resulting digest bits. Furthermore, it is computationally impossible to find two meaningful messages that have the same 160-bit digest. On the other hand, given a 160-bit message digest, it is also impossible to find a meaningful message with that digest.

5.5   Hashed Message Authentication Codes (HMAC)

The keyed-hashed Message Authentication Code (HMAC) is a key-dependent one-way hash function which provides both data integrity and data origin authentication for files sent between two users. HMACs have the same properties as the one-way hash functions discussed earlier in this chapter, but they also include a secret key. HMACs can be used to authenticate data or files between two users (data authentication). They can also be used by a single user to determine whether or not the user's files have been altered (data integrity).

To evaluate HMAC over the message or file, the following expression is required to compute.

$$\text{HMAC} = H\,[(K \oplus \text{opad})\|H\,[(K \oplus \text{ipad})\|M\,]]$$

where

   ipad = inner padding

      = 0x36 (repeated b times)

      = 00110110 (0x36) repeated 64 times (512 bits);

   opad = outer padding

      = 0x5c (repeated b times)

      = 01011100 (0x5c) repeated 64 times (512 bits);

b = block length of 64 bytes = 512 bits;
h = length of hash values, that is, h = 16 bytes = 128 bits for MD5
    and h = 20 bytes = 160 bits for SHA-1;
K = secret key of any length up to b = 512 bits;
H = hash function where message is hashed by iterating a basic key K.

The HMAC equation is explained as follows:

1. Append zeros to the end of K to create a b-byte string (i.e., if $K = 160$ bits in length and $b = 512$ bits, then K should be appended with 352 zero bits or 44 zero bytes 0x00, resulting in K′ = ($K \| 0x00$).
2. XOR (bitwise XOR) K' with ipad to produce the b-bit block computed in step 1.
3. Append M to the b-byte string resulting from step 2.
4. Apply H to the stream generated in step 3.
5. XOR (bitwise XOR) K' with opad to produce the b-byte string computed in step 1.
6. Append the hash result H from step 4 to the b-byte string resulting from step 5.
7. Apply H to the stream generated in step 6 and output the result.

Figure 5.14 illustrates the overall operation of HMAC, explaining these steps.

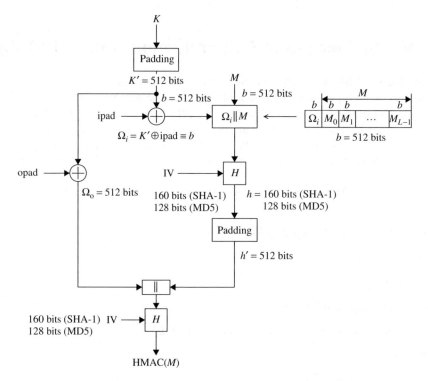

Figure 5.14 Overall operation of HMAC computation using either MD5 or SHA-1 (message length computation is based on $\Omega_i \| M$).

Example 5.8 Consider HMAC computation by using a hash function SHA-1. Assume that the message (M), the key (K), and the initialization vector (IV) are given as follows:

M : 0x1a7fd53b4c
K : 0x31fa7062c45113e32679fd1353b71264
IV : $A = 0x67452301$, $B = 0xefcdab89$, $C = 0x98badcfe$,
    $D = 0x10325476$, $E = 0xc3d2e1f0$

Referring to Figure 5.14, the HMAC–SHA-1 calculation proceeds with the steps shown below.

$K' = K \| (0x00 \ldots 00)(512 \text{ bits})$

= 31fa7062	c45113e3	2679fd13	53b71264
00000000	00000000	00000000	00000000
00000000	00000000	00000000	00000000
00000000	00000000	00000000	00000000

$\Omega_i = K' \oplus \text{ipad} = K' \oplus (0x3636 \ldots 36)$

= 07cc4654	f26725d5	104fcb25	65812452
36363636	36363636	36363636	36363636
36363636	36363636	36363636	36363636
36363636	36363636	36363636	36363636

$M' = $ 1a7fd53b	4c800000	00000000	00000000
00000000	00000000	00000000	00000000
00000000	00000000	00000000	00000000
00000000	00000000	00000000	00000028

$\Omega_i \| M'$:

07cc4654	f26725d5	104fcb25	65812452
36363636	36363636	36363636	36363636
36363636	36363636	36363636	36363636
36363636	36363636	36363636	36363636
1a7fd53b	4c800000	00000000	00000000
00000000	00000000	00000000	00000000
00000000	00000000	00000000	00000000
00000000	00000000	00000000	00000028

$h = H[(\Omega_i \| M'), \text{IV}] = \text{Inner SHA-1}$

$\quad = $ 9691eb0c d263a12f ab7e0e2f e60ced5f 546c857a

$\Omega_o = K' \oplus \text{opad} = K' \oplus (0x5c5c \ldots 5c)$

= 6da62c3e	980d4fbf	7a25a14f	0feb4e38
5c5c5c5c	5c5c5c5c	5c5c5c5c	5c5c5c5c
5c5c5c5c	5c5c5c5c	5c5c5c5c	5c5c5c5c
5c5c5c5c	5c5c5c5c	5c5c5c5c	5c5c5c5c

$h' =$ 9691eb0c	d263a12f	ab7e0e2f	e60ced5f
546c857a	80000000	00000000	00000000
00000000	00000000	00000000	00000000
00000000	00000000	00000000	000000a0

$\Omega_o \| h'$:

6da62c3e	980d4fbf	7a25a14f	0feb4e38
5c5c5c5c	5c5c5c5c	5c5c5c5c	5c5c5c5c
5c5c5c5c	5c5c5c5c	5c5c5c5c	5c5c5c5c
5c5c5c5c	5c5c5c5c	5c5c5c5c	5c5c5c5c
9691eb0c	d263a12f	ab7e0e2f	e60ced5f
546c857a	80000000	00000000	00000000
00000000	00000000	00000000	00000000
00000000	00000000	00000000	000000a0

$\text{HMAC}[\Omega_o \| h'] = \text{Outer SHA-1}$

$$= \text{c19e1236} \quad \text{ae346195} \quad \text{16594259} \quad \text{4c5202b3} \quad \text{4a85c5e}$$

The alternative operation for computation of either HMAC-MD5 or HMAC–SHA-1 is based on the following expression.

$\text{HMAC} = H[H[M, (IV)_i], (IV)_o]$

$(IV)_i \quad = f[(K' \oplus \text{ipad}), IV]$

$(IV)_o \quad = f[(K' \oplus \text{opad}), IV]$

$K' \qquad = K \| (0x00 \ldots 0) \quad (512 \text{ bits})$

f denotes a compression function representing either SHA-1 or MD5 hash function.
  The procedure can be explained in words as follows:

1. Append zeros to K to create a b-bit string K', where $b = 512$ bits.
2. XOR K' (padding with zero) with ipad to produce the b-bit block.

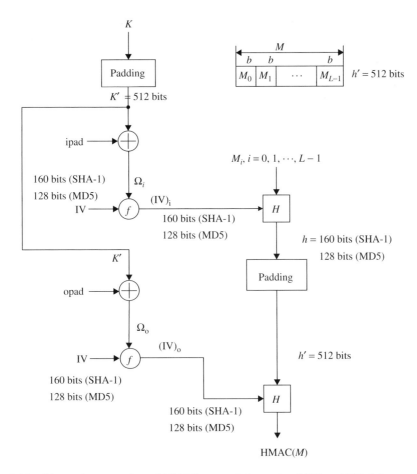

Figure 5.15 Alternative operation of HMAC computation using MD5 or SHA-1 (message length computation is based on M only).

3. Apply the compression function $F(K' \oplus \text{ipad, IV})$ to produce $(\text{IV})_i = 160$ bits for SHA-1.
4. Compute the hash code h with $(\text{IV})_i$ and M_i.
5. Raise the hash value computed from step 4 to a b-bit string.
6. XOR K' (padded with zeros) with opad to produce the b-bit block.
7. Apply the compression function $F(K' \oplus \text{opad, IV})$ to produce $(\text{IV})_o = 160$ bits for SHA-1.
8. Compute the HMAC with $(\text{IV})_o$ and the raised hash value resulted from step 5.

Figure 5.15 shows the alternative scheme based on these steps.

Example 5.9 Consider the HMAC computation by the alternative method. Assume that the message (M), the key (K), and the initialization vector (IV) are given as follows:

M : 0x 1a7fd53b4c
K : 0x 31fa7062c45113e32679fd1353b71264
IV : A = 0x67452301, B = 0xefcdab89, C = 0x98badcfe,
   D = 0x10325476, E = 0xc3d2e1f0

Referring to Figure 5.15, the HMAC–SHA-1 calculation proceeds through the steps shown below.

$K' = K \| (0x00 \dots 00)(512 \text{ bits})$

= 31fa7062	c45113e3	2679fd13	53b71264
00000000	00000000	00000000	00000000
00000000	00000000	00000000	00000000
00000000	00000000	00000000	00000000

$\Omega_i = K' \oplus \text{ipad} = K' \oplus (0x3636 \dots 36)$

= 07cc4654	f26725d5	104fcb25	65812452
36363636	36363636	36363636	36363636
36363636	36363636	36363636	36363636
36363636	36363636	36363636	36363636

$(\text{IV})_i = f(\Omega_i, \text{IV})$

  = c6edf676   ef938cee   84dd1b00   5b3b8996   cb172ad4

$M' =$ 1a7fd53b	4c800000	00000000	00000000
00000000	00000000	00000000	00000000
00000000	00000000	00000000	00000000
00000000	00000000	00000000	00000028

$h = H(M', (\text{IV})_i) = \text{Inner SHA-1}$

  = 613f6cbd   b336740e   8af4b185   367b1773   d260afce

$\Omega_o = K' \oplus \text{opad} = K' \oplus (0x5c5c \dots 5c)$

= 6da62c3e	980d4fbf	7a25a14f	0feb4e38
5c5c5c5c	5c5c5c5c	5c5c5c5c	5c5c5c5c
5c5c5c5c	5c5c5c5c	5c5c5c5c	5c5c5c5c
5c5c5c5c	5c5c5c5c	5c5c5c5c	5c5c5c5c

$$(IV)_o = f(\Omega_o, IV)$$

$$= a46e7eba \quad 64c80ca4 \quad c42317b3 \quad dd2b4f1e \quad 81c21ab0$$

$$HMAC(M) = H(h', (IV)_o) = \text{Outer SHA-1}$$

$$= af625840 \quad ed120ccd \quad ba408de3 \quad b259a95b \quad d4d98eda$$

The HMAC is a cryptographic checksum with the highest degree of security against attacks. HMACs are used to exchange information between two parties, where both have knowledge of the secret key. A digital signature does not require any secret key to be verified for authentication.

6

Asymmetric Public-Key Cryptosystems

Public-key cryptography became public soon after Whitefield Diffie and Martin Hellman (1976) proposed the innovative concept of an exponential key exchange scheme. Since 1976, numerous public-key algorithms have been proposed, but many of them have since been broken. Of the many algorithms that are still considered to be secure, most are impractical.

Only a few public-key algorithms are both secure and practical. Of these, only some are suitable for encryption. Others are only suitable for digital signatures. Among these numerous public-key cryptography algorithms, only four algorithms, RSA (1978), ElGamal (1985), Schnorr (1990), and ECC (1985), are considered to be suitable for both encryption and digital signatures. Another public-key algorithm that is designed to only be suitable for secure digital signatures is DSA (Digital Signature Algorithm; 1991). The designer should bear in mind that the security of any encryption scheme depends on the length of the key and the computational work involved in breaking a cipher.

6.1 Diffie–Hellman Exponential Key Exchange

In 1976, Diffie and Hellman proposed a scheme using the exponentiation modulo q (a prime) as a public key exchange algorithm. Exponential key exchange takes advantage of easy computation of exponentials in a finite field $GF(q)$ with a prime q compared with the difficulty of computing logarithms over $GF(q)$ with q elements $\{1, 2, \ldots, q - 1\}$. Let q be a prime number and α a primitive element of the prime number q. Then the powers of α generate all the distinct integers from 1 to $q - 1$ in some order. For any integer Y and a primitive element α of prime number q, a unique exponent X is found such that

$$Y \equiv \alpha^X \pmod{q}, 1 \leqslant X \leqslant q - 1$$

Then X is referred to as the discrete logarithm of Y to the base α over $GF(q)$:

$$X = \log \alpha^Y \text{ over } GF(q), 1 \leqslant Y \leqslant q - 1$$

Wireless Mobile Internet Security, Second Edition. Man Young Rhee.
© 2013 John Wiley & Sons, Ltd. Published 2013 by John Wiley & Sons, Ltd.

Calculation of Y from X is comparatively easy, using repeated squaring, but computation of X from Y is typically far more difficult.

Suppose the user i chooses a random integer X_i and the user j a random integer X_j. Then the user i picks a random number X_i from the integer set $\{1, 2, \ldots, q - 1\}$. The user i keeps X_i secret, but sends

$$Y_i \equiv \alpha^{X_i} \ (\text{mod } q)$$

to the user j. Similarly, the user j chooses a random integer X_j and sends

$$Y_j \equiv \alpha^{X_j} \ (\text{mod } q)$$

to the user i.

Both users i and j can now compute:

$$K_{ij} \equiv \alpha^{X_i X_j} \ (\text{mod } q)$$

and use K_{ij} as their common key.

The user i computes K_{ij} by raising Y_j to the power X_i:

$$K_{ij} \equiv Y_j^{X_i} \ (\text{mod } q)$$
$$\equiv (\alpha^{X_j})^{X_i} \ (\text{mod } q)$$
$$\equiv \alpha^{X_j X_i} \equiv \alpha^{X_i X_j} \ (\text{mod } q)$$

and the user j computes K_{ij} in a similar fashion:

$$K_{ij} \equiv Y_i^{X_j} \ (\text{mod } q)$$
$$\equiv (\alpha^{X_i})^{X_j} \equiv \alpha^{X_i X_j} \ (\text{mod } q)$$

Thus, both users i and j have exchanged a secret key. Since X_i and X_j are private, the only available factors are the public values q, α, Y_i, and Y_j. Therefore, the opponent is forced to compute a discrete logarithm which is considered to be unrealistic, particularly for large primes. Figure 6.1 illustrates the Diffie–Hellman (D-H) key exchange scheme.

When utilizing finite field GF(q), where q is either a prime or $q = 2^k$, it is necessary to ensure the $q - 1$ factor has a large prime, otherwise it is easy to find discrete logarithms in GF(q).

Example 6.1 Consider a prime field Z_q where q is a prime modulus. If α is a primitive root of the modulus q, then α generates the set of nonzero integer modulo q such that $\alpha, \alpha^2, \ldots, \alpha^{q-1}$. These powers of α are all distinct and are all relatively prime to q. Given $\alpha, 1 \leqslant \alpha \leqslant q - 1$, and $q = 11$, all the primitive elements of q are computed as shown in Table 6.1.

For the modulus $q = 11$, the primitive elements are $\alpha = 2, 6, 7$, and 8 whose order is 10, respectively.

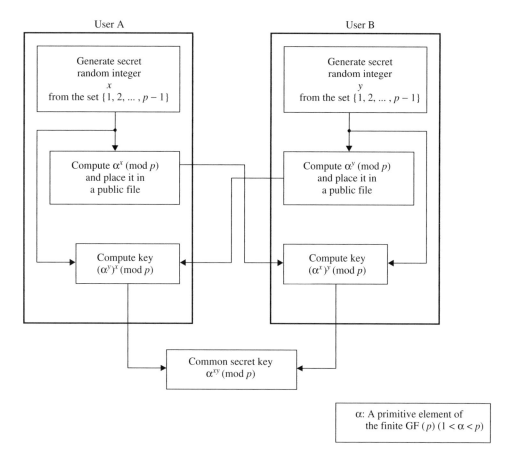

Figure 6.1 The Diffie–Hellman exponential key exchange scheme.

Table 6.1 Powers of primitive element α (over Z_{11})

α	α^2	α^3	α^4	α^5	α^6	α^7	α^8	α^9	α^{10}
1	1	1	1	1	1	1	1	1	1
2	4	8	5	10	9	7	3	6	1
3	9	5	4	1	3	9	5	4	1
4	5	9	3	1	4	5	9	3	1
5	3	4	9	1	5	3	4	9	1
6	3	7	9	10	5	8	4	2	1
7	5	2	3	10	4	6	9	8	1
8	9	6	4	10	3	2	5	7	1
9	4	3	5	1	9	4	3	5	1
10	1	10	1	10	1	10	1	10	1

Example 6.2 Consider a finite field GF(q) of a prime q. Choose a primitive element $\alpha = 2$ of the modulus $q = 11$.

Compute:

$2^{\lambda}(1 \leqslant \lambda \leqslant 10) : \quad 2^1 \quad 2^2 \quad 2^3 \quad 2^4 \quad 2^5 \quad 2^6 \quad 2^7 \quad 2^8 \quad 2^9 \quad 2^{10}$

$2^{\lambda}(\text{mod } 11) \qquad : 2 \quad 4 \quad 8 \quad 5 \quad 10 \quad 9 \quad 7 \quad 3 \quad 6 \quad 1$

To initiate communication, the user i chooses $X_i = 5$ randomly from the integer set $2^{\lambda}(\text{mod } 11) = \{1, 2, \ldots, 10\}$ and keeps it secret. The user i sends

$$Y_i \equiv \alpha^{X_i} \ (\text{mod } q)$$
$$\equiv 2^5 \ (\text{mod } 11) \equiv 10$$

to the user j. Similarly, the user j chooses a random number $X_j = 7$ and sends

$$Y_j \equiv \alpha^{X_j} \ (\text{mod } q)$$
$$\equiv 2^7 \ (\text{mod } 11) \equiv 7$$

to the user i.

Finally, compute their common key K_{ij} as follows:

$$K_{ij} \equiv Y_j^{X_i} \ (\text{mod } q)$$
$$\equiv 7^5 \ (\text{mod } 11) \equiv 10$$

and

$$K_{ji} \equiv Y_i^{X_j} \ (\text{mod } q)$$
$$\equiv 10^7 \ (\text{mod } 11) \equiv 10$$

Thus, each user computes the common key.

Example 6.3 Consider the key exchange problem in the finite field GF(2^m) for $m = 3$. The primitive polynomial $p(x)$ of degree $m = 3$ over GF(2) is $p(x) = 1 + x + x^3$. If α is a root of $p(x)$ over GF(2), then the field elements of GF(2^3) generated by $p(\alpha) = 1 + \alpha + \alpha^3 = 0$ are as shown in Table 6.2.

Suppose users i and j select $X_i = 2$ and $X_j = 5$, respectively. Both X_i and X_j are kept secret, but

$$Y_i \equiv \alpha^{X_i} \ (\text{mod } q) \equiv \alpha^2 \ (\text{mod } 7) \equiv 001$$
$$\text{and} \quad Y_j \equiv \alpha^{X_j} \ (\text{mod } q) \equiv \alpha^5 \ (\text{mod } 7) \equiv 111$$

Table 6.2 Field elements of $GF(2^3)$ for $q = 7$

Power	Polynomial	Vector
1	1	100
α	α	010
α^2	α^2	001
α^3	$1 + \alpha$	110
α^4	$\alpha + \alpha^2$	011
α^5	$1 + \alpha + \alpha^2$	111
α^6	$1 + \alpha^2$	101

are placed in the public file. User i can communicate with user j by taking $Y_j = 111$ from the public file and computing their common key K_{ij} as follows:

$$K_{ij} \equiv (Y_j)^{Xi} \pmod{q}$$

$$\equiv (\alpha^5)^2 \pmod{7} \equiv \alpha^{10} \pmod{7} \equiv \alpha^3 \equiv 110$$

User j computes K_{ij} in a similer fashion:

$$K_{ij} \equiv (Y_i)^{Xj} \pmod{q} \equiv (\alpha^2)^5 \pmod{7} \equiv \alpha^{10} \pmod{7} \equiv \alpha^3 \equiv 110$$

Thus two users i and j arrive at a key K_{ij} in common. These examples are extremely small in size and are intended only to illustrate the technique. So far, we have shown how to calculate the D-H key exchange, the security of which lies in the fact that it is very difficult to compute discrete logarithms for large primes.

This pioneering work relating to the key-exchange algorithm introduced a new approach to cryptography that met the requirements for public-key systems. The first response to the challenge was the development of the RSA scheme which was the only widely accepted approach to the public key encryption. The RSA cryptosystem will be examined in the next section.

6.2 RSA Public-Key Cryptosystem

In 1976, Diffie and Hellman introduced the idea of the exponential key exchange. In 1977 Rivest, Schamir, and Adleman invented the RSA algorithm for encryption and digital signatures which was the first public-key cryptosystem. Soon after the publication of the RSA algorithm, Merkle and Hellman devised a public-key cryptosystem for encryption based on the knapsack algorithm. The RSA cryptosystem resembles the D-H key exchange system in using exponentiation in modula arithmetic for its encryption and decryption, except that RSA operates its arithmetic over the composite numbers. Even though the cryptanalysis was researched for many years for RSA's security, it is still popular and reliable. The security of RSA depends on the problem of factoring large numbers. It is proved that 110-digit numbers are being factored with the power of current factoring technology. To keep RSA's level of security, more than 150-digit values for n will be required. The speed of RSA does not beat DES, because DES is about 100 times faster than RSA in software.

6.2.1   RSA Encryption Algorithm

Given the public key e and the modulus n, the private key d for decryption has to be found by factoring n. Choose two large prime numbers, p and q, and compute the modulus n which is the product of the two primes:

$$n = pq$$

Choose the encryption key e such that e and $\phi(n)$ are coprime, that is, gcd $(e, \phi(n)) = 1$, in which $\phi(n) = (p - 1)(q - 1)$ is called *Euler's totient function*.

Using euclidean algorithm, the private key d for decryption can be computed by taking the multiplicative inverse of e such that

$$d \equiv e^{-1} \pmod{\phi(n)}$$

or $ed \equiv 1 \pmod{\phi(n)}$

The decryption key d and the modulus n are also relatively prime. The numbers e and n are called the *public keys*, while the number d is called the *private key*.

To encrypt a message m, the ciphertext c corresponding to the message block can be found using the following encryption formula:

$$c \equiv m^e \pmod{n}$$

To decrypt the ciphertext c, c is raised to the power d in order to recover the message m as follows:

$$m \equiv c^d \pmod{n}$$

It is proved that

$$c^d \equiv (m^e)^d \equiv m^{ed} \equiv m \pmod{n}$$

due to the fact that $ed \equiv 1 \pmod{\phi(n)}$.

Because Euler's formula is $m^{\phi(n)} \equiv 1 \pmod{n}$, the message m is relatively prime to n such that gcd $(m, n) = 1$. Since $m^{\lambda\,\phi(n)} \equiv 1 \pmod{n}$ for some integer λ, it can be written $m^{\lambda\,\phi(n)+1} \equiv m \pmod{n}$, because $m^{\lambda\,\phi(n)+1} \equiv mm^{\lambda\,\phi(n)} \equiv m \pmod{n}$. Thus, the message m can be restored.

Figure 6.2 and Table 6.3 illustrate the RSA algorithm for encryption and decryption. Using Table 6.3, the following examples are demonstrated.

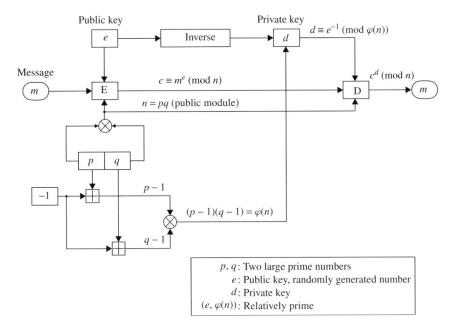

Figure 6.2   RSA public-key cryptosystem for encryption/decryption.

Table 6.3   RSA encryption algorithm

Public key e:

> n (product of two primes p and q (secret integers))
> e (encryption key, relatively prime to $\phi(n) = (p-1)(q-1)$)

Private key d:

> d (decryption key, $d = e^{-1} \pmod{\phi(n)}$)
> $ed \equiv 1 \pmod{\phi(n)}$

Encryption:

> $c \equiv m^e \pmod{n}$, where m is a plaintext.

Decryption:

> $m \equiv c^d \pmod{n}$, where c is a ciphertext.

Example 6.4   If $p = 17$ and $q = 31$ are chosen, then

$$n = pq = 17 \times 31 = 527$$

$$\phi(n) = (p-1)(q-1) = 16 \times 30 = 480$$

If $e = 7$ is chosen, then compute:

$$d \equiv e^{-1} \pmod{\phi(n)} \equiv 7^{-1} \pmod{480} \equiv 343$$

This decryption key d is calculated using the extended euclidean algorithm.

$ed \equiv 7 \times 343 \ (\text{mod } 480) \equiv 2401 \ (\text{mod } 480) \equiv 1$

The public key (e, n) is required for encryption of m. If $m = 2$, then the message m is encrypted as:

$c \equiv m^e \ (\text{mod } n)$

$\equiv 2^7 \ (\text{mod } 527) \equiv 128$

To decipher, the private key d is needed to compute the message as follows:

$m \equiv c^d \ (\text{mod } n)$

$\equiv 128^{343} \ (\text{mod } 527) \equiv 2$

Example 6.5   If $p = 47$ and $q = 71$, then compute

$n = pq = 47 \times 71 = 3337$

$\phi(n) = (p - 1)(q - 1) = 46 \times 70 = 3220$

Choose the encryption key $e = 79$ randomly such that $\gcd(e, \phi(n)) = \gcd(79, 3220) = 1$, that is, e and $\phi(n)$ are relatively prime. Using the extended euclidean algorithm (i.e., $\gcd(e, \phi(n)) = 1 = ed + \phi(n)s$), compute the decryption key d such that

$ed \equiv 1 \ (\text{mod } \phi(n))$

$79d \equiv 1 \ (\text{mod } 3220)$

$3220 = 79 \times 40 + 60$

$79 = 60 + 19$

$60 = 19 \times 3 + 3$

$19 = 3 \times 6 + 1 \rightarrow \gcd(79, 3220) = 1 \ (\text{coprime})$

$1 = 19 - 3 \times 6 = 19 - (60 - 19 \times 3) \times 6$

$= 19 \times 19 - 60 \times 6$

$1 = (79 - 60) \times 19 - 60 \times 6$

$= 79 \times 19 - 60 \times 25$

$1 = 79 \times 19 - (3220 - 79 \times 40) \times 25$

$= 79 \times 1019 - 3220 \times 25$

$(79)(1019) \equiv 1 \ (\text{mod } 3220)$

$d = 1019 \ (\text{privatekey})$

Table 6.4 Two-digit numbers representing each character

Blank	00	E	05	J	10	O	15	T	20	Y	25
A	01	F	06	K	11	P	16	U	21	Z	26
B	02	G	07	L	12	Q	17	V	22		
C	03	H	08	M	13	R	18	W	23		
D	04	I	09	N	14	S	19	X	24		

To encrypt a message $m = 688$ with $e = 79$, compute

$$c \equiv m^e \ (\text{mod } n) \equiv 688^{79} \ (\text{mod } 3337)$$

$$688^2 \ (\text{mod } 3337) \equiv 2827, 688^4 \ (\text{mod } 3337) \equiv 3151$$

$$688^8 \ (\text{mod } 3337) \equiv 1226, 688^{16} \ (\text{mod } 3337) \equiv 1426$$

$$688^{32} \ (\text{mod } 3337) \equiv 1243, 688^{64} \ (\text{mod } 3337) \equiv 18$$

$$c \equiv 688^{79} \ (\text{mod } 3337) \equiv 688^{64+8+4+2+1}$$

$$\equiv 18 \times 1426 \times 3151 \times 2827 \times 688 \ (\text{mod } 3337)$$

$$\equiv 1570 \ (\text{mod } 3337)$$

To decrypt a message, perform the same exponentiation process using the decryption key $d = 1019$ such that

$$m \equiv c^d \ (\text{mod } n) \equiv 1570^{1019} \ (\text{mod } 3337)$$

$$m = (1570)^{512} \times (1570)^{256} \times (1570)^{128} \times (1570)^{64} \times (1570)^{32}$$

$$\times (1570)^{16} \times (1570)^8 \times (1570)^2 \times (1570)$$

$$= 3925000 \ (\text{mod } 3337) \equiv 688$$

Thus, the message is recovered.

To encrypt the message m, break it into a series of m_i-digit blocks, $1 \leqslant i \leqslant n - 1$. Suppose each character in the message is represented by a two-digit number as shown in Table 6.4.

Example 6.6 Encode the message "INFORMATION SECURITY" using Table 6.4.

$$m = (09140615181301200915140019050321180920 25)$$

Choose $p = 47$ and $q = 71$. Then

$$n = pq = 47 \times 71 = 3337$$

$$\phi(n) = (p - 1)(q - 1) = 46 \times 70 = 3220$$

Break the message m into blocks of four digits each:

0914   0615   1813   0120   0915

1400   1905   0321   1809   2025

Choose the encryption key $e = 79$. Then the decryption key d becomes

$$d \equiv e^{-1} \pmod{\phi(n)} \equiv 79^{-1} \pmod{3220} \equiv 1019$$

The first block, $m_1 = 914$, is encrypted by raising it to the power $e = 79$ and dividing by $n = 3337$ and taking the remainder $c_1 = 3223$ as the first block of ciphertext:

$$c_1 \equiv m_1^e \pmod{n}$$
$$\equiv 914^{79} \pmod{3337}$$
$$\equiv 3223$$

Thus, the whole ciphertext blocks c_i, $1 \leqslant i \leqslant 10$, are computed as

3223   3155   1012   1712   1595

2653   0802   2360   0832   1369

To decrypt the first ciphertext $c_1 = 3223$, use the decryption key, $d = 1019$, and compute

$$m_1 \equiv c_1^d \pmod{n}$$
$$\equiv 3223^{1019} \pmod{3337} \equiv 914$$
$$m_2 \equiv c_2^d \pmod{n}$$
$$\equiv 3155^{1019} \pmod{3337} \equiv 615$$

$$\vdots$$

The recreated message of this example is computed as

0914   0615   1813   0120   0915

1400   1905   0321   1809   2025

6.2.2 RSA Signature Scheme

The RSA public-key cryptosystem can be used for both encryption and signatures. Each user has three integers e, d, and n, $n = pq$ with p and q large primes. For the key pair (e, d), $ed \equiv 1 \pmod{\phi(n)}$ must be satisfied. If sender A wants to send signed message c corresponding to message m to receiver B, A signs it using A's private key, computing $c \equiv m^{d_A} \pmod{n_A}$. First A computes

$$\varphi(n_A) \equiv 1 \ cm \ (p_A - 1, q_A - 1)$$

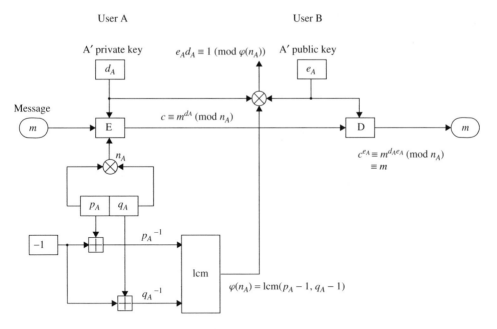

Figure 6.3 The RSA signature scheme.

where 1 cm stands for the least common multiple. The sender A selects his own key pair (e_A, d_A) such that

$$e_A \cdot d_A \equiv 1 \pmod{\varphi(n_A)}$$

The modulus n_A and the public key e_A are published., Figure 6.3 illustrates the RSA signature scheme.

Example 6.7 Choose $p = 11$ and $q = 17$. Then $n = pq = 187$.

Compute $\varphi(n) = 1 \text{ cm } (p - 1, q - 1)$

$$= 1 \text{ cm } (10, 16) = 80$$

Select $e_A = 27$. Then $e_A d_A \equiv 1 \pmod{\varphi(n_A)}$

$$27 d_A \equiv 1 \pmod{80}$$

$$d_A = 3$$

Suppose $m = 55$. Then the signed message is

$$c \equiv m^{d_A} \pmod{187}$$

$$\equiv 55^3 \pmod{187} \equiv 132$$

The message will be recreated as:

$m \equiv c^{eA} \pmod{n}$

$\equiv 132^{27} \pmod{187} \equiv 55$

Thus, the message m is accepted as authentic.

Next, consider a case where the message is much longer. The larger m requires more computation in signing and verification steps. Therefore, it is better to compute the message digest using an appropriate hash function, for example, the SHA-1 (Secure Hash Algorithm). Signing the message digest rather than the message often improves the efficiency of the process because the message digest is usually much smaller than the message.

When the message is assumed to be $m = 75\ 139$, the message digest h of m is computed using the SHA-1 as follows:

$h \equiv H(m) \pmod{n}$

$\equiv H(75\ 139) \pmod{187}$

$\equiv 86a0aab5631e729b0730757b0770947307d9f597$

$\equiv 768587753333627872847426508024461003561962698135$

$\quad \pmod{187}$ (decimal)

The message digest h is then computed as

$h \equiv H(75\ 139) \pmod{187} \equiv 11$

Signing h with A's private key d_A produces

$c \equiv h^{d_A} \pmod{n}$

$\equiv 11^3 \pmod{187} \equiv 22$

Thus, the signature verification proceeds as follows:

$h \equiv c^{eA} \pmod{n}$

$\equiv 22^{27} \pmod{187} \equiv 11$

which shows that verification is accomplished.

In hardware, RSA is about 1000 times slower than DES. RSA is also implemented in smartcards, but these implementations are slower. DES is about 100 times faster than RSA. However, RSA will never reach the speed of symmetric cipher algorithms.

It is known that the security of RSA depends on the problem of factoring large numbers. To find the private key from the public key e and the modulus n, one has to factor n. Currently, n must be larger than a 129 decimal digit modulus. Easy methods to break RSA have not yet been found. A brute-force attack is even less efficient than trying to factor n. RSA encryption and signature verification are faster if you use a low value for e, but can be insecure.

6.3 ElGamal's Public-Key Cryptosystem

ElGamal proposed a public-key cryptosystem in 1985. The ElGamal algorithm can be used for both encryption and digital signatures. The security of the ElGamal scheme relies on the difficulty of computing discrete logarithms over GF(p), where p is a large prime. Prime factorization and discrete logarithms are required to implement the RSA and the ElGamal cryptosystems.

In the RSA cryptosystems, each user has three integers e, d, and n, where $n = pq$ with two large primes p and q, and $ed \equiv 1 (\mathrm{mod}\ \phi(n))$, ϕ being Euler's totient function. User A has a public key consisting of the pair (e_A, n_A) and a private key d_A; similarly, user B has (e_B, n_B) and d_B. To encrypt the message m to B, A uses B's public key for computing the encrypted message (or ciphertext) such that $c \equiv m^{e_B} (\mathrm{mod}\ n_B)$. If A wants to send the signed message to B, A signs the message m using his own private key d_A such that $c \equiv m^{d_A} (\mathrm{mod}\ n_A)$.

To describe the ElGamal system, choose a prime number p and two random numbers, g and x, such that both $g < p$ and $x < p$, where x is a private key. The random number g is a primitive root modulo p. The public key is defined by y, g, and p. Then we compute $y \equiv g^x (\mathrm{mod}\ p)$. To encrypt the message m, $0 < m \leqslant p - 1$, first pick a random number k such that gcd $(k, p - 1) = 1$. The encrypted message (or ciphertext) can be expressed by the pair (r, s) as follows:

$$r \equiv g^k\ (\mathrm{mod}\ p)$$

$$s \equiv (y^k m\ (\mathrm{mod}\ p))\ (m\ (\mathrm{mod}\ p - 1))$$

To decrypt m, divide s by r^x such that $s/r^x \equiv m\ (\mathrm{mod}\ p - 1)$. To sign a given message m, first choose a random number k such that gcd $(k, p - 1) = 1$, and compute $m \equiv xr + ks\ (\mathrm{mod}\ p - 1)$ using the extended euclidean algorithm to solve s. The basic technique for encryption and signature using the ElGamal algorithm as a two-key cryptosystem is described in the following section.

6.3.1 ElGamal Encryption

To generate a key pair, first choose a prime p and two random numbers g and x such that $g < p$ and $x < p$. Then compute

$$y \equiv g^x\ (\mathrm{mod}\ p)$$

The public key is (y, g, p) and the private key is $x < p$.

To encrypt the message $m, 0 \leqslant m \leqslant p - 1$, first choose a random number k such that gcd $(k, p - 1) = 1$. The encrypted message (or ciphertext) is then the following pair (r, s):

$$r \equiv g^k\ (\mathrm{mod}\ p)$$

$$s \equiv (y^k\ (\mathrm{mod}\ p))\ (m (\mathrm{mod}\ p - 1))$$

Note that the size of the ciphertext is double the size of the message. To decrypt the message, divide s by r^x, as shown below:

$$r^x \equiv (g^k)^x \pmod{p}$$

$$s/r^x \equiv y^k m/(g^k)^x \equiv (g^x)^k m/(g^k)^x \equiv m \pmod{p-1}$$

The ElGamal encryption scheme is plotted in Figure 6.4 and Table 6.5.

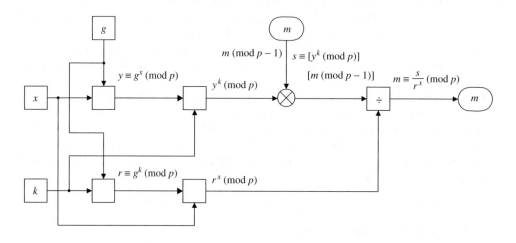

Figure 6.4 The ElGamal encryption scheme.

Table 6.5 The ElGamal encryption algorithm

Public key:

    p (a prime number)
    $g, x < p$ (two random numbers)
    $y \equiv g^x \pmod{p}$
    y, g and p: public key

Private key:

    $x < p$

Enciphering:

    k: a random number such that gcd $(k, p-1) = 1$
    $r \equiv g^k \pmod{p}$
    $s \equiv (y^k \pmod{p})(m \pmod{p-1})$

Deciphering:

    $m \equiv s/r^x \pmod{p}, 0 \leqslant m \leqslant p-1$

Example 6.8   Choose:

$p = 11$ (a prime)

$g = 4$ (a random number such that $g < p$)

$x = 8$ (a private key such that $x < p$)

Then compute:

$y \equiv g^x \pmod{p} \equiv 4^8 \pmod{11} \equiv 9$

The public key is $y = 9, g = 4$, and $p = 11$. The private key $x = 8$ is given above. To encrypt the message $m = 5$, first choose a random number $k = 7$ such that gcd $(k, p - 1) = \gcd(7, 10) = 1$ and then compute:

$r \equiv g^k \pmod{p} \equiv 4^7 \pmod{11} \equiv 5$

$s \equiv (y^k \pmod{p}) \, (m \pmod{p - 1})$

$\quad \equiv (9^7 \pmod{11}) \, (5 \pmod{10}) \equiv 4 \times 5 \equiv 20$

To decipher the message m, first compute:

$r^x \pmod{p} \equiv 5^8 \pmod{11} \equiv 4$

and take the ratio:

$m = s/r^x \pmod{p} \equiv 20/4 \equiv 5$

It thus proves that the message m is completely restored using the ElGamal encryption algorithm (Table 6.5).

6.3.2   ElGamal Signatures

To sign a message m, first choose a random number k such that gcd $(k, p - 1) = 1$ (relatively prime). The public key is described by

$y \equiv g^x \pmod{p}$

where the private key is $x < p$. Let m be a message to be signed, $0 \leqslant m \leqslant p - 1$. Choose first a random number k such that gcd $(k, p - 1) = 1$ (relatively prime). Then compute

$r \equiv g^k \pmod{p}$

The signature for m is the pair (r, s), $0 \leqslant r, s < p - 1$.

$g^m \equiv y^r r^s \pmod{p}$

$\quad \equiv (g^x)^r \, (g^k)^s \pmod{p}$

$\quad \equiv g^{xr + ks} \pmod{p}$

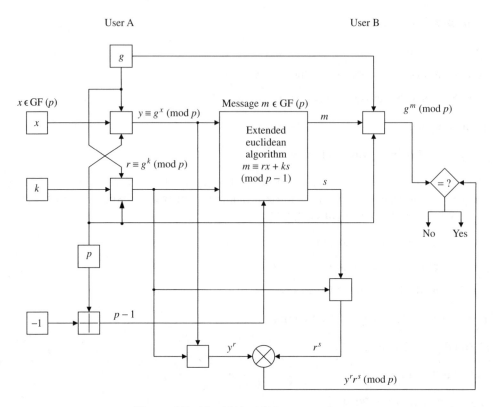

Figure 6.5   The ElGamal signature scheme.

from which

$$m \equiv xr + ks \pmod{p-1}$$

Use the extended euclidean algorithm to solve s. The signature for m is the pair (r, s). The random number s should be kept secret. To verify a signature, confirm that:

$$y^r r^s \pmod{p} \equiv g^m \pmod{p}$$

Figure 6.5 illustrates the ElGamal signature scheme based on Table 6.6.

Example 6.9   To sign a message m, first choose a prime $p = 11$ and two random numbers $g = 7$ and $x = 3$, where $x < p$ is a private key.
Compute:

$$y \equiv g^x \pmod{p} \equiv 7^3 \pmod{11} \equiv 2$$

The public key is $y = 2$, $g = 7$, and $p = 11$.

Table 6.6   The ElGamal signature algorithm

Public key:

> p (a prime number)
> $g < p$ (a random number)
> $y \equiv g^x \pmod{p}$ where $x < p$

Private key:
> k: a random number
> $r \equiv g^k \pmod{p}$
> s: compute from $m \equiv xr + ks \pmod{p-1}$

Verifying:

> Accept as valid if
> $y^r r^s \pmod{p} \equiv g^m \pmod{p}$

To authenticate $m = 6$, choose a random number $k = 7$ such that gcd $(k, p-1) =$ gcd$(7, 10) = 1$. Compute:

$r \equiv g^k \pmod{p} \equiv 7^7 \pmod{11} \equiv 6$

$m \equiv xr + ks \pmod{p-1}$ (euclidean algorithm to solve for s)

$6 \equiv 3 \times 6 + 7s \pmod{10}$

$7s \equiv -2 \pmod{10} \equiv 28 \pmod{10}$

$\ s \equiv 4 \pmod{10}$

The signature is the pair of $r = 6$ and $s = 4$.
    To verify a signature, it must be confirmed that

$$y^r r^s \pmod{p} \equiv g^m \pmod{p}$$

$$(2^6)\,(6^4) \pmod{11} \equiv 7^6 \pmod{11}$$

$$81 \pmod{11} \equiv 15 \pmod{11}$$

$$4 \pmod{11} \equiv 4 \pmod{11}$$

6.3.3 ElGamal Authentication Scheme

The ElGamal signature or authentication scheme looking at another angle is described in the following.
    The sender chooses a finite field GF(p) where p is a prime. Let g be a primitive element of GF(p). First choose two random integers g and x such that $g < p$ and $x < p$. A key x is kept secret by both the sender and the receiver. Let m denote a message which is relatively prime to p. Then compute:

$u \equiv g^m \pmod{p}$

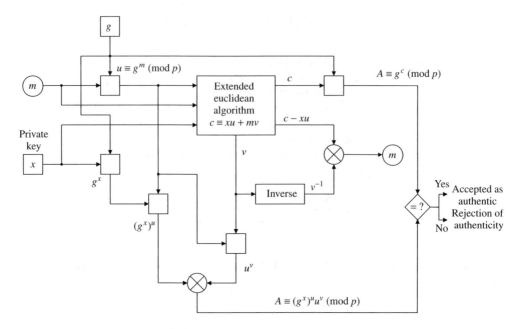

Figure 6.6   The ElGamal authentication scheme.

Let c denote a ciphertext such that gcd $(c, p) = 1$.

Using the extended euclidean algorithm, the following congruence is to solve for v:

$$c \equiv xu + mv \pmod{p - 1}$$

or $v \equiv m^{-1} (c - xu) \pmod{p - 1}$

To authenticate the ciphertext c, the signed cryptogram (c, u, v) is transmitted to the receiver. Upon receipt of (c, u, v), the receiver computes

$$A \equiv (g^x)^u u^v \pmod{p}$$
$$\equiv g^{c-mv} (g^m)^v \pmod{p}$$
$$\equiv g^c \pmod{p}$$

Thus, the ciphertext c is accepted as authentic if $A \equiv g^c \pmod{p}$. Once this ciphertext has been accepted, the message m is recovered by:

$$m \equiv v^{-1} (c - xu) \pmod{p - 1}$$

The ElGamal authentication scheme is shown in Figure 6.6. The ElGamal authentication algorithm given in Table 6.7 is illustrated by the following example.

Example 6.10   Take the finite field GF(11). Then the set of primitive elements of GF(11) is $\{2, 6, 7, 8\}$. Choose a primitive element $g = 7$ from the set. Define the public

Table 6.7   The ElGamal authentication algorithm

Sender

    p (a prime integer)
    $g < p$ (a primitive element of GF(p))
    $u \equiv g^m$ (mod p) where $m < p$ is a message
    $x < p$ (a private key)
    c (ciphertext)
    $c \equiv xu + mv$: solve for v
    (c, u, v): (the signed cryptogram to be transmitted)

Receiver

    $A \equiv (g^x)^u u^v$ (mod p)

Verifying:

    Accept as valid if and only if $A \equiv g^c$ (mod p)

Decryption:

    $m \equiv v^{-1}(c - xu)$ (mod $p - 1$)

key as $(g, p) = (7, 11)$ and $x = 5$ as the chosen private key which is shared by both the sender and the receiver. If the sender now wants to transmit a message $m = 3$ such that $\gcd(m, p) = \gcd(3, 11) = 1$, then compute first:

$$u \equiv g^m \text{ (mod } p) \equiv 7^3 \text{ (mod 11)} \equiv 2$$

Next, compute v by solving the following congruence:

$$c \equiv xu + mv \text{ (mod } p - 1)$$

$$7 \equiv 5 \times 2 + 3v \text{ (mod 10)}$$

$$3v \equiv 7 \text{ (mod 10)}$$

$$v \equiv 9 \text{ (mod 10)}$$

where $c = 7$ is assumed.

Send the signed cryptogram $(c, u, v) = (7, 2, 9)$ to the receiver. At the receiving end, compute:

$$A \equiv (g^x)^u u^v \text{ (mod } p)$$
$$\equiv (7^5)^2 2^9 \text{(mod 11)}$$
$$\equiv (10^2)(2^9) \text{ (mod 11)} \equiv 6$$

and $A \equiv g^c \text{ (mod } p) \equiv 7^7 \text{ (mod 11)} \equiv 6$

Thus, the cryptogram $(7, 2, 9)$ is accepted, and $c = 7$ is authentic. Finally, the message is restored in the following manner:

$m \equiv v^{-1} (c - xu) \pmod{p - 1}$

$\equiv 9^{-1}(7 - 5 \times 2)(\bmod\ 10)$

$\equiv (9^{-1}) (7) \pmod{10} \equiv 3$

The message $m = 3$ has been completely recovered.

6.4 Schnorr's Public-Key Cryptosystem

In 1990, Schnorr introduced his authentication and signature schemes based on discrete logarithms.

6.4.1 Schnorr's Authentication Algorithm

First choose two primes, p and q, such that q $(1 < q < p - 1)$ is a prime factor of $p - 1$. To generate a public key, choose $a \neq 1$ such that $a \equiv h^{(p-1)/q} \pmod{p}$, that is, $a^q \equiv h^{p-1} \pmod{p}$. If h is relatively prime to p, by Fermat's theorem it can then be written as $h^{p-1} \equiv 1 \pmod{p}$. As a result, we have $a^q \equiv 1 \pmod{p}, 1 < a < p - 1$. All these numbers, p, q, and a, can be freely published and shared with a group of users. To generate a key pair, choose a random number $s < q$ which is used as the private key. Next, compute $\lambda \equiv a^{-s} \pmod{p}$ which is the public key.

Now, user A picks a random number $r < q$ and computes $x \equiv a^r \pmod{p}$. User B picks a random number t and sends it to user A, where $t \in (0, 1, 2, \ldots, 2^v - 1)$ indicates the security level. Schnorr recommends the value of $v = 72$ for sufficient security. User A computes $y \equiv r + st \pmod{q}$ and sends it to user B. Thus, user B tests verification of authenticity such that $x \equiv a^y \lambda^t \pmod{p}$. Figure 6.7 illustrates Schnorr's authentication scheme, and Table 6.8 shows the related algorithm.

Example 6.11   Choose two primes $p = 23$ and $q = 11$ such that $q = 11$ is a prime factor of $p - 1 = 22$. Choose $a = 3$ satisfying $a^q \equiv 1 \pmod{p}$, that is, $3^{11} \equiv 1 \pmod{23}$. Choose $s = 8 < q$ as the private key and compute the public key such that $\lambda \equiv a^{-s} \pmod{p} \equiv 3^{-8} \pmod{23}$. Compute the multiplicative inverse of $a = 3 : aa^{-1} \equiv 1 \pmod{p}$, $3a^{-1} \equiv 1 \pmod{23}$ from which $a^{-1} = 8$. Thus, $\lambda \equiv 8^8 \pmod{23} \equiv 4$.

The sender picks $r = 5 < q$ and computes:

$x \equiv a^r \pmod{p}$

$\equiv 3^5 \pmod{23} \equiv 13$

The receiver sends $t = 15$ to the sender and the sender computes:

$y \equiv r + st \pmod{q}$

$\equiv (5 + 8 \times 15)(\bmod\ 11)$

$\equiv 125 \pmod{11} \equiv 4$

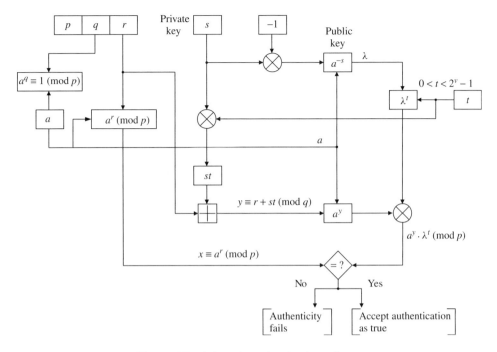

Figure 6.7   Schnorr's authentication scheme.

Table 6.8   Schnorr's authentication algorithm

Preprocessing:

   Choose two primes, p and q, such that q is a prime factor of $p - 1$
   Choose a such that $a^q \equiv 1 \pmod{p}$

Key generation:

   Choose a random number $s < q$ (private key)
   Compute $\lambda \equiv a^{-s} \pmod{p}$ (public key)

User A	*User B*
Choose a random number $r < q$	
Compute $x \equiv a^r \pmod{p}$	Pick a random number t such that $0 < t < 2^v - 1$
$\leftarrow$	Send t to user A
Compute $y \equiv r + st \pmod{q}$	
Send y to user B $\rightarrow$	Verify that $x \equiv a^y \lambda^t \pmod{p}$

To verify $x \equiv a^y \cdot \lambda^t \pmod{p} \equiv 13$, compute:

$x \equiv (3^4)(4^{15}) \pmod{23}$

$\equiv 12 \times 3 \pmod{23} \equiv 13$

Since $a^r \pmod{p} \equiv a^y \lambda^t \pmod{p} \equiv 13$, the authentication is accepted.

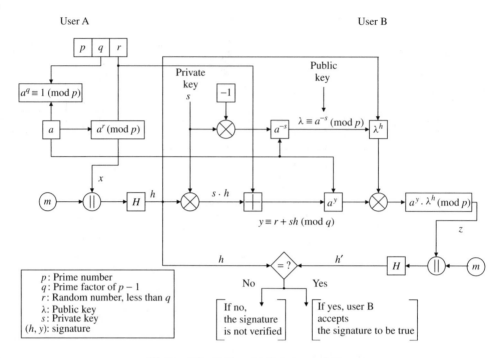

Figure 6.8　Schnorr's signature scheme.

6.4.2　Schnorr's Signature Algorithm

For a digital signature, user A concatenates the message m and x and computes the hash code:

$$h \equiv H(m\|x)$$

User A sends the signature (h, y) to user B. User B computes $z \equiv a^y \lambda^h \pmod{p}$ and confirms whether hashing the concatenation of m and z yields:

$$h' \equiv H(m\|z)$$

If $h = h'$, then user B accepts the signature as valid.

For the same level of security, Schnorr's signature algorithms are shorter than the RSA ones. Also, Schnorr's signatures are much shorter than ElGamal signatures. Figure 6.8 and Table 6.9 illustrate Schnorr's signature algorithm.

Example 6.12　First choose two primes $p = 29$ and $q = 7$ such that $q|p - 1$, that is, q is a prime factor of $p - 1$. Determine $a = 7$ in order to meet the requirement of $a^q \equiv 1$

Table 6.9   Schnorr's signature algorithm

Preprocessing stage and the two key pair are the same.

User A	User B
Choose $r < q$ (a random number)	
Compute $x \equiv a^r \pmod{p}$	
Concatenate m and x, that is, $m\|x$ and hash	
such that $h = H(m\|x)$	
Compute $y = r + sh \pmod{q}$	
Send the signature (h, y) to user B $\rightarrow$	Compute $z \equiv a^y \lambda^h \pmod{p}$
	Concatenate m and z and hash:
	$h' = H(m\|z)$
	If the two hash values match ($h = h'$),
	then user B accepts the
	signature as valid

$\pmod{p}$ such that $7^7 \equiv 823\ 543 \equiv 1 \pmod{29}$. Pick a private key $s = 4$ such that $s < q$ and compute the public key as follows:

$$\lambda \equiv a^{-s} \pmod{p}$$

$$\equiv 7^{-4} \pmod{29} \equiv 24$$

User A

Choose a random number $r = 5 < q$ and then compute:

$$x \equiv a^r \pmod{p}$$

$$\equiv 7^5 \pmod{29} \equiv 16$$

Concatenate m and x and hash $m\|x$ such that

$$h \equiv H(m\|x) = H(12\ 345\|16)$$

where the message $m = 12\ 345$ is assumed. To produce the message digest $h = H(m\|x)$, use the SHA which is closely modeled on MD4. Utilizing SHA for h yields a 160-bit message digest as the output, as follows:

$$h \equiv H\ (m\|x) \pmod{q} \equiv H(12\ 345\|16) \pmod{7}$$

$$= \text{a11784b83ea003cd66491c7e1de07296d9d9242c (hexadecimal)}$$

$$= 919671992759145855242593220263016201851705566252$$

$$\pmod 7 \text{ (decimal)}$$

$$\equiv 5$$

User A computes $y \equiv r + sh \pmod{q}$:

$y \equiv (5 + 4 \times 5) \pmod 7 \equiv 25 \pmod 7 \equiv 4$

Send signature $(h, y) = (5, 4)$ to user B. User B first computes:

$z \equiv a^y \cdot \lambda^h \pmod p$

$\equiv 7^4 \times 24^5 \pmod{29}$

$\equiv (23 \times 7) \pmod{29}$

$\equiv 16$

Concatenate $m = 12\ 345$ and z and hash it as follows:

$h' \equiv H(m\|z) \pmod q$

$\equiv H(12\ 345\|16) \pmod 7$

$\equiv 5$

which is identical to h. Therefore, user B accepts the signature as valid because $h = h'$.

The next example demonstrates how to solve the problem, making use of the MD5 algorithm in order to compute the 128-bit message digest. The source code of the MD5 program can be obtained from ftp.funet.fi:/pub/crypt/hash/mds/md5.

Example 6.13   If two primes $p = 23$ and $q = 11$ are given, then $a = 9$ is determined. Choose a private key $s = 4$, a random number $r = 7$, and the message $m = 135$.

Key generation

Private key: $s = 4$

Public key: $\lambda \equiv a^{-s} \pmod p$

$\equiv 9^{-4} \pmod{23} \equiv 4$

User A

Compute $x \equiv a^r \pmod p$

$\equiv 9^7 \pmod{23} \equiv 4$

Using the MD5 algorithm, compute the message digest:

$h \equiv H(m\|x) \pmod q$

$\equiv H(135\|4) \pmod{11}$

$h \equiv af4732711661056eadbf798ba191272a$ (hexadecimal)

$\equiv 232984575419504758889249578349365372714 \pmod{11}$

$\equiv 0$

Using $h = 0, y \equiv r + sh \pmod{q}$ becomes $y \equiv 7 \pmod{11}$.

Send the signature $(h, y) = (0, 7)$ to user B.

User B

When user B receives the signature (h, y), compute:

$$z \equiv a^y \lambda^h \pmod{p}$$

$$\equiv 9^7 \pmod{23} \equiv 4$$

Applying MD5 to $h' \equiv H(m \| z) \pmod{q} \equiv H(135 \| 4) \pmod{11}$, we have

$$h' = af\,4732711661056eadbf\,798ba\,191272a$$

Thus, user B confirms verification of $h' \pmod{11} \equiv h \pmod{11} \equiv 0$.

6.5 Digital Signature Algorithm

In 1991, the National Institute of Standards and Technology (NIST) proposed the DSA for federal digital signature applications. The proposed new Digital Signature Standard (DSS) uses a public-key signature scheme to verify to a recipient the integrity of data received and the identity of the sender of the data.

DSA provides smartcard applications for digital signature. Key generation in DSA is faster than that in RSA. Signature generation has the same level of speed as RSA, but signature verification is much slower than RSA.

Many software companies, such as IBM, Microsoft, Novell, and Apple, that have already licensed the RSA algorithm protested against the DSS. Many companies wanted NIST to adopt ISO/IEC 9796 for use instead of RSA as the international digital signature standard.

The DSA is based on the difficulty of computing discrete logarithms, and originated from schemes presented by ElGamal and Schnorr. The public key consists of three parameters, p, q, and g, and is common to a group of users. Choose q of a 160-bit prime number and select a prime number p with $512 < p < 1024$ bits such that q is a prime factor of $p - 1$. Next, choose $g > 1$ to be of the form $h'^{(p-1)/q} \pmod{p}$ such that h' is an integer between 1 and $p - 1$.

With these three numbers, each user chooses a private key x in the range $1 < x < q - 1$ and the public key y is computed from x as $y \equiv g^x \pmod{p}$. Recall that determining x is computationally impossible because the discrete logarithm of y to the base $g \pmod{p}$ is difficult to calculate.

To sign a message m, the sender computes two parameters, r and s, which are functions of $(p, q, g,$ and $x)$, the message digest $H(m)$, and a random number $k < q$. At the receiver, verification is performed as shown in Table 6.10. The receiver generates a quantity v that is a function of parameters $(x, y, r, s^{-1},$ and $H(m))$.

When a one-way hash function H operates on a message m of any length, a fixed-length message digest (hash code) h can be produced such that $h = H(m)$. The message digest h to the DSA input computes the signature for the message m. Signing the message digest

Table 6.10 DSA signatures

Key pair generation:

p: a prime number between 512 and 1024 bits long
q: a prime factor of $p - 1$, 160 bits long
$g \equiv h^{'(p-1)/q}$ (mod p) > 1, and $h^{'} < p - 1$
$(p, q$ and $g)$: public parameters
$x < q$: the private key, 160 bits long
$y \equiv g^x$ (mod p): the public key, 160 bits long

Signing process (sender):

$k < q$: a random number
$r \equiv (g^k \bmod p)$ (mod q)
$s \equiv k^{-1} (h + xr)$ (mod q), $h = H(m)$ is a one-way hash function of the message m
(r, s): signature

Verifying signature (receiver):

$w \equiv s^{-1}$ (mod q)
$u1 \equiv h \times w$ (mod q)
$u2 \equiv r \times w$ (mod q)
$v \equiv (g^{u1}y^{u2}$ (mod p)) (mod q)

If $v = r$, then the signature is verified.

---

rather than the message itself often improves the efficiency of the signature process, because the message digest h is usually much smaller than the message m. The SHA is called *secure* because it is designed to be computationally impossible to recover a message corresponding to a given message digest. Any change to a message in transit will result in a different message digest, and the signature will fail to verify. The structure of the DSA algorithm is illustrated in Figure 6.9.

Example 6.14 Choose $p = 23$ and $q = 11$ such that q is a prime factor of $p - 1$. Choose $h^{'} = 16 < p - 1$ such that $g \equiv 16^2$ (mod 23) $\equiv 3 > 1$. Choose the private key $x = 7 < q$ and compute the public key $y \equiv g^x$ (mod p) $\equiv 3^7$ (mod 23) $\equiv 2$.
Sender: (signing)
Choose $k = 5$ such that $k < q = 11$ and compute the signatures (r, s) as follows:

$$r \equiv (g^k \bmod p) \text{ (mod } q)$$

$$\equiv (3^5 \bmod 23) \text{ (mod } 11) \equiv 13 \text{ (mod } 11) \equiv 2$$

Assume that $h = H(m) = 10$ and compute:

$$s \equiv k^{-1} (h + xr) \text{ (mod } q)$$

$$\equiv 5^{-1} (10 + 7 \times 2) \text{ (mod } 11) \equiv (9 \times 24) \text{ (mod } 11) \equiv 216 \text{ (mod } 11) \equiv 7$$

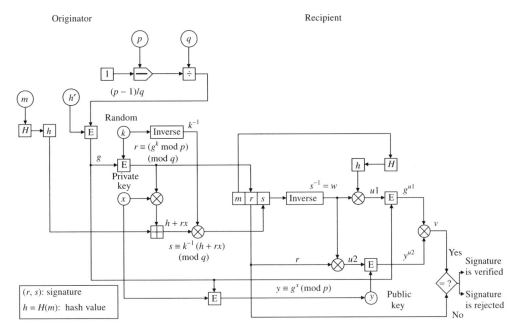

Figure 6.9   DSA digital signature scheme.

where the multiplicative inverse k^{-1} is:

$k \cdot k^{-1} \equiv 1 \pmod{q}$

$5k^{-1} \equiv 1 \pmod{11}$ from which $k^{-1} = 9$

Receiver: (verifying)
Compute:

$w \equiv s^{-1} \pmod{q}$

$\equiv 7^{-1} \pmod{11} \equiv 8$

$u1 \equiv (h \times w) \pmod{q}$

$\equiv (10 \times 8) \pmod{11} \equiv 3$

$u2 \equiv (r \times w) \pmod{q}$

$\equiv (2 \times 8) \pmod{11} \equiv 5$

$v \equiv ((g^{u1} \times y^{u2}) \bmod p) \pmod{q}$

$\equiv ((3^3 \times 2^5) \bmod 23) \pmod{11}$

$\equiv (864 \pmod{23}) \pmod{11} \equiv 13 \pmod{11} \equiv 2$

Since $v = r = 2$, the signature is verified.

6.6 The Elliptic Curve Cryptosystem (ECC)

The Elliptic Curve Cryptosystem (ECC) was introduced by Neal Koblitz and Victor Miller in 1985. The elliptic curve (EC) discrete logarithm problem appears to be substantially more difficult than the existing discrete logarithm problem. Considering they have equal levels of security, ECC uses smaller parameters than the conventional discrete logarithm systems.

In this section, we first present the concept of an EC and then discuss its applications to existing public-key algorithms. Finally, we will look at cryptographic algorithms with ECs over the prime or finite fields.

6.6.1 Elliptic Curves

ECs have been studied for many years. ECs over the prime field Z_p or the finite field $GF(2^m)$ are particularly interesting because they provide a way of constructing cryptographic algorithms. ECs have the potential to provide faster public-key cryptosystem with smaller key sizes.

Elliptic Curves over Prime Field Z_p

Figure 6.10 shows the EC $y^2 = x^3 + ax + b$ defined over Z_p where $a, b \in Z_p$. Z_p is called a *prime field* if and only if $p > 3$ is an odd prime. An EC can be made into an abelian group with all points on an EC, including the point at infinity O under the condition of $4a^3 + 27b^2 = 0 \pmod{p}$.

Computation over Prime Field Z_p

If the points on an EC $y^2 = x^3 + ax + b$ over Z_p are represented by $P(x, y)$, $Q(x_2, y_2)$, and $R(x_3, y_3)$, then two cases can be considered:

- $P \neq Q$ (addition of two points on EC).
- $P = Q$ (doubling of two points on EC).

Consider the curve where $P \neq Q$ as shown in Figure 6.10

$P + Q = R$

First draw a linear curve $y = \alpha x + c$ (where $\alpha = (y_2 - y_1)/(x_2 - x_1)$); $c = y_1 - \alpha x_1$) passing through two points P and Q and find the intersection point R. As shown in Figure 6.10, draw a straight line vertically (perpendicularly) across the x-axis so that it meets the third point $R(x_3, y_3)$, which is the sum of P and Q.

EC: $y^2 = x^3 + ax + b$ expresses

$$(\alpha x + c)^2 = x^3 + ax + b \tag{6.1}$$

$$x^3 - \alpha^2 x^2 + (a - 2\alpha c)x + (b - c^2) = 0 \tag{6.2}$$

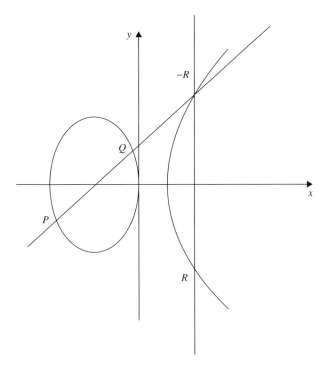

Figure 6.10   An elliptic curve.

Consider a cubic equation with three roots:

$$(x - x_1)(x - x_2)(x - x_3) = 0 \tag{6.3}$$

From which

$$x^3 - (x_1 + x_2 + x_3)x^2 + (x_1x_2 + x_2x_3 + x_3x_1)x - x_1x_2x_3 = 0 \tag{6.4}$$

Comparing Equations 6.1 and 6.2 results in

$$\alpha^2 = x_1 + x_2 + x_3$$
$$x^3 = \alpha_2 - x_1 - x_2$$

Using

$$y_3 = \alpha x_3 + c$$
$$= \alpha x_3 + (y_1 - \alpha x_1)$$
$$= \alpha(x_3 - x_1) + y_1$$
$$-y_3 = \alpha(x_1 - x_3) - y_1$$

If x does have a constant value A, EC becomes

$$y^2 = A^3 + aA + b = B$$

or

$$y^2 - B = 0 \quad (B \text{ is another constant value})$$

Thus, y will have two roots:

$$-R = (x_3, y_3)$$
$$R = (x_3, -y_3)$$

Since the third point $R = (x_3, y_3)$, it results in

$$-y_3 = y_3$$

Finally, we have

$$y_3 = \alpha(x_1 - x_3) - y_1$$

This can be summed up as

$$\boxed{\begin{aligned} &P \neq Q : \alpha = \frac{y_2 - y_1}{x_2 - x_1} \\ &x_3 = \alpha - x_1 - x_2 \\ &y_3 = \alpha(x_1 - x_3) - y_1 \end{aligned}}$$

Consider the case $P = Q$ as shown in Figure 6.11
$P = Q$ (doubling of two points on EC)

$$y^2 = x^3 + ax + b$$

Differentiating with respect to x yields

$$2y\frac{dy}{dx} = 3x^2 + a$$

$$\frac{dy}{dx} = \frac{3x^2 + a}{2y}$$

Considering $\beta = \frac{dy}{dx}$, we have $\beta = \frac{3x^2 + a}{2y}$
Since $y = \beta x + c$,

$$(\beta x + c)^2 = x^3 + ax + b \tag{6.5}$$

$$x^3 - \beta^2 x^2 + (a - 2\beta c)x + b - c^2 = 0 \tag{6.6}$$

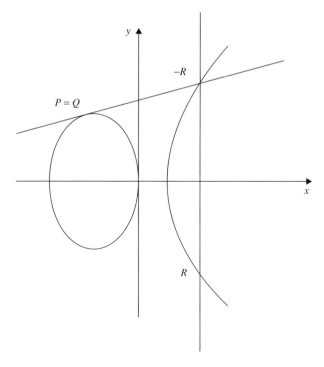

Figure 6.11   The doubling of an elliptic curve point.

Comparing Equation 6.6 with Equation 6.4 results in $\beta^2 = x_1 + x_2 + x_3 = 2x_1 + x_3$ (by doubling)

$x_3 = \beta^2 - 2x_1$

Since $y_3 = \beta x_3 + c = \beta x_3 + (y_1 - \beta x_1)$

$y_3 = \beta(x_3 - x_1)y_1$

Thus, the case of $P = Q$ can be summed up as shown below:

$$\beta = \frac{3x_1^2 + a}{2y_1}$$

$x_3 = \beta^2 - 2x_1$

$y_3 = \beta(x_1 - x_3) - y_1$

Example 6.15   Let $p = 17$. Choose $a = 1$ and $b = 5$ such that the EC over Z_{17} becomes $y^2 \equiv x^3 + x + 5 \pmod{17}$.

$4a^3 + 27b^2 = 4 + 675 = 679 \equiv 16 \pmod{17}$

Hence, the given equation is indeed an EC.

1. Let $P = (3, 1)$ and $Q = (8, 10)$ be two points on the EC. Then $P + Q = R(x_3, y_3)$ is computed as follows:

$$P + Q = (3, 1) + (8, 10)$$

$$x_3 = \left(\frac{y_2 - y_1}{x_2 - x_1} \right)^2 - x_1 - x_2$$

$$= \left(\frac{9}{5} \right)^2 - 3 - 8$$

Since 9×5^{-1} (mod 17) $= 9 \times 7$ (mod 17) $= 12$, it gives

$$x_3 = (12^2 - 3 - 8) \text{ (mod 17)} \equiv 14$$

$$y_3 = -1 + \left(\frac{9}{5} \right) \times (3 - 14) = -1 + 12 \times (-11) = -133 \text{ (mod 17)} \equiv 3$$

Hence, $P + Q = R(14, 3)$.

2. Let $P = (3, 1)$. Then $2P = P + P = (x_3, y_3)$ is computed as follows:

$$2P = (3, 1) + (3, 1)$$

$$x_3 = \left(\frac{3x_1^2 + a}{2y_1} \right)^2 - 2x_1$$

$$= \left(\frac{27 + 1}{2} \right)^2 - 6$$

$$= 14^2 - 6 = 196 - 6 = 190 \text{(mod 17)} \equiv 3$$

and

$$y_3 = -y_1 + \left(\frac{3x_1^2 + a}{2y_1} \right)(x_1 - x_3)$$

$$= -1 + 14(3 - 3) = -1 \text{(mod 17)} \equiv 16$$

Hence, $2P = (3, 16)$.

If P is an odd prime, $0 < z < p$, and $\gcd(z, p) = 1$, then z is called a *quadratic residue* modulo p if and only if $y^2 \equiv z$ (mod p) has a solution for some y; otherwise, z is called a *quadratic nonresidue*.

For example, the quadratic residues modulo 13 are determined as follows:

$$Z_{13}^* = \{1, 2, 3, \ldots, 12\}$$

The square of the integers in Z_{13}^* for modulo 13 is computed as:

$$\{1^2, 2^2, 3^2, \ldots, 11^2, 12^2\} \pmod{13} = \{1, 3, 4, 9, 10, 12\}$$

Hence, the quadratic nonresidues modulo 13 are $\{2, 5, 6, 7, 8, 11\}$. Now you can see that the set $Z_{13}^* = \{1, 2, 3, \ldots, 12\}$ is equally divided into quadratic residues and nonresidues. In general, there are precisely $(p-1)/2$ quadratic residues and $(p-1)/2$ quadratic non-residues of p.

Euler's Criterion

Let p be an odd prime and $\gcd(z, p) = 1$. Using Fermat's theorem $z^{p-1} \equiv 1 \pmod{p}$, or $z^{p-1} - 1 \equiv 0 \pmod{p}$, it gives $[z^{(p-1)/2} - 1][z^{(p-1)/2} + 1] \equiv 0 \pmod{p}$ from which z is a quadratic residue of p if $z^{(p-1)/2} \equiv 1 \pmod{p}$ and a quadratic nonresidue of p if and only if $z^{(p-1)/2} \equiv -1 \pmod{p}$.

Legendre Symbol (z/p)

If $p > 2$ is a prime, $0 < z < p$, and $\gcd(z, p) = 1$, the Legendre symbol (z/p) is a characteristic function of the set of quadratic residues modulo p as follows:

$$\left(\frac{z}{p}\right) = \begin{cases} 1 & \text{if } z \text{ is a quadratic residue of } p \\ -1 & \text{if } z \text{ is a quadratic nonresidue of } p \end{cases}$$

Example 6.16   Let $p = 17, a = 6$, and $b = 5$. Then the EC is defined as $y^2 \equiv x^3 + 6x + 5$ over Z_{17}. Note that $4a^3 + 27b^2 = 1539 \pmod{17} \equiv 9$, so the given EC is indeed an EC. The points in $EC(Z_{17})$ are $\{0\} \cup \{(2, 5), (2, 12), \ldots, (16, 10)\}$. Let us first determine the points on EC. Compute $y^2 = x^3 + 6x + 5 \pmod{17}$ for each possible $x \in Z_{17}$. It will be necessary to check whether or not $z \equiv x^3 + 6x + 5 \pmod{17}$ is a quadratic residue for a given value of x. If z is a quadratic residue, then y can be computed by solving $y^2 \equiv z \pmod{17}$.

For $x = 0$, $z = 5$. Hence, $5^{(p-1)/2} \pmod{z} \equiv 5^8 \pmod{17} \equiv 16 \pmod{17} \equiv -1$ (quadratic nonresidue).

For $x = 1$, $z = 12$. Hence, $12^8 \pmod{17} \equiv 16 \pmod{17} \equiv -1$ (quadratic nonresidue).

For $x = 2$, $z = 25$. Hence, $25^8 \pmod{17} \equiv 1$ (quadratic residue).

Then, solving $y^2 \equiv 25 \pmod{17}$, we obtain $y = 5$ and $y = 12$.

Two points on the EC are found as (x, y): (2, 5) and (2, 12).

Check: $5^2 \pmod{17} = 25 \pmod{17} \equiv 8$ and $12^2 \pmod{17} = 144 \pmod{17} \equiv 8$.

Hence, $y = 5$ and $y = 12$ are checked as two solutions.

Continuing in this way, the quadratic residues and the remaining points on the EC can be computed as shown in Table 6.11.

Let EC be an elliptic curve over z_p. Hasse states that the number of points on an EC, including the point at infinity O, is $\#EC(Z_p) = p + 1 - t$ where $|t| \leqslant 2\sqrt{p}$. $\#EC(Z_p)$ is called the *order of EC* and t is called the *trace of EC*.

Table 6.11 Quadratic residues and points on EC $y^2 = x^3 + 6x + 5 = z$ over z_{17}

x	z (mod 17)	Quadratic residue $z^{(p-1)/2} \equiv 1$ or $(z/p) = 1$	Point (x, y) on EC
0	5	−1	–
1	12	−1	–
2	8	1	(2, 5) (2, 12)
3	16	1	(3, 4) (3, 13)
4	8	1	(4, 5) (4, 12)
5	7	−1	–
6	2	1	(6, 6) (6, 11)
7	16	1	(7, 4) (7, 13)
8	4	1	(8, 2) (8, 15)
9	6	−1	–
10	11	−1	–
11	8	1	(11, 2) (11, 15)
12	3	−1	–
13	2	1	(13, 6) (13, 11)
14	11	−1	–
15	2	1	(15, 6) (15, 11)
16	15	1	(16, 7) (16, 10)

Example 6.17 Let EC be the elliptic curve $y^2 \equiv x^3 + x + 6$ over Z_{11}. All points on EC can be determined as:

$$EC(z_{11}) = \{(2,4), (2,7), (3,5), (3,6), (5,2), (5,9),$$
$$(7,2), (7,9), (8,3), (8,8), (10,2), (10,9)\} \cup \{O\}$$

Any point other than the point at infinity can be a generator G of EC. If we pick $G = (8,3)$ as the generator, the multiples of G can be computed as follows:

When $P = Q, 2G = (8,3) + (8,3)$. Using $x_3 = \beta^2 - 2x_1$ and $y_3 = -y_1 + \beta(x_1 - x_3)$ where $\beta = [(3x_1^2 + a)/(2y_1)](\text{mod } p)$, $2G(x_3, y_3)$ is computed as follows:

Since $\beta = \dfrac{3 \times 8^2 + 1}{2 \times 3}(\text{mod } 11) \equiv 1, x_3 = 1^2 - 16 \ (\text{mod } 11) \equiv 7$

and $y_3 = -3 + 1(8 - 7) \ (\text{mod } 11) \equiv 9$.

Hence, $2G = (7,9)$.

For $3G = 2G + G = (7,9) + (8,3)$, it may be expressed as $P = 2G$ and $Q = G$. Since $P \neq Q$, we use $x_3 = \beta^2 - x_1 - x_2$ and $y_3 = -y_1 + \beta(x_1 - x_3)$ where $\beta = (y_2 - y_1)/(x_2 - x_1)$. Compute β first as: $\beta = [(9 - 3)/(7 - 8)](\text{mod } 11) \equiv 5$. Thus, $x_3 = 5^2 - 7 - 8$ (mod 11) $\equiv 10$ and $y_3 = -9 + 5 \ (7 - 10) \ (\text{mod } 11) \equiv 9$. Hence, $3G = (10,9)$.

Continuing computation in this way, the remaining multiples are evaluated as shown below:

$G = (8, 3)$ $2G = (7, 9)$ $3G = (10, 9)$ $4G = (2, 4)$ $5G = (5, 2)$ $6G = (3, 6)$
$7G = (3, 5)$ $8G = (5, 9)$ $9G = (2, 7)$ $10G = (10, 2)$ $11G = (7, 2)$ $12G = (8, 8)$

The generator $G = (8, 3)$ is called a *primitive element* that generates the multiples.

Elliptic Curve over Finite Field GF(2^m)

An EC over GF(2^m) is defined by the following equation:

$$y^2 + xy = x^3 + ax^2 + b$$

where $a, b \in$ GF(2^m) and $b \neq 0$. The set of EC over GF(2^m) consists of all points (x, y), $x, y \in$ GF(2^m), that satisfy the above defining equation, together with the point at infinity O.

Addition

Adding points on an EC over GF(2^m) will give a third EC point. The set of EC points forms a group with O (point at infinity) serving as its identity. The algebraic formula for the sum of two points and the doubling point are defined as follows:

1. If $P \in$ EC(GF(2^m)), then $P + (-P) = O$, where $P = (x, y)$ and $-P = (x, x + y)$ are indeed the points on the EC.
2. If P and Q (but $P \neq Q$) are the points on the EC(GF(2^m)), then $P + Q = P(x_1, y_1) + Q(x_2, y_2) = R(x_3, y_3)$, where $x_3 = \lambda^2 + \lambda + x_1 + x_2 + a$ and $y_3 = \lambda(x_1 + x_3) + x_3 + y_1$, where $\lambda = (y_1 + y_2)/(x_1 + x_2)$.
3. If P is a point on the EC (GF(2^m)), but ($P \neq -P$), then the point of doubling is $2P = R(x_3, y_3)$, where

$$x_3 = x_1^2 + \frac{b}{x_1^2} \quad \text{and} \quad y_3 = x_1^2 + \left(x_1 + \frac{y_1}{x_1}\right)x_3 + x_3$$

Example 6.18   Consider GF(2^4) whose primitive polynomial is $p(x) = x^4 + x + 1$ of degree 4. If α is a root of $p(x)$, then the field elements of GF(2^4) generated by $p(x)$ are shown in Table 6.12. Since $p(\alpha) = \alpha^4 + \alpha + 1 = 0$, that is, $\alpha^4 = \alpha + 1$, the field elements of GF(2^4) are expressed by four-tuple vectors such as $1 = (1000), \alpha = (0100), \alpha^2 = (0010), \ldots, \alpha^{14} = (1001)$.
Choosing $a = \alpha^4$ and $b = 1$, the EC equation over GF(2^4) becomes

$$y^2 + xy = x^3 + \alpha^4 x^2 + 1$$

Table 6.12 Field elements of $GF(2^4)$ using $\alpha^4 = \alpha + 1$

$\alpha^i, 0 \leqslant i \leqslant 14$	Polynomial expression	Vector form			
α^0	1	1	0	0	0
α^1	α	0	1	0	0
α^2	α^2	0	0	1	0
α^3	α^3	0	0	0	1
α^4	$1 + \alpha$	1	1	0	0
α^5	$\alpha + \alpha^2$	0	1	1	0
α^6	$\alpha^2 + \alpha^3$	0	0	1	1
α^7	$1 + \alpha + \alpha^3$	1	1	0	1
α^8	$1 + \alpha^2$	1	0	1	0
α^9	$\alpha + \alpha^3$	0	1	0	1
α^{10}	$1 + \alpha + \alpha^2$	1	1	1	0
α^{11}	$\alpha + \alpha^2 + \alpha^3$	0	1	1	1
α^{12}	$1 + \alpha + \alpha^2 + \alpha^3$	1	1	1	1
α^{13}	$1 + \alpha^2 + \alpha^3$	1	0	1	1
α^{14}	$1 + + \alpha^3$	1	0	0	1

Check whether one element (α^3, α^8) satisfies the EC equation over $GF(2^4)$.

$$(\alpha^8)^2 + (\alpha^3)(\alpha^8) = (\alpha^3)^3 + \alpha^4(\alpha^3)^2 + 1$$

$$\alpha^{16} + \alpha^{11} = \alpha^9 + \alpha^{10} + 1$$

$$(0100) + (0111) = (0101) + (1110) + (1000)$$

$$(0011) = (0011)$$

Thus, the points on the $EC(GF(2^4))$ are O (point at infinity) and the following 15 elements:

$(0, 1)$ $(1, \alpha^6)$ $(1, \alpha^{13})$ (α^3, α^8) (α^3, α^{13})

(α^5, α^3) (α^5, α^{11}) (α^6, α^8) (α^6, α^{14}) (α^9, α^{10})

(α^9, α^{13}) (α^{10}, α) (α^{10}, α^8) $(\alpha^{12}, 0)$ $(\alpha^{12}, \alpha^{12})$

Example 6.19 Consider the EC $y^2 + xy = x^3 + \alpha^4 x^2 + 1$ over $GF(2^4)$ used in Example 6.18. Then the point addition $P(\alpha^6, \alpha^8) + Q(\alpha^3, \alpha^{13}) = R(x_3, y_3)$ is computed as follows:

Since $\lambda = \frac{\alpha^8 + \alpha^{13}}{\alpha^6 + \alpha^3} = \alpha$, we have $x_3 = \lambda^2 + \lambda + x_1 + x_2 + a = \alpha^2 + \alpha + \alpha^6 + \alpha^3 + \alpha^4 = 1$ and $y_3 = \lambda(x_1 + x_3) + x_3 + y_1 = \alpha(\alpha^6 + 1) + 1 + \alpha^8 = \alpha(\alpha^{13}) + \alpha^2 = \alpha^{13}$.

Hence, $P + Q = R(1, \alpha^{13})$.

Next, the point-doubling problem of $2P = P + P = R(x_3, y_3)$ is considered as shown below:

$x_3 = x_1^2 + \frac{b}{x_1^2} = \alpha^{12} + \frac{1}{\alpha^{12}} = \alpha^{12} + \alpha^3 = \alpha^{10}$ (Take the inverse of α^i to be $\alpha^{-i} = \alpha^{-i+15 \ (\text{mod} 15)}$)

and $y_3 = x_1^2 + \left(x_1 + \dfrac{y_1}{x_1}\right) x_3 + x_3$

$$= \alpha^{12} + \left(\alpha^6 + \dfrac{\alpha^8}{\alpha^6}\right)\alpha^{10} + \alpha^{10}$$

$$= \alpha^{12} + \alpha^{13} + \alpha^{10} = (1010) = \alpha^8$$

Hence, $2P = R(x_3, y_3) = (\alpha^{10}, \alpha^8)$

6.6.2 Elliptic Curve Cryptosystem Applied to the ElGamal Algorithm

As an application problem to ECC, consider the ElGamal public-key cryptosystem based on the EC defined over the prime field z_p. The ElGamal crypto-algorithm is based on the discrete logarithm problem. Referring to Table 6.5 for the ElGamal encryption algorithm, choose a prime p such that the discrete logarithm problem in z_p is intractable, and let α be a primitive element of z_p^*. The values of p, α, and y are public, and x is secret.

$y \equiv \alpha^x \pmod{p}$

Choose a random number k such that $\gcd(k, p-1) = 1$. Then the encryption process of the message $m, 0 \leqslant m \leqslant p-1$, is accomplished by the following pair (r, s):

$r \qquad \equiv \alpha^k \pmod{p}$

$s \equiv (m \pmod{p-1})\, (y^k \pmod{p})$

For $r, s \in Z_p^*$, the decryption is defined as:

$m \equiv \dfrac{s}{r^x} \pmod{p}$

Elliptic Curve Cryptosystem by the ElGamal Algorithm

User A		User B
Let X be the plaintext and k a random number. Choose X and k	$\leftarrow$	Generate B's private key e_B and a public base point G. The public key is represented by $(G, e_B G)$
Compute $Y = (x, y)$ where $x = kG$ and $y = X + k(e_B G)$ Send Y to user B	$\rightarrow$	Receive $Y = (x, y) = (kG, X + k(e_B G))$ Decryption yields $X = y - e_B x$

Many public-key algorithms, such as D-H, ElGamal, and Schnorr, can be implemented in ECs over finite fields.

Example 6.20 Suppose user B generates a private key $e_B = 10$ and picks a base point $G = (8, 3)$ as a generator on the EC $y^2 \equiv x^3 + x + 6$ over z_{11}. Then B's public key becomes $(G, e_B G) = ((8, 3), 10(8, 3)) = ((8, 3), (10, 2))$.

User A wishes to send the plaintext $X = (2, 4)$ and chooses a random number $k = 5$. Compute the ciphertext $Y = (x, y)$, $x, y \in EC$

Where $x = kG = 5(8, 3) = (5, 2)$,

$$y = X + k(e_B G) = (2, 4) + 5(10, 2)$$

$$= (2, 4) + (7, 2) = (7, 9)$$

Send $Y = (x, y) = ((5, 2), (7, 9))$ to B.

B receives Y and decrypts it as follows:

$$X = y - e_B x$$

$$= (7, 9) - 10(5, 2) = (7, 9) + (7, 9) = (2, 4)$$

Thus, the correct plaintext X is recovered by decryption.

6.6.3   Elliptic Curve Digital Signature Algorithm

In the author's recent book "Mobile Communications and Security," John-Wiley (2009), Chapter 9 presents some cryptography algorithms for providing the data security in the air interface security layer. One of the topics is ECC. Some examples on Elliptic Curve Digital Signature Algorithm (ECDSA) over the finite binary extension field GF(2^m) is analyzed with the Weierstrass equations without derivation of addition and doubling formula. The purpose of this presentation is twofold: derivation of addition and doubling formula applicable to EC(GF(2^m)) and pointing out some innovative derivation in an example.

Analysis of the effect of different extension field of ECDSA is examined. Some practical examples of EC(GF(2^m)) field are presented to illustrate performances based on changing some of the parameters. The results examine the relationship between analyses and predict any future performances of DSA.

Definition

The Weierstrass equations are described by

$$E : y^2 + a_1 xy + a_3 y = x^3 + a_2 x^2 + a_4 x + a_6$$

where $a_i \in$ GF(2^m) (from Rosen, K.H., *Handbook of Elliptic and Hyperelliptic Curve Cryptography*. Boca Raton, FL: Chapman & Hall/CRC).

Arithmetic of EC over GF(2^m)

This section shows the basic arithmetic of EC over GF(2^m), especially for

$$EC : y^2 + xy = x^3 + ax^2 + b$$

where $a, b \in$ GF(2^m).

Addition

The addition on an EC is defined as the point addition of the two given points $P(x_1, y_1)$ and $Q(x_2, y_2)$ on an EC to obtain another point $R(x_3, y_3)$ on the same EC, which is expressed as

$$R(x_3, y_3) = P(x_1, y_1) + Q(x_2, y_2)$$

where $P(x_1, y_1) \neq Q(x_2, y_2)$.

$R(x_3, y_3)$ is defined as the additive inverse of the third crossing point on the EC by the line determined by $P(x_1, y_1)$ and $Q(x_2, y_2)$; that is, the third crossing point is $-R(x_3, y_3)$. Now, we are going to find x_3 and y_3 from the definition.

Let, $y = \lambda x + c$ be the line determined by $P(x_1, y_1)$ and $Q(x_2, y_2)$ and we can get $R(x_3, y_3)$ by solving the following simultaneous equations with $P(x_1, y_1)$ and $Q(x_2, y_2)$.

$$\begin{cases} y^2 + xy = x_3 + ax^2 + b \\ y = \lambda x + c \end{cases}$$

$$(6.7)$$
$$(6.8)$$

where

$$\lambda = \frac{y_2 - y_1}{x_2 - x_1} = \frac{y_2 + y_1}{x_2 + x_1}$$

Substituting y in Equation 6.7 results in

$$(\lambda x + c)^2 + x(\lambda x + c) = x^3 + ax^2 + b \tag{6.9}$$

$$\Rightarrow x^3 + (\lambda^2 + \lambda + a)x^2 + (c + 2\lambda c)x + b + c^2 = 0 \tag{6.10}$$

If x_1, x_2, and x_3 are the roots of the equation, then they should be expressed as

$$(x - x_1)(x - x_2)(x - x_3) = 0 \tag{6.11}$$

Comparing coefficients of like terms of Equations 6.9 and 6.11 yields

$$x_1 + x_2 + x_3 = \lambda^2 + \lambda + a \tag{6.12}$$

$$x_1 x_2 + x_2 x_3 + x_3 x_1 = c + 2\lambda c \tag{6.13}$$

$$x_1 x_2 x_3 = b + c^2 \tag{6.14}$$

From Equation 6.12, we get x_3 as

$$x_3 = x_1 + x_2 + \lambda^2 + \lambda + a \tag{6.15}$$

Next we are going to find y_3. Since $R(x_3, y_3)$ and $-R(x_3, y_3)$ are additive inverse, it should be $-R(x_3, y_3) = (x_3, x_3 + y_3)$(details may be required). Since $P(x_1, y_1)$ is on the straight line l, we get the following equations.

$$\begin{cases} y_1 = \lambda x_1 + c \\ x_3 + y_3 = \lambda x_3 + c \end{cases}$$

$$(6.16)$$
$$(6.17)$$

which results in

$$y_3 = \lambda(x_1, x_3) + x_3 + y_1 \tag{6.18}$$

In summary, the addition on the EC

EC: $y^2 + xy = x_3 + ax^2 + b$

is defined as

$$R(x_3, y_3) = P(x_1, y_1) + Q(x_2, y_2)$$

and the resulting point is calculated as

$$R(x_3, y_3) = (x_1 + x_2 + \lambda^2 + \lambda + a, (x_1 + x_3) + x_3 + y_1)$$

where

$$\lambda = \frac{y_2 + y_1}{x_2 + x_1}$$

$$x_3 = x_1 + x_2 + \lambda^2 + \lambda + a$$

$$y_3 = \lambda(x_1 + x_3) + x_3 + y_1$$

Verification of

$$-y_3 = x_3 + y_3$$

Doubling

When $P(x_1, y_1) = Q(x_2, y_2)$, we cannot get λ in Equation 6.5 because $\lambda = (y_2 - y_1)/(x_2 - x_1) = (y_2 + y_1)/(x_2 + x_1) = (2y_1)/(2x_1) = 0/0$. Therefore, we need another way to get λ in Equation 6.6. Let us differentiate the Equation (6.4) with respect to x and letting $\lambda = dy/dx$, then we get the following.

$$2y\frac{dy}{dx} + x\frac{dy}{dx} + y = 3x^2 + 2ax$$

$$\Rightarrow \frac{dy}{dx}(2y + x) + y = 3x^2 + 2ax$$

and

$$\lambda(2y + x) + y = 3x^2 + 2ax$$

$$\Rightarrow \lambda(2y + x) = 3x^2 + 2ax + y$$

$$\Rightarrow \lambda = \frac{3x^2 + 2ax + y}{2y + x} = \frac{3x^2 + 0 + y}{0 + x} = \frac{x^2 + y}{x} = x + \frac{y}{x}$$

In short,

$$\lambda = x + \frac{y}{x} \tag{6.19}$$

By solving Equations 6.7 and 6.8 with $P(x_1, y_1) = Q(x_2, y_2)$, we get x_3 as

$$x_3 = \frac{b + c^2}{x_1 x_2} = \frac{b + c^2}{x^2} = \frac{b}{x_1^2} + \frac{c^2}{x_1^2} = \frac{b}{x_1^2} + \left(\frac{c}{x_1}\right)^2 = \frac{b}{x_1^2} + \left(\frac{y_1 - \lambda x_1}{x_2}\right)^2$$

$$= \frac{b}{x_1^2} + \left(\frac{y_1 - \lambda x_1}{x_1}\right)^2$$

$$= \frac{b}{x_1^2} + \left(\frac{y_1}{x_1} + \lambda\right)^2$$

$$= \frac{b}{x_1^2} + \left(\frac{y_1}{x_1} + x_1 + \frac{y_1}{x_2}\right)^2$$

$$= \frac{b}{x_1^2} + x_1^2$$

In short,

$$x_3 = \frac{b}{x_1^2} + x_1^2 \tag{6.20}$$

Now, we are going to find y_3. Again, from the straight line through $P(x_1, y_1)$ and $-R(x_3, y_3)$, we get the following equations.

$$y_1 = \lambda x_1 + c$$

$$x_3 + y_3 = \lambda x_3 + c$$

By substituting $y = x + (y/x)$ with $(x, y) = (x_1, y_1)$, we get

$$x_3 + y_3 = \lambda x_3 + y_1 + \lambda x_1$$

$$= \left(x_1 + \frac{y_1}{x_1}\right) x_3 + y_1 + \left(x_1 + \frac{y_1}{x_1}\right) x_1$$

$$= \left(x_1 + \frac{y_1}{x_1}\right) x_3 + y_1 + x_1^2 + y_1$$

$$= \left(x_1 + \frac{y_1}{x_1}\right) x_3 + x_1^2$$

Therefore,

$$y_3 = \left(x_1 + \frac{y_1}{x_1}\right) x_3 + x_1^2 + x_3 \tag{6.21}$$

In summary, the doubling on the EC

$$EC: y^2 + xy = x^3 + ax^2 + b$$

is defined as

$$R(x_3, y_3) = P(x_1, y_1) + Q(x_2, y_2) = 2P(x_1, y_1) \text{ for } P(x_1, y_1) = Q(x_2, y_2)$$

And the resulting point is calculated as

$$R(x_3, y_3) = \left(\frac{b}{x_1^2} + x_1^2, \left(x_1 + \frac{y_1}{x_1} \right) x_3 + x_1^2 + x_3 \right)$$

$$x_3 = x_1^2 + \frac{b}{x_1^2}$$

$$y_3 = \left(x_1 + \frac{y_1}{x_1} \right) x_3 + x_1^2 + x_3$$

6.6.4 ECDSA Signature Computation

ECCs are viewed as EC analogs to the conventional discrete logarithm cryptosystems in which the subgroup of Z_p is replaced by the group of point on an EC over a finite field. In that sense, ECDSA is the EC analog of DSA. The ECDSA signature and verification algorithms are presented in this section. In addition, the ECDSA examples are presented to clarify the algorithms.

ECDSA Signature Scheme over GF(2^m)

The domain parameters for ECDSA consist of a proper EC defined over an extension field of GF(2^m) of characteristic 2. A set of EC domain parameters is composed of

$$D = (q, \text{FR}, a, b, G, n, \lambda)$$

where

1. $q =$ a field of 2^m;
2. FR $=$ field representation used for elements of GF(2^m);
3. $a, b \in$ GF(2^m) $=$ two field elements that define an EC over GF(2^m)

   $$EC: y^2 + xy = x^3 + ax^2 + b;$$

4. $G =$ the basic point, $G \in$ GF(2^m);
5. $n =$ the order of the point G;
6. $\lambda =$ the cofactor is defined to be $\lambda = \#EC(GF(2^m))/ n$.

Algorithm

1. Choose an arbitrary bit string E of length $g \geqslant 160$ bits.
2. Compute the hash code $h = \text{SHA-1(E)}$ and let b_0 be the bit string of length v bits obtained by taking the v rightmost bits of h.
3. Let z be the integer whose binary expansion is the g-bit stream E.
4. For i from 1 to s do: Compute $b_i = \text{SHA} - 1(s_i)$, where s_i is the g-bit string of the integer $(z + i) \bmod 2^y$.
5. Let b be the field element obtained by concatenation as $b = b_0 \| b_1 \| \ldots \| b_s$.
6. If $b = 0$, then go to step 1.
7. Let a be an arbitrary element of $\text{GF}(2^m)$.
8. Output (E, a, b).
9. Let b' be the field element such that $b' = b_0 \| b_1 \| \ldots \| b_s$.
10. If $b' = b$, then accept. Otherwise reject.

Key Pair Generation

An ECDSA key pair is associated with a particular set of EC domain parameters $D = (q, \text{FR}, a, b, G, n, \lambda)$ that must be valid before key generation.

User A selects a random integer k for $1 \leqslant k \leqslant n - 1$ and computes $Q = kG$, where Q is A's public key and k is A's private key.

- Choose that $Q \neq 0$.
- Check whether a public key $Q = (x_Q, y_Q)$ is properly represented by the elements of Z_p over $(0, p - 1)$ and m-bit string over $\text{GF}(2^m)$ of 2^m.
- Check that Q lies on the EC defined by a and b.
- Check $nQ = 0$.
- If any check fails, then Q is invalid; otherwise Q is valid.

Example 6.21   User A uses the EC $y^2 \equiv x^3 + x + 6$ over Z_{11}. Choose the key pair (d, Q) in which $d = 2$ (A's private key), $Q = (7, 9)$ (A's public key), and $k = 5$ (a random integer). $G = (8, 3)$. Compute the following steps:

User A

$kG = 5(8, 3) = (5, 2)$ from which $r = x_1 = 5$.

$k^{-1} = 8$ is the multiplicative inverse of $k \equiv 5 \pmod{13}$.

Suppose the message digest $h = \text{SHA-1}(m) = 8$ is a converted integer e.

Compute $s \equiv k^{-1}(e + dr) \pmod{13}$

$$\equiv 8(8 + 2 \times 5) \pmod{13} \equiv 8(18) \pmod{13} \equiv 1$$

Thus, A's signature for m is $(r, s) = (5, 1)$.

To verify A's signature (r, s) on m, the following computations are required:

User B

$w \equiv s^{-1} \equiv 1^{-1} \pmod{13} \equiv 1$

$u_1 \equiv ew \pmod{13} \equiv 8 \times 1 \pmod{13} \equiv 8$

$u_2 \equiv rw \pmod{13} \equiv 5 \times 1 \pmod{13} \equiv 5$

$X = u_1 G + u_2 Q = 8(8, 3) + 5(7, 9) = (5, 9) + (10, 2) = (5, 2)$

Since $v = 5 = r$, the signature is accepted.

Example 6.22   User A uses the elliptic curve EC $y^2 + xy = x^3 + \alpha^4 x^2 + 1$ over GF(2^4) and the primitive polynomial of GF(2^4) is $p(x) = x^4 + x + 1$ whose root is α.

The number of points on EC(GF(2^4)) is $n = 16$, which is the order, that is, for given point X on EC(GF(2^4)), $X = 16X$.

User A selects the base point $G = (\alpha^{12}, 0)$

Signature Generation at User A

Choose key pair (d, Q), where $d = 2$ is the private key and Q is the public key, $Q = (\alpha^5, \alpha^3)$.

User A chooses a random integer $k = 3$ between $1 \leqslant k \leqslant 15$. User A computes $kG = (x_1, y_1) = 3(\alpha^{12}, 0) = (\alpha^6, \alpha^{14})$.

User A converts $x_1 = \alpha^6$ into an integer by means of the following integer conversion

$r \quad \leftrightarrow x_1$

$1 \quad \leftrightarrow \alpha^0$

$2 \quad \leftrightarrow \alpha^1$

$\quad \vdots$

$15 \leftrightarrow \alpha^{14}$

For example, when α^6 is converted into an integer, it would be 7, that is, $r = 7$.

Suppose the message digest of message m is $h = 9$:

User A computes $s = k^{-1}(h + dr) \pmod{n}$

$$= 11(9 + 2 \times 7) \pmod{16} = 13$$

User A's signature for the message is $(r, s) = (7, 13)$.
User A sends the signature $(r, s) = (7, 13)$ to user B.

Signature Verification at User B

User B computes:

$$w = s^{-1} \ (\text{mod } n) = 13^{-1} \ (\text{mod } 16) = 5$$

$$u_1 = hw \ (\text{mod } n) = 9 \times 5 \ (\text{mod } 16) = 13$$

$$u_2 = rw \ (\text{mod } n) = 7 \times 5 \ (\text{mod } 16) = 3$$

Finally, user B computes

$$X = u_1 G + u_2 Q$$
$$= 13(\alpha^{12}, 0) + 3(\alpha^5, \alpha^3)$$
$$= (\alpha^6, \alpha^{14})$$

We can again convert α^6 into an integer $v = 7$. Since $v = r = 7$, the ECDSA signature is accepted.

7

Public-Key Infrastructure

This chapter presents the profiles related to public-key infrastructure (PKI) for the Internet. The PKI manages public keys automatically through the use of public-key certificates. It provides a basis for accommodating interoperation between PKI entities. A large-scale PKI issues, revokes, and manages digital signature public-key certificates to allow distant parties to reliably authenticate each other. A sound digital signature PKI should provide the basic foundation needed for issuing any kind of public-key certificate.

The PKI provides a secure binding of public keys and users. The objective is how to design an infrastructure that allows users to establish certification paths which contain more than one key. Creation of certification paths, commonly called *chains of trust*, is established by Certification Authorities (CAs). A certification path is a sequence of CAs. CAs issue, revoke, and archive certificates. In the hierarchical model, trust is delegated by a CA when it certifies a subordinate CA. Trust delegation starts at a root CA that is trusted by every node in the infrastructure. Trust is also established between any two CAs in peer relationships (cross-certification).

The CAs will certify a PKI entity's identity (a unique name) and that identity's public key. A CA performs user authentication and is responsible for keeping the user's name and the associated public key. Hence, each CA must be a trusted entity, at least to the extent described in the Policy Certification Authority (PCA) policies. The CAs will need to certify public keys, create certificates, distribute certificates, and generate and distribute Certificate Revocation Lists (CRLs). The PCA is a special purpose CA which creates a policy-setting responsibility: that is, how the CA's and PCA's functions and responsibilities are defined and how they interact to determine the nature of the infrastructure. Therefore, PKI tasks are centered on researching and developing these functions, responsibilities, and interactions.

This chapter presents the interoperability functional specifications that are carried out by CA entities at all levels. It describes what the PAA (Policy Approval Authority), PCAs, and CAs perform. It also describes the role of an Organizational Registration Authority (ORA) that acts as an intermediary between the CA and a prospective certificate subject.

Wireless Mobile Internet Security, Second Edition. Man Young Rhee.
© 2013 John Wiley & Sons, Ltd. Published 2013 by John Wiley & Sons, Ltd.

In the long run, the goal of the Internet PKI is to satisfy the requirements of identification, authentication, access control, and authorization functions.

7.1 Internet Publications for Standards

The Internet Activities Board (IAB) is the body responsible for coordinating Internet design, engineering, and management. The IAB has two subsidiary task forces:

- The Internet Engineering Task Force (IETF), which is responsible for short-term engineering issues including Internet standards.
- The Internet Research Task Force (IRTF), which is responsible for long-term research.

The IETF working groups meet three times annually at large conventions to discuss standards development, but the development process is conducted primarily via open e-mail exchanges. Participants of IETF are individual technical contributors, rather than formal organizational representatives.

The most important series of Internet publications for all standards specifications appear in the Internet Request for Comments (RFCs) document series. Anyone interested in learning more about current developments on Internet standards can readily track their progress via e-mail. Another important series of Internet publications are the Internet Drafts. These are working documents prepared by the IETF, its working groups, or other groups or individuals working on Internet technical topics. Internet Drafts are valid for a maximum of 6 months and may be updated, replaced, or rendered obsolete by other documents at any time. Specifications that are destined to become Internet standards evolve through a set of maturity level as the standards evolve, which has three recognized levels: Proposed Standard, Draft Standard, and Refined Standard.

To review the complete listing of current Internet Drafts, Internet standards associated with PKI will be briefly summarized in the following.

A public directory service or repository that can distribute certificates is particularly attractive. The X.500 standard specifies the directory service. A comprehensive online directory service has been developed through the ISO/ITU standardization processes. These directory standards provide the basis for constructing a multipurpose distributed directory service by interconnecting computer systems belonging to service providers, governments, and private organizations. In this way, the X.500 directory can act as a source of information for private people, communications network components, or computer applications. When the X.500 standards were first developed in 1984–1988, the use of X.500 directories for distributing public-key certificates was recognized. Therefore, the standards include full specifications of data items required for X.500 to fulfill this role. Since the X.500 technology is somewhat complex, adoption of X.500 was slower than expected until the mid-1990s. Nevertheless, deployment of X.500 within large enterprises is increasing and some organizations are finding this repository a useful means of public-key certificate distribution.

The Internet Lightweight Directory Access Protocol (LDAP) is a protocol which can access information stored in a directory, including access to stored public-key certificates.

LDAP is an access protocol which is compatible with the X.500 directory standards. However, LDAP is much simpler and more effective than the standard X.500 protocols.

The X.509 certificate format describes the authentication service using the X.500 directory. The certificate format specified in the Privacy Enhanced Mail (PEM) standards is the 1988 version of the X.509 certificate format. The certificate format specified in the American National Standards Institute (ANSI) X9.30 standards is based on the 1992 version of the X.509 certificate format. The ANSI X9.30 standard requires that the issuer unique identifier field be filled in. This field will contain information that allows the private key to sign the certificate and be uniquely identified.

The certificate format used with the Message Security Protocol (MSP) is also based on the 1988 X.509 certificate format, but it does not include the issuer unique identifier or the subject unique identifier fields that are found in the 1992 version of the X.509 format.

The ISO/IEC/ITU X.509 standard defines a standard CRL format. The X.509 CRL format has evolved somewhat since first appearing in 1988. When the extension fields were added to the X.509 v3 certificate format, the same type of mechanism was added to the CRL to create the X.509 v2 CRL format. Of the various CRL formats studied, the PEM CRL format best meets the requirements of the PKI CRL format. ITU-T X.509 (formerly CCITT X.509) and ANSI X9.30 CRL formats are compared with the PEM CRL format to show where they differ. For example, the ANSI X9.30 CRL format is based on the PEM format, but the former adds one reason code field to each certificate entry within the list of revoked certificates.

All CAs are assumed to generate CRLs. The CRLs may be generated on a periodic basis or every time a certificate revocation occurs. These CRLs will include certificates that have been revoked because of key compromises, and changes in a user's affiliation. All entities are responsible for requesting the CRLs that they need from the directory, but to keep querying the directory is impractical. Any CA which generates a CRL is responsible for sending its latest CRL to the directory. However, CRL distribution is the biggest cost driver associated with the operation of the PKI. CAs certifying fewer users result in much smaller CRLs because each CRL requested carries far less unwanted information. The delta CRL indicator is a critical CRL extension that identifies a delta CRL. The use of delta CRLs can significantly improve processing time for applications that store revocation information in a format other than the CRL structure. This allows changes to be added to the local database while ignoring unchanged information that is already in the local database.

7.2 Digital Signing Techniques

Since user authentication is so important for the PKI environment, it is appropriate to discuss the concept of digital signature at an early stage in this chapter. Digital signing techniques are employed to provide sender authentication, message integrity, and sender nonrepudiation, provided that private keys are kept secret and the integrity of public keys is preserved. Provision of these services is furnished with the proper association between the users and their public/private-key pairs.

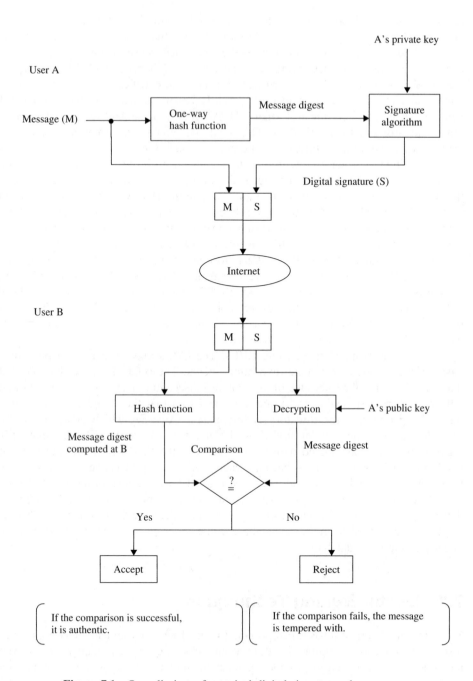

Figure 7.1 Overall view of a typical digital signature scheme.

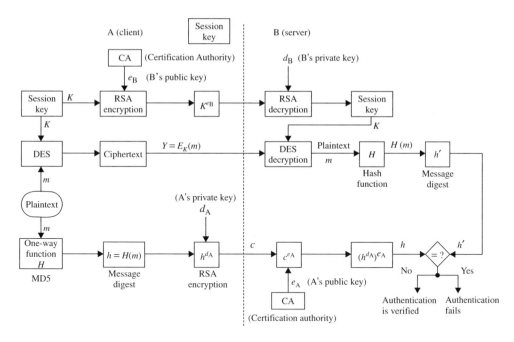

Figure 7.2 Signature and authentication with DES/RSA/MD5 (compatible with PEM method).

When two users A and B communicate, they can use their public keys to keep their messages confidential. If A wishes to hide the contents of a message to B, A encrypts it using B's public key. If A wishes to sign a document, he or she must use the private key available only to him or her. When B receives a digitally signed message from A, B must verify its signature. B needs A's public key for this verification. A should have high confidence in the integrity of that key.

The scenario of a typical signature scheme is described in Figure 7.1. The following example is presented to illustrate one practical system (Figure 7.2) applicable to the digital signature computation for user authentication. The combination of SHA-1 (or MD5) and RSA provides an effective digital signature scheme. As an alternative, signatures can also be generated using DSS/SHA-1.

For digital signatures, the content of a message m is reduced to a message digest with a hash function (such as MD5). An octet string containing the message digest is encrypted with the RSA private key of the signer. The message and the encrypted message digest are represented together to yield a digital signature. This application is compatible with the PEM method. For digital envelopes, the message is first encrypted under a DES key with a DES algorithm and then the DES key (message-encryption key) is encrypted with the RSA public key of the recipients of the message. The encrypted message and the encrypted DES key are represented together to yield a digital envelope. This application is also compatible with PEM methods.

Example 7.1 Utilizing the practical signature/authentication scheme shown in Figure 7.2, the analytic solution is as follows:

Client A

1. DES encryption of message m:

   The 64-bit message m is

   $m = 785ac3a4bd0fe12d$

   The 56-bit DES session key K is

   $K = ba0c2b3c484ff9$ (hexadecimal)

   The 64-bit ciphertext Y (output of 16-round DES) is

   $Y = a78791c0c8f0b444$

2. RSA encryption of K:

   $K = 52367725502681081$ (decimal)

   Split K into blocks of two digits:

   $K = 05\ 23\ 67\ 72\ 55\ 02\ 68\ 10\ 81$

   Obtain B's public key $e_B = 79$ from CA and choose public modulo $n = 3337$. Encrypt every 2-bit block of K as follows:

   $5^{79}(\mathrm{mod}\ 3337) \equiv 270$
   $23^{79}(\mathrm{mod}\ 3337) \equiv 2524$

   $\vdots$

   $81^{79}(\mathrm{mod}\ 3337) \equiv 3198$

   Encrypted $K = 0270\ 2524\ 1479\ 0285\ 1773\ 3139\ 2753\ 3269\ 3198$
   This encrypted symmetric key is called the *digital envelope*.
   Send this encrypted key (digital envelope) K to B.

3. Computation of hash code using MD5:
   Compute the hash value h of m:

   $h = H(m) = H(785ac3a4bd0fe12d)$

   $\qquad\qquad = 6a26ee0ed9ce3963ec8b0f98ebda8476$ (hexadecimal)

   $h = 141100303223912907143183747760118203510$ (decimal)

   Choose $d_A = 13$ (A's private key) and compute:

   $c = h^{d_A}$

Let us break the hash code into two decimal numbers as follows:

$h = 1$ 41   10   03   03   22   39   12   90   71

       43   18   37   47   76   01   18   20   35   10

Using $d_A = 13$ and $n = 851$, compute the RSA signature:

$1^{13} (\text{mod } 851) \equiv 1$

$41^{13} (\text{mod } 851) \equiv 545$

$$\vdots$$

$10^{13} (\text{mod } 851) \equiv 333$

$c = h^{d_A} = 001$ 669   084   400   400   091   348   719   157   303

                    635   439   333   047   089   001   439   520   466   084

Send c to B.

A → B

Send (ciphertext Y, encrypted value of K, and signed hash code c) to B.

Server B

1. Decryption of secret session key K:
   Received encryption key K:

   $K = 0270\ 2524\ 1479\ 0285\ 1773\ 3139\ 2753\ 3669\ 3198$

   Choose $d_B = 1019$ (B's private key) and decrypt K block by block:

   $270^{1019} (\text{mod } 3337) \equiv 5$

   $2524^{1019} (\text{mod } 3337) \equiv 23$

   $$\vdots$$

   $3198^{1019} (\text{mod } 3337) \equiv 81$

   $K = 05\ 23\ 67\ 72\ 55\ 02\ 68\ 10\ 81$

   or

   $K = 52367725502681081$ (decimal)

   $\quad = ba0c2b3c484ff9$ (hexadecimal)

2. Decryption of m using DES:

   Ciphertext $Y = a78791c0c8f0b444$

   Restored DES key $K = ba0c2b3c484ff9$

Using Y and K, the message m can be recreated:

$$m = 785ac3a4bd0fe12d$$

3. Computation of hash code and verification of signature:
   Apply MD5 algorithm to the restored message in order to compute the hash code:

$$h' = H(m) = H(785ac3a4bd0fe12d)$$
$$= 6a26ee0ed9ce3963ec8b0f98ebda8476$$

Obtain A's public key $e_A = 61$ from CA and apply e_A to the signed hash value c:

$$c = 001 \quad 669 \quad 084 \quad 400 \quad 400 \quad 091 \quad 348 \quad 719 \quad 157 \quad 303$$
$$635 \quad 439 \quad 333 \quad 047 \quad 089 \quad 001 \quad 439 \quad 520 \quad 466 \quad 084$$

Using e_A, compute $h = c^{e_A}$ as follows:

$$1^{61}(\text{mod } 851) \equiv 1$$
$$669^{61}(\text{mod } 851) \equiv 41$$
$$\vdots$$
$$084^{61}(\text{mod } 851) \equiv 10$$

Hence, $h = 1 \quad 41 \quad 10 \quad 03 \quad 03 \quad 22 \quad 39 \quad 12 \quad 90 \quad 71$
$$43 \quad 18 \quad 37 \quad 47 \quad 76 \quad 01 \quad 18 \quad 20 \quad 35 \quad 10$$

Convert it to the hexadecimal number:

$$h = 6a26ee0ed9ce3963ec8b0f98ebda8476$$

Thus, we can easily check $h = h'$.

Digital signing techniques are used in a number of applications. Since digital signature technology has grown in demand, its extensive use and development will be expected to continue in the future. Several applications are considered in the following.

- *Electronic-mail security.* E-mail is needed to sign digitally, especially in cases where sensitive information is being transmitted and security services such as authentication, integrity, and nonrepudiation are desired. Signing an e-mail message assures all recipients that the sender of the information is the person who he or she claims to be, thus authenticating the sender. For example, the DSS is using MOSAIC to provide security services for e-mail messages. The digital signature algorithm (DSA) has been incorporated into MOSAIC and is used to digitally sign e-mails as well as public-key certificates. Pretty Good Privacy (PGP) provides security services as well as data integrity services for messages and data files by using digital signatures, encryption, compression (zip), and radix-64 conversion (ASCII Armor). MIME defines a format for text messages being sent using e-mail. MIME is actually intended to address some of the problems and limitations of the use of SMTP. S/MIME is a security enhancement to the MIME Internet e-mail format, based on technology from RSA Data Security. Although both PGP and S/MIME are on an IETF standards track, it appears likely that PGP will remain the choice for personal e-mail security for many users, while S/MIME will emerge as the industry standard for commercial and organizational use.

- *Financial transactions.* This encompasses a number of areas in which money is being transferred directly or in exchange for services and goods. One area of financial transactions whichcould benefit especially from the use of digital signatures is Electronic Funds Transfer (EFT). Digitally signing EFTs is a way of providing security services such as authentication, integrity, and nonrepudiation.

  Secure Electronic Transaction (SET) is the most important protocol related to e-commerce. SET introduced a new concept of digital signature called *dual signatures*. A dual signature is generated by creating the message digest of two messages: order digest and payment digest. The SET protocol for payment processing utilizes cryptography to provide confidentiality of information and to ensure payment integrity and identity authentication.

- *Electronic filing.* Contracting requirements expect certain mandated certificates to be submitted from contractors. This requirement is often filed through the submission of a written form and usually requires a handwritten signature. If filings are digitally signed and electronically filed, digital signatures may be used to replace written signatures and to provide authentication and integrity services.

  One of the largest information submission processes is perhaps the payment of taxes, and the request for tax-related information will require signatures. In fact, the IRS in the United States is converting many of these processes electronically and is considering the use of digital signatures. The IRS has several prototypes under development that utilize digital signatures generated using DSA. At present, individuals send their tax forms to the IRS in bulk transactions. The IRS will require them to sign the bulk transactions digitally to provide added assurances. In future, the electronically generated tax returns may be digitally signed. The taxpayer may send the digitally signed electronic form to the IRS directly or through a tax accountant or adviser.

- *Software protection.* Digital signatures are also used to protect software. By signing the software, the integrity of the software is assured when it is distributed. The signature may be verified when the software is installed to ensure that it was not modified during the distribution process.

- *Signing and authenticating.* Signing is the process of using the sender's private key to encrypt the message digest of a document. Anyone with the sender's public key can decrypt it. A person who wants to sign the data has only to encrypt the message digest to ensure that the data originated from the sender. Authentication is provided when the sender encrypts the hash value with the sender's private key. This assures the receiver that the message originated from the sender.

Digital signatures can be used in cryptography-based authentication schemes to sign either the message being authenticated or the authentication challenge used in the scheme. The X.509 strong authentication is an example of an authentication scheme that utilizes digital signatures.

Careful selection and appropriate protection of the prime numbers p and q, of the primitive element g of p, and of the private and public components x and y of each key are at the core of security in digital signatures. Therefore, whoever generates these keys and their parameters is a vital concern for security. PCAs are responsible for defining who should generate these numbers.

When generating the key for itself and its CA, each PCA needs to specify the acceptable algorithms used to generate the prime numbers and parameters. For example, a larger p means more security, but requires more computation in the signing and verification steps. Thus, the size of p allows a trade-off between security and performance. Each PCA must specify the range of p for itself, its CAs, and its end users. The range of p is largest for the PCA and smallest for the end user.

One-way hash functions and DSAs are used to sign certificates and CRLs. They are used to identify OIDs (object identifiers) for public keys contained in a certificate. SHA-1 is the preferred one-way function for use in the Internet PKI. It was developed by the US government for use with both the RSA signature algorithm and DSA. However, MD5 is used in other legacy applications, but it is still reasonable to use MD5 to verify existing signatures. RSA and DSA are the most popular signature algorithms used in the Internet. They combine RSA with either MD5 or SHA-1 one-way hash functions; DSA is used in conjunction with the SHA-1 one-way hash function. The signature algorithm with the MD5 and RSA encryption algorithm is defined in PKCS#1 (RFC 2437). The signature algorithm with the SHA-1 and RSA encryption algorithms is implemented using the padding and encoding mechanisms also described in PKCS#1 (RFC 2437).

7.3   Functional Roles of PKI Entities

This section describes the functional roles of the whole entities at all levels within the PKI. It also describes how the PAA, PCAs, CAs, and ORAs perform.

7.3.1   Policy Approval Authority

The PAA is the root of the certificate management infrastructure. This authority is known to all entities at all levels in the PKI and creates the overall guidelines that all users, CAs, and subordinate policy-making authorities must follow.

The PAA approves policies established on behalf of subclasses of users or communities of interest. It is also responsible for supervising other policy-making authorities.

Figure 7.3 illustrates the PAA functions and their performances. Each PAA performs the following functions:

- Publishes the PAA's public key.
- Sets the policies and procedures that entities (PCAs, CAs, ORAs, and users) of the infrastructure should follow.

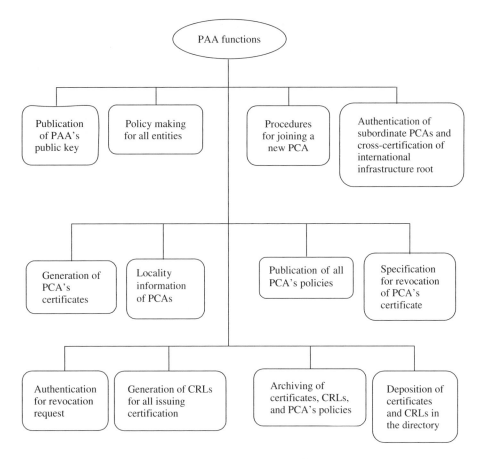

Figure 7.3 Illustration of PAA functions.

- Sets the policies and procedures, if any, for a new PCA to join the PKI.
- Carries out identification and authentication of each of its subordinate PCAs and national or international infrastructure roots and judges the proper measures to be taken for cross-certification.
- Generates certificates of subordinate PCAs and of national or international infrastructure roots to be cross-certified.
- Publishes identification and locality of subordinate PCAs such as directory name, e-mail address, postal address, phone number, and fax number.
- Receives and publishes policies of all subordinate PCAs.
- Specifies information required from subordinate PCAs for a revocation request of the PCA's certificate.
- Receives and authenticates revocation requests concerning certificates it has generated.
- Generates CRLs for all the certificates it has issued.
- Archives certificates, CRLs, audit files, and PCA's policies.
- Deposits the certificates and the CRLs it generates in the directory.

7.3.2 Policy Certification Authority

PCAs are formed by all entities at the second level of the infrastructure. Each PCA describes the users whom it serves. All PCAs have both policy and certification responsibilities, and must publish their security policies, procedures, legal issues, fees, or any other subjects they may consider necessary. For PCAs, the users may be people who are affiliated to an organization or part of a specific community, or a nonhuman entity. All PCA security policies are published and stored on an end user's local database. Each PCA performs the following functions, as illustrated in Figure 7.4.

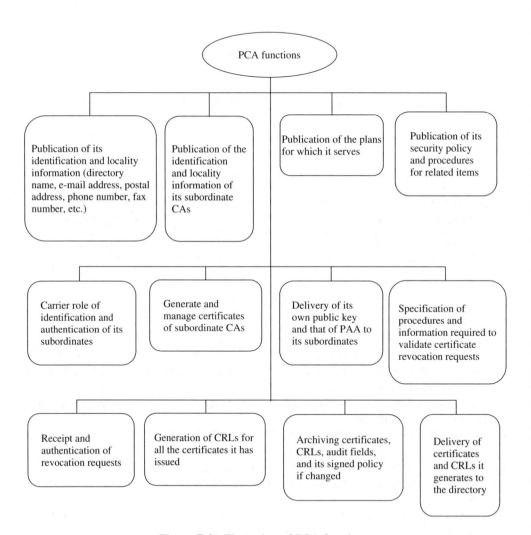

Figure 7.4 Illustration of PCA functions.

- Publishes its identification and locality information, such as directory name, e-mail address, postal address, phone number, and fax number.
- Publishes identification and locality information of its CAs.
- Publishes who it plans to serve.
- Publishes its security policy and procedures which specify the following items:

  - Who generates key variables p, q, g, x, and y.
  - The ranges of allowed sizes of p for itself, its CAs, and end users.
  - Identification and authentication requirements for the PCA, CAs, ORAs, and end users.
  - Security controls at the PCA and CA systems that generate certificates and CRLs.
  - Security controls at ORA systems.
  - Security controls for every user's private key.
  - The frequency of CRL issuance.
  - The constraints it imposes on naming schemes.
  - Audit procedures.

- Carries out identification and authentication of each of its subordinates.
- Generates and manages certificates of subordinate CAs.
- Delivers its own public key and that of PAA to its subordinates.
- Specifies procedures and information required to validate certificate revocation requests.
- Receives and authenticates revocation requests concerning certificates it has generated.
- Generates CRLs for all the certificates it has issued.
- Archives certificates, CRLs, audit files, and its signed policy if changed.
- Delivers the certificates and CRLs it generates to the directory.

7.3.3 Certification Authority

CAs form the next level below the PCAs. The PKI contains many CAs with no policy-making responsibilities. The majority are plain CAs. A few are CAs that are associated with PCAs. A CA has any combination of users and ORAs whom it certifies.

The primary function of the CA is to generate, publish, revoke, and archive the public-key certificates that bind the user's identity with the user's public key. A better and trusted way of distributing public keys is to use a CA. CAs are expected to certify the public keys of users or of other CAs according to PCA and PAA policies. The CAs ensure that all key parameters are in the range specified by the PCA. Thus, CAs either create key pairs that satisfy the PCA regulations or they examine user-generated keys to ascertain whether they fit within the required range assignment. Referring to Figure 7.5, a CA performs the following functions:

- Publishes and augments PCA policy.
- Carries out identification and authentication of each of its subordinates.
- Generates and manages certificates of subordinates.
- Delivers its own public key and its predecessor's public keys.
- Verifies ORA certification requests.

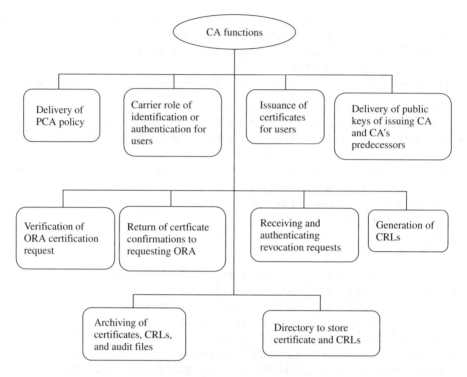

Figure 7.5 Functions of certificate authority (CA).

- Returns certificate creation confirmations or new certificates to requesting ORA.
- Receives and authenticates revocation requests concerning certificates it has generated.
- Generates CRLs for all the certificates it has issued.
- Archives certificates, CRLs, and audit files.
- Delivers the certificates and the CRLs it generates to the directory.

7.3.4 Organizational Registration Authority

The ORA is the interface between a user and a CA. The prime function that an ORA performs is user identification and authentication on behalf of a CA and it delivers the CA-generated certificate to the end user. After authenticating a user, an ORA transmits a signed request for a certificate to the appropriate CA. In response to an ORA request for key certification, the CA returns a certificate to the ORA. The ORA passes the certificate on to the user. Thus, an ORA's sole task is to help a user who is far from the user's CA to register with that CA and to obtain a public-key certificate. ORAs must pass certificate revocation reports timely and accurately to a CA. In order to verify the signature on the information at a future time, ORAs must archive the public key or the certificate associated with the signer. The ORA uses a signed message to inform the CA of the need to revoke

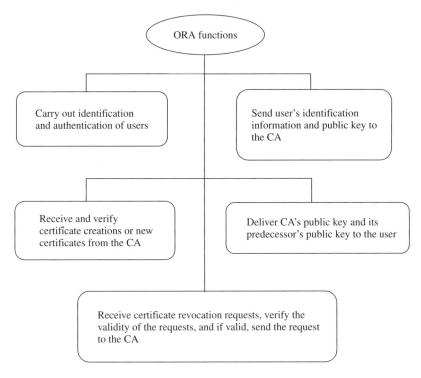

Figure 7.6 Illustration of ORA functions.

the certificate and to issue a new one. Nowadays, RA is preferred for simple use rather than ORA. An ORA performs the following functions that are illustrated in Figure 7.6:

- Carries out identification and authentication of users.
- Sends user identification information and the user's public key to the CA in a signed message.
- Receives and verifies certificate creations or new certificates from the CA.
- Delivers the CA's public key and its predecessor's public keys as well as the certificate to the user if returned.
- Receives certificate revocation requests, verifies the validity of the requests, and if valid, sends the request to the CA.

7.4 Key Elements for PKI Operations

This section describes the operational concepts of the PKI. In order to comprehend the overall PKI operation, one must understand how it conducts its various activities. Each activity is broken down into functional steps. The resources required for each functional step within each activity must be defined. The resources required for an activity are presented in relation to the entities such as User, Key Generator (KG), CA, ORA, PCA, or Directory. The steps associated with PKI activities are applied to all PKI relationships: User–CA, User–ORA, ORA–CA, CA–PCA, and PCA–PAA.

This section also presents the architectural structures for the PKI certificate management infrastructure. These structures should allow users to establish chains of trust that are no more than a few certificates in length. The functions and responsibilities of the CAs and PCAs are briefly reviewed, followed by how the CAs are interconnected to permit establishment of reliable certification paths. Some major activities associated with the PKI operations are presented subsequently.

7.4.1 Hierarchical Tree Structures

Chains of trust follow a strict tree hierarchy with a root CA (PAA or PCA) to which all trust is referenced. Each CA certifies the public keys of its users and the public key of the root CA is distributed to all PKI entities. Thus, every entity is linked to the root CA via a unique trust path. Figure 7.7 depicts such a tree structure. A number of hierarchies may be joined together by cross-certifying their root CA directly or using bridge CAs. Figure 7.8 illustrates a bridge-type scheme joining a hierarchical tree structure to a mesh structure.

With a mesh structure, entities may be connected via several chains of trust. PGP is a PKI that uses a mesh structure, with every entity acting as their own CA. Gateway structures are new structures appearing in VPN (Virtual Private Network) applications. In a gateway structure, each domain is separated and relies on its gateway to provide external PKI services. Figure 7.9 depicts a gateway structure with three cross-certified gateways through which the trust of the network is channeled. Horizontal structures offer improved robustness to penetration by distributing the trust path horizontally. Multiple platform structures can be used to introduce redundancy into a PKI structure and thus

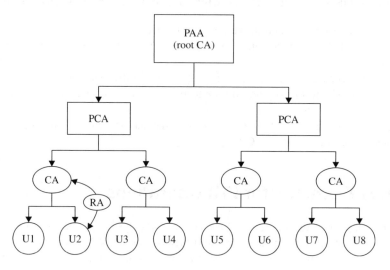

Figure 7.7 Hierarchical tree structure.

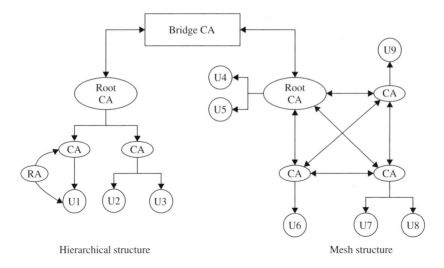

Hierarchical structure   Mesh structure

Figure 7.8   A mixed structure using a bridge CA.

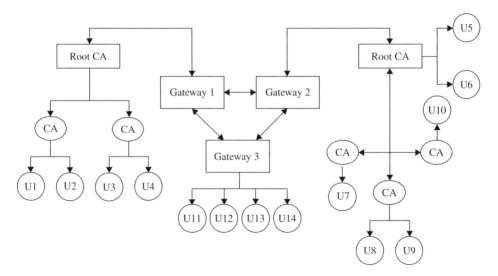

Figure 7.9   A gateway structure.

reduce risk. The public key of each user is authenticated in each platform. This is a particular advantage with hierarchical structures because it can remove a single point of failure.

7.4.2   Policy-Making Authority

Chains of trust are based on appropriate policies at all levels in the infrastructure. Associated with the entire PKI is a policy-establishing authority which will create the overall guidelines and security policies that all users, CAs, and subordinate policy-making authorities must follow.

- The PAA has the responsibility of supervising other policy-making authorities. The PAA will approve policies established on behalf of subclasses of users or of communities of interest.

- The PCAs will create policy details that expand or extend the overall PAA policies. Each PCA establishes policy for a single organization or for a single community of interest. PCAs must publish their security policies, procedures, any legal issues, any fees, or any other subjects that they consider necessary.

- The CAs are expected to certify the public key of end users or of other CAs in accordance with PCA and PAA policies. The CA must ensure that all key parameters are in the range specified by the PCA. Therefore, the CA either creates key pairs according to the PCA regulations or examines the user-generated keys to ascertain that they satisfy the requirements of the range. A few CAs are associated with PCAs, but the majority are plain CAs at all points in the infrastructure.

- The ORA submits a certificate request on behalf of an authenticated entity. The CA returns the signed certificate or an error message to the ORA. The ORA or certificate holder requests revocation of a certificate to the issuing CA. The CA responds with acceptance or rejection of the revocation request. CRLs contain all revoked certificates that CAs have issued and have not expired. The CA returns the signed certificate and its certificate or an error message to the end user. The CA posts a new certificate and CRL to the repository.

7.4.3   Cross-Certification

Suppose the CA has its private/public-key pair and the X.509 certificate issued by the CA. If a user knows the CA's public key, then the user can decrypt the certificate with the CA's public key and verify the X.509 certificate signed by the CA. Thus the user can recover his or her public key contained in the X.509 certificate; the user's public key is verified as illustrated in Figure 7.10.

The signature algorithm and one-way hash function used to sign a certificate are indicated by use of an algorithm identifier or an OID. The one-way hash functions commonly used are SHA-1 and MD5. RSA and DSA are the most popular signature algorithms used in the X.509 PKI (PKIX).

Because no one can modify the certificate, it can be placed in a directory without any special effort made to protect the certificate. A user can transmit his or her certificate directly to other users. In the case when a CA encompasses several users, there must be a common trust of that CA. These users' certificates can be stored in the directory for access by all users.

When all users in a large community subscribe to the same CA, it may not be practical for these users. With many users, it is more desirable to have a limited number of participating CAs, each CA securely providing its public key to the subordinate users. Since the CA signs the certificates, each user must have a copy of the CA's public key

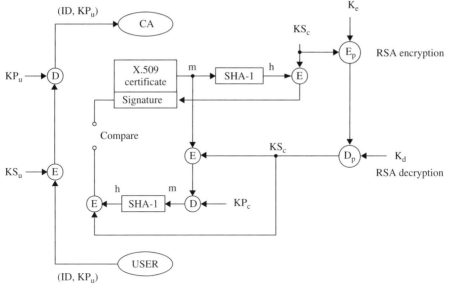

Figure 7.10 Certification of the user's public key.

to verify signatures. The CA should provide its public key to each user in an absolutely secure way so that the user has confidence in the associated certificates.

Suppose there are two users A and B. A certificate is defined in the following notation:

$$X \ll A \gg$$

which means the certificate of user A issued by the CA X. Consider Figure 7.11a which depicts a simple example, where X_1 and X_2 represent two CAs. User A uses a chain of certificates to obtain user B's public key. The chain of certificates is expressed as:

$$X_1 \ll X_2 \gg X_2 \ll B \gg$$

Similarly, user B can obtain A's public key with the reverse chain such that:

$$X_2 \ll X_1 \gg X_1 \ll A \gg$$

This scheme need not be limited to a chain of two certificates. An arbitrarily long path of CAs can produce a chain. All the certificates of CAs by CAs need to appear in the directory, and the user needs to know how they are linked to follow a path to another

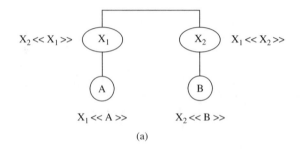

(a)

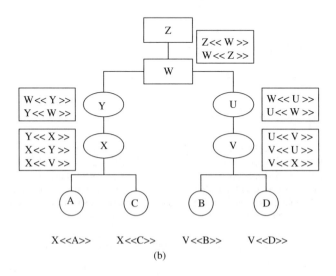

(b)

Figure 7.11   X.509 hierarchical scheme for a chain of certificates.

user's public-key certificate. X.509 suggests that CAs be arranged in a hierarchy so that tracing is straightforward.

Figure 7.11b is an example of such a hierarchy. The connected ellipses and circles indicate the hierarchical relationship among CAs; the associated boxes indicate certificates maintained in the directory for each CA entry. Four users are indicated by circles. In this example, user A can acquire the following certificates from the directory to establish a certification path to user B:

$$X \ll Y \gg Y \ll W \gg W \ll U \gg U \ll V \gg V \ll B \gg$$

When A has obtained these certificates, A can unwrap the certification path in sequence to recover a trusted copy of B's public key. Using this public key, A can send encrypted messages to B. If A wishes to receive encrypted messages back from B, or to sign messages sent to B, then B will require A's public key, which can be obtained from the following certification path:

$$V \ll U \gg U \ll W \gg W \ll Y \gg Y \ll X \gg X \ll A \gg$$

B can obtain this set of certificates from the directory, or A can provide them as part of the initial message to B.

CAs may issue certificates to other CAs with appropriate constraints. Each CA determines the appropriate constraints for path validation by its users. After obtaining the other CA's public key, the CA generates the certificate and posts it to the repository.

The procedure for certifying path validation for the PKIX describes the verification process for binding both the subject distinguished name (DN) and the subject public key. The binding is limited by constraints that are specified in the certificates which comprise the path.

7.4.4 X.500 Distinguished Naming

X.509 v1 and v2 certificates employ X.500 names exclusively to identify subjects and issuers. The information stored in X.500 directories comprises a set of entries. Each entry is associated with a person, an organization, or a device which has a DN. The directory entry for an object contains values of a set of attributes pertaining to that object. For example, an entry for a person might contain values of attributes of type common name, telephone number, e-mail address, and job title. All X.500 entries have the unambiguous naming structure called the *Directory Information Tree* (*DIT*), as shown in Figure 7.12. The DIT has a single conceptual root and unlimited further vertices with DNs. The DN

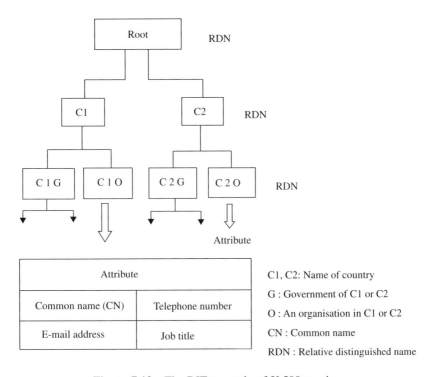

Figure 7.12   The DIT example of X.500 naming.

for an entry is constructed by joining the DN of its immediate superior entry in the tree with a relative distinguished name (RDN).

Suppose a staff member of an organization has an X.500 name. If this person leaves the corporation and a new staff member joins the corporation and is reassigned the same X.500 name, this may cause authorization ambiguities in the access control of X.500 data objects. The idea of the unique identifier fields in the X.509 v2 certificate format is that a new value could be put in this field whenever an X.500 name is reused. Unfortunately, unique identifiers do not contribute a very reliable solution to this problem due to the managing difficulty. A much better approach is to systematically ensure that all X.500 names are unambiguous. This can be achieved by an RDN and a new attribute value, ensuring that employee numbers are not reused over time.

7.4.5   Secure Key Generation and Distribution

Each user must assure the integrity of the received key and must rely on the PKI to supply the public keys generated from associated certificates.

Consider a scenario in which a user's public/private-key pair can be generated, certified, and distributed. There are two ways to consider:

- The user generates his or her own public/private-key pair. In this way, the user is responsible for ensuring that he or she used a good method for generating the key pair. The user is also responsible for having his or her public key certified by a CA. The advantage for the user in generating the key pair is that the user's private key is never released to another entity. This allows for the provision of true nonrepudiation services. The user must store his or her private key in a tamperproof secure location such as on a smart card, floppy disk, or a PCMCIA (Personal Computer Memory Card International Association) card.

- A trusted third party generates the key pair for the user. This method assumes that security measures are employed by the third party to prevent tampering. To obtain a key pair from another entity such as a centralized KG, the user goes to the KG and requests the KG to generate a key pair. This KG will be collocated with either a CA or an ORA. The KG generates the key pair and gives the public and private keys to the user. The private key must certainly be transmitted to the user in a secure manner such as on a token which might be a smart card, a PCMCIA card, or an encrypted diskette. It is not appropriate for the KG to send the user's public key to the CA for certification. It must give the copy of the public key to the user so that he or she can be properly identified during the certificate generating procedure. The KG also automatically destroys the copy of the user's private key once it has been to the user.

If key generation is conducted by a trusted third party on behalf of the user, it is necessary to assure the integrity of the public key and the confidentiality of the private key. Therefore, the generation and distribution of key pairs must be done in a secure fashion.

CA keys are generated by the CA itself. Thus, the PAA, PCAs, and CAs all generate their own key pairs. An ORA can generate its own key pair or have it generated by a third party depending upon the PCA policy. A PCA has its public key certified by the

PAA. At that time, it can obtain the PAA's public key. A CA's public key is certified by the appropriate PCA.

Besides these elements, other important key elements for PKI operations are X.509 certificates, CRLs, and certification path validation. These subjects are covered in the following sections.

7.5 X.509 Certificate Formats

These formats are described in this section, and an algorithm for X.509 certificate path validation is also discussed. The specification profiles the format of certificates and CRLs for the Internet PKIX. Procedures are described for processing certification paths in the Internet environment. Encryption and authentication rules are provided with well-known cryptographic algorithms.

X.500 specifies the directory service. X.509 describes the authentication service using the X.500 directory. A standard certificate format of X.509 which was defined by the ITU-T X.509 (formerly CCITT X.509) or the ISO/IEC/ITU 9594-8 was first published in 1988 as part of the X.500 directory recommendations. The certificate format in the 1988 standard is called the *version 1* (*v1*) format. When X.509 was revised in 1993, two more fields were added, resulting in the version 2 (v2) format. These two fields are used to support directory access control.

The Internet PEM, published in 1993, includes specifications for a PKI based on the X.509 v1 certificate (RFC 1422). Experience has shown that the X.509 v1 and v2 certificate formats are not adequate enough in several aspects. It was found that more fields were needed to contain necessary information for PEM design and implementation. In response to these new requirements, ISO/IEC/ITU and ANSI X9 developed the X.509 v3 certificate format. It extends the v2 format by including additional fields. Standardization of the basic format of X.509 v3 was completed in June 1996.

The standard extensions for use in the v3 extensions field can convey data such as subject identification information, key attribute information, policy information, and certification path constraints. In order to develop interoperable implementations of X.509 v3 systems for Internet use, it is necessary to specify a profile for use of the X.509 v3 extensions for the Internet.

X.509 defines a framework for the provision of authentication services by the X.500 directory to its users. X.509 is an important standard because the certificate structure and authentication protocols defined in X.509 are used in various areas. The X.509 certificate format is used in S/MIME for e-mail security, IPsec (IP security) for network-level security, SSL/TLS (Secure Sockets Layer/Transport-Layer Security) for transport-level security, and SET for secure payment systems.

7.5.1 X.509 v1 Certificate Format

As stated above, the X.509 certificate format has evolved through three versions: version 1 in 1988, version 2 in 1993, and version 3 in 1996. We start by describing the v1 format.

Certificate fields	Interpretation of contents
Version	Version of certificate format
Serial number	Certificate serial number
Signature algorithm	Signature algorithm identifier for certificate issuer's signature
Issuer	CA's X.500 name
Validity period	Start and expiry dates/times
Subject name	Subject X.500 name
Subject public-key information	Algorithm identifier and subject public-key value
Issuer's signature	CA's digital signature

Figure 7.13   X.509 version 1 certificate format.

This format contains information associated with the subject of the certificate and the CA who issued it. The certificate (value equals 0) contains a version number, a serial number, the CA signature algorithm, the names of the subject and issuer, a validity period, a public key associated with the subject, and an issuer's signature. These basic fields are as shown in Figure 7.13. The certificate fields are interpreted as follows:

- *Version.* In this field, the format of the certificate is identified as the indicator of version 1, 2, or 3 format. The 1988 X.509 certificate v1 format is used only when basic fields are present. The value of this field in a v1 format is assigned as "0". The v2 certificate format is assigned the value "1". The value of this field is 2, signifying a v3 certificate.

- *Serial number.* This is an integer assigned by the CA to each certificate. In other words, this field contains a unique identifying number for this certificate, assigned by the issuing CA. The issuer must ensure that it never assigns the same serial number to two distinct certificates.

- *Signature.* The algorithm used by the issuer in order to sign the certificate is specified. The signature field contains the algorithm identifier for the algorithm used to sign the certificate.

- *Issuer.* This field provides a globally unique identifier of the authority signing the certificate. The syntax of the issuer name is an X.500 distinct name. This field contains the X.500 name of the issuer that generated and signed the certificate. The DN is composed of attribute type–attribute value pairs.

- *Validity.* This field denotes the start and expiry dates/times for the certificate. The validity field indicates the dates on which the certificate becomes valid (not before) and on which the certificate ceases to be valid (not after). In other words, it contains two time and date indications that denote the start and the end of the time period for which the certificate is valid. The validity field always uses UTCTime (Coordinated Universal Time), which is expressed in Greenwich Mean Time (Zulu).

- *Subject.* The purpose of the subject field is to provide a unique identifier of the subject of the certificate. The syntax of the subject name will be an X.500 DN. This field contains the name of the entity for whom the certificate is being generated. The field denotes the X.500 name of the holder of the private key, for which the corresponding public key is being certified.

- *Subject public-key information.* This field contains the value of a public key of the subject together with an identifier of the algorithm with which this public key is to be used. It includes the subject public-key field and an algorithm identifier field with algorithm and parameters subfields.

- *Issuer's signature.* This field denotes the CA's signature for which the CA's private key is used. The actual signature on the certificate is defined by the use of a sequence of the data being signed, an algorithm identifier, and a bit string which is the actual signature. The algorithm identifier is used to sign the certificate. Although this algorithm identifier field includes a parameter field that can be utilised to pass the parameters used by the signature algorithm, it is not itself a signed object. The parameter field of the certificate signature is not to be used to pass parameters. When parameters are used to validate a signature, they may be obtained from the subject public-key information field of the issuing CA's certificate.

Experience has shown that the X.509 v1 certificate format is deficient in several respects. The v2 format extends the v1 format by including two more identifier fields.

7.5.2   X.509 v2 Certificate Format

RFC 1422 uses the X.509 v1 certificate format, which imposes several structural restrictions on clearly associating policy information and restricts the utility of certificates. The X.509 v2 format imposed by RFC 1422 can be addressed using two more fields – issuer and subject unique identifiers, which are illustrated in Figure 7.14. These two added fields are interpreted as follows:

- *Issuer unique identifier.* This field is present in the certificate to deal with the possibility of reuse of issuer names over time. In this field, an optional bit string is used to make the issuer's name unambiguous in the event that the same name has been reassigned to different entities over time.

- *Subject unique identifier.* This field is present in the certificate to deal with the possibility of reuse of subject names over time. This field is an optional bit string used to make the subject name unambiguous in the event that the same name has been reassigned to different entities over time.

Certificate fields	Interpretation of two more added fields
v1 = v2 (for seven fields)	Version, serial number, signature algorithm, issuer, validity period, subject name, subject public-key information
Issuer unique identifier	To handle the possibility of reuse of issuer and/or subject names through time
Subject unique identifier	
v1 = v2 (for the last field)	Issuer's signature

Figure 7.14   X.509 version 2 certificate format.

Submissive CAs do not issue certificates that include these unique identifiers. Submissive PKI clients are not required to process certificates that include these unique identifiers. However, if they do not process these fields, they are required to reject certificates that include these fields.

7.5.3   X.509 v3 Certificate Format

The Internet PEM RFCs, published in 1993, include specifications for a PKI based on X.509 v1 certificates. The experience gained from RFC 1422 indicates that the v1 and v2 certificate formats are deficient in several respects. In response to the new requirements and to overcome the deficiencies, ISO/IEC/ITU and ANSI X9 developed the X.509 v3 certificate format. This format extends the v2 format by including provision for additional extension fields. The addition of these extension fields is the principal change introduced to the v3 certificate.

Although the revision to ITU-T X.509 that specifies the v3 format is not yet published, the v3 format has been widely adopted and is specified in ANSI X 9.55–1995, and the IETF's Internet PKI working document (PKIX1). In June 1996, standardization of the basic X.509 v3 was completed. The v3 certificate includes the 11 fields, as shown in Figure 7.15. The version field describes the version of the encoded certificate. The value of this field is 2, signifying a version 3 certificate.

ISO/IEC/ITU and ANSI X9 have also developed standard extensions for use in the v3 extensions field. These extensions can convey data such as additional subject identification information, key attribute information, policy information, and certification path

Certificate fields	Interpretation of contents
v1 = v2 = v3 (for seven fields)	Version, serial number, signature algorithm, issuer, validity period, subject name, subject public-key information
v2 = v3 (for two fields)	Issuer unique identifier Subject unique identifier
Extensions (v3)	Key and policy information Subject and issuer attributes Certification path constraints Extensions related to CRLs
v1 = v2 = v3 (for the last field)	Issuer's signature

Figure 7.15   X.509 version 3 certificate format.

constraints. In order to develop interoperable implementations of X.509 v3 systems for Internet use, it will be necessary to specify a profile for use of the v3 extensions tailored for the Internet.

The extensions defined for the v3 certificates provide methods for associating additional attributes with users or public keys and for managing the certification hierarchy. The v3 format also allows communities to define private extensions to carry information unique to those communities.

Each extension includes an OID and an ASN.1 structure. When an extension appears in a certificate, the OID appears as the field extnID and the corresponding ASN.1 encoded structure is the value of the octet-string extnValue.

Conforming CAs must support such extensions as authority and subject key identifiers, key usage, certification policies, subject alternative name, basic constraint, and name and policy constraints. The format and content of certificate extensions in the Internet PKI are described in the following.

The standard extensions can be divided into the following groups:

- Key and policy information.
- Subject and issuer attributes.
- Certification path constraints.
- Extensions related to CRLs.

Key and Policy Information Extensions

The key and policy information extensions convey additional information about the subject and issuer keys. The extensions also convey indicators of certificate policy. The extensions facilitate the implementation of PKI and allow administrators to limit the purposes for which certificates and certified keys are used.

Authority key identifier extension

The authority key identifier extension provides a mean of identifying the public key corresponding to the private key used to sign a certificate. This extension is used where an issuer has multiple signing keys. The identification is based on either the subject key identifier in the issuer's certificate or the issuer name and serial number.

The key identifier field of the authority key identifier extension must be included in all certificates generated by conforming CAs to facilitate chain building. The value of the key identifier field should be derived from the public key used to verify the certificate's signature or a method that generates unique values. This field helps the correct certificate for the next CA in the chain to be found.

Subject key identifier extension

The subject key identifier extension provides a means of identifying certificates that contain a particular public key.

To facilitate chain building, this extension must appear in all conforming CA certificates including the basic constraints extension. The value of the subject key identifier is the value placed in the key identifier field of the authority key identifier extension of certificates issued by the subject of the certificate.

For CA certificates, subject key identifiers should be derived from the public key or a method that generates unique values. There are two common methods for generating key identifiers from the public key.

- The key identifier is composed of the 160-bit SHA-1 hash value of the bit string of the subject public key.
- The key identifier is composed of a 4-bit-type field with 0100 followed by the least significant 60 bits of the SHA-1 hash value of the bit string of the subject public key.

For end-entity certificates, the subject key identifier extension provides a means of identifying certificates containing the particular public key used in an application. For an end entity which has obtained multiple certificates from multiple CAs, the subject key identifier provides a mean to quickly identify the set of certificates containing a particular public key.

Key usage extension

This extension defines the key usage for encryption, signature, and certificate signing with the key contained in the certificate. When a key which is used for more than one operation is to be restricted, the usage restriction is required to be employed. An RSA key should be used only for signing; the digital signature and/or nonrepudiation bits would be asserted. Likewise, when an RSA key is used only for key management, the key encryption bit would be asserted. Bits in the key usage type are used as follows:

```
Key Usage :: = Bit String {
  digital signature bit        (0)
  nonrepudiation bit           (1)
  key encryption bit           (2)
  data encryption bit          (3)
  key certificate sign bit     (4)
  key agreement sign bit       (5)
  CRL sign bit                 (6)
  encipher only bit            (7)
  decipher only bit            (8) }
```

- The digital signature bit is asserted when the subject public key is used with a digital signature mechanism to support security services other than nonrepudiation (bit 1), certificate signing (bit 5), or revocation information signing (bit 6). Digital signature mechanisms are often used for entity authentication and data origin authentication with integrity.
- The nonrepudiation bit (bit 1) is asserted when the subject public key is used to verify digital signatures used to provide a nonrepudiation service. This service protects against the signing entity falsely denying some action, excluding certificate or CRL signing.
- The key encryption bit (bit 2) is asserted when the subject public key is used for key transport. For example, when an RSA key is used for key management, then this bit will be asserted.
- The data encryption bit (bit 3) is asserted when the subject public key is used to encipher user data, other than cryptographic keys.
- The key agreement bit (bit 4) is asserted when the subject public key is used for key agreement. For example, when the Diffie–Hellman exchange is used for key management, then this bit will be asserted.
- The key certificate signing bit (bit 5) is asserted when the subject public key is used to verify a signature on certificates. This bit is only asserted in CA certificates.
- The CRL sign bit (bit 6) is asserted when the subject public key is used to verify a signature on revocation information.
- The encipher only bit (bit 7) is undefined in the absence of the key agreement bit. When this bit is asserted and the key agreement bit is also set, the subject public key can be used only to encipher data while performing key agreement.
- The decipher only bit (bit 8) is undefined in the absence of the key agreement bit. When the decipher only bit is asserted and the key agreement bit is also set, the subject public key can be used only to decipher data while performing key agreement.

This profile does not restrict the combinations of bits that may be set in an instantiation of the key usage extension.

Private-key usage period extension

This extension allows the certificate issuer to specify a different validity period for the private key than the certificate. The extension is intended for use with digital signature keys and consists of two optional components, "not before" and "not after." The private key associated with the certificate should not be used to sign objects before or after the times specified by the two components, respectively. CAs conforming to this profile must not generate certificates with private-key usage period extensions unless at least one of the two components is present.

Certificate policies extension

This extension contains a sequence of one or more policy information terms, each of which consists of an OID and optional qualifiers. These policy information terms indicate the policy under which the certificate has been issued and the purposes for which it may be used. Optional qualifiers are not expected to change the definition of the policy.

Applications with specific policy requirements are expected to list those policies which they will accept and to compare the policy OIDs in the certificate with that list. If the certificate policies extension is critical, the path validation software must be able to interpret this extension, or must reject the certificate. To promote interoperability, this profile recommends that policy information terms consist only of an OID.

Policy mappings extension

This extension is used in CA certificates. It lists one or more pairs of OIDs. Each pair includes an issuer domain policy and a subject domain policy. The pairing indicates that the issuing CA considers its issuer domain policy equivalent to the subject CA's subject domain policy. The issuing CA's users may accept an issuer domain policy for certain applications. The policy mapping tells the issuing CA's users which policies associated with the subject CA are comparable with the policy they accept. This extension may be supported by CAs and/or applications, and it must be noncritical.

Subject and Issuer Attributes Extensions

These extensions support alternative names for certificate subjects and issuers. They can also convey additional attribute information about the subject to help a certificate user gain confidence that the certificate applies to a particular person, organization, or device. These extensions are described in the following.

Subject alternative name extension

This extension allows additional identities to be bound to the subject of the certificate. Defined options include an Internet e-mail or EDI address, a DNS name, an IP address, and a uniform resource identifier (URI).

Whenever such identities are bound into a certificate, the subject alternative name (or issuer alternative name) extension must be used.

Since the subject alternative name is considered to be definitively bound to the public key, all parts of the subject alternative name must be verified by the CA.

Issuer alternative name extension

As with the previous section, this extension field contains one or more alternative names for the certificate issuer. The name forms are the same as for the subject alternative name extension. This extension is used to associate Internet-style identities with the certificate issuer. This field provides for CAs that are accessed via the Web or e-mail.

Subject directory attributes extension

This extension field conveys any desired X.500 attribute values to the subject of the certificate. It provides a general means of conveying additional identifying information about the subject beyond what is conveyed in the name field. This extension is not recommended as an essential part of this profile, but it may be used in local environments. The extension must be noncritical.

Certification Path Constraints Extensions

These extensions help different organizations link their infrastructures together. When one CA certifies another CA, it can include, in the certificate, information advising certificate users of restrictions on the types of certification paths that can stem from this point. These extensions are as follows:

Basic constraints extension

This indicates whether the certificate subject acts as a CA or is an end entity only. This indicator is important to prevent end users from fraudulently emulating CAs. If the subject acts as a CA, a certification path length constraint may also be specified on how deep a certification path may exist through that CA. For example, this extension field may indicate that certificate users must not accept certification paths that extend more than one certificate from this certificate.

Name constraints extension

This extension must be used only in a CA certificate. The extension indicates a name space within which all subject names in subsequent certificates in a certification path are located. Restrictions apply only when the specified name form, either the subject DN or subject alternative name, is present. In other words, if no name of this type is in the certificate, the certificate is acceptable. Restrictions are defined in terms of permitted or excluded name subtrees. Any name matching a restriction in the excluded subtrees field is invalid regardless of the information appearing in the permitted subtrees.

For URIs, the constraint applies to a host or a domain. Examples would be "foo.bar.com" and ".xyz.com". When the constraint begins with a full stop, the constraint ".xyz.com" can be expanded with one or more subdomains such as "abc.xyz.com" and "abc.def.xyz.com". When the constraint does not begin with a full stop, it specifies a host.

For a name constraint for Internet mail addresses, it specifies a particular mailbox, all addresses at a particular host, or all mailboxes in a domain. To indicate a particular mailbox, the constraint is the complete address. For example, "root@xyz.com" indicates the root mailbox on the host "xyz.com". To indicate all Internet mail addresses on a particular host, the constraint is specified as the host name.

DNS name restrictions are expressed as "foo.bar.com". Any DNS name constructed by simply adding to the left-hand side of the name satisfies the name constraint. For example, "www.foo.bar.com" would satisfy the constraint.

Policy constraints extension

The policy constraints extension is used in certificates issued to CAs. This extension constrains path validation in two ways:

- *Inhibited policy-mapping field.* This field can be used to prohibit policy mapping. If the inhibited policy-mapping field is present, the value indicates the number of additional certificates that may appear in the path before policy mapping is no longer permitted. For example, a value of one indicates that policy mapping is processed in certificates issued by the subject of this certificate, but not in additional certificates in the path.

- *Required explicit policy field.* This field can be used to require that each certificate in a path contains an acceptable policy identifier. If the required explicit policy field is present, subsequent certificates will include an acceptable policy identifier. The value of this explicit field indicates the number of additional certificates that may appear in the path before an explicit policy is required. An acceptable policy identifier is the identifier of a policy required by the user of the certification path or one which has been declared equivalent through policy mapping.

Conforming CAs must not issue certificates where policy constraints form a null sequence. At least one of the inhibited policy-mapping field or the required explicit policy field must be present.

Extended key usage field

This field indicates one or more purposes for which the certified public key can be used in place of the basic purposes in the key usage extension field. Key purposes can be defined by any organization. OIDs used to identify key purposes are assigned in accordance with IANA (Internet Assigned Numbers Authority) or ISO/IEC/ITU 9834-1.

This extension at the option of the certificate issuer is either critical or noncritical. If the extension is flagged as critical, then the certificate must be used only for one of the purposes indicated. If the extension is flagged as noncritical, then it indicates the intended purpose or purposes of the key and can be used to find the correct key/certificate of an entity that has multiple keys/certificates. It is an advisory field and does not imply that usage of the key is restricted by the CA to the purpose indicated.

If a certificate contains both a critical key usage field and a critical extended key usage field, then both fields must be processed independently and the certificate must only be used for a purpose consistent with both fields. If there is no purpose consistent with both fields, then the certificate must not be used for any purpose.

CRL distribution points extension

The CRL distribution points extension identifies how CRL information is obtained. The extension should be noncritical, but CAs and applications must support it. If this extension contains a distribution point name of type URL, the URI is a pointer to the CRL. When the subject alternative name extension contains a URI, the name must be stored in the URI (an IA5String).

Private Internet Extensions

This section defines one new extension for use in the Internet PKI. This extension may be used to direct applications to identify an online validation service supporting the issuing CA. As the information may be available in multiple forms, each extension is a sequence of IA5String values, each of which represents a URI. The URI implicitly specifies the location and format of the information. It also specifies the method for obtaining the information.

An OID is defined for the private extension. The OID associated with the private extension is defined under the arc id-pe within the id-pkix name space. Any future extensions defined for the Internet PKI will also be defined under the arc id-pe.

Authority information access extension

This extension indicates how to access CA information and services for the issuer of the certificate in which the extension appears. Information and services may include online validation services and CA policy data.

Each entry in this information access syntax describes the format and location of additional information about the CA who issued the certificate. The information type and format are specified by the access method field, while the access location field specifies the location of the information. The retrieval mechanism may be implied by the access method or specified by the access location.

This profile defines one OID for the access method. The id-ad-caIssuers OID is used when the additional information lists CAs that have issued certificates superior to the CA that issued the certificate containing this extension. The referenced CA issuers description is intended to help certificate users select a certification path that terminates at a point trusted by the certificate user.

When id-ad-caIssuers appears as the access information type, the access location field describes the referenced description server and the access protocol to obtain the referenced description. The access location field is defined as a general name, which can take several forms.

Where the information is available via http, ftp, or ldap, the access location must be a URI. Where the information is available via the Directory Access Protocol (dap), the access location must be a directory name. When the information is available via e-mail, the access location must be an RFC 2822 name.

7.6   Certificate Revocation List

CRLs are used to list unexpired certificates that have been revoked. Certificates may be revoked for a variety of reasons, ranging from routine administrative revocations to situations where the private key is compromised.

CRLs are used in a wide range of applications and environments covering a broad spectrum of interoperability goals and an even broader spectrum of operational and assurance requirements.

The ISO/IEC/ITU X.509 standard also defines the X.509 CRL format that, like the certificate format, has evolved somewhat since first appearing in 1998. In fact, when the extensions field was added to the certificate to create the X.509 v3 certificate format, the same type of mechanism was added to the CRL to create the X.509 v2 CRL format. The main elements of the X.509 v2 CRL are shown in Figure 7.16. The X.509 v2 CRL format is augmented by several optional extensions, similar in concept to those defined for certificates. CAs are able to generate X.509 v2 CRLs.

7.6.1   CRL Fields

The following items describe the use of the X.509 v2 CRL:

- *Version.* This optional field describes the version of the encoded CRLs. The integer value of this field is 1, indicating a v2 CRL. When extensions are used, this field must be present and must specify the v2 CRL.

- *Signature.* This field contains the algorithm identifier for the algorithm used to sign the CRL. The signature algorithm and one-way hash function used to sign a certificate or CRL is indicated by the use of an algorithm identifier. The algorithm identifier is an OID, and possibly includes associated parameters. RSA and DSA are the most popular signature algorithms used in the Internet PKI. The one-way hash functions commonly used are MD5 and SHA-1.

- *Issuer name.* This identifies the entity which has signed and issued the CRL. The issuer identity is carried in the issuer name field. Alternative name forms may also appear in the issuer alternative name extension. The issuer name is an X.500 DN. The issuer name field is defined as the X.501 type name and must follow the encoding rules for the issuer name field in the certificate.

Version (optional)	X.509 CRL format This field is present only if extensions are used
Signature	For CRL issuer's signature, signature algorithm (RSA or DSA), and hash function (MD5 or SHA-1)
Issuer name	CRL issuer (X.500 distinguished name)
This update	Issue the data of CRL (date/time)
Next update	Issue the CRLs with a next update time equal to or later than all previous CRLs (date/time)
Revoked certificates	A list of certificates have been revoked: • Identify uniquely by certificate serial number • Date on the revocation occurrence is specified • Optional CRL entry extensions: - Give the reason for revoked certificate - State the data for invalidity - State the name of CA issuing the revoked certificate
CRL extensions	Authority key identifier Issuer alternative name CRL number, delta CRL indicator, issuing distribution point
CRL entry extensions	Reason code Hold instruction code Invalidity date certificate issuer
CRL issuer's digital signature	—

Figure 7.16 X.509 v2 CRL format.

- *This update.* This field indicates the issue date of the CRL. The update field may be encoded as UTCTime or GeneralisedTime. CAs conforming to this field that issue CRLs must encode this update as UTCTime for dates to the year 2049 and as GeneralisedTime for dates to the year 2050 or later. For this specification, where encoded as UTCTime, the update field must be specified and interpreted as defined in the rules for the certificate validity field.

- *Next update.* This field indicates the date by which the next CRL will be issued. It could be issued before the indicated date, but it will not be issued any later than that date. CAs should issue CRLs with a next update time equal to or later than all previous CRLs. The next update field may be encoded as UTCTime or GeneralisedTime.

  This profile requires inclusion of the next update field in all CRLs issued by conforming CAs. Note that the ASN.1 syntax of TBCCertList described this field as optional, which is consistent with the ASN.1 structure defined in X.509. CAs conforming to this profile that issue CRLs must encode the next update as UTCTime for dates to the year 2049 and as GeneralisedTime for dates to the year 2050 or later. For this specification, the next update field should follow the rules for the certificate validity field.

- *Revoked certificates.* This field is a list of the certificates that have been revoked. Each revoked certificate listed contains the following:

  - The revoked certificates are identified by their serial numbers. Certificates revoked by the CA are uniquely identified by the certificate serial number.
  - The date on which the revocation occurred is specified. The time for revocation must be encoded as UTCTime or GeneralisedTime.
  - The optional CRL entry extensions may give the reason why the certificate was revoked, state the date when the invalidity is believed to have occurred, and may state the name of the CA that issued the revoked certificate, which may be a different CA from the one issuing the CRL. Note that the CA that issued the CRL is assumed to be the one that issued the revoked certificate unless the certificate issuer CRL entry extension is included.

7.6.2   CRL Extensions

The extensions defined by ANSI X9 and ISO/IEC/ITU for X.509 v2 CRLs provide methods for associating additional attributes with CRLs. The X.509 v2 CRL format also allows communities to define private extensions to carry information unique to those communities. Each extension in a CRL is designated as critical or noncritical. A CRL validation must fail if it encounters a critical extension which it does not know how to process. However, an unrecognized noncritical extension may be ignored. The extensions used within Internet CRLs will be presented in the following:

- *Authority key identifier.* This extension provides a means of identifying the public key corresponding to the private key used to sign a CRL. The identification can be based on either the key identifier or the issuer name and serial number. This extension is particularly useful where an issuer has more than one signing key, either due to multiple concurrent key pairs or due to changeover.

- *Issuer alternative name.* This extension is a noncritical CRL extension that allows additional identities to be associated with the issuer of the CRL. Defined options include an e-mail address, a DNS name, an IP address, and a URI. Multiple instances of a name and multiple name forms may be included. Whenever such identities are used, the issuer alternative name extension must also be used. CAs are capable of generating this extension in CRLs, but clients are not required to process it.

- *CRL number.* This field is a noncritical CRL extension which conveys a monotonically increasing sequence number to each CRL issued by a CA. This extension allows users to easily determine when a particular delete CRL is replaced by another CRL. CAs conforming to this profile must include this extension in all CRLs.

- *Delta CRL indicator.* This is a critical CRL extension that identifies a delta CRL. The use of delta CRLs can significantly improve processing time for applications which store revocation information in a format other than the CRL structure. This allows changes to be added to the local database while ignoring unchanged information that is already in the local database. When a delta CRL is issued, the CAs must also issue a complete CRL.

The value of the base CRL number identifies the CRL number of the base CRL that was used as the starting point in the generation of this delta CRL. The delta CRL contains the changes between the base CRL and the current CRL issued along with the delta CRL. It is the decision of a CA as to whether to provide delta CRLs. Again, a delta CRL must not be issued without a corresponding complete CRL. The value of the CRL number for both the delta CRL and the corresponding complete CRL must be identical.

A CRL user constructing a locally held CRL from delta CRLs must consider the constructed CRL as incomplete and unusable if the CRL number of the received delta CRL is more than one greater than the CRL number of the delta CRL last processed.

- *Issuing distribution point.* The issuing distribution point is a critical CRL extension that identifies the CRL distribution point for a particular CRL, and it indicates whether the CRL covers revocation for end-entity certificates only, CA certificates only, or a limited set of reason codes that have been revoked for a particular reason. Although the extension is critical, conforming implementations are not required to support this extension. The CRL is signed using the CA's private key. CRL distribution points do not have their own key pairs. If the CRL is stored in the X.500 directory, it is stored in the directory entry corresponding to the CRL distribution point, which could be different from the directory entry of the CA.

  The reason codes associated with a distribution point are specified in onlySomeReasons. A ReasonsFlag bit string indicates the reasons for which certificates are listed in the CRL. If onlySomeReasons does not appear, the distribution point contains revocations for all reason codes. CAs may use the CRL distribution point to partition the CRL on the basis of compromise and routine revocation. The revocations with reason code keyCompromise (used to indicate compromise or suspected compromise) and cACompromise (used to indicate that the certificate has been revoked because of a CA key compromise) appear in one distribution point, and the revocations with other reason codes appear in another distribution point.

7.6.3 CRL Entry Extensions

The CRL entry extensions already defined by ANSI X9 and ISO/IEC/ITU for X.509 v2 CRLs provide methods for associating additional attributes with CRL entries. The X.509 v2 CRL format also allows communities to define private CRL entry extensions to carry information unique to those communities. Each extension in a CRL entry is designated as critical or noncritical. A CRL validation must fail if it encounters a critical CRL entry extension which it does not know how to process. However, an unrecognized noncritical CRL entry extension may be ignored. The following list presents recommended extensions used within Internet CRL entries and standard locations for information.

All CRL entry extensions used in this specification are noncritical. Support for these extensions is optional for conforming CAs and applications. However, CAs that issue CRLs must include reason codes and invalidity dates whenever this information is available.

- *Reason code.* This is a noncritical CRL entry extension that identifies the reason for revocation of the certificate. CAs are strongly encouraged to include meaningful reason codes in CRL entries. However, the reason code CRL entry extension must be absent instead of using the unspecified reason code value (0). The following enumerated reasonCode values are defined:

  - unspecified (0) should not be used.
  - all keyCompromise (1) indicates compromise or suspected compromise.
  - cACompromise (2) indicates that the certificate has been revoked because of a CA key compromise. It is only used to revoke CA certificates.
  - affiliationChanged (3) indicates that the certificate was revoked because of a change of affiliation in the certificate subject.
  - superseded (4) indicates that the certificate has been replaced by a more recent certificate.
  - cessationOfOperation (5) indicates that the certificate is no longer needed for the purpose for which it was issued, but there is no reason to suspect that the private key has been compromised.
  - certificateHold (6) indicates that the certificate will not be used at this time. When clients process a certificate that is listed in a CRL with a reasonCode of certificate-Hold, they will fail to validate the certification path.
  - removeFromCRL (7) is used only with delta CRLs and indicates that an existing CRL entry should be removed.

- *Hold instruction code.* This code is a noncritical CRL entry extension that provides a registered instruction identifier. This identifier indicates the action to be taken after encountering a certificate that has been placed on hold.

- *Invalidity date.* This is a noncritical CRL entry extension that provides the date on which it is known or suspected that the private key was compromised or that the certificate otherwise became invalid. The invalidity date is the date at which the CA processed the revocation, but it may be earlier than the revocation date in the CRL entry. When a revocation is first posted by a CA in a CRL, the invalidity date may precede the date of issue of earlier CRLs. However, the revocation date should not precede the date of issue of earlier CRLs. Whenever this information is available, CAs are strongly encouraged to share it with CRL users. The generalized time values included in this field must be expressed in Greenwich Mean Time (Zulu).

- *Certificate issuer.* This CRL entry extension identifies the certificate issuer associated with an entry in an indirect CRL (i.e., a CRL that has the indirect CRL indicator set in its issuing distribution point extension). If this extension is not present on the first entry of an indirect CRL, the certificate issuer defaults to the CRL issuer. On subsequent entries of an indirect CRL, if this extension is not present, the certificate issuer for the entry is the same as the issuer of the preceding CRL entry.

7.7 Certification Path Validation

The certification path validation procedure for the Internet PKI describes the verification process for the binding between the subject DN and/or subject alternative name and subject public key. The binding is limited by constraints that are specified in the certificates which comprise the path.

This section describes an algorithm for validating certification paths. For basic path validation, all valid paths begin with certificates issued by a single most-trusted CA. The algorithm requires the public key of the CA, the CA's name, the validity period of the public key, and any constraints upon the set of paths which may be validated using this key. Depending on policy, the most-trusted CA could be a root CA in a hierarchical PKI, the CA that issued the verifier's own certificate, or any other CA in a network PKI. The path validation procedure is the same regardless of the choice of the most-trusted CA.

This section also describes extensions to the basic path validation algorithm. Two specific cases are considered: (i) the case where paths are begun with one of several trusted CAs and (ii) the case where compatibility with the PEM architecture is required.

7.7.1 Basic Path Validation

It is assumed that the trusted public-key and related information are contained in a self-signed certificate in order to simplify the description of the path processing procedure. Note that the signature on the self-signed certificate does not provide any security services.

The goal of path validation is to verify the binding between a subject DN or subject alternative name and subject key, as represented in the end-entity certificate, based on the public key of the most-trusted CA. This requires obtaining a sequence of certificates which support that binding.

A certification path is a sequence of n certificates where, for all x in $\{1, (n-1)\}$, the subject of certificate x is the issuer of certificate $x + 1$. Certificate $x = 1$ is the self-signed certificate, and certificate $x = n$ is the end-entity certificate.

The inputs that are provided to the path processing logic are assumed as follows:

- A certificate path of length n.
- A set of initial policy identifiers which identifies one or more certificate policies.
- The current date and time.
- The time T for which the validity of the path must be determined.

From the inputs, the procedure initializes five state variables:

- *Acceptable policy set.* A set of certificate policy identifiers comprising the policy or policies recognized by the public-key user together with policies considered equivalent through policy mapping.
- *Constrained subtrees.* A set of root names defining a set of subtrees within which all subject names in subsequent certificates in the certification path will fall.
- *Excluded subtrees.* A set of root names defining a set of subtrees within which no subject name in subsequent certificates in the certification path may fall.

- *Explicit policy.* An integer that indicates if an explicit policy identifier is required. The integer indicates the first certificate in the path where this requirement is imposed.
- *Policy mapping.* An integer which indicates if policy mapping is permitted. The integer indicates the last certificate on which policy mapping can be applied.

The actions performed by the path processing software for each certificate $x = 1$ to n are described below. The self-signed certificate is $x = 1$ and the end-entity certificate is $x = n$.

- Verify the basic certificate information:

  - The certificate was signed using the subject public key from certificate $x - 1$. For the special case $x = 1$, this step is omitted.
  - The certificate validity period includes time T.
  - The certificate had not been revoked at time T and is not currently on hold, a status that commenced before time T.
  - The subject and issuer names chain correctly; that is, the issuer of this certificate was the subject of the previous certificate.

- Verify that the subject name and subject alternative name extension are consistent with the constrained subtree state variables.
- Verify that the subject name and subject alternative name extension are consistent with the excluded subtree state variables.
- Verify that policy information is consistent with the initial policy set:

  - If the explicit policy state variable is less than or equal to x, a policy identifier in the certificate should be in the initial policy set.
  - If the policy-mapping variable is less than or equal to x, the policy identifier may not be mapped.

- Verify that policy information is consistent with the acceptable policy set:

  - If the certificate policies extension is marked as critical, the intersection of the policies extension and the acceptable policy set will be non-null.
  - The acceptable policy set is assigned the resulting intersection as its new value.

- Verify that the intersection of the acceptable policy set and the initial policy set is non-null.
- Recognize and process any other critical extension present in the certificate.
- Verify that the certificate is a CA certificate as specified in a basic constraints extension or as verified out of band.
- If permittedSubtrees is present in the certificate, set the constrained subtree state variable to the intersection of its previous value and the value indicated in the extension field.
- If excludedSubtrees is present in the certificate, set the excluded subtree state variable to the union of its previous value and the value indicated in the extension field.
- If a policy constraints extension is included in the certificate, modify the explicit policy and policy-mapping state variable as follows:

  - If the required explicit policy is present and has value r, the explicit policy state variable is set to the minimum of its current value and the sum of r and x.

- If the inhibited policy mapping, whose value is q, is present, the policy-mapping state variable is set to the minimum of its current value and the sum of q and x.
- If a key usage extension is marked as critical, ensure that the KeyCertSign bit is set.

If any one of the above checks fails, the procedure terminates, returning a failure indication and an appropriate reason. If none of the above checks fail on the end-entity certificate, the procedure terminates, returning a success indication together with the set of all policy qualifier values encountered in the set of certificates.

7.7.2 Extending Path Validation

The path validation algorithm presented in Section 7.7.1 is based on a simplifying assumption, that is, a single trusted CA that starts all valid paths. This algorithm can be extended for multiple trusted CAs by providing a set of self-signed certificates to the validation module. In this case, a valid path could begin with any one of the self-signed certificates. Limitations in the trust paths for any particular key may be incorporated into the self-signed certificate's extensions. In this way, the self-signed certificates permit the path validation module to automatically incorporate local security policy and requirements.

It is also possible to specify an extended version of the above certification path processing procedure which results in a default behavior identical to the rules of PEM of REC 1422. In this extended version, additional inputs to the procedure are a list of one or more PCA names and an indicator of the position in the certification path where the PCA is expected. At the nominated PCA position, if the CA name is found, then a constraint of SubordinateToCA is implicitly assumed for the remainder of the certification path and processing continues. If no valid PCA name is found, and if the certification path cannot be validated on the basis of identified policies, then the certification path is considered invalid.

The PKI scheme discussed in this chapter is chiefly embodied in the US scheme of PKI. After the appearance of the US version, several countries devised their own PKI systems, mostly derived from many of the principles and system architectures originating from the US PKI scheme. These systems are:

USA:	Federal Public Key Infrastructure (FPKI)
Europe:	European Trusted Service (ETS) and Internetworking Public Key Certification of Europe (ICE-TEL)
Australia:	Public Key Authentication Framework (PKAF)
Canada:	Government of Canada Public Key Infrastructure (GoC-PKI)
Korea:	GPKI for government sector and NPKI for Civilian sector

It will be worthwhile for readers to examine each country's PKI system through its web site.

8

Network Layer Security

TCP/IP communication can be made secure with the help of cryptography. Cryptographic methods and protocols have been designed for different purposes in securing communication on the Internet. These include the SSL and TLS for HTTP Web traffic, S/MIME and PGP for e-mail, and IPsec (Internet Protocol security) for network layer security. This chapter mainly addresses security only at the IP layer and describes various security services for traffic offered by IPsec.

8.1 IPsec Protocol

IPsec is designed to protect communication in a secure manner by using TCP/IP. The IPsec protocol is a set of security extensions developed by the IETF and it provides privacy and authentication services at the IP layer by using modern cryptography.

To protect the contents of an IP datagram, the data is transformed using encryption algorithms. There are two main transformation types that form the basics of IPsec, the Authentication Header (AH) and the Encapsulating Security Payload (ESP). Both AH and ESP are two protocols that provide connectionless integrity, data origin authentication, confidentiality, and an antireplay service. These protocols may be applied alone or in combination to provide a desired set of security services for the IP layer. They are configured in a data structure called a Security Association (SA).

The basic components of the IPsec architecture are explained in terms of the following functionalities:

- Security protocols for AH and ESP.
- SAs for policy management and traffic processing.
- Manual and automatic key management for the Internet Key Exchange (IKE), the Oakley key determination protocol, and the Internet Security Association and Key Management Protocol (ISAKMP).
- Algorithms for authentication and encryption.

Wireless Mobile Internet Security, Second Edition. Man Young Rhee.
© 2013 John Wiley & Sons, Ltd. Published 2013 by John Wiley & Sons, Ltd.

The set of security services provided at the IP layer includes access control, connectionless integrity, data origin authentication, protection against replays, and confidentiality. The modularity which is designed to be algorithm independent permits selection of different sets of algorithms without affecting the other parts of the implementation.

A standard set of default algorithms is specified to facilitate interoperability in the global Internet. The use of these algorithms in conjunction with IPsec traffic protection and key management protocols is intended to permit system and application developers to deploy high-quality, Internet layer, cryptographic security technology. Thus, the suite of IPsec protocols and associated default algorithms is designed to provide high-quality security for Internet traffic.

An IPsec implementation operates in a host or a security gateway environment, affording protection to IP traffic. The protection offered is based on requirements defined by a Security Policy Database (SPD) established and maintained by a user or system administrator.

IPsec provides security services at the IP layer by enabling a system to select the required security protocols, determine the algorithms to use for the services, and put in place any cryptographic keys required to provide the requested service. IPsec can be used to protect one or more paths between a pair of hosts, between a pair of security gateways (routers or firewalls), or between a security gateway and a host.

8.1.1 IPsec Protocol Documents

This section will discuss the protocols and standards which apply to IPsec. The set of IPsec protocols is divided into seven groups as illustrated in Figure 8.1.

In November 1998, the Network Working Group of the IETF published RFC 2411 for IP Security Document Roadmap. This document is intended to provide guidelines for the development of collateral specifications describing the use of new encryption and authentication algorithms used with the AH protocol as well as the ESP protocol. Both these protocols are part of the IPsec architecture. The seven-group documents describing the set of IPsec protocols are explained in the following:

- *Architecture.* The main architecture document covers the general concepts, security requirements, definitions, and mechanisms defining IPsec technology.
- *ESP.* This document covers the packet format and general issues related to the use of the ESP for packet encryption and optional authentication. This protocol document also contains default values, if appropriate, and dictates some of the values in the Domain of Interpretation (DOI).
- *AH.* This document covers the packet format and general issues related to the use of AH for packet authentication. This document also contains default values, such as the default padding contents, and dictates some of the values in the DOI document.
- *Encryption algorithm.* This is a set of documents that describes how various encryption algorithms are used for ESP. Specifically,
  - specification of the key sizes and strengths for each algorithm;
  - any available estimates on performance of each algorithm;

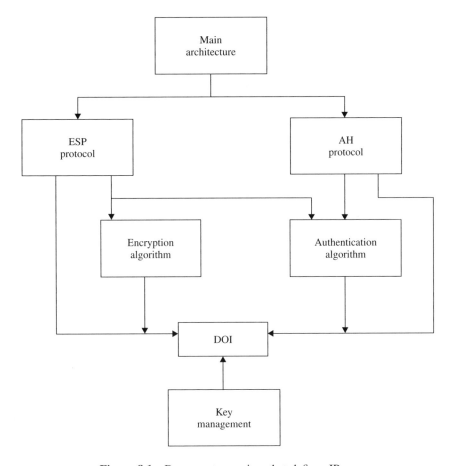

Figure 8.1 Document overview that defines IPsec.

- general information on how this encryption algorithm is to be used in ESP; features of this encryption algorithm to be used by ESP, including encryption and/or authentication.

When these encryption algorithms are used for ESP, the DOI document has to indicate certain values, such as an encryption algorithm identifier, so these documents provide input to the DOI.

- *Authentication algorithm.* This is a set of documents that describes how various authentication algorithms are used for AH and for the authentication option of ESP. Specifically,
  - specification of operating parameters such as number of rounds and input or output block format;
  - implicit and explicit padding requirements of this algorithm;
  - identification of optional parameters/methods of operation;
  - defaults and mandatory ranges of the algorithm;
  - authentication data comparison criteria for the algorithm.

- *Key management.* This is a set of documents that describes key management schemes. These documents also provide certain values for the DOI. Currently, the key management represents the Oakley, ISAKMP, and Resolution protocols.
- *DOI.* This document contains values needed for the other documents to relate each other. These include identifiers for approved encryption and authentication algorithms, as well as operational parameters such as key lifetime.

8.1.2   Security Associations (SAs)

An SA is fundamental to IPsec. Both AH and ESP make use of SAs. Thus, the SA is a key concept that appears in both the authentication and confidentiality mechanisms for IPsec. An SA is a simplex connection between a sender and receiver that affords security services to the traffic carried on it. If both AH and ESP protection are applied to a traffic stream, then two SAs are required for two-way secure exchange.

An SA is uniquely identified by three parameters as follows:

- *Security Parameter Index. (SPI)* This is assigned to each SA, and each SA is identified through an SPI. A receiver uses the SPI to identify the SA for a packet. Before a sender uses IPsec to communicate with a receiver, the sender must know the index value for a particular SA. The sender then places the value in the SPI field of each outgoing datagram. The SPI is carried in AH and ESP headers to enable the receiver to select the SA under which a received packet is processed. However, index values are not globally specified. A combination of the destination address and SPI is needed to identify an SA.
- *IP destination address.* Because, at present, unicast addresses are only allowed by IPsec SA management mechanisms, this is the address of the destination endpoint of the SA. The destination endpoint may be an end user system or a network system such as a firewall or router.
- *Security protocol identifier.* This identifier indicates whether the association is an AH or an ESP SA.

There are two nominal databases in a general model for processing IP traffic relative to SAs, namely, the SPD and the Security Association Database (SAD). To ensure interoperability and to provide a minimum management capability that is essential for productive use of IPsec, some external aspects for the processing standardization are required.

The SPD specifies the policies that determine the disposition of all IP traffic inbound or outbound from a host or security gateways, while the SAD contains parameters that are associated with each SA.

Security Policy Database

The SPD, which is an essential element of SA processing, specifies what services are to be offered to IP datagrams and in what fashion. The SPD is used to control the flow of all traffic (inbound and outbound) through an IPsec system, including security and key management traffic (i.e., ISAKMP). The SPD contains an ordered list of policy entries. Each policy entry is keyed by one or more selectors that define the set of all IP traffic encompassed by this entry. Each entry encompasses every indication mechanism for bypassing, discarding, or IPsec processing. The entry for IPsec processing includes SA (or SA bundle) specification, limiting the IPsec protocols, modes, and algorithms to be employed.

Security Association Database

The SAD contains parameters that are associated with each SA. Each SA has an entry in the SAD. For outbound processing, entries are pointed out by entries in the SPD. For inbound processing, each entry in the SAD is indexed by a destination IP address, IPsec protocol type, and SPI.

Transport Mode SA

There are two types of SAs to be defined: a transport mode SA and a tunnel mode SA. A transport mode provides protection primarily for upper-layer protocols, that is, a TCP packet or UDP segment or an Internet Control Message Protocol (ICMP) packet, operating directly above the IP layer. A transport mode SA is an SA between two hosts. When a host runs AH or ESP over IPv4, the payload is the data that normally follows the IP header. For IPv6, the payload is the data that normally follows both the IP header and any IPv6 extension headers. In the case of AH, AH in transport mode authenticates the IP payload and the protection is also extended to selected portions of the IP header, selected portions of IPv6 extension headers, and the selected options.

In the case of ESP, ESP in transport mode primarily encrypts and optionally authenticates the IP payload but not the IP header. A transport mode SA provides security services only for higher-layer protocols, not for the IP header or any extension headers proceeding the ESP header.

Tunnel Mode SA

Tunnel mode provides protection to the entire IP packet. A tunnel mode SA is essentially an SA applied to an IP tunnel. Whenever either end of an SA is a security gateway, the SA must be tunnel mode, as is an SA between a host and a security gateway. Note that a host must support both transport and tunnel modes, but a security gateway is required to support only the tunnel mode. If a security gateway supports transport mode, it should be used as an acting host. But in this case, the security gateway is not acting as a gateway.

When the entire inner (original) packet travels through a tunnel from one point of the IP network to another, routers along the path are unable to examine the inner IP header because the original inner packet is encapsulated. As a result, the new larger packet will have totally different source and destination addresses. When the AH and ESP fields are added to the IP packet, the entire packet plus the security field (AH or ESP) is treated as the new outer IP packet with a new outer IP header.

ESP in tunnel mode encrypts and optionally authenticates the entire inner IP packet, including the inner IP header. AH in tunnel mode authenticates the entire inner IP packet and selected portions of the outer IP header.

8.1.3 Hashed Message Authentication Code (HMAC)

A mechanism that provides a data integrity check based on a secret key is usually called the *Message Authentication Code* (MAC). An HMAC (Hashed Message Authentication Code) mechanism can be used with any iterative hash functions in combination with a secret key. MACs are used between two parties (e.g., client and server) that share a secret key in order to validate information transmitted between them. An MAC mechanism based on a cryptographic hash function is called *HMAC*. MD5 and SHA-1 are examples of such hash functions. HMAC uses a secret key for computation and verification of the message authentication values. The MAC mechanism should allow for easy replacement of the embedded hash function in case faster or more secure hash functions are found or required; that is, if it is desired to replace a given hash function in an HMAC implementation, all that is required is simply to remove the existing hash function module and replace it with the new, more secure module. HMAC can be proven as secure, provided the underlying hash function has some reasonable cryptographic strengths.

Current candidates for secure hash functions include SHA-1, MD5, and RIPEMD-160. Hash functions such as MD5 and SHA-1 are generally known to execute faster in software than symmetric block ciphers such as DES–CBC (Data Encryption Standard-Cipher Block Chaining). There has been a number of proposals for the incorporation of a secret key into an existing hash function. MD5 has been recently shown to be vulnerable to collision search attacks. Therefore, it seems that MD5 does not compromise its use within HMAC because it does not rely on a secret key. However, SHA-1 appears to be a cryptographically stronger function.

HMAC Structure

HMAC is a secret-key authentication algorithm which provides both data integrity and data origin authentication for packets sent between two parties. Its definition requires a cryptographic hash function H and a secret key K. H denotes a hash function where the message is hashed by iterating a basic compression function on data blocks. Let b denote the block length of 64 bytes or 512 bits for all hash functions such as MD5 and SHA-1. h denotes the length of hash values, that is, $h = 16$ bytes or 128 bits for MD5 and 20 bytes or 160 bits for SHA-1. The secret key K can be of any length up to $b = 512$ bits.

To compute HMAC over the message, the HMAC equation is expressed as follows:

$$\text{HMAC} = H\left[(K \oplus \text{opad}) \| H\left[(K \oplus \text{ipad}) \| M\right]\right]$$

where

  ipad $= 00110110(0\text{x}36)$ repeated 64 times (512 bits);
  opad $= 01011100\ (0\text{x}5\text{c})$ repeated 64 times (512 bits);
  ipad is inner padding and opad is outer padding.

The following explains the HMAC equation:

1. Append 0s to the end of K to create a b-byte string (i.e., if $K = 160$ bits in length and $b = 512$ bits, then K will be appended with 352 0 bits or 44 0 bytes 0x00).

2. XOR (bitwise exclusive-OR) K with ipad to produce the b-bit block computed in step 1.
3. Append M to the b-byte string resulting from step 2.
4. Apply H to the stream generated in step 3.
5. XOR (bitwise exclusive-OR) K with opad to produce the b-byte string computed in step 1.
6. Append the hash result H from step 4 to the b-byte string resulting from step 5.
7. Apply H to the stream generated in step 6 and output the result.

Figure 8.2 illustrates the overall operation of HMAC–MD5.

Example 8.1

HMAC–MD5 computation using the RFC method:

Data: 0x 2143f501 f014a713 c1059e23 7123fd68

Key: 0x 31fa7062 c45113e3 2679fd13 53b71264

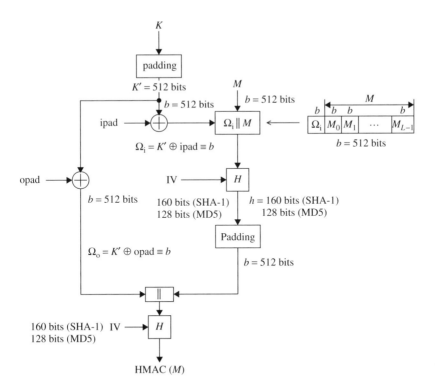

Figure 8.2 Overall operation of HMAC computation using either MD5 or SHA-1 (message length computation based on $\Omega_i \| M$).

	A	B	C	D
IV	67452301	efcdab89	98badcfe	10325476
$H[(K \oplus \text{ipad})\|M]$	4f556d1d	62d021b7	6db31022	00219556
$H[(K \oplus \text{opad})\| H[(K \oplus \text{ipad})\|M]]$	b1c3841c	73b63dff	1a22d4bd	f468e7b4

HMAC–MD5 = 0 x b1c3841c 73b63dff 1a22d4bd f468e7b4

The alternative operation for computation of either HMAC–MD5 or HMAC–SHA-1 is described in the following:

1. Append 0s to K to create a b-bit string K', where $b = 512$ bits.
2. XOR K' (padding with zero) with ipad to produce the b-bit block.

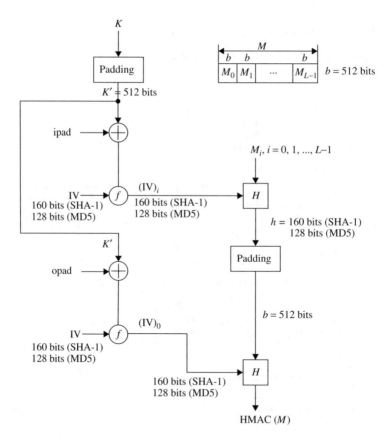

Figure 8.3 Alternative operation of HMAC computation using either MD5 or SHA-1 (message length computation based on M only).

3. Apply the compression function $f(\text{IV}, K' \oplus \text{ipad})$ to produce $(\text{IV})_i = 128$ bits.
4. Compute the hash code h with $(\text{IV})_i$ and M_i.
5. Raise the hash value computed from step 4 to a b-bit string.
6. XOR K' (padded with zeros) with opad to produce the b-bit block.
7. Apply the compression function $f(\text{IV}, K' \oplus \text{opad})$ to produce $(\text{IV})_o = 128$ bits.
8. Compute the HMAC with $(\text{IV})_o$ and the raised hash value resulting from step 5.

Figure 8.3 shows the alternative scheme based on these steps.

Example 8.2

HMAC–SHA-1 computation using alternative method:

Data: 0x 7104f218 a3192e65 1cf7025d 8011bf79 4a19

Key: 0x 31fa7062 c45113e3 2679fd13 53b71264

	A	B	C	D	E
IV	67452301	efcdab89	98badcfe	10325476	c3d2e1f0
$f[(K \oplus \text{ipad}), \text{IV}] = (\text{IV})_i$	c6edf676	ef938cee	84dd1b00	5b3be996	cb172ad4
$H[M, (\text{IV})_i]$	f75ebdde	df6b486e	796daefd	e9cadc38	6bb33c7d
$f[(K \oplus \text{opad}), \text{IV}] = (\text{IV})_o$	a46e7eba	64c80ca4	c42317b3	dd2b4f1e	81c21ab0
$H[H[M, (\text{IV})_i], (\text{IV})_o]$	ee70e949	d7439e60	7865108b	6325235f	e220024e

HMAC–SHA-1 = 0x ee70e949 d7439e60 7865108b 6325235f e220024e

8.2 IP Authentication Header

The IP AH is used to provide data integrity and authentication for IP packets. It also provides protection against replays. The AH provides authentication for the IP header, as well as for upper-level protocol (TCP, UDP) data. But some IP header fields may change in transit and the sender may not be able to predict the value of these fields when the packet arrives at the receiver. Thus, the protection provided to the IP header by AH is somewhat piecemeal. The AH can be used in conjunction with ESP or with the use of tunnel mode. Security services can be provided between a pair of hosts, between a pair of security gateway, or between a security gateway and a host. The ESP provides a confidentiality service. The primary difference between the authentication provided by ESP and that by AH is the extent of the coverage. ESP does not protect any IP header fields unless these fields are encapsulated by ESP (tunnel mode). The current key management options required for both AH and ESP are manual keying and automated keying via IKE. Authentication is based on the use of an MAC or the Integrity Check Value (ICV) computation so that two hosts must share a secret key.

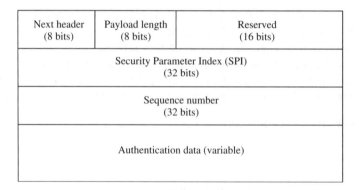

Figure 8.4 IPsec AH format.

8.2.1 AH Format

The IPsec AH format is shown in Figure 8.4. The following six fields comprise the AH format:

- *Next header* (*8 bits*). This field identifies the type of the next payload after the AH. The value of this field is chosen from the set of IP numbers defined in the Internet Assigned Numbers Authority (IANA).

- *Payload length* (*8 bits*). This field specifies the length of the AH in 32-bit words, minus 2. The default length of the authentication data field is 96 bits, or three 32-bit words. With a three-word fixed header, there are a total of six words in the header, and the Payload Length field has a value of 4.

- *Reserved* (*16 bits*). This field is reserved for future use. It must be set to "0".

- *SPI* (*32 bits*). This field uniquely identifies the SA for this datagram, in combination with the destination IP address and security protocol (AH).
    The set of SPI values in the range of 1–255 is reserved by the IANA for future use. The SPI value of zero (0) is reserved for local, implementation-specific use. A key management implementation may use the zero SPI value to mean "No SA Exists" during the period when the IPsec implementation has requested that its key management entity establishes a new SA, but the SA has not yet been established.

- *Sequence number* (*32 bits*). This field contains the monotonically increasing counter value which provides an antireplay function. Even if the sender always transmits this field, the receiver need not act on it, that is, processing of the sequence number field is at the discretion of the receiver. The sender's counter and the receiver's counter are initialized to 0 when an SA is established. The first packet sent using a given SA will have a sequence number of 1. The sender increments the sequence number for this SA and inserts the new value into the sequence number field.
    If antireplay is enabled, the sender checks to ensure that the counter has not cycled before inserting the new value in the sequence number field. If the counter

has cycled, the sender will set up a new SA and key. If the antireplay is disabled, the sender does not need to monitor or reset the counter. However, the sender still increments the counter and when it reaches the maximum value, the counter rolls over to zero.

- *Authentication data* (*variable*). This field is a variable-length field that contains the ICV or MAC for this packet. This field must be an integral multiple of 32-bit words. It may include explicit padding. This padding is included to ensure that the length of AH is an integral multiple of 32 bits (IPv4) or 64 bits (IPv6).

8.2.2 AH Location

Either AH or ESP is employed in two ways: transport mode or tunnel mode. The transport mode is applicable only to host implementations and provides protection for upper-layer protocols. In the transport mode, AH is inserted after the IP header and before an upper-layer protocol (TCP, UDP, or ICMP), or before any other IPsec header that may have already been inserted.

In the IPv4 context, AH is placed after the original IP header and before the upper-layer protocol, that is, TCP or UDP. Note that an ICMP message may be sent using either the transport mode or the tunnel mode. Authentication covers the entire packet, excluding mutable fields in the IPv4 header that are set to zero for MAC computation. The positioning of AH transport mode for an IPv4 packet is illustrated in Figure 8.5a.

In the IPv6 context, AH should appear after hop-to-hop, routing, and fragmentation extension headers. The destination options extension header(s) could appear either before or after AH, depending on the semantics desired. Authentication again covers the entire packet, excluding mutable fields that are set to zero for MAC computation. The positioning of AH transport mode for an IPv6 packet is illustrated in Figure 8.5b.

Tunnel mode AH can be employed in either hosts or security gateways. When AH is implemented in a security gateway to protect transit traffic, tunnel mode must be used. In tunnel mode, the *inner* IP header carries the ultimate source and destination addresses, while an *outer* IP header may contain different IP addresses (i.e., addresses of firewalls or other security gateways). In tunnel mode, AH protects the entire inner IP packet, including the entire inner IP header. The position of AH in tunnel mode, relative to the outer IP header, is the same as for AH in transport mode. Figure 8.5c illustrates AH tunnel mode positioning for typical IPv4 and IPv6 packets.

8.3 IP ESP

The ESP header is designed to provide security services in IPv4 and IPv6. ESP can be applied alone, in combination with the IP AH, or through the use of tunnel mode. Security services are provided between a pair of hosts, between a pair of security gateways, or between a security gateway and a host.

The ESP header is inserted after the IP header and before the upper-layer protocol header (transport mode) or before an encapsulated IP header (tunnel mode).

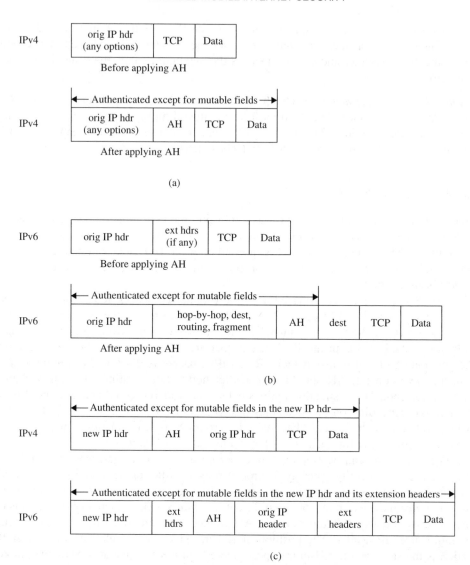

Figure 8.5 Transport mode and tunnel mode for AH authentication. (a) AH transport mode for an IPv4 packet, (b) AH transport mode for an IPv6 packet, and (c) AH tunnel mode for typical IPv4 and IPv6 packets.

ESP is used to provide confidentiality (encryption), data authentication, integrity and antireplay service, and limited traffic flow confidentiality. Confidentiality could be selected independent of all other services. However, use of confidentiality without integrity/authentication may subject traffic to certain forms of active attacks that undermine the confidentiality service. Data authentication and integrity are joint services offered as an option with confidentiality. The antireplay service is chosen only if data

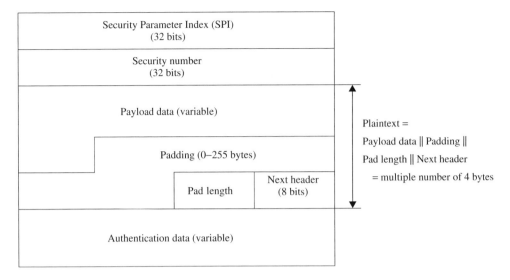

Figure 8.6 IPsec ESP format.

origin authentication is selected, and the service is effective only if the receiver checks the sequence number. The current key management options required for both AH and ESP are manual keying and automated keying via IKE.

8.3.1 ESP Packet Format

Figure 8.6 shows the format of an ESP packet and the fields in the header format are defined in the following.

- *SPI* (*32 bits*). The SPI is an arbitrary 32-bit value that uniquely identifies an SA for this datagram. The set of SPI values in the range 1–255 is reserved by the IANA for future use. The SPI field in the ESP packet format is mandatory and always present.

- *Sequence number* (*32 bits*). This field contains a monotonically increasing counter value. This provides an antireplay function. It is mandatory and is always present even if the receiver does not elect to enable the antireplay service for a specific SA. If antireplay is enabled, the transmitted sequence number must not be allowed to cycle. Thus, the sender's counter and the receiver's counter must be reset prior to the transmission of the 2^{32}nd packet on an SA.

- *Payload data* (*variable*). This variable-length field contains data described by the next header field. The field is an integral number of bytes in length. If the algorithm requires an initialization vector (IV) to encrypt payload, then this data may be carried explicitly in the payload field. Any encryption algorithm that requires such IP data must indicate the length, structure, and location of this data by specifying how the algorithm is used with ESP. For some IP-based modes of operation, the receiver treats the IP as the start of the ciphertext, feeding it into the algorithm directly.

- *Padding.* This field for encryption requires several factors:
  - If an encryption algorithm requires the plaintext to be a multiple number of bytes, the padding field is used to fill the plaintext to the size required by the algorithm. The plaintext consists of the payload data, pad length, and next header field, as well as the padding (Figure 8.6).
  - Padding is also required to ensure that the ciphertext terminates on a 32-bit boundary. Specifically, the pad length and next header fields must be right-aligned within a 32-bit word to ensure that the authentication data field is aligned on a 32-bit boundary.

  The sender may add 0–255 bytes of padding. Inclusion of the padding field in an ESP packet is optional, but all implementations must support the generation and consumption of padding. For the purpose of ensuring that either the bits to be encrypted are a multiple of the algorithm's blocksize or the authentication data is aligned on a 32-bit boundary, the padding is applied to the payload data exclusive of the IV, the pad length, and next header fields.

  The padding bytes are initialized with a series of integer values such that the first padding byte appended to the plaintext is numbered 1, with subsequent padding bytes following a monotonically increasing sequence: 1, 2, 3, ... When this padding scheme is employed, the receiver should inspect the padding field. Any encryption algorithm requiring padding must define the padding contents, while any required receiver must process these padding bytes in specifying how the algorithm is used with ESP. In such circumstances, the encryption algorithm and mode selected will determine the content of the padding field. Subsequently, a receiver must inspect the padding field and inform senders how the receiver will handle the padding field.

- *Pad length.* This field indicates the number of pad bytes immediately preceding it. The range of valid values is 0–255, where a value of 0 indicates that no padding bytes are present. This field is mandatory.

- *Next header (8 bits).* This field identifies the type of data contained in the payload data field, that is, an extension header in IPv6 or an upper-layer protocol identifier. The value of this field is chosen from the set of IP numbers defined by the IANA. The next header field is mandatory.

- *Authentication data (variable).* This is a variable-length field containing an ICV computed over the ESP packet minus the authentication data. The length of this field is specified by the authentication function selected. The field is optional and is included only if the authentication service has been selected for the SA in question. The authentication algorithm must specify the length of the ICV and the comparison rules and processing steps for validation.

8.3.2   ESP Header Location

Like AH, ESP is also employed in the transport or the tunnel mode. The transport mode is applicable only to host implementations and provides protection for upper protocols, but not the IP header. In the transport mode, ESP is inserted after the IP header and before an upper-layer protocol (TCP, UDP, or ICMP), or before any other IPsec headers that have already been inserted.

In the IPv4 context, ESP is placed after the IP header, but before the upper-layer protocol. Note that an ICMP message may be sent using either the transport mode or the tunnel mode. Figure 8.7a illustrates ESP transport mode positioning for a typical IPv4 packet, on a *before and after* basis. The ESP trailer encompasses any padding, plus the pad length, and next header fields.

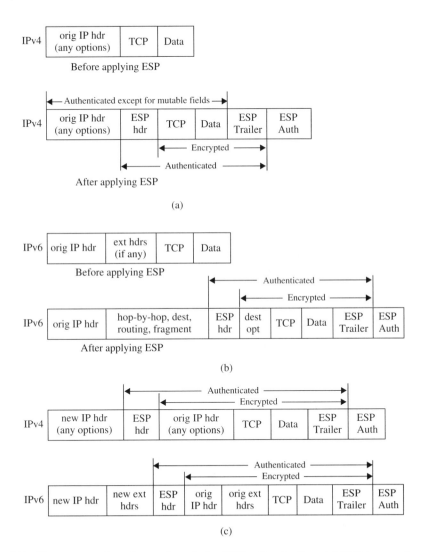

Figure 8.7 Transport mode and tunnel mode for ESP authentication. (a) ESP transport mode for an IPv4 packet, (b) ESP transport mode for an IPv6 packet, and (c) ESP tunnel mode for typical IPv4 and IPv6 packets.

In the IPv6 context, the ESP appears after hop-by-hop, routing, and fragmentation extension headers. The destination options extension header(s) could appear either before or after the ESP header depending on the semantics desired. However, since ESP protects only fields after the ESP header, it is generally desirable to place the destination options header(s) after the ESP header. Figure 8.7b illustrates ESP transport mode positioning for a typical IPv6 packet.

Tunnel mode ESP can be employed in either hosts or security gateways. When ESP is implemented in a security gateway to protect subscriber transit traffic, tunnel mode must be used. In tunnel mode, the *inner* IP header carries the ultimate source and destination addresses, while an *outer* IP header may contain different IP addresses such as addresses of security gateways. In tunnel mode, ESP protects the entire inner IP packet, including the entire inner IP header. The position of ESP in tunnel mode, relative to the outer IP header, is the same as for ESP in transport mode. Figure 8.7c illustrates ESP tunnel mode positioning for typical IPv4 and IPv6 packets.

8.3.3   Encryption and Authentication Algorithms

ESP is applied to an outbound packet associated with an SA that calls for ESP processing. The encryption algorithm employed is specified by the SA, as is the authentication algorithm.

Encryption

ESP is designed for use with symmetric algorithms like a triple DES in CBC mode. However, a number of other algorithms have been assigned identifiers in the DOI document. These algorithms for encryption are RC5, IDEA (International Data Encryption Algorithm), CAST, and Blowfish.

For encryption to be applied, the sender encapsulates the ESP payload field, adds any necessary padding, and encrypts the result (i.e., payload data, padding, pad length, and next header). The sender encrypts the fields (payload data, padding, pad length, and next header) using the key, encryption algorithm, algorithm mode indicated by the SA and an IV (cryptographic synchronization data). If the algorithm to be encrypted requires an IV, then this data is carried explicitly in the payload field. The payload data field is an integral number of bytes in length. Since ESP provides padding for the plaintext, encryption algorithms employed by ESP exhibit either block or stream mode characteristics.

The encryption is performed before the authentication and does not encompass the authentication data field. The order of this processing facilitates rapid detection and rejection of replayed or bogus packets by the receiver, prior to decrypting the packet. Therefore, it will reduce the impact of service attacks. At the receiver, parallel processing of packets is possible because decryption can take place in parallel with authentication. Since the authentication data is not protected by encryption, a keyed authentication algorithm must be employed to compute the ICV.

Referring to Figure 8.8, the 3DES–CBC mode requires an IV that is the same size as the block size. The IV is XORed with the first plaintext block before it is encrypted. For successive blocks, the previous ciphertext block is XORed with the current plaintext

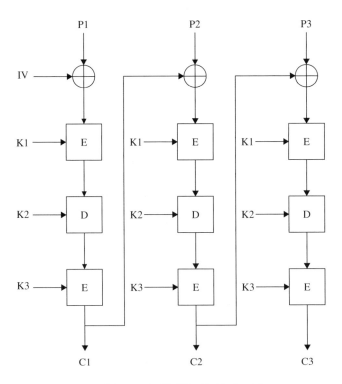

Figure 8.8 DES–EDE3–CBC algorithm.

before it is encrypted. Triple DES, known as *DES–EDE3*, processes each block three times, each time with a different key. Therefore, the triple DES algorithm has 48 rounds. In DES–EDE3-CBC, an IV is XORed with the first 64-bit plaintext block (P1).

Some cipher algorithms allow for a variable-sized key (RC5), while others only allow a specific key size (DES, IDEA).

Decryption

The receiver decrypts the ESP payload data, padding, pad length, and next header using the key, encryption algorithm, algorithm mode, and IV data. If explicit IV data is indicated, it is taken from the payload field and input to the decryption algorithm. If implicit IV data is indicated, a local version of the IV is constructed and input to the decryption algorithm.

The exact steps for reconstructing the original datagram depend on the mode (transport or tunnel) and are described in the Security Architecture document. The receiver processes any padding as given in the encryption algorithm specification. For transport mode, the receiver reconstructs the original IP datagram from the original IP header plus the original upper-layer protocol information in the ESP payload field. For tunnel mode, the receiver reconstructs the tunnel IP header plus the entire IP datagram in the ESP payload field.

If authentication has been computed, verification and decryption are performed serially or in parallel. If performed serially, then ICV or MAC verification should be performed first. If performed in parallel, verification must be completed before the decrypted packet is passed on for further processing. This order of processing facilitates rapid detection and rejection of replayed or bogus packets by the receiver.

Authentication

The authentication algorithm employed for the ICV computation is specified by the SA. For communication between two points, suitable authentication algorithms include Keyed MACs based on symmetric encryption algorithms (i.e., DES) or on one-way hash function (i.e., MD5 or SHA-1). For multicast communication, one-way hash algorithms combined with asymmetric signature algorithms are appropriate.

If authentication is selected for the SA, the sender computes the ICV over the ESP packet minus the authentication data. As stated previously, the fields of payload data, padding, pad length, and next header are all in ciphertext form because encryption is performed prior to authentication. Thus, the SPI, sequence numbers, and these four fields are all encompassed by the ICV computation.

ICV

Once the SA selects the authentication algorithm, the sender computes the ICV over the ESP packet minus the authentication data. The ICV is an MAC or a truncated value of a code produced by an MAC algorithm. As with AH, ESP supports the use of an MAC with a default length of 96 bits. The current specification for use of the HMAC computation must support:

HMAC–MD5–96
HMAC–SHA-1–96

8.4 Key Management Protocol for IPsec

The key management mechanism of IPsec involves the determination and distribution of a secret key. Key establishment is at the heart of data protection that relies on cryptography. A secure key distribution for the Internet is an essential part of packet protection.

Prior to establishing a secure session, the communicating parties need to negotiate the terms that are defined in the SA. An automated protocol is needed in order to establish the SAs for making the process feasible on the Internet. This automated process is the IKE. IKE combines ISAKMP with the Oakley key exchange.

We begin our discussion with an overview of Oakley and then look at ISAKMP.

8.4.1 OAKLEY Key Determination Protocol

The Diffie–Hellman (D-H) Key Exchange algorithm provides a mechanism that allows two users to agree on a shared secret key without requiring encryption. This shared key is immediately available for use in encrypting subsequent data transmission. Oakley is not only a refinement of the D-H Key Exchange algorithm but also a method to establish an authentication key exchange. The Oakley protocol is truly used to establish a shared key with an assigned identifier and associated authenticated identities for the two parties. Oakley can be used directly over the IP or the UDP using a well-known port number assignment available.

It is worth to note that Oakley uses the cookies for two purposes: anticlogging (denial of service) and key naming. The anticlogging tokens provide a form of source address identification for both parties. The construction of the cookies prevents an attacker from obtaining a cookie using a real IP address and UDP port.

Creating the cookie is to produce the result of a one-way function applied to a secret value, the IP source and destination addresses, and the UDP source and destination ports. Protection against anticlogging always seems to be one of the most difficult to address. A cookie or anticlogging token is aimed for protecting the computing resources from attack without spending excessive CPU resources to determine its authenticity. Absolute protection against anticlogging is impossible, but this anticlogging token provides a technique for making it easier to handle.

Oakley employs *nonces* to ensure against replay attacks. Each nonce is a pseudorandom number which is generated by the transmitting entity. The Nonce Payload contains this random data used to guarantee liveness during a key exchange and protect against replay attacks. If nonces are used by a particular key exchange, the use of the Nonce Payload will be dictated by the key exchange. The nonces may be transmitted as part of the key exchange data.

All the Oakley message fields correspond to ISAKMP message payloads. The relevant payload fields are the SA payload, the authentication payload, the certification payload, and the exchange payload. Oakley is the actual instantiation of ISAKMP framework for IPsec key and SA generation. The exact mapping of the Oakley message fields to ISAKMP payloads is in progress at this time.

8.4.2   ISAKMP

ISAKMP defines a framework for SA management and cryptographic key establishment for the Internet. This framework consists of defined exchange, payloads, and processing guidelines that occur within a given DOI. ISAKMP defines procedures and packet formats to establish, negotiate, modify, and delete SAs. It also defines payloads for exchanging key generation and authentication data. These payload formats provide a consistent framework for transferring key and authentication data which is independent of the key generation technique, encryption algorithm, and authentication mechanism.

ISAKMP is intended to support the negotiation of SAs for security protocols at all layers of the network stack. By centralizing the management of the SAs, ISAKMP reduces the amount of duplicated functionality within each security protocol.

ISAKMP Payloads

ISAKMP payloads provide modular building blocks for constructing ISAKMP messages. The presence and ordering of payloads in ISAKMP is defined by and dependent upon the Exchange Type field located in the ISAKMP Header.

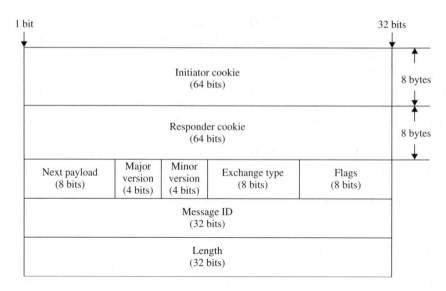

Figure 8.9   ISAKMP Header format.

ISAKMP Header

The ISAKMP header fields are defined as shown in Figure 8.9.

- Initiator cookie (64 bits)
  This field is the cookie of entity that initiated SA establishment, SA notification, or SA deletion.
- Responder cookie (64 bits)
  This field is the cookie of entity that is corresponded to an SA establishment request, SA notification, or SA deletion.
- Next payload (8 bits)
  This field indicates the type of the first payload in the message.
- Major version (4 bits)
  This field indicates the Major version of the ISAKMP in use. Set the Major version to 1 according to ISAKMP Internet Draft.
- Minor version (4 bits)
  This field indicates the Minor version of ISAKMP in use. Set the Minor version to 0 according to implementations based on the ISAKMP Internet Draft.
- Exchange type (8 bits)
  This field indicates the type of exchange being used. This dictates the message and payload orderings in the ISAKMP exchanges.
- Flags (8 bits)
  This field indicates specific options that are set for the ISAKMP exchange. The Flags are specified in the Flags field, beginning with the least significant bit: the encryption bit is bit 0 of the Flags field, the commit bit is bit 1, and the authentication only bit is bit 2 of the Flags field. The remaining bits of the Flags field must be set to 0 prior to transmission.

- All payloads following the header are encrypted using the encryption algorithm identified in the ISAKMP SA. The encryption should begin after both parties have exchanged Key Exchange Payloads.
- The commit bit is used to signal key exchange synchronization. In addition to synchronizing key exchange, the commit bit can be used to protect against loss of transmissions over unreliable networks and to guard against the need for multiple retransmissions.
- The authentication only bit is intended for use with the Information Exchange with a Notify Payload and will allow the transmission of information with integrity checking, but no encryption.

- Message ID (32 bits)
  Message ID is used to identify protocol state during Phase 2 negotiations. This value is randomly generated by the initiator of the Phase 2 negotiation. During Phase 1 negotiation, this value must be set to 0.
- Length (32 bits)
  Length of total message (header ∥ payload) is 32 bits. Encryption can expand the size of an ISAKMP message.

Generic Payload Header

Each ISAKMP payload begins with a generic header which provides a payload chaining capability and clearly defines the boundaries of a payload.
  The Generic Payload Header fields in 32 bits are defined as follows:

- Next payload (8 bits)
  This field is the identifier for the payload type of the next payload in the message. If the current payload is the last in the message, then this field will be 0. This field provides the chaining capability.
- Reserved (8 bits)
  This field is not used and set to 0.
- Payload length (16 bits)
  This field indicates the length in bytes of the current payload, including the generic payload header.

Payload Types for ISAKMP

ISAKMP defines several types of payloads that are used to transfer information such as SA data or key exchange data in DOI-defined formats. RFC 2408 presented by Maughan *et al.* is a good coverage of ISAKMP payloads.

Security Association Payload

The SA Payload is used to negotiate security attributes and to identify the DOI (32 bits) under which negotiation is taking place. A DOI value of 0 during a Phase 1 exchange specifies a Generic ISAKMP which can be used for any protocol during the Phase 2 exchange. A DOI value of 1 is assigned to the IPsec DOI.

The SA Payloads are defined as follows:

The Next Payload field (8 bits) is the identifier for the payload type of the next payload in the message. This field has a value of 0 if this is the last payload in the message.
The Reserved field (8 bits) is unused, set to 0.
The Payload Length field (16 bits) indicates the length in octets of the entire SA payload, including the SA payload, all Proposal payloads, and all Transform payloads associated with the proposed SA.
The Situation field (variable length) is a DOI-specific field that identifies the situation under which negotiation is taking place. The Situation field defines policy decisions regarding the security attributes being negotiated.

Proposal Payload

The Proposal Payload is used to build ISAKMP message for the negotiation and establishment of SAs. The Proposal Payload field contains information used during SA negotiation for securing the communications channel. The payload type for the Proposal Payload is 2.
    The Proposal Payload fields are defined as follows:

The Next Payload field (8 bits) is the identifier for the payload type of the next payload in the message. This field must only contain the value 2 or 0. This field will be 2 for additional Proposal payloads in the message and 0 when the current Proposal Payload is the last within the SA proposal.
The Reserved field (8 bits) is set to 0 and is reserved for the future use.
The Payload Length field (16 bits) is the length in octets of the entire Proposal Payload, including Generic Payload Header, the Proposal Payload, and all Transform Payloads associated with this proposal.
The Proposal # field (8 bits) identifies the proposal number for the current payload.
The Protocol-id field (8 bits) specifies the protocol identifier for the current negotiation. Examples are IPsec ESP, IPsec AH, OSPF, and TLS.
The SPI Size (8 bits) denotes the length in octets of the SPI. In the case of ISAKMP, the Initiator and Responder cookie pair from the ISAKMP Header is the ISAKMP SPI. The SPI size may be from 0 to 16. If the SPI size is nonzero, the content of the SPI field must be ignored. The DOI will dictate the SPI Size for other protocols.
# of Transform (8 bits) specifies the number of transforms for the proposal. Each of these is contained in a Transform Payload.
SPI field (variable) is the sending entity's SPI. In the event of the SPI size not being a multiple of 4 octets, there is no padding applied to the payload.

Transform Payload

The Transform Payload contains information used during SA negotiation. The Transform Payload consists of a specific security mechanism to be used to secure the communications channel. The Transform Payload also contains the SA attributes associated with the specific transform. These SA attributes are DOI-specific. The Transform Payload allows the initiating entity to present several possible supported transforms for that proposed protocol.

The Transform Payload fields are defined as follows:

The Next Payload ficld (8 bits) is the identifier for the payload type of the next payload in the message. This field must only contain the value 3 or 0. This field is 3 when there are additional Transform Payloads in the proposal. This field is 0 when the current Transform Payload is the last within the proposal.

The Reserved field (8 bits) is for unused, set to 0.

The Transform # field (8 bits) identifies the Transform number for the current payload. If there is more than one transform within the Proposal Payload, then each Transform Payload has a unique Transform number.

The Transform-id field (8 bits) specifies the Transform identifier for the protocol within the current proposal.

The Reserved 2 field (16 bits) is for unused, set to 0.

The SA Attributes field (variable length) contains the SA attributes as defined for the transform given in the Transform-id field. The SA Attributes should be represented using the Data Attributes format. These Data Attributes are not an ISAKMP payload, but are contained within ISAKMP payloads. The format of the Data Attributes provides the flexibility for representation of many different types of information. There may be multiple Data Attributes within a payload. The length of the Data Attributes will either be 4 octets or defined by the Attribute Length field (16 bits). If the SA Attributes are not aligned on 4-byte boundaries, then subsequent payloads will not be aligned and any padding will be added at the end of the message to make the message 4-byte aligned.

The payload type for the Transform Payload is 3.

Key Exchange Payload

The Key Exchange Payload supports a variety of key exchange techniques. Example key exchanges are Oakley, D-H, the enhanced D-H key exchange, and the RSA-based key exchange used by PGP.

The Key Exchange Payload fields are defined as follows:

The Next Payload field (8 bits) is the identifier for the payload type of the next payload in the message. If the current payload is the last in the message, then this field will be 0.

The Reserved field (8 bits) is unused for future use, set to 0.

The Payload Length field (16 bits) is the length in octets of the current payload, including the generic payload header.

The Key Exchange Data field (variable length) is the data required to generate a session key. The interpretation of this data is specified by the DOI and the associated Key Exchange algorithm. This field may also contain preplaced key indicators.

Identification Payload

The Identification Payload contains DOI-specific data used to exchange identification information. This information is used for determining the identities of communication partners and may be used for determining authenticity of information.

The Identification Payload fields are described as follows:

The Next Payload field (8 bits) is the identifier for the payload type of the Next Payload in the message. If the current payload is the last in the message, then this field will be 0. The Reserved field (8 bits) is not used, but set to 0.
The Payload Length field (16 bits) is the length in octets of the current payload, including the generic payload header.
The ID type field (8 bits) specifies the type of identification being used. This field is DOI dependent.
The DOI-specific ID Data field (24 bits) contains DOI-specific identification data. If unused, this field must be set to 0.
The Identification Data field (variable length) contains identity information. The values for this field are DOI specific, and the format is specified by the ID Type field. Specific details for the IETF IPsec DOI identification data are detailed in RFC 2407.
The payload type for the Identification Payload is 5.

Certificate Payload

The Certificate Payload provides a means to transport certificates via ISAKMP and can appear in any ISAKMP message. Certificate Payloads should be included in an exchange whenever an appropriate directory service is not available to distribute certificates. The Certificate Payload must be accepted at any point during an exchange.

The Certificate Payload fields are defined as follows:

The Next Payload field (8 bits) is the identifier for the Payload type of the next payload in the message. If the current payload is the last in the message, then this field will be 0. The Reserved field (8 bits) is unused, set to 0.
The Payload Length field (16 bits) is the length in octets of the current payload, including the generic payload header.
The Certificate Encoding field (8 bits) indicates the type of certificate or certificate-related information contained in the Certificate Data field.

Certificate Type	Value
NONE	0
PKCS #7 wrapped X.509 certificate	1
PGP Certificate	2
DNS Signed Key	3
X.509 Certificate-Signature	4
X.509 Certificate-Key Exchange	5
Kerberos Tokens	6
Certificate Revocation List (CRL)	7
Authority Revocation List (ARL)	8
SPKI Certificate	9
X.509 Certificate-Attribute	10
Reserved	11–255

The Certificate Data field (variable length) denotes actual encoding of Certificate Data. The type of certificate is indicated by the Certificate Encoding field.

The Payload type for the Certificate Payload is 6.

Certificate Request Payload

The Certificate Request Payload provides a means to request certificate via ISAKMP and can appear in any message. Certificate Request Payloads should be included in an exchange whenever an appropriate directory service is not available to distribute certificates. The Certificate Request Payload must be accepted at any point during the exchange. The responder to the Certificate Request Payload must send its certificate, if certificates are based on the values contained in the payload. If multiple certificates are required, then multiple Certificate Request Payloads should be transmitted.

The Certificate Request Payload fields are defined as follows:

The Next Payload field (8 bits) is the identifier for the payload type of the next payload in the message. If the current payload is the last in the message, then this field will be 0.

The Reserved field (8 bits) is not used, set to 0.

The Payload Length field (16 bits) is the length in octets of the current payload, including the generic payload header.

The Certificate Type field (8 bits) contains an encoding of the type of certificate requested. Acceptable values are listed in the Certificate Payload fields.

The Certificate Authority field (variable length) contains an encoding of an acceptable Certificate Authority for the type of certificate requested. As an example, for an X.509 certificate, this field would contain the Distinguished Name encoding of the Issuer Name of an X.509 Certificate Authority acceptable to the sender of this payload. This may assist the responder in determining how much of the certificate chain would need to be sent in response to this request. If there is no specific Certificate Authority requested, this field should not be included.

The payload type for the Certificate Request Payload is 7.

Hash Payload

The Hash Payload contains data generated by the hash function over some part of the message and/or the ISAKMP state. This payload can possibly be used to verify the integrity of the data in an ISAKMP message or for authentication of the negotiating entities.

The Hash Payload fields are defined as follows:

The Next Payload field (8 bits) is the identifier for the payload type of the next payload in the message. If the current payload is the last in the message, then this field will be 0.

The Reserved field (8 bits) is not used, set to 0.

The Payload Length field (16 bits) is the length in octets of the current payload, including the generic payload header.

The Hash Data field (variable length) is the data that results from applying the hash routine to the ISAKMP message and/or state.

The payload type for the Hash Payload is 8.

Signature Payload

The Signature Payload contains data generated by the digital signature function over some part of the message and/or ISAKMP state. This payload is used to verify the integrity of the data in the ISAKMP message, and may be of use for nonrepudiation services.

The Signature Payload fields are defined as follows:

The Next Payload field (8 bits) is the identifier for the payload type of the next payload in the message. If the current payload is the last in the message, then this field will be 0.
The Reserved field (8 bits) is not used, but set to 0.
The Payload Length field (16 bits) is the length in octets of the current payload, including the generic payload header.
The Signature Data field (variable length) is the data that results from applying the digital signature function to the ISAKMP message and/or state.
The payload type for the Signature Payload is 9.

Nonce Payload

The Nonce Payload contains random data used to guarantee liveness during an exchange and protect against replay attacks. If nonce is used by a particular key exchange, the use of the Nonce Payload will be dictated by the key exchange. The nonces may be transmitted as part of the key exchange data, or as a separate payload. However, this is defined by the key exchange, not by ISAKMP.

The Nonce Payload fields are defined as follows:

The Next Payload field (8 bits) is the identifier for the payload type of the next payload in the message. If the current payload is the last in the message, then this field will be 0.
The Reserved field (8 bits) is unused, but set to 0.
The Payload Length field (16 bits) is the length in octets of the current payload, including the generic payload header.
The Nonce Data field (variable length) contains the random data generated by the transmitting entity.
The Payload type for the Nonce Payload is 10.

Notification Payload

The Notification Payload can contain both ISAKMP and DOI-specific data and is used to transmit information data, such as error conditions, to an ISAKMP peer. It is possible to send multiple Notification Payloads in a single ISAKMP message. Notification which occurs during a Phase 1 negotiation is identified by the Initiator and Responder cookie pair in the ISAKMP Header. Notification which occurs during a Phase 2 negotiation is identified by the Initiator and Responder cookie pair in the ISAKMP header and the Message ID and SPI associated with the current negotiation.

The Notification Payload fields are defined as follows:

The Next Payload field (8 bits) is the identifier for the payload type of the next payload in the message. If the current payload is the last in the message, then this field will be 0.

The Reserved field (8 bits) is unused, but set to 0.

The Payload Length field (16 bits) is the length in octets of the current payload, including the generic payload header.

The DOI field (32 bits) identifies the DOI under which this notification is taking place. For ISAKMP this value is 0 and for the IPsec DOI it is 1.

The Protocol-id field (8 bits) specifies the protocol identifier for the current notification. Examples are ISAKMP, IPsec ESP, IPsec AH, OSPF, and TLS.

The SPI Size field (8 bits) is the length in octets of the SPI as defined by the protocol-id. In the case of ISAKMP, the Initiator and Responder cookie pair from the ISAKMP Header is the ISAKMP SPI. Therefore, the SPI size is irrelevant and may be from 0 to 16. If the SPI size is nonzero, the content of the SPI field must be ignored. The DOI will dictate the SPI size for other protocols.

The Notify Message Type field (16 bits) specifies the type of notification message. Additional text, if specified by the DOI, is placed in the Notification Data field.

The SPI field has the variable length. The length of this field is determined by the SPI Size field and is not necessarily aligned to a 4-octet boundary. During the SA establishment, an SPI must be generated. ISAKMP is designed to handle various-sized SPIs. This is accomplished by using the SPI Size field within the Proposal Payload during SA establishment.

The Notification Data field (variable length) is informational or error data transmitted in addition to the Notify Message Type. Values for this field are DOI specific.

The payload type for the Notification Payload is 11.

Delete Payload

The Delete Payload contains a protocol-specific SA identifier that the sender has removed from its SA database. Therefore, the sender is no longer valid. It is possible to send multiple SPIs in a Delete Payload. But each SPI must be for the same protocol.

The Delete Payload fields are defined as follows:

The Next Payload field (8 bits) is the identifier for the payload type of the next payload in the message. If the current payload is the last in the message, then this field will be 0.

The Reserved field (8 bits) is unused, but set to 0.

The Payload Length field (16 bits) is the length in octets of the current payload, including the generic payload header.

The DOI field (32 bits) identifies the DOI under which this deletion is taking place. For ISAKMP, this value is 0 and for the IPsec DOI it is 1.

The Protocol-id field (8 bits) specifies that ISAKMP can establish SAs for various protocols, including ISAKMP and IPsec. This field identifies which SA database to apply for the delete request.

The SPI Size field (8 bits) is the length in octets of the SPI as defined by the Protocol-id. In the case of ISAKMP, the Initiator and Responder cookie pair is the ISAKMP SPI. In this case, the SPI Size would be 16 bytes for each SPI being deleted.

The # of SPIs field (16 bits) is the number of SPIs contained in the Delete Payload. The size of each SPI is defined by the SPI Size field.

The SPI field (variable length) identifies the specific SAs to delete. Values for this field are DOI and protocol specific. The length of this field is determined by the SPI Size and # of SPIs fields.

The Payload type for the Delete Payload is 12.

Vendor ID Payload

The Vendor ID Payload contains a vendor-defined constant. The constant is used by vendors to identify and recognize remote instances of their implementations. This mechanism allows a vendor to experiment with new features while maintaining backwards compatibility. However, this is not a general extension facility of ISAKMP.

If a Vendor ID Payload is sent, it must be sent during the Phase 1 negotiation. Reception of a familiar Vendor ID Payload in the Phase 1 negotiation allows an implementation to make use of Private Use Payload numbers for vendor-specific extension during Phase 2 negotiation.

The Vendor ID Payload fields are defined as follows:

The Next Payload field (8 bits) is the identifier for the payload type of the next payload in the message. If the current payload is the last in the message, then this field will be 0.

The Reserved field (8 bits) is unused, but set to 0.

The Payload Length field (16 bits) is the length in octets of the current payload, including the generic payload header.

The Vendor ID field (variable length) contains the choice of hash and text to hash. Vendors could generate their vendor ID by taking a keyless hash of a string containing the product name and the version of the product.

The Payload type for the Vendor ID Payload is 13.

ISAKMP Exchanges

ISAKMP supplies the basic syntax of a message exchange. ISAKMP allows the creation of exchanges for SA establishment and key exchange. There are currently five default exchange types defined for ISAKMP. Exchanges define the content and ordering of ISAKMP messages during communications between peers. Most exchanges include all the basic payload types: SA (SA Payload), KE (Key Exchange Payload), ID (Identity Payload), and SIG (Signature Payload). The primary difference between exchange types is the ordering of messages and the payload ordering within each message.

The defined exchanges are not meant to satisfy all DOI and key exchange protocol requirements. If the defined exchanges meet the DOI requirements, then they can be used as outlined. If the defined exchanges do not meet the security requirements defined by the DOI, then the DOI must specify new exchange type(s) and the valid sequences of payloads that make up a successful exchange, and how to build and interpret those payloads.

Base Exchange

The Base Exchange is designed to allow the Key Exchange and Authentication-related information to be transmitted together. Combining the Key Exchange and

Authentication-related information into one message reduces the number of round-trips at the expense of not providing identity protection.

Identity Protection Exchange

The Identity Protection Exchange is designed to separate the Key Exchange information from the Identity and Authentication-related information. Separating the Key Exchange from the Identity and Authentication-related information protects the communicating identities at the expense of two additional messages. Identities are exchanged under the protection of a previously established common shared secret.

Authentication Only Exchange

The authentication only exchange is designed to allow only Authentication-related information to be transmitted. The benefit of this exchange is the ability to perform only authentication without the computational expense of computing keys. Using this exchange during negotiation, none of the transmitted information will be encrypted. But the Authentication Only Exchange will be encrypted by the ISAKMP SA, negotiated in the first phase.

Aggressive Exchange

The Aggressive Exchange is designed to allow the SA, Key Exchange, and Authentication-related payloads to be transmitted together. Combining these SA, KE, and Auth information into one message reduces the number of round-trips at the expense of not providing identity protection. Identity protection is not provided because identities are exchanged before a common shared secret has been established.

Informational Exchange

The Information Exchange is designed as a one-way transmittal of information that can be used for SA management. If the Information Exchange occurs prior to the exchange of keying material during an ISAKMP Phase 1 negotiation, there will be no protection provided for the Information Exchange. Once keying material has been exchanged or an ISAKMP SA has been established, the Information Exchange must be transmitted under the protection provided by the keying material or the ISAKMP SA.

ISAKMP Payload Processing

The ISAKMP Payloads are used in the exchanges described in the section "ISAKMP Exchanges" and can be used in exchanges defined for a specific DOI. This section describes the processing for each of the payloads.

General Message Processing

Every ISAKMP message has basic processing applied to ensure protocol reliability and to minimize threats such as denial of services and replay attacks. All processing should include packet length checks to insure that the packet received is at least as long as the length given in the ISAKMP Header. If the ISAKMP message length and the value in the Payload Length field of the ISAKMP Header are not the same, then the ISAKMP message must be rejected.

ISAKMP Header Processing

When an ISAKMP message is created at the transmitting entity, the initiator (transmitter) must create the respective cookie, determine the relevant security characteristics of the session, construct an ISAKMP Header with fields, and transmit the message to the destination host (responder).

When an ISAKMP is received at the receiving entity, the responder (receiver) must verify the Initiator and Responder cookies, check the Next Payload field to confirm it is valid, check the Major and Minor version fields to confirm they are correct, check the Exchange Type field to confirm it is valid, check the Flags field to ensure it contains correct values, and check the Message ID field to ensure it contains correct values.

Thus, processing of the ISAKMP message continues using the value in the Next Payload field.

Generic Payload Header Processing

When any of the ISAKMP Payloads are created, a Generic Payload Header is placed at the beginning of these payloads.

When creating the Generic Payload Header, the transmitting entity (initiator) must place the value of the Next Payload in the Next Payload field, place the value 0 in the Reserved field, place the length (in octets) of the payload in the Payload Length field, and construct the payloads.

When any of the ISAKMP Payloads are received, the receiving entity (responder) must check the Next Payload field to confirm it is valid, verify that the Reserved field contains the value 0, and process the remaining payloads as defined by the Next Payload field.

Security Association Payload Processing

When an SA Payload is created, the transmitting entity (initiator) must determine the DOI for which this negotiation is being preformed, determine the situation within the determined DOI for which this negotiation is being formed, determine the proposal(s) and transform(s) within the situation, construct an SA payload, and transmit the message to the receiving entity (responder).

When an SA Payload is received, the receiving entity (responder) must determine if the DOI is supported, determine if the given situation can be protected, and process the remaining payloads (Proposal, Transform) of the SA Payload. If the SA Proposal is not accepted, then the Invalid Proposal event may be logged in the appropriate system audit file. An Information Exchange with a Notification Payload containing the No-Proposal-Chosen message type may be sent to the transmitting entity (initiator). This action is dictated by a system security policy.

Proposal Payload Processing

When a Proposal Payload is created, the transmitting entity (initiator) must determine the protocol for this proposal, determine the number of proposals to be offered for this proposal and the number of transform for each proposal, generate a unique pseudorandom SPI, and construct a Proposal Payload.

When a Proposal Payload is received, the receiving entity (responder) must determine if the proposal is supported and if the Protocol-ID field is invalid, determine whether the SPI is valid or not, ensure whether or not proposals are formed correctly, and then process the Proposal and Transform payloads as defined by the Next Payload field.

Transform Payload Processing

When creating a Transform Payload, the transmitting entity (initiator) must determine the Transform # for this transform, determine the number of transforms to be offered for this proposal, and construct a Transform Payload.

When a Transform Payload is received, the receiving entity (responder) must do as follows: determine if the Transform is supported. If the Transform-ID field contains an unknown or unsupported value, then that Transform Payload must be ignored. Ensure Transforms are presented according to the details given in the Transform Payload and SA Establishment. Finally, process the subsequent Transform and Proposal payloads as defined by the Next Payload field.

Key Exchange Payload Processing

When creating a Key Exchange Payload, the transmitting entity (initiator) must determine the Key Exchange to be used as defined by the DOI, determine the usage of Key Exchange Data field as defined by the DOI, and construct a Key Exchange Payload. Finally, it must transmit the message to the receiving entity (responder).

When a Key Exchange Payload is received, the receiving entity (responder) must determine if the Key Exchange is supported. If the Key Exchange determination fails, the message is discarded and the following actions are taken:

The event of Invalid Key Information may be logged in the appropriate system audit file. An Information Exchange with a Notification Payload containing the Invalid-Key-Information message type may be sent to the transmitting entity. This action is dictated by a system security policy.

Identification Payload Processing

When an Identification Payload is created, the transmitting entity (initiator) must determine the Identification information to be used as defined by the DOI, determine the usage of the Identification Data field as defined by the DOI, construct an Identification Payload, and finally transmit the message to the receiving entity.

When an Identification Payload is received, the receiving entity (responder) must determine if the Identification Type is supported. This may be based on the DOI and Situation.

If the Identification determination fails, the message is discarded. An Information Exchange with a Notification Payload containing the Invalid-ID-Information message type is sent to the transmitting entity (initiator).

Certificate Payload Processing

When a Certificate Payload is created, the transmitting entity (initiator) must determine the Certificate Encoding which is specified by the DOI, ensure the existence of a certificate formatted as defined by the Certificate Encoding, construct a Certificate Payload, and then transmit the message to the receiving entity (responder).

When a Certificate Payload is received, the receiving entity (responder) must determine if the Certificate Encoding is supported. If the Certificate Encoding is not supported, the payload is discarded. The responder then processes the Certificate Data field. If the Certificate Data is improperly formatted, the payload is discarded.

Certificate Request Payload Processing

When creating a Certificate Request Payload, the transmitting entity (initiator) must determine the type of Certificate Encoding to be requested, determine the name of an acceptable Certificate Authority, construct a Certificate Request Payload, and then transmit the message to the receiving entity (responder).

When a Certificate Request Payload is received, the receiving entity (responder) must determine if the Certificate Encoding is supported. If the Certificate Encoding is invalid, the payload is discarded. The responder must determine if the Certificate Authority is supported for the specified Certificate Encoding. If the Certificate Authority is improperly formatted, the payload is discarded. Finally, the responder must process the Certificate Request. If a requested Certificate Type with the specified Certificate Authority is not available, then the payload is discarded.

Hash Payload Processing

When creating a Hash Payload, the transmitting entity (initiator) must determine the Hash function to be used as defined by the SA negotiation, determine the usage of the Hash Data field as defined by the DOI, construct a Hash Payload, and then transmit the message to the receiving entity (responder).

When a Hash Payload is received, the receiving entity (responder) must determine if the Hash is supported. If the Hash determination fails, the message is discarded. The responder also performs the Hash function as outlined in the DOI and/or Key Exchange protocol documents. If the Hash function fails, the message is discarded.

Signature Payload Processing

When a Signature Payload is created, the transmitting entity(initiator) must determine the Signature function to be used as defined by the SA negotiation, determine the usage of the Signature Data field as defined by the DOI, construct a Signature Payload, and finally transmit the message to the receiving entity (responder).

When a Signature Payload is received, the receiving entity must determine if the Signature is supported. If the Signature determination fails, the message is discarded. The responder must perform the Signature function as outlined in the DOI and/or Key Exchange protocol documents. If the Signature function fails, the message is discarded.

Nonce Payload Processing

When creating a Nonce Payload, the transmitting entity (initiator) must create a unique random value to be used as a nonce, construct a Nonce Payload, and transmit the message to the receiving entity.

When a Nonce Payload is received, the receiving entity (responder) must do as follows: there are no specific procedures for handling Nonce Payloads. The procedures are defined by the exchange types and possibly the DOI and Key Exchange descriptions.

Notification Payload Processing

During communication, it is possible that errors may occur. The Information Exchange with a Notify Payload provides a controlled method of informing a peer entity that error has occurred during protocol processing. It is recommended that Notify Payloads be sent in a separate Information Exchange rather than appending a Notify Payload to an existing exchange.

When a Notification Payload is created, the transmitting entity (initiator) must determine the DOI for this Notification, determine the Protocol-ID for this Notification, determine the SPI size based on the Protocol-ID field, determine the Notify Message Type based on the error or status message desired, determine the SPI which is associated with this notification, determine if additional Notification Data is to be included, construct a Notification Payload, and finally, transmit the messages to the receiving entity.

When a Notification Payload is received, the receiving entity (responder) must determine if the Information Exchange has any protection applied to it by checking the encryption bit and authentication only bit in the ISAKMP Header, determine if the DOI is supported, determine if the protocol-ID is supported, determine if the SPI is valid, determine if the Notify Message Type is valid, and then process the Notification Payload, including additional Notification Data, and take appropriate action according to local security policy.

Delete Payload Processing

During communication, it is possible that hosts may be compromised or that information may be interrupted during transmission. If it is discovered that transmissions are being compromised, then it is necessary to establish a new SA and delete the current SA.

When a Delete Payload is created, the transmitting entity (initiator) must determine the DOI for this Deletion, determine the Protocol-ID for this Deletion, determine the SPI size based on the Protocol-id field, determine the # of SPIs to be deleted for this protocol, determine the SPI(s) which is (are) associated with this deletion, construct a Delete Payload, and then transmit the message to the receiving entity.

When a Delete Payload is received, the receiving entity (responder) must do as follows:

- Since the Information Exchange is protected by authentication for an Auth-Only SA and encryption for other exchange, the message must have these security services applied using the ISAKMP SA. Any errors that occur during the Security Service processing will be evident when checking information in the Delete Payload.
- Determine if the DOI is supported.
- Delete if the Protocol-ID is supported.
- Determine if the SPI is valid for each SPI included in the Delete Payload.
- Process the Delete Payload and take appropriate action, according to local security policy.

The ISAKMP is a well-designed protocol provided for Internet security services. ISAKMP provides the ability to establish SAs for multiple security protocols and applications. ISAKMP establishes the common base that allows all other security protocols to interoperate.

ISAKMP's SA feature coupled with authentication and key establishment provides the security and flexibility that will be needed for future growth and security diversity. As the Internet grows and evolves, new payloads to support new security functionality can be added without modifying the entire protocol.

9

Transport Layer Security: SSLv3 and TLSv1

Secure Sockets Layer version 3 (SSLv3) was introduced by Netscape Communications Corporation in 1995. SSLeay implements both SSLv2 and SSLv3 and TLSv1 as of the release of SSLeay-0.9.0. SSLv3 was designed with public review and input from industry and was published as an Internet-Draft document. After reaching a consensus of opinion to Internet standardization, the Transport Layer Security (TLS) Working Group was formed within Internet Engineering Task Force (IETF) in order to develop an initial version of TLS as an Internet standard. The first version of TLS is very closely compatible with SSLv3. The TLSv1 protocol provides communications privacy and data integrity between two communicating parties over the Internet. Both the SSL and TLS protocols allow client/server applications to communicate in such a way that they prevent eavesdropping, tampering, or message forgery. The SSL (or TLS) protocol is composed of two layers: the SSL (or TLS) Record Protocol and the SSL (or TLS) Handshake Protocol.

This chapter is devoted to a full discussion of the protocols of both SSLv3 and TLSv1.

9.1 SSL Protocol

SSL is a layered protocol. It is not a single protocol but rather two layers of protocols. At the lower level, the SSL Record Protocol is layered on top of some reliable transport protocol such as TCP. The SSL Record Protocol is also used to encapsulate various higher-level protocols. A higher-level protocol can layer on top of the SSL protocol transparently. For example, the Hypertext Transfer Protocol (HTTP), which provides a transfer service for Web client/server interaction, can operate on top of the SSL Record Protocol.

The SSL Record Protocol takes the upper-layer application message to be transmitted, fragments the data into manageable blocks, optionally compresses the data, applies an

Wireless Mobile Internet Security, Second Edition. Man Young Rhee.
© 2013 John Wiley & Sons, Ltd. Published 2013 by John Wiley & Sons, Ltd.

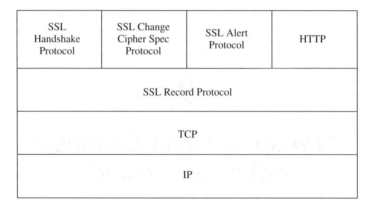

Figure 9.1   Two-layered SSL protocols.

MAC, encrypts, adds a header, and transmits the result to TCP. The received data is decrypted, verified, decompressed, reassembled, and then delivered to higher-level clients. Figure 9.1 illustrates the overview of the SSL protocol stack.

9.1.1   Session and Connection States

There are two defined specifications: SSL session and SSL connection.

SSL Session

An SSL session is an association between a client and a server. Sessions are created by the Handshake Protocol. They define a set of cryptographic security parameters, which can be shared among multiple connections. Sessions are used to avoid the expensive negotiation of new security parameters for each connection. An SSL session coordinates the states of the client and the server. Logically, the state is represented twice as the current operating state and pending state. When the client or server receives a *change cipher spec* message, it copies the pending read state into the current read state. When the client or server sends a *change cipher spec* message, it copies the pending write state into the current write state. When the handshake negotiation is completed, the client and server exchange *change cipher spec* messages, and they then communicate using the newly agreed-upon cipher spec.

The session state is defined by the following elements:

- *Session identifier.* This is a value generated by a server that identifies an active or a resumable session state.
- *Peer certificate.* This is an X.509 v3 certificate of the peer. This element of the state may be null.
- *Compression method.* This is the algorithm used to compress data prior to encryption.

- *Cipher spec.* This specifies the bulk data encryption algorithm (such as null and DES) and a hash algorithm (such as MD5 or SHA-1) used for MAC computation. It also defines cryptographic attributes such as the hash size.
- *Master secret.* This is a 48-byte secret shared between the client and the server. It represents secure secret data used for generating encryption keys, MAC secrets, and IVs.
- *Is resumable.* This designates a flag indicating whether the session can be used to initiate new connections.

SSL Connection

A connection is a transport (in the Open Systems Interconnect (OSI) layering model definition) that provides a suitable type of service. For SSL, such connections are peer-to-peer relationships. The connections are transient. Every connection is associated with one session.

The connection state is defined by the following elements:

- *Server and client random.* These are byte sequences that are chosen by the server and client for each connection.
- *Server write MAC secret.* This indicates the secret key used in MAC operations on data sent by the server.
- *Client write MAC secret.* This represents the secret key used in MAC operations on data sent by the client.
- *Server write key.* This is the conventional cipher key for data encrypted by the server and decrypted by the client.
- *Client write key.* This is the conventional cipher key for data encrypted by the client and decrypted by the server.
- *Initialization vectors.* When a block cipher in cipher block chaining (CBC) mode is used, an IV is maintained for each key. This field is first initialized by the SSL Handshake Protocol. Thereafter, the final ciphertext block from each record is preserved for use as the IV with the following record. The IV is XORed with the first plaintext block prior to encryption.
- *Sequence numbers.* Each party maintains separate sequence numbers for transmitted and received messages for each connection. When a party sends or receives a change cipher spec message, the appropriate sequence number is set to zero. Sequence numbers may not exceed $2^{64} - 1$.

9.1.2 SSL Record Protocol

The SSL Record Protocol provides basic security services to various higher-layer protocols. Three upper-layer protocols are defined as part of SSL: the Handshake Protocol, the Change Cipher Spec Protocol, and the Alert Protocol. Two layers of SSL protocols are shown in Figure 9.1. The SSL Record Layer receives data from higher layers in blocks of arbitrary size.

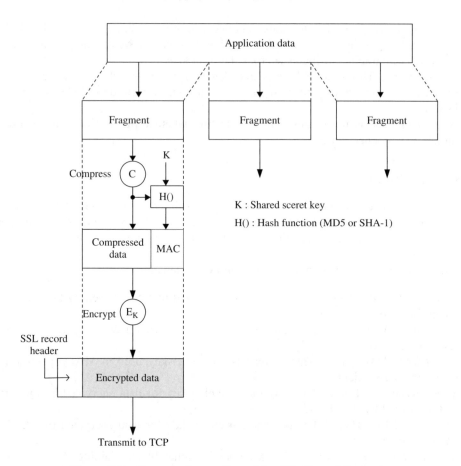

Figure 9.2   The overall operation of the SSL Record Protocol.

The SSL Record Protocol takes an application message to be transmitted, fragments the data into manageable blocks, optionally compresses the data, applies an MAC, encrypts, adds a header, and transmits the result in a TCP segment. The received data is decrypted, verified, decompressed, reassembled, and then delivered to higher-level clients. The overall operation of the SSL Record Protocol is shown in Figure 9.2.

- *Fragmentation.* A higher-layer message is fragmented into blocks (SSLPlaintext records) of 2^{14} bytes or less.
- *Compression and decompression.* All records are compressed using the compression algorithm defined in the current session state. The compression algorithm translates an SSLPlaintext structure into an SSLCompressed structure. Compression must be lossless and may not increase the current length by more than 1024 bytes. If the decompression function encounters an SSLCompressed.fragment that would decompress to a length in excess of $2^{14} = 16\,348$ bytes, it should issue a fatal decompression-failure alert.

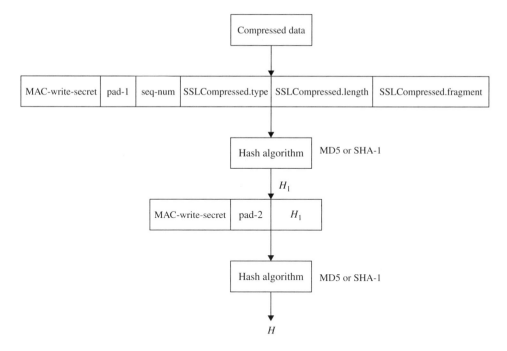

Figure 9.3 Computation of MAC over the compressed data.

Compression is essentially useful when encryption is applied. If both compression and encryption are required, compression should be applied before encryption. The compression processing should ensure that an SSLPlaintext structure is identical after being compressed and decompressed. Compression is optionally applied in the SSL Record Protocol, but, if applied, it must be done before encryption and MAC computation.

- *MAC.* The MAC is computed before encryption. The computation of an MAC over the compressed data is illustrated in Figure 9.3. Using a shared secret key, the calculation is defined as follows:

$H_1 = \text{hash(MAC-write-secret} \parallel \text{pad-1} \parallel \text{seq-num} \parallel \text{SSLCompressed.type} \parallel$

$\qquad \text{SSLCompressed.length} \parallel \text{SSLCompressed.fragment)}$

$H \ = \text{hash(MAC-write-secret} \parallel \text{pad-2} \parallel H_1)$

where

MAC-write-secret:	Shared secret key
Hash (H_1 and H):	Cryptographic hash algorithm; either MD5 or SHA-1
Pad-1:	The byte 0x36 (0011 0110) repeated 48 times (384 bits) for MD5 and 40 times (320 bits) for SHA-1

Pad-2:	The byte 0x5C (0101 1100) repeated 48 times for MD5 and 40 times for SHA-1
Seq-num:	The sequence number for this message
SSLCompressed.type:	The higher-level protocol used to process this fragment
SSLCompressed.length:	The length of the compressed fragment
SSLCompressed.fragment:	The compressed fragment (the plaintext fragment if not compressed)
‖:	concatenation symbol

The compressed message plus the MAC are encrypted using symmetric encryption. The block ciphers being used as encryption algorithms are:

DES(56), Triple DES(168), IDEA(128),

RC5(variable), and Fortezza(80)

where the number inside the brackets indicates the key size. Fortezza is a Personal Computer Memory Card International Association (PCMCIA) card that provides both encryption and digital signing.

For block encryption, padding is added after the MAC prior to encryption. The total size of the data (plaintext plus MAC plus padding) to be encrypted must be a multiple of the cipher's block length. Padding is added to force the length of the plaintext to be a multiple of the block cipher's block length. Padding is formed by appending a single '1' bit to the end of the message and then '0' bits are added, as many as needed. The last 64 bits of the total size of padded data are reserved for the original message length.

For stream encryption, the compressed message plus the MAC are encrypted. Since the MAC is computed before encryption takes place, it is encrypted along with the compressed plaintext.

- *Append SSL record header.* The final processing of the SSL Record Protocol is to append an SSL record header. The composed fields consist of

  - *Content type (8 bits).* This field is the higher-layer protocol used to process the enclosed fragment.
  - *Major version (8 bits).* This field indicates the major version of SSL in use. For SSLv3, the value is 3.
  - *Minor version (8 bits).* This field indicates the minor version of SSL in use. For SSLv3, the value is 0.
  - *Compressed length (16 bits).* This field indicates the length in bytes of the plaintext fragment or compressed fragment if compression is required. The maximum value is $2^{14} + 2048$.

Figure 9.4 illustrates the SSL Record Protocol format.

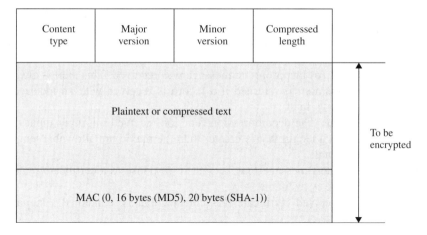

Content type	Major version	Minor version	Compressed length

Plaintext or compressed text

MAC (0, 16 bytes (MD5), 20 bytes (SHA-1))

To be encrypted

Figure 9.4   SSL Record Protocol format.

The SSL-specific protocols consist of the Change Cipher Spec Protocol, the Alert Protocol, and the Handshake Protocol, as shown in Figure 9.1. The contents of these three protocols are presented in what follows.

9.1.3   SSL Change Cipher Spec Protocol

The Change Cipher Spec Protocol is the simplest of the three SSL-specific protocols. This protocol consists of a single message, which is compressed and encrypted under the current CipherSpec. The message consists of a single byte of value 1. The change cipher spec message is sent by both the client and server to notify the receiving party that subsequent records will be protected under the just-negotiated CipherSpec and keys. Reception of this message causes the pending state to be copied into the current state, which updates the cipher suite to be used on this connection. The client sends a change cipher spec message following handshake key exchange and certificate verify messages (if any), and the server sends one after successfully processing the key exchange message it received from the client.

9.1.4   SSL Alert Protocol

One of the content types supported by the SSL Record Layer is the alert type. Alert messages convey the severity of the message and a description of the alert. Alert messages consist of 2 bytes. The first byte takes the value warning or fatal to convey the seriousness of the message. If the level is fatal, SSL immediately terminates the connection. In this case, other connections on the same session may continue, but the session identifiers must be invalidated, preventing the failed session from being used to establish new connections. The second byte contains a code that indicates the specific alert. As with other applications that use SSL, alert messages are compressed and encrypted, as specified by the current connection state.

A specification of SSL-related alerts that are always fatal is listed in the following:

- *unexpected-message.* An inappropriate message was received. This alert is always fatal.
- *bad-record-mac.* This alert is returned if a record is received with an incorrect MAC. This message is always fatal.
- *decompression-failure.* The decompression function received improper input (i.e., data that would expand to a length that is greater than the maximum allowable length). This message is always fatal.
- *no-certificate.* This alert message may be sent in response to a certificate request if no appropriate certificate is available.
- *bad-certificate.* A received certificate was corrupt, that is, contained a signature that did not verify correctly.
- *unsupported certificate.* The type of the received certificate is not supported.
- *certificate-revoked.* A certificate has been revoked by its signer.
- *certificate-expired.* A certificate has expired or is not currently valid.
- *certificate-unknown.* This means some other unspecified issue arose in processing the certificate, rendering it unacceptable.
- *illegal-parameter.* A field in the handshake was out of range or inconsistent with other fields. This is always fatal.
- *close-notify.* This message notifies the recipient that the sender will not send any more messages on this connection. The session becomes unresumable if any connection is terminated without proper close-notify messages with level equal to warning. Each party is required to send a close-notify alert before closing the write side of the connection. Either party may initiate a close-notify alert. Any data received after a closure alert is ignored.

9.1.5   SSL Handshake Protocol

The SSL Handshake Protocol being operated on top of the SSL Record Layer is the most important part of SSL. This protocol provides three services for SSL connections between the server and client. The Handshake Protocol allows the client/server to agree on a protocol version, to authenticate each other by forming an MAC, and to negotiate an encryption algorithm and cryptographic keys for protecting data sent in an SSL record before the application protocol transmits or receives its first byte of data.

The Handshake Protocol consists of a series of messages exchanged by the client and server. Figure 9.5 shows the exchange of handshake messages needed to establish a logical connection between client and server. The contents and significance of each message are presented in detail in the following sections.

Phase 1: Hello Messages for Logical Connection

The client sends a client hello message to which the server must respond with a server hello message, or else a fatal error will occur and the connection will fail. The client

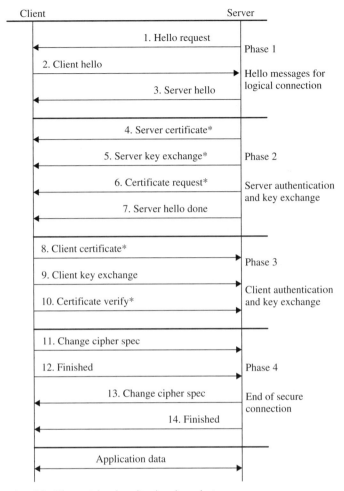

Asterisks (*) are optional or situation-dependent
messages that are not always sent.

Figure 9.5   SSL Handshake Protocol.

hello and server hello are used to establish security enhancement capabilities between
client and server. The client hello and server hello establish the following attributes:
protocol version, random values (ClientHello.random and ServerHello.random), session
ID, cipher suite, and compression method.

Hello messages

The hello phase messages are used to exchange security enhancement capabilities between
client and server.

- *Hello request.* This message is sent by the server at any time, but may be ignored by the client if the Handshake Protocol is already underway. A client who receives a hello request while in a handshake negotiation state should simply ignore the message.
- *Client hello.* The exchange is initiated by the client. A client sends a client hello message using the session ID of the session to be resumed. The server then checks its session cache for a match. If a match is found, the server will send a server hello message with the same session ID value. The client sends a client hello message with the following parameters:

  - *Client version.* This is the version of the SSL protocol in which the client wishes to communicate during this session. This should be the most recent (highest-valued) version supported by the client. The value of this version will be 3.0.
  - *Random.* This is a client-generated random structure with 28 bytes generated by a secure random number generator.
  - *Session ID.* This is the identity of a session when the client wishes to use this connection. A nonzero value indicates that the client wishes to update the parameters of an existing connection or create a new connection in this session. A zero value indicates that the client wishes to establish a new connection in a new session.
  - *Cipher suites.* This is a list of the cryptographic options supported by the client, with the client's first preference first. The single cipher suite is an element of a list selected by the server from the list in ClientHello.cipher_suites. For a resumed session, this field is the value from the state of the session being resumed.
  - *Compression method.* This is a list of the compression methods supported by the client, sorted by client preference. If the session ID field is not empty, it must include the compression method from that session.

- *Server hello.* The server will send the server hello message in response to a client hello message when it has found an acceptable set of algorithms. If it is unable to find such a match, it will respond with a handshake failure alert. The structure of this message consists of server version, random, session ID, cipher suite, and compression method.

  - *Server version.* This field will contain the lower-valued version suggested by the client in the client hello and the highest-valued version supported by the server. The value of this version is 3.0.
  - *Random.* This structure is generated by the server and must be different from Client-breakHello.random.
  - *Session ID.* This field represents the identity of the session corresponding to this connection. If the ClientHello.session_id is nonempty, the server will look in its session cache for a match. If a match is found and the server is willing to establish the new connection using the specified session state, the server will respond with the same value as was supplied by the client. This indicates a resumed session and dictates that the parties must proceed directly to the finished messages.
  - *Cipher suite.* This is the single cipher suite selected by the server from the list in ClientHello.cipher_suites. For a resumed session, this field is the value from the state of the session being resumed.

- *Compression method.* This is the single compression algorithm selected by the server from the list in ClientHello.compression_methods. For resumed sessions, this field is the value from the resumed session state.

Phase 2: Server Authentication and Key Exchange

Following the hello messages, the server begins this phase by sending its certificate if it needs to be authenticated. Additionally, a server key exchange message may be sent if it is required. If the server is authenticated, it may request a certificate from the client, if that is appropriate to the cipher suite selected. Then the server will send the server hello done message, indicating that the hello message phase of the handshake is complete. The server will then wait for a client response. If the server has sent a certificate request message, the client must send the certificate message.

- *Server certificate.* If the server is to be authenticated, it must send a certificate immediately following the server hello message. The certificate type must be appropriate for the selected cipher suite's key exchange algorithm, and is generally an X.509 v3 certificate. It must contain a key which matches the key exchange method. The signing algorithm for the certificate must be the same as the algorithm for the certificate key.
- *Server key exchange message.* The server key exchange message is sent by the server only when it is required. This message is not used if the server certificate contains Diffie–Hellman (DH) parameters, or RSA key exchange is to be used for a signature-only RSA.

  - *params.* The server's key exchange parameters.
  - *signed-params.* For nonanonymous key exchange, a hash of the corresponding params value, with the signature appropriate to that hash applied.

  As usual, a signature is created by taking the hash of the message and encrypting it with the sender's public key. Hence, the hash is defined as:

  md5-hash : MD5(ClientHello.random‖ServerHello.random‖serverParams)

  sha-hash : SHA(ClientHello.random‖ServerHello.random‖serverParams)

  $\qquad$ enum{anonymous, rsa, dsa}SignatureAlgorithm

  For a Digital Signature Standard (DSS) signature, the hash is performed using the SHA-1 algorithm. In the case of an RSA signature, both an MD5 and an SHA-1 hash are calculated, and the concatenation of the two hashes is encrypted with the server's public key.
- *Certificate request message.* A nonanonymous server can optionally request a certificate from the client, if appropriate for the selected cipher suite. This message includes two parameters, certificate type and certificate authorities. Its structure is as follows:

```
enum{
    rsa_sign(1), des_sign(2), rsa_fixed_dh(3),
```

```
    dss_fixed_dh(4),
    rsa_ephemeral_dh(5), dss_ephemeral_dh(6),
    fortezza_dms(20), (255)
} ClientCertificateType;
opaque DistinguishedName<1..2^16-1>;
struct {
        ClientCertificateType certificate_types<1..2^8-1>;
        DistinguishedName certificate_authorities<3..2^16-1>
} CertificateRequest;
```

- *certificate_types*. This field is a list of the types of certificates requested, sorted in order of the server's preference.
- *certificate_authorities*. This is a list of the distinguished names of acceptable certificate authorities. These distinguished names may specify a desired distinguished name for a root CA or for a subordinate CA; thus, this message can be used to describe both known roots and a desired authorization space.
Note that DistinguishedName is derived from X.509 and that it is a fatal handshake_ failure alert for an anonymous server to request client identification.
- *Server hello done message.* This message is sent by the server to indicate the end of the server hello and associated messages. After sending this message, the server will wait for a client response. This message means that server has finished sending messages to support the key exchange, and the client can proceed with its phase of the key exchange. Upon receipt of the server hello done message, the client should verify that the server provided a valid certificate if required and check that the server hello parameters are acceptable. If all is satisfactory, the client sends one or more messages back to the server.

Phase 3: Client Authentication and Key Exchange

If the server has sent a certificate request message, the client must send the certificate message. The client key exchange message is then sent, and the content of that message will depend on the public key algorithm selected between the client hello and the server hello. If the client has sent a certificate with signing ability, a digitally signed certificate verify message is sent to explicitly verify the certificate.

- *Client certificate message.* This is the first message the client can send after receiving a server hello done message. This message is sent only when the server requests a certificate. If no suitable certificate is available, the client should send a certificate message containing no certificates. If client authentication is required by the server for the handshake to continue, it may respond with a fatal handshake failure alert. The same message type and structure will be used for the client's response to a certificate request message. Note that a client may send no certificates if it does not have an appropriate certificate to send in response to the server's authentication request. The client's DH certificates must match the server-specified DH parameters.

- *Client key exchange message.* This message is always sent by the client. It will immediately follow the client certificate message, if it is sent. Otherwise, it will be the first message sent by the client after it receives the server hello done message. With this message, the premaster secret is set, either through direct transmission of the RSA-encrypted secret, or by transmission of DH parameters which will allow each side to agree upon the same premaster secret. When the key exchange method is DH–RSA or DH–DSS, client certification has been requested, and the client was able to respond with a certificate which contained a DH public key whose parameters matched those specified by the server in its certificate; this message will not contain any data.

- *Certificate verify message.* This message is used to provide explicit verification of a client certificate. The message is only sent following any client certificate that has signing capability (i.e., all certificates except those containing fixed DH parameters). When sent, it will immediately follow the client key exchange message. This message signs a hash code based on the preceding messages, and its structure is defined as follows:

```
struct{
      Signature signature;
} CertificateVerify;
CertificateVerify.signature.md5_hash
    MD5(master_secret||pad2||MD5(handshake-message||
          master_secret||pad1))
Certificate.signature.sha_hash
    SHA(master_secret||pad2||SHA(handshake-message||
          master_secret||pad1))
```

where pad1 and pad2 are the values defined earlier for the MAC, handshake messages refer to all Handshake Protocol messages sent or received starting at client hello but not including this message, and master_secret is the calculated secret. If the user's private key is DSS, then it is used to encrypt the SHA-1 hash. If the user's private key is RSA, it is used to encrypt the concatenation of the MD5 and SHA-1 hashes.

Phase 4: End of Secure Connection

At this point, a change cipher spec message is sent by the client, and the client copies the pending CipherSpec into the current CipherSpec. The client then immediately sends the finished message under the new algorithms, keys, and secrets. In response, the server will send its own change cipher spec message, transfer the pending CipherSpec to the current one, and then send its finished message under the new CipherSpec. At this point, the handshake is complete and the client and the server may begin to exchange application layer data (Figure 9.5).

- *Change cipher spec messages.* The client sends a change cipher spec message and copies the pending CipherSpec in the current CipherSpec. This message is immediately sent after the certificate verify message that is used to provide explicit verification of a client certificate. It is essential that a change cipher spec message is received between the other handshake messages and the finished message. It is a fatal error if a change cipher spec message is not preceded by a finished message at the appropriate point in the handshake.

- *Finished message.* This is always sent immediately after a change cipher spec message to verify that the key exchange and authentication processes were successful. The content of the finished message is the concatenation of two hash values:

```
MD5(master_secret||pad2||MD5(handshake_messages||Sender||
    master_secret||pad1))
SHA(master_secret||pad2||SHA(handshake_messages||Sender||
    master_secret||pad1))
```

where 'Sender' is a code that identifies that the sender is the client and "handshake_ messages" is code that identifies the data from all handshake messages up to but not including this message.

The finished message is first protected with just-negotiated algorithms, keys, and secrets. Recipients of finished messages must verify that the contents are correct. Once a side has sent its finished message and received and validated the finished message from its peer, it may begin to send and receive application data over the connection. Application data treated as *transparent data* is carried by the Record Layer and is fragmented, compressed, and encrypted based on the current connection state.

9.2 Cryptographic Computations

The key exchange, authentication, encryption, and MAC algorithms are determined by the cipher suite selected by the server and revealed in the server hello message. The compression algorithm is negotiated in the hello messages, and the random values are exchanged in the hello messages. The creation of a shared master secret by means of the key exchange and the generation of cryptographic parameters from the master secrete are of interest to study as two further items.

9.2.1 Computing the Master Secret

For all key exchange methods, the same algorithm is used to convert the premaster secret into the master secret. In order to create the master secret, a premaster secret is first exchanged between two parties and then the master secret is calculated from it. The master secret is always exactly 48 bytes (384 bits) shared between the client and server. But the length of the premaster secret is not fixed and will vary depending on the key exchange method. There are two ways for the exchange of the premaster secret.

- *RSA.* When RSA is used for server authentication and key exchange, a 48-byte premaster secret is generated by the client, encrypted with the server's public key, and sent to the server. The server decrypts the ciphertext (of the premaster secret) using its private key to recover the premaseter secret. Both parties then convert the premaster secret into the master secret as specified below.

- *Diffie–Hellman.* A conventional DH computation is performed. Both client and server generate a DH common key. This negotiated key is used as the premaster secret and is converted into the master secret, as specified below.

The client and server then compute the master secret as follows:

```
master_secret = MD5(pre_master_secret||SHA('A'||
                   pre_master_secret||ClientHello.random||
                     ServerHello.random))||
                 MD5(pre_master_secret||SHA('BB'||
                   pre_master_secret||ClientHello.random||
                     ServerHello.random))||
                 MD5(pre_master_secret||SHA('CCC'||
                   pre_master_secret||ClientHello.random||
                     ServerHello.random))
```

where ClientHello.random and ServerHello.random are the two nonce values exchanged in the initial hello messages.

The generation of the master secret from the premaster secret is shown in Figure 9.6.

9.2.2 Converting the Master Secret into Cryptographic Parameters

CipherSpec specifies the bulk data encryption algorithm and a hash algorithm used for MAC computation, and defines cryptographic attributes such as the hash size.

To generate the key material, the following is computed.

```
key_block = MD5(master_secret||SHA('A'||master_secret||
               ServerHello.random||ClientHello.random))||
             MD5(master_secret||SHA('BB'||master_secret||
               ServerHello.random||ClientHello.random))||
             MD5(master_secret||SHA('CCC'||master_secret||
               ServerHello.random||ClientHello.random))||...
```

until enough output has been generated. Note that the generation of the key block from the master secret uses the same format for generation of the master secret from the premaster secret. Figure 9.7 illustrates the steps for generation of the key block from the master secret.

9.3   TLS Protocol

The TLSv1 protocol itself is based on the SSLv3 protocol specification as published by Netscape. Many of the algorithm-dependent data structures and rules are very close so that the differences between TLSv1 and SSLv3 are not dramatic. The current work on

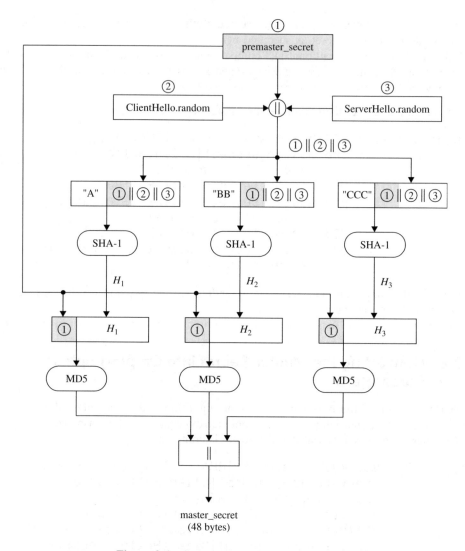

Figure 9.6 Computation of the master secret.

TLS is aimed at producing an initial version as an Internet standard. It is recommended that readers examine the comparative studies between the TLSv1 of RFC 2246 and SSLv3 of Netscape. In this section, we will not repeat every detailed step of identical protocol contents, but only highlight the differences.

9.3.1 HMAC Algorithm

A Keyed-hashing Message Authentication Code (HMAC) is a secure digest of some data protected by a secret. Forging the HMAC is infeasible without knowledge of the MAC secret. HMAC can be used with a variety of different hash algorithms, namely, MD5 and SHA-1, denoting these as HMAC_MD5(secret, data) and HMAC_SHA-1(secret, data).

There are two differences between the SSLv3 and TLSMAC schemes. TLS makes use of the HMAC algorithm defined in RFC 2104. HMAC was fully discussed in Chapters 4 and 7 and defined as

$$\text{HMAC} = H[(K \oplus \text{opad})\|H[(K \oplus \text{ipad})\|M]]$$

where

ipad $= 00110110$ (0x36) repeated 64 times (512 bits)
opad $= 01011100$ (0x5c) repeated 64 times (512 bits)
  $H =$ one-way hash function for TLS (either MD5 or SHA-1)
  $M =$ message input to HMAC
  $K =$ padded secret key equal to the block length of the hash code
      (512 bits for MD5 and SHA-1)

The following explains the HMAC equation.

1. Append zeros to the end of K to create a b-byte string (i.e., if $K = 160$ bits in length and $b = 512$ bits, then K will be appended with 352 zero bits or 44 zero bytes 0x00).
2. XOR (bitwise exclusive OR) K with ipad to produce the b-bit block computed in step 1.
3. Append M to the b-byte string resulting from step 2.
4. Apply H to the stream generated in step 3.
5. XOR (bitwise exclusive OR) K with opad to produce the b-byte string computed in step 1.
6. Append the hash result H from step 4 to the b-byte string resulting from step 5.
7. Apply H to the stream generated in step 6 and output the result.

Figure 9.8 illustrates the overall operation of HMAC–MD5 or HMAC–SHA-1.

Example 9.1 HMAC–SHA-1 computation using RFC method:

Data : 0x 7104f218 a3192e65 1cf7025d 8011bf79 4a19
Key : 0x 31fa7062 c45113e3 2679fd13 53b71264

–	A	B	C	D	E
IV	67452301	efcdab89	98badcfe	10325476	c3d2e1f0
$H[(K \oplus \text{ipad})\|M]$	8efeef30	f64b360f	77fd8236	273f0784	613bbd4b
$H[(K \oplus \text{opad})\|H[(K \oplus \text{ipad})\|M]]$	31db10b8	ed346850	d0f0b7dd	50fd71f4	2dacd24c

HMAC–SHA-1 $= 0x$ 31 db10b8 ed346850 d0f0b7dd 50fd71f4 2dacd24c

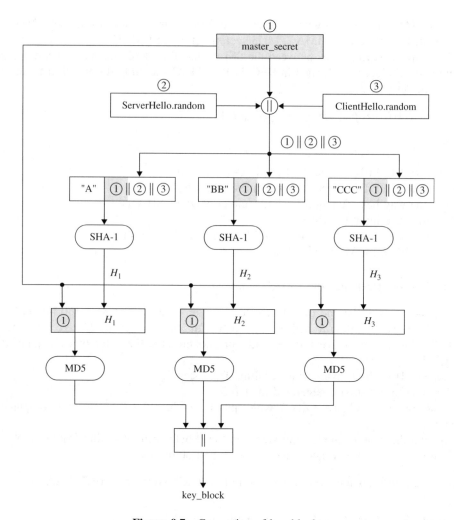

Figure 9.7 Generation of key block.

The alternative operation for computation of either HMAC–MD5 or HMAC–SHA-1 is described in the following.

1. Append zeros to K to create a b-bit string K', where $b = 512$ bits.
2. XOR K' (padding with zero) with ipad to produce the b-bit block.
3. Apply the compression function f(IV, $K' \oplus$ ipad) to produce (IV)$_i$ = 128 bits.
4. Compute the hash code h with (IV)$_i$ and M_i.
5. Raise the hash value computed from step 4 to a b-bit string.
6. XOR K' (padded with zeros) with opad to produce the b-bit block.
7. Apply the compression function f(IV, $K' \oplus$ opad) to produce (IV)$_o$ = 128 bits.
8. Compute the HMAC with (IV)$_o$ and the raised hash value resulting from step 5.

Figure 9.9 illustrates the overall operation of HMAC–MD5.

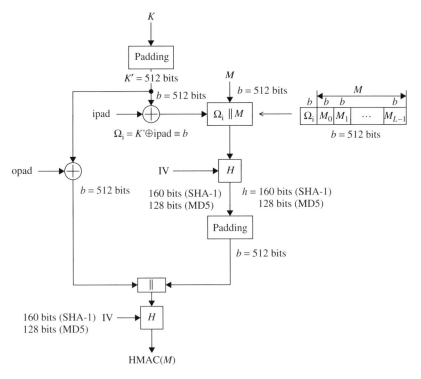

Figure 9.8 Overall operation of HMAC computation using either MD5 or SHA-1 (message length computation based on $\Omega_i \| M$.)

Example 9.2 HMAC–MD5 computation using alternative method:

Data: 0x 2143f501 f014a713 c1059e23 7123fd68
Key: 0x 31fa7062 c45113e3 2679fd13 53b71264

–	A	B	C	D
IV	67452301	efcdab89	98badcfe	10325476
f[$(K \oplus \text{ipad}), IV] = (IV)_i$	13fbaf34	034879ab	35e73505	526a8d28
$H[M, (IV)_i]$	90c6d9b0	0f281bc8	94d04b33	7f0f4265
f[$(K \oplus \text{opad}), IV] = (IV)_o$	5f8647d7	fa8e9afa	bffa4989	3cd471d1
$H[H[M, (IV)_i], (IV)_o]$	2c47cd5b	68830268	7d255059	45c7bef0

HMAC–MD5 = 0x 2c47cd5b 68830268 7d255059 45c7bef0

For TLS, the MAC computation encompasses the fields indicated in the following expression.

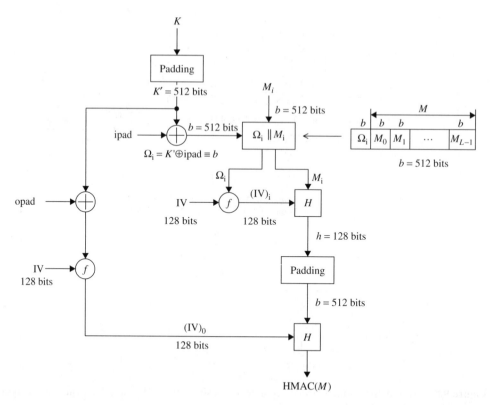

Figure 9.9 Alternative operation of HMAC computation using MD5 (message length computation is based on *M* only).

```
HMAC_hash(MAC_write_secret, seq_num||TLScompressed.type||
  TLSCompressed.version||TLSCompressed.length||
  TLSCompressed.fragment)
```

Note that the MAC calculation includes all of the fields covered by the SSLv3 computation, plus the field TLSCompressed.version, which is the version of the protocol being employed.

9.3.2 Pseudo-random Function

TLS utilizes a pseudo-random function (PRF) to expand secrets into blocks of data for the purposes of key generation or validation. The PRF takes relatively small values such as a secret, a seed, and an identifying label as input and generates an output of arbitrary longer blocks of data.

The data expansion function, P_hash(secret, data), uses a single hash function to expand a secret and seed into an arbitrary quantity of output. The data expansion function is defined as follows:

```
P_hash(secret, seed) = HMAC_hash (secret, A(1)||seed) ||
                       HMAC_hash (secret, A(2)||seed) ||
                       HMAC_hash (secret, A(3)||seed) ||...
```

where A() is defined as

A(0) = seed
A(i) = HMAC_hash(secret, A(i-1)) and ‖ indicates concatenation.

Applying A(i), $i = 0, 1, 2, \ldots$, to P_hash, the resulting sketch can be depicted as shown in Figure 9.10. As you can see, P_hash is iterated as many times as necessary to produce the required quantity of data. Thus, the data expansion function makes use of the HMAC algorithm with either MD5 or SHA-1 as the underlying hash function. As an example, consider SHA-1 whose value is 20 bytes (160 bits). If P_SHA-1 is used to create 64 bytes (512 bits) of data, it will have to be iterated four times up to A(4), creating $20 \times 4 = 80$ bytes (640 bits) of output data. Hence, the last 16 bytes (128 bits) of the final iteration A(4) must be discarded, leaving $(80 - 16) = 64$ bytes of output data. On the other hand, MD5 produces 16 bytes (128 bits). In order to generate an 80-byte output, P_MD5 should exactly be iterated through A(5), while P_SHA-1 will only iterate through A(4) as described above. In fact, alignment to a shared 64-byte output will be required to discard the last 16 bytes from both P_SHA-1 and P_MD5.

TLS's PRF is created by splitting the secret into two halves (S1 and S2) and using one half to generate data with P_MD5 and the other half to generate data with P_SHA-1. These two results are then XORed to produce the output. S1 is taken from the first half of the secret and S2 from the second half. Their length is respectively created by rounding up the length of the overall secret divided by 2. Thus, if the original secret is an *odd* number of bytes long, the last bytes of S1 will be the same as the first byte of S2.

L_S = length in bytes of secret

L_S1 = L_S2 = ceil(L_S/2)

The PRF is then defined as the result of mixing the two pseudo-random streams by XORing them together. The PRF is defined as

PRF(secret, label, seed) = P_MD5(S1, label‖seed) $\oplus$ P_SHA $-$ 1(S2, label‖seed)

The label is an ASCII string. Figure 9.11 illustrates the PRF generation scheme to expand secrets into blocks of data.

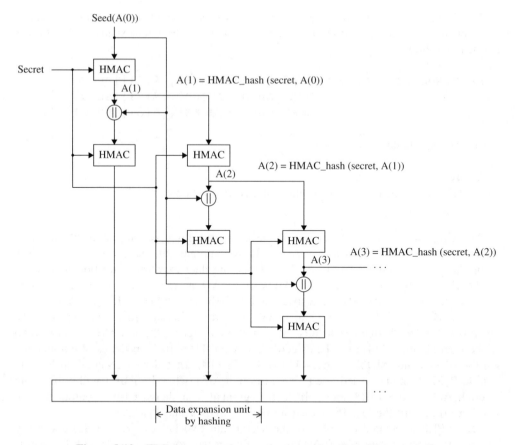

Figure 9.10   TLS data expansion mechanism using P_hash(secret,seed).

Example 9.3   Refer to Figure 9.11. Suppose the following parameters are given:

seed = 0x 80 af 12 5c 7e 36 f3 21

label = rocky mountains = 0x 72 6f 63 6b 79 20 6d 6f 75 6e 74 61 69 6e 73

secret = 0x 35 79 af 12 c4

Then

label∥seed = 0x 72 6f 63 6b 79 20 6d 6f 75 6e 74 61 69 6e 73 80 af 12 5c 7e

            36 f3 21

        = A(0)

      S1 = 0x 35 79 af for P_MD5,   S2 = 0x af 12 c4 for P_SHA − 1

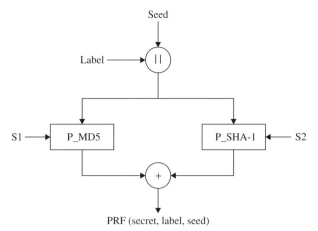

S1: First half of the secret
S2: Second half of the secret

P_MD5: Data expansion function to expand a secret
 S1 and (seed‖secret) using MD5
P_SHA-1: Data expansion function to expand a secret
 S2 and (seed‖secret) using SHA-1

Figure 9.11 A pseudorandom function (PRF) generation scheme.

Data expansion by P_MD5

$A(1) = \text{HMAC\_MD5}(S1, A(0))$

 = d0 de 36 53 79 78 04 a0 21 b8 6f f8 29 60 d5 f7

$\text{HMAC\_MD5}(S1, A(1)\|A(0))$

 = 32 fd b3 70 eb 36 11 70 a4 3b 50 a9 fb ea 2a ec

$A(2) = \text{HMAC\_MD5}(S1, A(1))$

 = 8c ce 5b 50 02 af 75 91 e7 20 cd 86 d9 3e 67 9d

$\text{HMAC\_MD5}(S1, A(2)\|A(0))$

 = 1f a8 4c af 5d e1 20 01 ea b0 38 6a a5 76 f9 8e

$A(3) = \text{HMAC\_MD5}(S1, A(2))$

 = 45 48 5d 00 4e 64 07 45 eb 2c 18 60 7c e6 fa 1f

$\text{HMAC\_MD5}(S1, A(3)\|A(0))$

 = f0 23 29 d9 5e 89 4b 70 cc 45 f8 aa 1f 58 8e 55

$A(4) = \text{HMAC\_MD5}(S1, A(3))$

 = 87 39 c6 d3 7a b f8 e3 29 79 3a ae 63 24 6a ff

HMAC_MD5(S1, A(4)||A(0))

   = 2e 0c 27 26 d0 b4 78 85 09 a2 69 1c 1b 1b d7 8d

A(5) = HMAC_MD5(S1, A(4))

   = 3a 2c aa d8 b3 ec 2e 5d 40 1c 39 bd 3e 48 1a d9

HMAC_MD5(S1, A(5)||A(0))

   = 92 f2 63 5d 88 3a dd bf 8d ec e1 cf 0c 5c 8f 4c

where S1 = 0x 35 79 af = first half of the secret, and

A(0) = label||seed

Thus, P_MD5 equals:

32	fd	b3	70	eb	36	11	70	a4	3b	50	a9	fb	ea	2a	ec	
1f	a8	4c	af	5d	e1	20	01	ea	b0	38	6a	a5	76	f9	8e	
f0	23	29	d9	5e	89	4b	70	cc	45	f8	aa	1f	58	8e	55	
2e	0c	27	26	d0	b4	78	85	09	a2	69	1c	1b	1b	d7	8d	
92	f2	63	5d	88	3a	dd	bf	8d	ec	e1	cf	0c	5c	8f	4c	(80 bytes)

Data expansion by P_SHA-1

A(1) = HMAC_SHA1(S2, A(0))

   = aa ea 46 1b a6 ad 43 34 51 f8 c6 ef 70 dd f4 60 ca b9 40 2f

HMAC_SHA1(S2, A(1)||A(0))

   = d0 8a d5 07 e0 b8 30 78 70 d9 c8 bb dd ba f5 a3 d0 77 49 e8

A(2) = HMAC_SHA1(S2, A(1))

   = 33 fd 23 41 01 ce 06 f8 c0 2b b3 e6 54 21 1c f4 6c 88 ab da

HMAC_SHA1(S2, A(2)||A(0))

   = 64 b5 cc 3f 79 31 5b 5d e6 e4 4f eb 98 a8 bf 3f 97 13 38 e1

A(3) = HMAC_SHA1(S2, A(2))

   = 86 1f a3 a5 37 58 41 71 f1 9f a5 f3 48 2e 5d 84 7c a8 b6 52

HMAC_SHA1(S2, A(3)||A(0))

   = 03 26 11 02 ce 69 74 4a 21 f4 76 55 13 af 77 80 2d fb 2f 36

A(4) = HMAC_SHA1(S2, A(3))

   = 9c 4d 01 3a 8c 48 54 42 68 07 4d f1 f0 a9 78 c3 6f ab d8 b4

HMAC_SHA1(S2, A(4)‖A(0))

$\quad$ = 48 56 04 b5 b4 5f 9b d8 c7 2f 28 f6 9e 1d 8a c4 72 9a b9 32

where S2 = 0x af 12 c4 = second half of the secret, and

A(0) = label‖seed

Thus, P_SHA1 equals:

d0	8a	d5	07	e0	b8	30	78	70	d9	c8	bb	dd	ba	f5	a3
d0	77	49	e8	64	b5	cc	3f	79	31	5b	5d	e6	e4	4f	eb
98	a8	bf	3f	97	13	38	e1	03	26	11	02	ce	69	74	4a
21	f4	76	55	13	af	77	80	2d	fb	2f	36	48	56	04	b5
b4	5f	9b	d8	c7	2f	28	f6	9e	1d	8a	c4	72	9a	b9	32 (80 bytes)

Finally, P_MD5 $\oplus$ P_SHA $-$ 1 equals:

e2	77	66	77	0b	8e	21	08	d4	e2	98	12	26	50	df	4f
cf	df	05	47	39	54	ec	3e	93	81	63	37	43	92	b6	65
68	8b	96	e6	c9	9a	73	91	cf	63	e9	a8	d1	31	fa	1f
0f	f8	51	73	c3	1b	0f	05	24	59	46	2a	53	4d	d3	38
26	ad	f8	85	4f	15	f5	49	13	f1	6b	0b	7e	c6	36	7e (80 bytes)

9.3.3 Error Alerts

The Alert Protocol is classified into the closure alert and the error alert. One of the content types supported by the TLS Record Layer is the alert type. Alert messages convey the severity of the message and a description of the alert. Alert messages with a fatal level result in the immediate termination of the connection.

The client and the server must share knowledge that the connection is ending in order to avoid a truncation attack. Either party may initiate a close by sending a close_notify alert. This message notifies the recipient that the sender will not send any more messages on this connection.

Error handling in the TLS Handshake Protocol is very simple. When an error is detected, the detecting party sends a message to the other party. Upon transmission or receipt of a fatal alert message, both parties immediately close the connection.

TLS supports all of the error alerts defined in SSLv3 with the exception of additional alert codes defined in TLS. The additional error alerts are described in the following.

- *decryption_failed*. A TLS ciphertext is decrypted in an invalid way: either it was not an even multiple of the block length or its padding values, when checked, were incorrect. This message is always fatal.

- *record_overflow.* A TLS record was received with a ciphertext whose length exceeds $2^{14} + 2048$ bytes, or the ciphertext decrypted to a TLS compressed record with more than $2^{14} + 1024$ bytes. This message is always fatal.
- *unknown_ca.* A valid certificate chain or partial chain was received, but the certificate was not accepted because the CA certificate could not be located or could not be matched with a known, trusted CA. This message is always fatal.
- *access_denied.* A valid certificate was received, but when access control was applied, the sender decided not to proceed with the negotiation. This message is always fatal.
- *decode_error.* A message could not be decoded because a field was out of its specified range or the length of the message was incorrect. This message was incorrect. It is always fatal.
- *decrypt_error.* A handshake cryptographic operation failed, including being unable to verify a signature, decrypt a key exchange, or validate a finished message.
- *export_restriction.* A negotiation not in compliance with export restrictions was detected; for example, attempting to transfer a 1024-bit ephemeral RSA key for the RSA_EXPORT handshake method. This message is always fatal.
- *protocol_version.* The protocol version the client has attempted to negotiate is recognized but not supported due to the fact that old protocol versions might be avoided for security reasons. This message is always fatal.
- *insufficient_security.* Returned instead of hanshake_failure when a negotiation has failed specifically because the server requires ciphers more secure than those supported by the client. This message is always fatal.
- *internal_error.* An internal error unrelated to the peer or the correctness of the protocol, such as a memory allocation failure, makes it impossible to continue. This message is always fatal.
- *user_canceled.* This handshake is being cancelled for some reason unrelated to a protocol failure. If the user cancels an operation after the handshake is complete, just closing the connection by sending a close_notify is more appropriate. This alert should be followed by a close_notify. This message is generally a warning.
- *no_renegotiation.* This is sent by the client in response to a hello request or by the server in response to a client hello after initial handshaking. Either of these messages would normally lead to renegotiation, but this alert indicates that the sender is not able to renegotiate. This message is always a warning.

For all errors where an alert level is not explicitly specified, the sending party may determine at its discretion whether this is a fatal error or not; if an alert with a level of warning is received, the receiving party may decide at its discretion whether to treat this as a fatal error or not. However, all messages which are transmitted with a level of fatal must be treated as fatal messages.

9.3.4 Certificate Verify Message

Recall that the hash computations for SSLv3 are included with the master secret, the handshake message, and pads. In the TLS certificate verify message, the MD5 and SHA-1 hashes are calculated only over handshake messages as shown below.

```
CertificateVerify.signature.md5_hash
      MD5(handshake_message)
CertificateVerify.signature.sha_hash
      SHA(handshake_message)
```

Here handshake messages refer to all handshake messages sent or received starting at client hello up to, but not including, this message, including the type and length fields of the handshake messages.

9.3.5 Finished Message

A finished message is always sent immediately after a change cipher spec message to verify that the key exchange and authentication processes were successful. It is essential that a change cipher spec message be received between the other handshake messages and the finished message. As with the finished message in SSLv3, the finished message in TLS is a hash based on the shared master_secret, the previous handshake messages, and a label that identifies client and server. However, the TLS computation for verify_data is somewhat different from that of the SSL calculation as shown below.

```
PRF(master_secret, finished_label, MD5(handshake_message)||
      SHA-1(handshake_message))
```

where

- The finished_label indicates either the string "client finished" sent by the client or the string "server finished" sent by the server, respectively.
- The handshake_message includes all handshake messages starting at client hello up to, but not including, this finished message. This is only visible at the handshake layer and does not include record layer headers. In fact, this is the concatenation of all the handshake structures exchanged thus far. This may be different from handshake messages for SSL because it would include the certificate verify message. Also, the handshake_message for the finished message sent by the client will be different from that for the finished message sent by the server.

Note that change cipher spec messages, alters and any other record types are not handshake messages and are not included in the hash computations.

9.3.6 Cryptographic Computations (for TLS)

In order to begin connection protection, the TLS Record Protocol requires specification of a suite of algorithms, a master secret, and the client and server random values. The authentication, encryption, and MAC algorithms are determined by the cipher_suite selected by the server and revealed in the server hello message. The compression algorithm is negotiated in the hello messages, and the random values are exchanged in the hello messages.

All that remains is to compute the master secret and the key block. The premaster_secret for TLS is calculated in the same way as in SSLv3. The presmater_secret should be deleted from memory once the master_secret has been computed. As in SSLv3, the master_secret in TLS in calculated as a hash function of the premaster_secret and two hello random numbers. The TLS master_secret computation is different from that of SSLv3 and is defined as follows:

```
master_secret = PRF(premaster_secret, ''master secret'',
                    ClientHello.random||ServerHello.random)
```

The master_secret is always exactly 48 bytes (384 bits) in length. The length of the premaster_secret will vary depending on key exchange method.

- *RSA*. When RSA is used for server authentication and key exchange, a 48-byte premaster_secret is generated by the client, encrypted with the server's public key, and sent to the server. The server uses its private key to decrypt the premaster_secret. Both parties then convert the premaster_secret into the master_secret, as specified above.
- *DH*. A conventional DH computation is performed. The negotiated key Z is used as the premaster_secret, and is converted into the master_secret, as specified above.

The computation of the key block parameters (MAC secret keys, session encryption keys, and IVs) is defined as follows:

```
key_block = PRF(master_secret, ''key expansion'',
                SecurityParameters.server_random||
                    SecurityParameters.client_random)
```

This is computed until sufficient output has been generated. As with SSLv3, key_block is a function of the master_secret and the client and the server random numbers, but for TLS, the actual algorithm is different.

On leaving this chapter, it is recommended that readers search for and find any other small differences between SSL and TLS.

10

Electronic Mail Security: PGP, S/MIME

Pretty Good Privacy (PGP) was invented by Philip Zimmermann who released version 1.0 in 1991. Subsequent versions 2.6.x and 5.x (or 3.0) of PGP have been implemented by an all-volunteer collaboration under the design guidance of Zimmermann. PGP is widely used in the individual and commercial versions that run on a variety of platforms throughout the computer community. PGP uses a combination of symmetric secret-key and asymmetric public-key encryption to provide security services for electronic mail and data files. It also provides data integrity services for messages and data files by using digital signature, encryption, compression (zip), and radix-64 conversion (ASCII Armor). With the explosively growing reliance on e-mail and file storage, authentication and confidentiality services have become increasing demands.

MIME is an extension to the RFC 2822 framework which defines a format for text messages being sent using e-mail. MIME is actually intended to address some of the problems and limitations of the use of SMTP. Secure/Multipurpose Internet Mail Extension (S/MIME) is a security enhancement to the MIME Internet e-mail format standard, based on technology from RSA Data Security.

Although both PGP and S/MIME are on an IETF standards track, it appears likely that PGP will remain the choice for personnel e-mail security for many users, while S/MIME will emerge as the industry standard for commercial and organizational use. Two schemes of PGP and S/MIME are discussed in this chapter.

10.1 PGP

Before looking at the operation of PGP in detail, it is convenient to confirm the notation. In the forthcoming analyses for security and data integrity services, the following symbols are generally used:

Wireless Mobile Internet Security, Second Edition. Man Young Rhee.
© 2013 John Wiley & Sons, Ltd. Published 2013 by John Wiley & Sons, Ltd.

K_s = session key H = hash function
KP_a = public key of user A KP_b = public key of user B
KS_a = private key of user A KS_b = private key of user B
E = conventional encryption D = conventional decryption
E_p = public-key encryption D_p = public-key decryption
Z = compression using zip algorithm Z^{-1} = decompression
$\|$ = concatenation

10.1.1   Confidentiality via Encryption

PGP provides confidentiality by encrypting messages to be transmitted or data files to be stored locally using a conventional encryption algorithm such as IDEA, 3DES, or CAST-128. In PGP, each symmetric key, known as a *session key*, is used only once. A new session key is generated as a random 128-bit number for each message. Since it is used only once, the session key is bound to the message and transmitted with it. To protect the key, it is encrypted with the receiver's public key. Figure 10.1 illustrates the sequence, which is described as follows:

- The sender creates a message.
- The sending PGP generates a random 128-bit number to be used as a session key for this message only.
- The session key is encrypted with RSA, using the recipient's public key.
- The sending PGP encrypts the message, using CAST-128 or IDEA or 3DES, with the session key. Note that the message is also usually compressed.
- The receiving PGP uses RSA with its private key to decrypt and recover the session key.
- The receiving PGP decrypts the message using the session key. If the message was compressed, it will be decompressed.

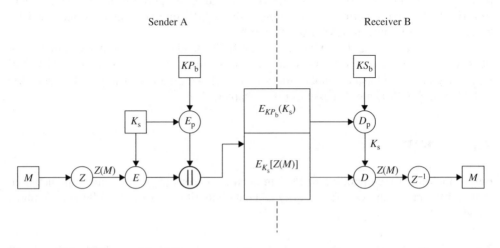

Figure 10.1 PGP confidentiality computation scheme with compression/decompression Algorithms.

Instead of using RSA for key encryption, PGP may use a variant of Diffie–Hellman (known as *ElGamal*) that does provide encryption/decryption. In order for the encryption time to reduce, the combination of conventional and public-key encryption is used in preference to simply using RSA or ElGamal to encrypt the message directly. In fact, CAST-128 and other conventional algorithms are substantially faster than RSA or ElGamal. Since the recipient is able to recover the session key that is bound to the message, the use of the public-key algorithms solves the session key exchange problem. Finally, to the extent that the entire scheme is secure, PGP should provide the user with a range of key size options from 768 to 3072 bits.

Both digital signature and confidentiality services may be applied to the same message. First, a signature is generated from the message and attached to the message. Then the message plus signature are encrypted using a symmetric session key. Finally, the session key is encrypted using public-key encryption and prefixed to the encrypted block.

10.1.2 Authentication via Digital Signature

The digital signature uses a hash code of the message digest algorithm and a public-key signature algorithm. Figure 10.2 illustrates the digital signature service provided by PGP. The sequence is as follows:

- The sender creates a message.
- SHA-1 is used to generate a 160-bit hash code of the message.
- The hash code is encrypted with RSA using the sender's private key and a digital signature is produced.
- The binary signature is attached to the message.
- The receiver uses RSA with the sender's public key to decrypt and recover the hash code.
- The receiver generates a new hash code for the received message and compares it with the decrypted hash code. If the two match, the message is accepted as authentic.

The combination of SHA-1 and RSA provides an effective digital signature scheme. As an alternative, signatures can be generated using DSS/SHA-1. The National Institute

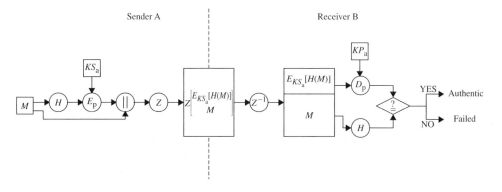

Figure 10.2 PGP authentication computation scheme using compression algorithm.

of Standards and Technology (NIST) has published FIPS PUB 186, known as the *Digital Signature Standard* (DSS). The DSS uses an algorithm that is designed to provide only the digital signature function. Although DSS is a public-key technique, it cannot be used for encryption or key exchange. The DSS approach for generating digital signatures was fully discussed in Chapter 5. The DSS makes use of the secure hash algorithm (SHA-1) described in Chapter 4 and presents a new digital signature algorithm (DSA).

10.1.3 Compression

As a default, PGP compresses the message after applying the signature but before encryption. The placement of Z for compression and Z^{-1} for decompression is shown in Figures 10.1 and 10.2. This compression algorithm has the benefit of saving space both for e-mail transmission and for file storage. However, PGP compression technique will present a difficulty.

Referring to Figure 10.1, message encryption is applied after compression to strengthen cryptographic security. In reality, cryptanalysis will be more difficult because the compressed message has less redundancy than the original message.

Referring to Figure 10.2, signing an uncompressed original message is preferable because the uncompressed message together with the signature are directly used for future verification. On the other hand, for a compressed message, one may consider two cases, either to store a compressed message for later verification or to recompress the message when verification is required. Even if a recompressed message were recovered, PGP's compression algorithm would present a difficulty due to the fact that different trade-offs in running speed versus compression ratio produce different compressed forms.

PGP makes use of a compression package called *ZIP* which is functionally equivalent to PKZIP developed by PKWARE, Inc. The zip algorithm is perhaps the most commonly used cross-platform compression technique.

Two main compression schemes, named after Abraham Lempel and Jakob Ziv, were first proposed by them in 1977 and 1978, respectively. These two schemes for text compression (generally referred to as *lossless compression*) are broadly used because they are easy to implement and also fast.

In 1982 James Storer and Thomas Szymanski presented their scheme, LZSS, based on the work of Lempel and Ziv. In LZSS, the compressor maintains a window of size N bytes and a lookahead buffer. Sliding-window-based schemes can be simplified by numbering the input text characters mod N, in effect creating a circular buffer. Variants of sliding-window schemes can be applied for additional compression to the output of the LZSS compressor, which include a simple variable-length code (LZB), dynamic Huffman coding (LZH), and Shannon–Fano coding (ZIP 1.x). All of them result in a certain degree of improvement over the basic scheme, especially when the data is rather random and the LZSS compressor has little effect.

Recently an algorithm was developed which combines the idea behind LZ77 and LZ78 to produce a hybrid called *LZFG* . LZFG uses the standard sliding window, but stores the data in a modified tree data structure and produces as output the position of the text in the tree. Since LZFG only inserts complete phrases into the dictionary, it should run faster than other LZ77-based compressors.

Huffman compression is a statistical data compression technique which reduces the average code length used to represent the symbols of an alphabet. Huffman code is an example of a code which is optimal when all symbols probabilities are integral powers of 1/2. A technique related to Huffman coding is Shannon–Fano coding. This coding divides the set of symbols into two equal or almost equal subsets based on the probability of occurrence of characters in each subset. The first subset is assigned a binary 0, the second a binary 1. Huffman encoding always generates optimal codes, but Shannon–Fano sometimes uses a few more bits.

Decompression of LZ77-compressed text is simple and fast. Whenever a (position, length) pair is encountered, one goes to that *position* in that window and copies *length* bytes to the output.

10.1.4   Radix-64 Conversion

When PGP is used, usually part of the block to be transmitted is encrypted. If only the signature service is used, then the message digest is encrypted (with the sender's private key). If the confidentiality service is used, the message plus signature (if present) are encrypted (with a one-time symmetric key). Thus, part or all of the resulting block consists of a stream of arbitrary 8-bit octets. However, many electronic mail systems only permit the use of blocks consisting of ASCII text. To accommodate this restriction, PGP provides the service of converting the raw 8-bit binary octets to a stream of printable 7-bit ASCII characters, called *radix-64 encoding* or *ASCII Armor*. Therefore, to transport PGP's raw binary octets through unreliable channels, a printable encoding of these binary octets is needed.

The scheme used for this purpose is radix-64 conversion. Each group of 3 octets of binary data is mapped into four ASCII characters. This format also appends a Cyclic Redundancy Check (CRC) to detect transmission errors. This radix-64 conversion is a wrapper around the binary PGP messages and is used to protect the binary messages during transmission over nonbinary channels, such as Internet e-mail.

Table 10.1 shows the mapping of 6-bit input values to characters. The character set consists of the upper- and lower-case letters, the digits 0–9, and the characters '+' and '/'. The '=' character is used as the padding character. The hyphen "-" character is not used.

Thus, a PGP text file resulting from ASCII characters will be immune to the modifications inflicted by mail systems. It is possible to use PGP to convert any arbitrary file to ASCII Armor. When this is done, PGP tries to compress the data before it is converted to radix-64.

Example 10.1   Consider the mapping of a 24-bit input (a block of 3 octets) into a four-character output consisting of the 8-bit set in the 32-bit block.

Suppose the 24-bit raw text is:

10110010   01100011   00101001

The hexadecimal representation of this text sequence is b2 63 29.

Table 10.1 Radix-64 encoding

6-Bit value	Character encoding	6-Bit value	Character encoding	6-Bit value	Character encoding	6-Bit value	Character encoding
0	A	16	Q	32	g	48	w
1	B	17	R	33	h	49	x
2	C	18	S	34	i	50	y
3	D	19	T	35	j	51	z
4	E	20	U	36	k	52	0
5	F	21	V	37	l	53	1
6	G	22	W	38	m	54	2
7	H	23	X	39	n	55	3
8	I	24	Y	40	o	56	4
9	J	25	Z	41	p	57	5
10	K	26	a	42	q	58	6
11	L	27	b	43	r	59	7
12	M	28	c	44	s	60	8
13	N	29	d	45	t	61	9
14	O	30	e	46	u	62	+
15	P	31	f	47	v	63	/
						(pad)	=

Arranging this input sequence in blocks of 6 bits yields:

101100   100110   001100   101001

The extracted 6-bit decimal values are 44, 38, 12, and 41.
Referring to Table 10.1, the radix-64 encoding of these decimal values produces the following characters:

smMp

If these characters are stored in 8-bit ASCII format with zero parity, we have them in hexadecimal as follows:

73 6d 4d 70

In binary representation, this becomes:

01110110   01101101   01001101   01110000

ASCII Armor Format

When PGP encodes data into ASCII Armor, it puts specific headers around the data, so PGP can construct the data later. PGP informs the user about what kind of data is encoded in ASCII Armor through the use of the headers.

Concatenating the following data creates ASCII Armor: an Armor head line, Armor headers, a blank line, ASCII-Armored data, Armor checksum, and Armor tail. Specifically, an explanation for each item is as follows:

- *An Armor head line.* This consists of the appropriate header line text surrounded by five dashes ("-", 0x2D) on either side of the header line text. The header line text is chosen based upon the type of data that is being encoded in Armor, and how it is being encoded. Header line texts include the following strings:
  - BEGIN PGP MESSAGE – used for signed, encrypted, or compressed files.
  - BEGIN PGP PUBLIC KEY BLOCK – used for armoring public keys.
  - BEGIN PGP PRIVATE KEY BLOCK – used for armoring private keys.
  - BEGIN PGP MESSAGE, PART X/Y – used for multipart messages, where the armour is divided among Y parts, and this is the Xth part out of Y.
  - BEGIN PGP MESSAGE, PART X – used for multipart messages, where this is the Xth part of an unspecified number of parts and requires the MESSAGE-ID Armor header to be used.
  - BEGIN PGP SIGNATURE – used for detached signatures, PGP/MIME signatures, and natures following clear-signed messages. Note that PGP 2.xs BEGIN PGP MESSAGE is used for detached signatures.

- *Armor headers.* There are pairs of strings that can give the user or the receiving PGP implementation some information about how to decode or use the message. The Armor headers are a part of the armor, not a part of the message, and hence are not protected by any signatures applied to the message. The format of an Armor header is that of a (key, value) pair. A colon (":" 0x38) and a single space (0x20) separate the key and value. PGP should consider improperly formatted Armor headers to be corruptions of ASCII Armor. Unknown keys should be reported to the user, but PGP should continue to process the message.

  Currently defined Armor header keys include:
  - *Version.* This states the PGP version used to encode the message.
  - *Comment.* This is a user-defined comment.
  - *MessageID.* This defines a 32-character string of printable characters. The string must be the same for all parts of a multipart message that uses the "PART X" Armor header. MessageID string should be unique enough that the recipient of the mail can associate all the parts of a message with each other. A good checksum or cryptographic hash function is sufficient.
  - *Hash.* This is a comma-separated list of hash algorithms used in the message. This is used only in clear-signed messages.
  - *Charset.* This is a description of the character set that the plaintext is in. PGP defines text to be in UTF-8 by default. An implementation will get the best results by translating into and out of UTF-8 (see RFC 2279). However, there are many instance where this is easier *said* than *done*. Also, there are communities of users who have no need for UTF-8 because they are all satisfied with a character set like ISO Latin-5 or a Japanese one. In such instances, an implementation may override the UTF-8 default by using this header key.

- *A blank line.* This indicates zero length or contains only white space.

- *ASCII-Armored data.* An arbitrary file can be converted to ASCII-Armored data by using Table 10.1.

- *Armor checksum.* This is a 24-bit CRC converted to four characters of radix-64 encoding by the same MIME base 64 transformation, preceded by an equals sign (=). The CRC is computed by using the generator 0x864cfb and an initialization of 0xb704ce. The accumulation is done on the data before it is converted to radix-64, rather than on the converted data. The checksum with its leading equals sign may appear on the first line after the base 64 encoded data.

- *Armor tail.* The Armor tail line is composed in the same manner as the Armor header line, except the string "BEGIN" is replaced by the string "END".

Encoding Binary in Radix-64

The encoding process represents three 8-bit input groups as output strings of four encoded characters. These 24 bits are then treated as four concatenated 6-bit groups, each of which is translated into a single character in the radix-64 alphabet. Each 6-bit group is used as an index. The character referenced by the index is placed in the output string.

Special processing is performed if fewer than 24 bits are available at the end of the data being encoded. There are three possibilities:

1. The last data group has 24 bits (3 octets). No special processing is needed.
2. The last data group has 16 bits (2 octets). The first two 6-bit groups are processed as above. The third (incomplete) data group has two zero-value bits added to it, and is processed as above. A pad character (=) is added to the output.
3. The last data group has 8 bits (1 octet). The first 6-bit group is processed as above. The second (incomplete) data group has four zero-value bits added to it, and is processed as above. Two pad characters (=) are added to the output.

Radix-64 printable encoding of binary data is shown in Figure 10.3.

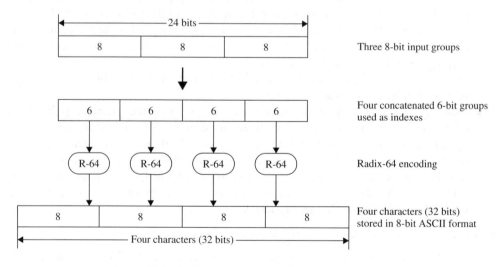

Figure 10.3   Radix-64 printable encoding of binary data.

Example 10.2 Consider the encoding process from 8-bit input groups to the output character string in the radix-64 alphabet.

1. Input raw text: 0x 15 d0 2f 9e b7 4c

8-bit octets:	00010101 11010000 00101111 10011110 10110111 01001100
6-bit index:	000101 011101 000000 101111 100111 101011 011101 001100
Decimal:	5 29 0 47 39 43 29 12
Output character:	F d A v n r d M
(radix-64 encoding)	
ASCII format (0x):	46 64 41 76 6e 72 64 4d
Binary:	01000110 01100100 01000001 01110110
	01101110 01110010 01100100 01001101

2. Input raw text: 0x 15 d0 2f 9e b7

8-bit octets:	00010101 11010000 00101111 10011110 10110111
6-bit index:	000101 011101 000000 101111 100111 101011 011100
	Pad with 00 (=)
Decimal:	5 29 0 47 39 43 28
Output character:	F d A v n r c =

3. Input raw text: 0x 15 d0 2f 9e

8-bit octets:	00010101 11010000 00101111 10011110
6-bit index:	000101 011101 000000 101111 100111 100000
	Pad with 0000 (==)
Decimal:	5 29 0 47 39 32
Output character:	F d A v n g ==

10.1.5 Packet Headers

A PGP message is constructed from a number of packets. A packet is a chunk of data which has a tag specifying its meaning. Each packet consists of a packet header of variable length, followed by the packet body.

The first octet of the packet header is called the *packet tag* as shown in Figure 10.4. The MSB is "bit 7" (the leftmost bit) whose mask is 0x80 (10000000) in hexadecimal. PGP 2.6.x only uses old format packets. Hence, software that interoperates with PGP 2.6.x must only use old format packets. These packets have 4 bits of content tags, but new format packets have 6 bits of content tags.

Packet Tags

The packet tag denotes what type of packet the body holds. The defined tags (in decimal) are:

0 –Reserved
1 –Session key packet encrypted by public key

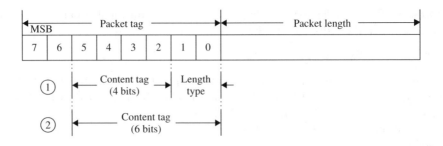

Figure 10.4   Packet header.

2 –Signature packet
3 –Session key packet encrypted by symmetric key
4 –One-pass signature packet
5 –Secret-key packet
6 –Public-key packet
7 –Secret-subkey packet
8 –Compressed data packet
9 –Symmetrically encrypted data packet
10 –Marker packet
11 –Literal data packet
12 –Trust packet
13 –User ID packet
14 –Public subkey packet
60 ∼ 63–Private or experimental values.

Old-format Packet Lengths

The meaning of the length type in old-format packets is:

0 –The packet has a 1-octet length. The header is 2 octets long.
1 –The packet has a 2-octet length. The header is 3 octets long.
2 –The packet has a 4-octet length. The header is 5 octets long.
3 –The packet is of indeterminate length. An implementation should not use indeterminate length packets except where the end of data will be clear from the context. It is better to use a new-format header described below.

New-format Packet Lengths

New-format packets have four possible ways of encoding length.

- *One-octet lengths.* A 1-octet body length header encodes packet lengths from 0 to 191 octets. This type of length header is recognized because the 1-octet value is less than 192. The body length is equal to:

bodyLen = 1st_octet

- *Two-octet lengths.* A 2-octet body length header encodes a length from 192 to 8383 octets. It is recognized because its first octet is in the range 192–223. The body length is equal to:

bodyLen = ((1st_octet − 192) ≪ 8) + (2nd_octet) + 192

- *Five-octet lengths.* A 5-octet body length header encodes packet lengths of up to 4 294 967 295(0xffffffff) octets in length. This header consists of a single octet holding the value 255, followed by a 4-octet scalar. The body length is equal to:

bodyLen = (2nd_octet ≪ 24)|(3rd_octet ≪ 16)|(4th_octet ≪ 8)|5th_octet

- *Partial body lengths.* A partial body length header is 1 octet long and encodes the length of only part of the data packet. This length is a power of 2, from 1 to 1 073 741 824 (2 to the 30th power). It is recognized by its 1-octet value that is greater than or equal to 224 and less than 255. The partial body length is equal to:

partialBodyLen = 1 ≪ (1st_octet   0x1f)

Each partial body length header is followed by a portion of the packet body data. The header specifies this portion's length. Another length header (of one of the three types: 1 octet, 2 octet, or partial) follows that portion. The last length header in the packet *must not* be a partial body length header. The latter headers may only be used for the nonfinal parts of the packet.

Example 10.3   Consider a packet with length 100. Compute its length encoded in 1 octet.
Now:

$$100 \text{ (decimal)} = 2^6 + 2^5 + 2^2 = 01100100 \text{(binary)} = 0x64 \text{ (hex)}$$

Thus, a packet with length 100 may have its length encoded in 1 octet: 0x64. This header is followed by 100 octets of data. Similarly, a packet with length 1723 may have its length encoded in 2 octets: 0xc5 and 0xfb. This header is followed by the 1723 octets of data. A packet with length 100 000 may have its length encoded in 5 octets: 0xff, 0x00, 0x01, 0x86, and 0xa0.

10.1.6   PGP Packet Structure

A PGP file consists of a message packet, a signature packet, and a session key packet.

Message Packet

This packet includes the actual data to be transmitted or stored as well as a header that includes control information generated by PGP such as a filename and a timestamp. A timestamp specifies the time of creation. The message component consists of a single literal data packet.

Signature Packet (Tag 2)

This packet describes a binding between some public key and some data. The most common signatures are a signature of a file or a block of text and a signature that is a certification of a user ID.

Two versions of signature packets are defined. PGP 2.6.x only accepts version 3 signature. Version 3 provides basic signature information, while version 4 provides an expandable format with subpackets that can specify more information about the signature. It is reasonable to create a v3 signature if an implementation is creating an encrypted and signed message that is encrypted with a v3 key.

At first, version 3 for basic signature information will be presented in the following. The signature packet is the signature of the message component, formed using a hash code of the message component and sender a's public key. The signature component consists of single signature packet.

The signature includes the following components:

- *Timestamp*. This is the time at which the signature was created.
- *Message digest* (*or hash code*). A hash code represents the 160-bit SHA-1 digest, encrypted with sender a's private key. The hash code is calculated over the signature timestamp concatenated with the data portion of the message component. The inclusion of the signature timestamp in the digest protects against replay attacks. The exclusion of the filename and timestamp portion of the message component ensures that detached signatures are exactly the same as attached signatures prefixed to the message. Detached signatures are calculated on a separate file that has none of the message component header fields.

If the default option of compression is chosen, then the block consisting of the literal data packet and the signature packet is compressed to form a compressed data packet:

- *Leading 2 octets of hash code*. These enable the recipient to determine if the correct public key was used to decrypt the hash code for authentication, by comparing the plaintext copy of the first 2 octets with the first 2 octets of the decrypted digest. Two octets also serve as a 16-bit frame-check sequence for the message.
- *Key ID of sender's public key*. This identifies the public key that should be used to decrypt the hash code and hence identifies the private key that was used to encrypt the hash code.

The message component and signature component (optional) may be compressed using ZIP and may be encrypted using a session key.

There are a number of possible meanings of a signature, which are specified in signature-type octets as shown below:

0x00: Signature of a binary document
0x01: Signature of a canonical text document
0x02: Stand-alone signature
0x10: Generic certification of a user ID and public-key packet
   (All PGP key signatures are of this type of certification.)
0x11: Personal certification of a user ID and public-key packet
   (The issuer has not carried out any verification of the claim.)
0x12: Casual certification of a user ID and public-key packet
   (The issuer has carried out some casual verification of the identity claim.)
0x13: Positive certification of a user ID and public-key packet
   (The issuer has carried out substantial verification of the identity claim.)
0x18: Subkey binding signature
   (This signature is a statement by the top-level signing key indicating that it owns the subkey.)
0x1f: Signature directly on a key
   (This signature is calculated directly on a key. It binds the information in the signature subpackets to the key.)
0x20: Key revocation signature
   (This signature is calculated directly when the key is revoked. A revoked key is not to be used.)
0x28: Subkey revocation signature
   (This signature is calculated directly when the subkey is revoked. A revoked subkey is not to be used.)
0x30: Certification revocation signature
   (This signature revokes an earlier user ID certification signature. It should be issued by the same key that issued the revoked signature or an authorised revocation key.)
0x40: Timestamp signature
   (This signature is only meaningful for the timestamp contained in it.)

The contents of the signature packets of version 3 (v3) and version 4 (v4) are illustrated in Table 10.2.

The signature calculation for version 4 signature is based on a hash of the signed data. The data being signed is hashed, and then the signature data from the version number to the hashed subpacket data is hashed. The resulting hash value is what is signed. The left 16 bits of the hash are included in the signature packet to provide a quick test to reject some invalid signatures.

Session Key Packets (Tag 1)

This component includes the session key and the identifier of the receiver's public key that was used by the sender to encrypt the session key. A public-key-encrypted session key packet, $E_{KP_b}(K_s)$, holds the session key used to encrypt a message. The symmetrically encrypted data packets are preceded by one public-key-encrypted session key packet for

Table 10.2 Signature packet format of version 3 and version 4

Content	Length in octets	
	v3	v4
Version number: v3(3), v4(4)	1	1
Signature type	1	1
Creation time	4	–
Signer's key ID	8	–
Public-key algorithm	1	1
Hash algorithm	1	1
Field holding left 16 bits of signed hash value	2	2
One or more MPIs comprising the signature	Algorithm specific *	Algorithm specific
Scalar octet count for hashed subpacket data	–	2
Hashed subpacket data	–	Zero or more subpackets
Scalar octet count for all of the unhashed subpackets	–	2
Unhashed subpacket data	–	Zero or more subpackets

*Algorithm-specific fields for RSA signature: MPI of RSA signature value m^d; algorithm-specific fields for DSA signature: MPI of DSA value r, MPI of DSA value s. (MPI = Multiprecision Integer)

each PGP 5.x key to which the message is encrypted. The message is encrypted with the session key, and the session key is itself encrypted and stored in the encrypted session key packet. The recipient of the message finds a session key that is encrypted to its public key, decrypts the session key, and then uses the session key to decrypt the message.

The body of this session key component consists of:

- A 1-octet version number which is 3.
- An 8-octet key ID of the public key that the session key is encrypted to.
- A 1-octet number giving the public key algorithm used.
- A string of octets that is the encrypted session key. This string's contents are dependent on the public-key algorithm used:
  - Algorithm-specific fields for RSA encryption: multiprecision integer (MPI) of RSA encrypted value m^e-mod n.
  - Algorithm-specific fields for ElGamal encryption: MPI of ElGamal value g^k mod p; MIP of ElGamal value my^k mod p. The value 'm' is derived from the session key.

If compression has been used, then conventional encryption is applied to the compressed data packet format from the compression of the signature packet and the literal data packet. Otherwise, conventional encryption is applied to the block consisting of the signature packet and the literal data packet. In either case, the ciphertext is referred to as a *conventional-key-encrypted data packet*.

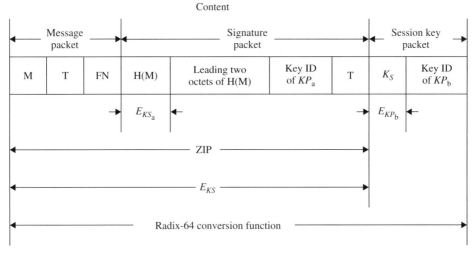

Figure 10.5   PGP message format.

As shown in Figure 10.5, the entire block of PGP message is usually encoded with radix-64 encoding.

10.1.7   Key Material Packet

A key material packet contains all the information about a public or private key. There are four variants of this packet type and two versions.

Key Packet Variants

There are:

- *Public-key packet (tag 6)*. This packet starts a series of packets that forms a PGP 5.x key.
- *Public subkey packet (tag 14)*. This packet has exactly the same format as a public-key packet, but denotes a subkey. One or more subkeys may be associated with a

top-level key. The top-level key provides signature services, and the subkeys provide encryption services. PGP 2.6.x ignores public-subkey packets.

- *Secret-key packet* (*tag 5*). This packet contains all the information that is found in a public-key packet, including not only the public-key materials but also the secret-key material after all the public-key fields.
- *Secret-subkey packet* (*tag 7*). A secret-subkey packet is the subkey analogous to the secret-key packet and has exactly the same format.

Public-key Packet Formats

There are two variants of version 3 packets and version 2 packets. Version 3 packets were originally generated by PGP 2.6. Version 2 packets are identical in format to version 3 packets, but are generated by PGP 2.5. However, v2 keys are deprecated and they must not be generated. PGP 5.0 introduced version 4 packets, with new fields and semantics. PGP 2.6.x will not accept key-material packets with versions greater than 3. PGP 5.x (or PGP3) implementation should create keys with version 4 format, but v4 keys correct some security deficiencies in v3 keys.

A v3 key packet contains:

- A 1-octet version number (3).
- A 4-octet number denoting the time that the key was created.
- A 2-octet number denoting the time in days that this key is valid.
- A 1-octet number denoting the public-key algorithm of this key.
- A series of MPIs comprising the key material: an MPI of RSA public module n and an MPI of RSA public encryption exponent e.

A key ID is an 8-octet scalar that identifies a key. For a v3 key, the 8-octet key ID consists of the low 64 bits of the public modulus of the RSA key. The *fingerprint* of a v3 key is formed by hashing the body (excluding the 2-octet length) of the MPIs that form the key material with MD5.

Note that MPIs are unsigned integers. An MPI consists of two parts: a 2-octet scalar that is the length of the MPI in bits followed by a string of octets that contain the actual integer.

Example 10.4 Suppose the string of octets [0009 01ff] forms an MPI. The length of the MPI in bits is [00000000 00001001] or 9 ($= 2^3 + 2^0$) in octets. The actual integer value of the MPI is:

$$[01\text{ff}] = 2^8 + 2^7 + 2^6 + 2^5 + 2^4 + 2^3 + 2^2 + 2^1 + 2^0 = 511$$

The MPI size is:

$$((\text{MPI.length} + 7)/8) + 2 = ((9 + 7)/8) + 2 = 4 \text{ octets}$$

which checks the given size of the MPI string.

The v4 format is similar to the v3 format except for the absence of a validity period. Fingerprints of v4 keys are calculated differently from v3 keys. A v4 fingerprint is the

160-bit SHA-1 hash of the 1-octet packet tag, followed by the 2-octet packet length, followed by the entire public-key packet starting with the version field. The key ID is the low-order 64 bits of the fingerprint.

A v4 key packet contains:

- A 1-octet version number (4).
- A 4-octet number denoting the time that the key was created.
- A 1-octet number denoting the public-key algorithm of this key.
- A series of MPIs comprising the key material:
  - Algorithm-specific fields for RSA public keys: MPI of RSA public modulus n and MPI of RSA public encryption exponent e.
  - Algorithm-specific fields for DSA public keys: MPI of DSA prime p; MPI of DSA group order q (q is a prime divisor of $p - 1$); MPI of DSA group generator g; and MPI of DSA public key value $y = g^x$ where x is secret.
  - Algorithm-specific fields for ElGamal public keys: MPI of ElGamal prime p; MPI of ElGamal group generator g; and MPI of ElGamal public key value $y = g^x$ where x is secret.

Secret-key Packet Formats

The secret-key and secret-subkey packets contain all the data of public-key and public-subkey packets in encrypted form, with additional algorithm-specific key data appended.

The secret-key packet contains:

- A public-key or public-subkey packet, as described above.
- One octet indicating string-to-key (S2K) usage conventions: 0 indicates that the secret-key data is not encrypted; 255 indicates that an S2K specifier is being given. Any other value specifies a symmetric-key encryption algorithm.
- If the S2K usage octet was 255, a 1-octet symmetric encryption algorithm (optional).
- If the S2K usage octet was 255, an S2K specifier (optional). The length of the S2K specifier is implied by its type, as described above.
- If secret data is encrypted, an 8-octet IV (optional).
- Encrypted MPIs comprising the secret-key data. These algorithm-specific fields are as described below.
- A 2-octet checksum of the plaintext of the algorithm-specific portion (sum of all octets, mod 2^{16} = mod 65 536):
  - Algorithm-specific fields for RSA secret keys: MPI of RSA secret exponent d; MPI of RSA secret prime value p; MPI of RSA secret prime value q ($p < q$); and MPI of u, the multiplicative inverse of p, mod q.
  - Algorithm-specific fields for DSA secret keys: MPI of DSA secret exponent x.
  - Algorithm-specific fields for ElGamal secret keys: MPI of ElGamal secret exponent x.

Simple S2K directly hashes the string to produce the key data:

Octet 0: 0x00
Octet 1: hash algorithm

It also hashes the *passphrase* to produce the session key. The hashing process to be done depends on the size of the session key and the size of the hash algorithm's output. If the hash size is greater than or equal to the session key size, the higher-order (leftmost) octets of the hash are used as the key. If the hash size is less than the key size, multiple instances are preloaded with 0, 1, 2, ... octets of zeros in order to produce the required key data.

S2K specifiers are used to convert passphrase strings into symmetric-key encryption/decryption keys. They are currently used in two ways: to encrypt the secret part of private keys in the private *keyring* and to convert passphrases to encryption keys for symmetrically encrypted messages.

Secret MPI values can be encrypted using a passphrase. If an S2K specifier is given, it describes the algorithm for converting the passphrase to a key, otherwise a simple MD5 hash of the passphrase is used. The cipher for encrypting the MPIs is specified in the secret-key packet.

Encryption/decryption of the secret data is done in CFB (Cipher Feedback) mode using the key created from the passphrase and IV from the packet. A different mode is used with v3 keys (which are only RSA) than with other key formats. With v3 keys, the prefix data (the first two octets) of the MPI is not encrypted; only the MPI nonprefix data is encrypted. Furthermore, the CFB state is resynchronized at the beginning of each new MPI value, so that the CFB block boundary is aligned with the start of the MPI data. With v4 keys, a simpler method is used: all secret MPI values are encrypted in CFB mode, including the MPI bitcount prefix.

The 16-bit checksum that follows the algorithm-specific portion is the algebraic sum, mod 65 536, of the plaintext of all the algorithm-specific octets (including the MPI prefix and data). With v4 keys, the checksum is encrypted like the algorithm-specific data. This value is used to check that the passphrase was correct.

Besides simple S2K, there are two more S2K specifiers currently supported:

- *Salted S2K.* This includes a *salt* value in the simple S2K specifier that hashes the passphrase to help prevent dictionary attacks:

  Octet 0: 0x01

  Octet 1: hash algorithm

  Octets 2–9: 8-octet salt value

  Salted S2K is exactly like simple S2K, except that the input to the hash function consists of the 8 octets of salt from the S2K specifier, followed by the passphrase.

- *Iterated and salted S2K.* This includes both a salt and an octet count. The salt is combined with the passphrase and the resulting value is hashed repeatedly. This further increases the amount of work an attacker would have to do.

  Octet 0: 0x03

  Octet 1: hash algorithm

  Octets 2–9: 8-octet salt value

  Octet 10: count, a 1-octet, coded value. (The count is coded into a 1-octet number.)

  Iterated–salted S2K hashes the passphrase and salt data multiple times. The total number of octets to be hashed is given in the encoded count in the S2K specifier. But the resulting count value is an octet count of how many octets will be hashed, not an iteration count. The salt followed by the passphrase data is repeatedly hashed until the number of octets specified by the octet count has been hashed. Implementations should use salted or iterated–salted S2K specifiers because simple S2K specifiers are more vulnerable to dictionary attacks.

10.1.8 Algorithms for PGP 5.x

This section describes the algorithms used in PGP 5.x.

Public-key Algorithms

ID	Algorithm
1	RSA (encrypt or sign)
2	RSA encryption only
3	RSA sign only
16	ElGamal (encrypt only)
17	DSA (DSS)
18	Reserved for elliptic curve
19	Reserved for ECDSA
20	ElGamal (encrypt or sign)
21	Reserved for Diffie–Hellman
100–110	Private/experimental algorithm

Symmetric-key Algorithms

ID	Algorithm
0	Plaintext or unencrypted data
1	IDEA
2	Triple DES (DES–EDE)
3	CAST 5 (128-bit key)
4	Blowfish (128-bit key, 16 rounds)
5	SAFER-SK128 (13 rounds)
6	Reserved for DES/SK

ID	Algorithm
7	Reserved for AES (128-bit key)
8	Reserved for AES (192-bit key)
9	Reserved for ASE (256-bit key)
100–110	Private/experimental algorithm

Compression Algorithm

ID	Algorithm
0	Uncompressed
1	ZIP (RFC 1951)
2	ZLIB (RFC 1950)
100–110	Private/experimental algorithm

Hash Algorithms

ID	Algorithm
1	MD5
2	SHA-1
3	RIPE-MD/160
4	Reserved for double-width SAH (experimental)
5	MD2
6	Reserved for TIGER/192
7	Reserved for HAVAL (5 pass, 160-bit)
100–110	Private/experimental algorithm

These tables are not an exhaustive list. An implementation may utilize an algorithm not on these lists.

10.2   S/MIME

S/MIME provides a consistent means to send and receive secure MIME data. S/MIME, based on the Internet MIME standard, is a security enhancement to cryptographic electronic messaging. Further, S/MIME not only is restricted to e-mail, but can be used with any transport mechanism that carries MIME data, such as HTTP. As such, S/MIME takes advantage of allowing secure messages to be exchanged in mixed-transport systems. Therefore, it appears likely that S/MIME will emerge as the industry standard for commercial and organizational use. This section describes a protocol for adding digital signature and encryption services to MIME data.

10.2.1   MIME

SMTP is a simple mail transfer protocol by which messages are sent only in NVT (Network Virtual Terminal) 7-bit ASCII format. NVT normally uses what is called *NVT ASCII*. This is an 8-bit character set in which the seven lowest-order bits are the same as ASCII and the highest-order bit is zero.

MIME was defined to allow transmission of non-ASCII data through e-mail. MIME allows arbitrary data to be encoded in ASCII and then transmitted in a standard e-mail message. It is a supplementary protocol that allows non-ASCII data to be sent through SMTP. However, MIME is not a mail protocol and cannot replace SMTP; it is only an extension to SMTP. In fact, MIME does not change SMTP or POP3, neither does it replace them.

The MIME standard provides a general structure for the content type of Internet messages and allows extensions for new content-type applications. To accommodate arbitrary data types and representations, each MIME message includes information that tells the recipient the type of the data and the encoding used. The MIME standard specifies that a content-type declaration must contain two identifiers, a content type and a subtype, separated by a slash.

MIME Description

MIME transforms non-ASCII data at the sender's site to NVT ASCII data and delivers it to the client SMTP to be sent through the Internet. The server SMTP at the receiver's site receives the NVT ASCII data and delivers it to MIME to be transformed back to the original non-ASCII data. Figure 10.6 illustrates a set of software functions that transforms non-ASCII data to ASCII data and vice versa.

MIME Header

MIME defines five headers that can be added to the original SMTP header section:

- MIME_Version
- Content_Type
- Content_Transfer_Encoding
- Content_Id
- Content_Description.

The MIMI header is shown in Figure 10.7 and described below.

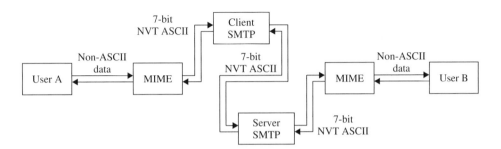

Figure 10.6   MIME showing a set of transforming functions.

Original header

MIME header
       MIME Version: 1.1
       Content Type: type/subtype
       Content Transfer Encoding: encoding type
       Content ID: message ID
       Content Description: textual explanation of non-textual contents

Mail message body

Figure 10.7   MIME header.

MIME_Version

This header defines the version of MIME used. The current version is 1.0.

Content_Type

This header defines the type of data used in the message body. The content type and the content subtype are separated by a slash. MIME allows seven different types of data:

- *Text.* The original message is in 7-bit ASCII format.

- *Multipart.* The body contains multiple, independent parts. The multipart header needs to define the boundary between each part. Each part has a separate content type and encoding.

  The multipart/signed content type specifies how to support authentication and integrity services via digital signature.

  Definition of multipart/signed:
  - MIME type name: multipart
  - MIME subtype name: signed
  - Required parameters: boundary, protocol, and micalg
  - Optional parameters: none
  - Security considerations: must be treated as opaque while in transit.

  The multipart/signed content type contains exactly two body parts. The first body part is the one over which the digital signature was created, including its MIME headers. The second body part contains the control information necessary to verify the digital signature.

  Definition of multipart/encrypted:
  - MIME type name: multipart
  - MIME subtype name: encrypted
  - Required parameters: boundary and protocol
  - Optional parameters: none
  - Security considerations: none.

Table 10.3 Five types of encoding

Type	Description
7 Bit	NVT ASCII characters and short lines
8 Bit	Non-ASCII characters and short lines
Binary	Non-ASCII characters with unlimited-length lines
Base64	6-Bit blocks of data encoded into 8-bit ASCII characters
Quoted-printable	Non-ASCII characters encoded as an equals sign followed by an ASCII code

The multipart/encrypted content type contains exactly two body parts. The first body part contains the control information necessary to decrypt the data in the second body part and is labeled according to the value of the protocol parameter. The second body part contains the data which was encrypted and is always labeled application/octet-stream.

- *Message.* In the message type, the body is itself a whole mail message, a part of a mail message, or a pointer to the message. Three subtypes are currently used: RFC 2822, partial body, or external body. The subtype RFC 2822 is used if the body is encapsulating another message. The subtype partial is used if the original message has been fragmented into different mail messages and this mail message is one of the fragments. The fragments must be reassembled at the destination by MIME. Three parameters must be added: ID, number, and total. The *id* identifies the message and is present in all the fragments. The *number* defines the sequence order of the fragment. The *total* defines the number of fragments that comprise the original message.

- *Image.* The original message is a stationary image, indicating that there is no animation. The two subtypes currently used are Joint Photographic Experts Group (JPEG), which uses image compression, and Graphics Interchange Format (GIF).

- *Video.* The original message is a time-varying image (animation). The only subtype is Motion Picture Experts Group (MPEG). If the animated image contains sound, it must be sent separately using the audio content type.

- *Audio.* The original message contains sound. The only subtype is basic, which uses 8-kHz standard audio data.

- *Application.* The original message is a type of data not previously defined. There are only two subtypes used currently: octet-stream and PostScript. Octet-stream is used when the data represents a sequence of binary data consisting of 8-bit bytes. PostScript is used when the data is in Adobe PostScript format for printers that support PostScript.

Content_Transfer_Encoding

This header defines the method to encode the messages into ones and zeros for transport. There are the five types of encoding: 7 bit, 8 bit, binary, Base64, and Quoted-printable. Table 10.3 describes the Content_Transfer_Encoding by the five types.

Note that lines in the header identify the type of the data as well as the encoding used.

- *7 Bit.* This is 7-bit NVT ASCII encoding. Although no special transformation is needed, the length of the line should not exceed 1000 characters.
- *8 Bit.* This is 8-bit encoding. Non-ASCII characters can be sent, but the length of the line still should not exceed 1000 characters. Since the underlying SMTP is able to transfer 8-bit non-ASCII characters, MIME does not do any encoding here. Base64 (or radix-64) and quoted-printable types are preferable.
- *Binary.* This is 8-bit encoding. Non-ASCII characters can be sent, and the length of the line can exceed 1000 characters. MIME does not do any encoding here; the underlying SMTP must be able to transfer binary data. Therefore, it is not recommended. Base64 (or radix-64) and quoted-printable types are preferable.
- *Base64.* This is a solution for sending data made of bytes when the highest bit is not necessarily zero. Base64 transforms this type of data of printable characters which can be sent as ASCII characters.
- *Quoted-printable.* Base64 is a redundant encoding scheme. The 24-bit non-ASCII data becomes four characters consisting of 32 bits. We have an overhead of 25%. If the data consists of mostly ASCII characters with a small non-ASCII portion, we can use quoted-printable encoding. If a character is ASCII, it is sent as it is; if a character is not ASCII it is sent as three characters.

Content_Id

This header uniquely identifies the whole message in a multiple message environment:

Content_Id: id =<content_id>

Content_Description

This header defines whether the body is image, audio, or video:
    Content_Description: <description>

Example 10.5   Consider an MIME message that contains a photograph in standard GIF representation. This GIF image is to be converted to 7-bit ASCII using Base64 encoding as follows:

    From: myrhee@tsp.snu.ac.kr
    To: kiisc2@kornet.net
    MIME_Version: 1.1
    Content_Type: image/gif
    Content_Transfer_Encoding: Base64

... data for the gif image ...

In this example, MIME_Version declares that the message was composed using version 1.1 of the MIME protocol. The MIME standard specifies that a Content_Type declaration must contain two identifiers, a content type and a subtype, separated by a slash. In this example, *image* is the content type, and *gif* is the subtype. Therefore, the Content_Type declares that the data is a GIF image. For the Content_Transfer_Encoding, the header declares that Base64 encoding was used to convert the image to ASCII. To view the image, a receiver's mail system must first convert from Base64 encoding back to binary, and then run an application that displays a GIF image on the user's screen.

MIME Security Multiparts

An Internet e-mail message consists of two parts: the headers and the body. The headers form a collection of field/value pairs, while the body is defined according to the MIME format. The basic MIME by itself does not specify security protection. Accordingly, a MIME agent must provide security services by employing a security protocol mechanism, by defining two security subtypes of the MIME multipart content type: signed and encrypted. In each of the security subtypes, there are exactly two related body parts: one for the protected data and one for the control information. The type and contents of the control information body parts are determined by the value of the protocol parameter of the enclosing multipart/signed or multipart/encrypted content type. A MIME agent should be able to recognize a security multipart body part and to identify its protected data and control information body part.

The multipart/signed content type specifies how to support authentication and integrity services via digital signature. The multipart/singed content type contains exactly two body parts. The first body part is the one over which the digital signature was created, including its MIME headers. The second body part contains the control information necessary to verify the digital signature. The Message Integrity Check (MIC) is the quantity computed over the body part with a message digest or hash function, in support of the digital signature service. The multipart/encrypted content type specifies how to support confidentiality via encryption. The multipart/encrypted content type contains exactly two body parts. The first body part contains the control information necessary to decrypt the data in the second body part. The second body part contains the data which was encrypted and is always labeled application/octet-stream.

MIME Security with OpenPGP

This subsection describes how the OpenPGP message format can be used to provide privacy and authentication using the MIME security content type. The integrating work on PGP with MIME suffered from a number of problems, the most significant of which was the inability to recover signed message bodies without parsing data structures specific to PGP. RFC 1847 defines security multipart formats for MIME. The security multiparts clearly separate the signed message body from the signature.

PGP can generate either ASCII Armor or a stream of arbitrary 8-bit octets when encrypting data, generating a digital signature, or extracting public-key data. The ASCII Armor output is the required method for data transfer. When the data is to be transmitted in many parts, the MIME message/partial mechanism should be used rather than the multipart ASCII Armor OpenPGP format.

Agents treat and interpret multipart/signed and multipart/encrypted as opaque, which means that the data is not to be altered in any way. However, many existing mail gateways will detect if the next hop does not support MIME or 8-bit data and perform conversion to either quoted-printable or Base64. This presents serious problems for multipart/signed where the signature is invalidated when such an operation occurs. For this reason all data signed according to this protocol must be constrained to 7 bits.

Before OpenPGP encryption, the data is written in MIME canonical format (body and headers). OpenPGP encrypted data is denoted by the *multipart/encrypted* content type, described in the Section MIME Security Multiparts, and must have a protocol parameter value of "application/pgp-encrypted". The multipart/encrypted MIME body must consist of exactly two body parts, the first with content type "application/pgp-encrypted." This body contains the control information. The second MIME body part must contain the actual encrypted data. It must be labeled with a content type of "application/octet-stream."

OpenPGP signed messages are denoted by the multipart/signed content type, described in the Section MIME Security Multiparts, with a protocol parameter which must have a value of "application/pgp-signature". The *micalg* parameter for the "application/pgp-signature" protocol must contain exactly one hash symbol of the format "pgp-<hash-identifier>" where <hash-identifier>identifies the MIC algorithm used to generate the signature. Hash symbols are contracted from text names or by converting the text name to lower case and prefixing it with the four characters "pgp-". Currently defined values are "pgp-md5," "pgp-sha1." "pgp-ripemd160," "pgp-tiger192," and "pgp-haval-5-160." The multipart/signed body must consist of exactly two parts. The first part contains the signed data in MIME canonical format, including a set of appropriate content headers describing the data. The second part must contain the OpenPGP digital signature. It must be labeled with a content type of 'application/pgp-signature.'

When the OpenPGP digital signature is generated:

- The data to be signed must first be converted to its content-type specific canonical form.
- An appropriate Content_Transfer_Encoding is applied. In particular, line endings in the encoded data must use the canonical <CR><LF> sequence where appropriate.
- MIME content headers are then added to the body, each ending with the canonical <CR><LF> sequence.
- Any trailing white space must be removed from the signed material.
- The digital signature must be calculated over both the data to be signed and its set of content headers.
- The signature must be generated as detached from the signed data so that the process does not alter the signed data in any way.

Note that the accepted OpenPGP convention is for signed data to end with a <CR><LF> sequence.

Upon receipt of a signed message, an application must:

- Convert line endings to the canonical <CR><LF>sequence before the signature can be verified.
- Pass both the signed data and its associated content headers along with the OpenPGP signature to the signature verification service.

Sometimes it is desirable both to digitally sign and then to encrypt a message to be sent. This encrypted and signed data protocol allows for two ways of accomplishing this task:

- The data is first signed as a multipart/signature body, and then encrypted to form the final multipart/encrypted body. This is most useful for standard MIME-compliant message forwarding.
- The OpenPGP packet format describes a method for signing and encrypting data in a single OpenPGP message. This method is allowed in order to reduce processing overheads and increase compatibility with non-MIME implementations of OpenPGP. The resulting data is formatted as a "multipart/encrypted" object. Messages which are encrypted and signed in this combined fashion are required to follow the same canonicalization rules as multipart/singed object. It is explicitly allowed for an agent to decrypt a combined message and rewrite it as a multipart/signed object using the signature data embedded in the encrypted version.

A MIME body part of the content type "application/pgp-keys" contains ASCII-Armoured transferable public-key packets as defined in RFC 2440.

Signatures of a canonical text document as defined in RFC 2440 ignore trailing white space in signed material. Implementations which choose to use signatures of canonical text documents will not be able to detect the addition of white space in transit.

10.2.2 S/MIME

S/MIME provides a way to send and receive 7-bit MIME data. S/MIME can be used with any system that transports MIME data. It can also be used by traditional mail user agents (MUAs) to add cryptographic security services to mail that is sent, and to interpret cryptographic security services in mail that is received. In order to create S/MIME messages, an S/MIME agent has to follow the specifications discussed in this section, as well as the specifications listed in the cryptographic message syntax (CMS).

The S/MIME agent represents user software that is a receiving agent, a sending agent, or both. S/MIME version 3 agents should attempt to have the greatest interoperability possible with S/MIME version 2 agents. S/MIME version 2 is described in RFC 2311 to RFC 2315 inclusively.

Before using a public key to provide security services, the S/MIME agent must certify that the public key is valid. S/MIME agents must use the Internet X.509 Public-Key Infrastructure (PKIX) certificates to validate public keys as described in the PKIX certificate and CRL profile.

Definitions

The following definitions are to be applied:

- *ASN.1.* Abstract Syntax Notation One, as defined in ITU-T X.680–689.
- *BER.* Basic Encoding Rules for ASN.1, as defined in ITU-T X.690.
- *DER.* Distinguished Encoding Rules for ASN.1, as defined in ITU-T X.690.

- *Certificate.* A type that binds an entity's distinguished name to a public key with a digital signature. This type is defined in the PKIX certificate and CRL profile. The certificate also contains the distinguished name of the certificate issuer (the signer), an issuer-specific serial number, the issuer's signature algorithm identifier, a validity period, and extensions also defined in that certificate.
- *CRL.* The Certificate Revocation List that contains information about certificates whose validity the issuer has prematurely revoked. The information consists of an issuer name, the time of issue, the next scheduled time of issue, a list of certificate serial numbers and their associated revocation times, and extensions as defined in Chapter 6. The CRL is signed by the issuer.
- *Attribute certificate.* An X.509 AC is a separate structure from a subject's PKIX certificate. A subject may have multiple X.509 ACs associated with each of its PKIX certificates. Each X.509 AC binds one or more attributes with one of the subject's PKIXs.
- *Sending agent.* Software that creates S/MIME CMS objects, MIME body parts that contains CMS objects, or both.
- *Receiving agent.* Software that interprets and processes S/MIME CMS objects, MIME parts that contain CMS objects, or both.
- *S/MIME agent.* User software that is a receiving agent, a sending agent, or both.

Cryptographic Message Syntax (CMS) Options

CMS allows for a wide variety of options in content and algorithm support. This subsection puts forth a number of support requirements and recommendations in order to achieve a base level of interoperability among all S/MIME implementations. CMS provides additional details regarding the use of the cryptographic algorithms.

DigestAlgorithmIdentifier

This type identifies a message digest algorithm which maps the message to the message digest. Sending and receiving agents must support SHA-1. Receiving agents should support MD5 for the purpose of providing backward compatibility with MD5-digested S/MIME v2 SignedData objects.

SignatureAlgorithmIdentifier

Sending and receiving agents must support id-dsa defined in DSS. Receiving agents should support rsaEncryption, defined in PRCS-1.

KeyEncryptionAlgorithmIdentifier

This type identifies a key encryption algorithm under which a content encryption key can be encrypted. A key-encryption algorithm supports encryption and decryption operations. The encryption operation maps a key string to another encrypted key string under the control of a key encryption key.

Sending and receiving agents must support Diffie–Hellman key exchange. Receiving agents should support rsaEncryption. Incoming encrypted messages contain symmetric keys which are to be decrypted with a user's private key. The size of the private key is determined during key generation. Sending agents should support rsaEncryption.

General syntax

The syntax is to support six different content types: data, signed data, enveloped data, signed-and-enveloped data, digested data, and encrypted data. There are two classes of content types: base and enhanced. Content types in the base class contain just *data* with no cryptographic enhancement, categorized as the data content type. Content types in the enhanced class contain content of some type (possibly encrypted), and other cryptographic enhancements. These types employ encapsulation, giving rise to the terms *outer* content containing the enhancements and *inner* content being enhanced.

CMS defines multiple content types. Of these, only the data, signed data and enveloped data types are currently used for S/MIME.

- *Data content type.* This type is arbitrary octet strings, such as ASCII text files. Such strings need not have any internal structure.
  The data content type should have ASN.1 type Data:

  Data:: = OCTET   STRING

  Sending agents must use the id-data content-type identifier to indicate the message content which has had security services applied to it.
- *Signed-data content type.* This type consists of any type and encrypted message digests of the content for zero or more signers. Any type of content can be signed by any number of signers in parallel. The encrypted digest for a signer is a digital signature on the content for that signer. Sending agents must use the signed-data content type to apply a digital signature to a message or in a degenerate case where there is no signature information to convey certificates. The syntax has a degenerate case in which there are no signers on the content. This degenerate case provides a means to disseminate certificates and certificate-revocation lists.

  The process to construct signed data is as follows. A message digest is computed on the content with a signer-specific message digest algorithm. A digital signature is formed by taking the message digest of the content to be signed and then encrypting it with the private key of the signer. The content plus signature are then encoded using Base64 encoding. A recipient verifies the signed-data message by decrypting the encrypted message digest for each signer with the signer's public key, then comparing the recovered message digest to an independently computed message digest. The signer's public key is either contained in a certificate included in the signer information, or referenced by an issuer distinguished name and an issuer-specific serial number that uniquely identify the certificate for the public key.
- *Enveloped-data content type.* An application/prcs7-mime subtype is used for the enveloped-data content type. This content type is used to apply privacy protection to a message. The type consists of encrypted content of any type and encrypted-content

encryption keys for one or more recipients. The combination of encrypted content and encrypted content-encryption key for a recipient is called a *digital envelope* for that recipient. Any type of content can be enveloped for any number of recipients in parallel. If a sending agent is composing an encrypted message to a group of recipients, that agent is forced to send more than one message.

The process by which enveloped data is constructed involves the following:

- A content-encryption key (a pseudorandom session key) is generated at random and is encrypted with the recipient's public key for each recipient.
- The content is encrypted with the content-encryption key. Content encryption may require that the content be padded to a multiple of some block size.
- The recipient-specific information values for all the recipients are combined with the encrypted content into an EnvelopedData value. This information is then encoded into Base64.

To cover the encrypted message, the recipient first strips off the Base64 encoding. The recipient opens the envelope by decrypting one of the encrypted content-encryption keys with the recipient's private key and decrypting the encrypted content with the recovered content-encryption key (the session key).

A sender needs to have access to a public key for each intended message recipient to use this service. This content type does not provide authentication.

- *Digested-data content type.* This type consists of content of any type and a message digest of the content. A typical application of the digested-data content type is to add integrity to content of the data content type, and the result becomes the content input to the enveloped-data content type. A message digest is computed on the content with a message digest algorithm. The message digest algorithm and the message digest are combined with the content into a DigestedData value.

  A recipient verifies the message digest by comparing the message digest to an independently computed message digest.

- *Encrypted-data content type.* This type consists of encrypted content of any type. Unlike the enveloped-data content type, the encrypted-data content type has neither recipients nor encrypted content-encryption keys. Keys are assumed to be managed by other means.

  It is expected that a typical application of the encrypted-data content type will be to encrypt content of the data content type for local storage, perhaps where the encryption key is a password.

10.2.3   Enhanced Security Services for S/MIME

The security services described in this section are extensions to S/MIME version 3. Some of the features of each service use the concept of a *triple wrapped* message. A triple wrapped message is one that has been signed, then encrypted, and then signed again. The signers of the inner and outer signatures may be different entities or the same entity. The S/MIME specification does not limit the number of nested encapsulations, so there may be more than three wrappings.

The inside signature is used for content integrity, nonrepudiation with proof of origin, and binding attributes to the original content. These attributes go from the originator to the recipient, regardless of the number of intermediate entities such as mail list agents that process the message. Signed attributes can be used for access control to the inner body. The encrypted body provides confidentiality, including confidentiality of the attributes that are carried in the inside signature.

The outside signature provides authentication and integrity for information that is processed hop by hop, where each hop is an intermediate entity such as a mail list agent. The outer signature binds attributes to the encrypted body. These attributes can be used for access control and routing decisions.

Triple Wrapped Message

The steps to create a triple wrapped message are as follows:

1. Start with the original content (a message body).
2. Encapsulate the original content with the appropriate MIME content-type headers.
3. Sign the inner MIME headers and the original content resulting from step 2.
4. Add an appropriate MIME construct to the signed message from step 3. The resulting message is called the *inside signature*.
   - If it is signed using multipart/signed, the MIME construct added consists of a content type of multipart/signed with parameters, the boundary, the step 2 result, a content type of application/pkcs7-signature, optional MIME headers, and a body part that is the result of step 3.
   - If it is instead signed using application/pkcs7-mime, the MIME construct added consists of a content type of application/pkcs7-mime with parameters, optional MIME headers, and the result of step 3.
5. Encrypt the step 4 result as a single block, turning it into an application/pkcs7-mime object.
6. Add the appropriate MIME headers: a content type of application/pkcs7-mime with parameters and optional MIME headers such as Content-Transfer-Encoding and Content-Disposition.
7. Sign the step 6 result (the MIME headers and the encrypted body) as a single block.
8. Using the same logic as in step 4, add an appropriate MIME construct to the signed message from step 7. The resulting message is called the *outside signature* and is also the triple wrapped message.

A triple wrapped message has many layers of encapsulation. The structure differs depending on the choice of format for the signed portions of the message. Because of the way that MIME encapsulates data, the layers do not appear in order, and the notion of layers becomes vague.

There is no need to use the multipart/signed format in an inner signature because it is known that the recipient is able to process S/MIME messages. A sending agent might choose to use the multipart/signed format in the outer layer so that a non-S/MIME agent could see that the next inner layer is encrypted. Because many sending agents always use multipart/signed structures, all receiving agents must be able to interpret either multipart/signed or application/pkcs7-mime signature structures.

Security Services with Triple Wrapping

This subsection briefly describes the relationship of each service with triple wrapping. If a signed receipt is requested for a triple wrapped message, the receipt request must be in the inside signature, not in the outside signature. A secure mailing list agent may change the receipt policy in the outside signature of a triple wrapped message when the message is processed by the mailing list.

A security label is included in the signed attributes of any SignedData object. A security label attribute may be included in either the inner signature or the outer signature, or both.

The inner security label is used for access control decisions related to the original plaintext content. The inner signature provides authentication and cryptographically protects the integrity of the original signer's security label that is in the inside body. The confidentiality security service can be applied to the inner security label by encrypting the entire inner SignedData block within an EnvelopedData block. The outer security label is used for access control and routing decisions related to the encrypted message.

Secure mail list message processing depends on the structure of S/MIME layers present in the message sent to the mail list agent. The agent never changes the data that was hashed to form the inner signature, if such a signature is present. If an outer signature is present, then the agent will modify the data that was hashed to form that outer signature.

Contain attributes should be placed in the inner or outer SignedData message. Some attributes must be signed, while signing is optional for others, and some attributes must not be signed.

Some security gateways sign messages that pass through them. If the message is of any type other than a SignedData type, the gateway has only one way to sign the message by wrapping it with a SignedData block and MIME headers. If the message to be signed by the gateway is a SignedData message already, the gateway can sign the message by inserting SignerInfo into the SignedData block.

Signed Receipts

Returning a signed receipt provides to the originator proof of delivery of a message and allows the originator to demonstrate to a third party that the recipient was able to verify the signature of the original message. This receipt is bound to the original message through the signature. Consequently, this service may be requested only if a message is signed. The receipt sender may optionally also encrypt a receipt to provide confidentiality between the sender and the recipient of the receipt.

The originator of a message may request a signed receipt from the message's recipients. The request is indicated by adding a receiptRequest attribute to the signedAttributes field of the SignerInfo object for which the receipt is requested. The receiving user agent software should automatically create a signed receipt when requested to do so, and return the receipt in accordance with mailing list expansion options, local security policies, and configuration options.

Receipts involve the interaction of two parties: the sender and the receiver. The sender is the agent that sent the original message that includes a request for a receipt. The receiver is the party that received that message and generated the receipt.

The interaction steps in a typical transaction are:

1. Sender creates a signed message including a receipt request attribute.
2. Sender transmits the resulting message to the recipient(s).
3. Recipient receives message and determines if there are a valid signature and receipt request in the message.
4. Recipient creates a signed receipt.
5. Recipient transmits the resulting signed receipt message to the sender.
6. Sender receives the message and validates that it contains a signed receipt for the original message.

Receipt Request Creation

Multilayer S/MIME messages may contain multiple SignedData layers. Receipts are requested only for the innermost SignedData layer in a multilayer S/MIME message such as a triple wrapped message. Only one receipt request attribute can be included in the signedAttributes of SignerInfo.

11

Internet Firewalls for Trusted Systems

A firewall is a device or group of devices that controls access between networks. A firewall generally consists of filters and gateway(s), varying from firewall to firewall. It is a security gateway that controls access between the public Internet and an intranet (a private internal network) and is a secure computer system placed between a trusted network and an untrusted Internet. A firewall is an agent that screens network traffic in some way, blocking traffic it believes to be inappropriate, dangerous, or both. The security concerns that inevitably arise between the sometimes hostile Internet and secure intranets are often dealt with by inserting one or more firewalls in the path connecting the Internet and the internal network. In reality, Internet access provides benefits to individual users, government agencies, and most organizations. But this access often creates a threat as a security flaw. The protective device that has been widely accepted is the firewall. When inserted between the private intranet and the public Internet, it establishes a controlled link and erects an outer security wall or perimeter. The aim of this wall is to protect the intranet from Internet-based attacks and to provide a choke point where security can be imposed.

Firewalls act as an intermediate server in handling SMTP and HTTP connections in either direction. Firewalls also require the use of an access negotiation and encapsulation protocol such as SOCKS to gain access to the Internet, the intranet, or both. Many firewalls support tri-homing, allowing use of a demilitarized zone (DMZ) network. It is possible for a firewall to accommodate more than three interfaces, each attached to a different network segment.

Firewalls can be classified into three main categories: packet filters, circuit-level gateways, and application-level gateways.

11.1 Role of Firewalls

The firewall imposes restrictions on packets entering or leaving the private network. All traffic from inside to outside, and vice versa, must pass through the firewall, but only authorized traffic will be allowed to pass. Packets are not allowed through unless they

Wireless Mobile Internet Security, Second Edition. Man Young Rhee.
© 2013 John Wiley & Sons, Ltd. Published 2013 by John Wiley & Sons, Ltd.

conform to a filtering specification or unless there is negotiation involving some sort of authentication. The firewall itself must be immune to penetration.

Firewalls create checkpoints (or choke points) between an internal private network and an untrusted Internet. Once the choke points have been clearly established, the device can monitor, filter, and verify all inbound and outbound traffic.

The firewall may filter on the basis of IP source and destination addresses and TCP port number. Firewalls may block packets from the Internet side that claim a source address of a system on the intranet, or they may require the use of an access negotiation and encapsulation protocol like SOCKS to gain access to the intranet.

The means by which access is controlled relate to using network layer or transport layer criteria such as IP subnet or TCP port number, but there is no reason that this must always be so. A growing number of firewalls control access at the application layer, using user identification as the criterion. In addition, firewalls for ATM networks may control access based on the data link layer criteria.

The firewall also enforces logging and provides alarm capacities as well. By placing logging services at firewalls, security administrators can monitor all access to and from the Internet. Good logging strategies are one of the most effective tools for proper network security.

Firewalls may block TELNET or RLOGIN connections from the Internet to the intranet. They also block SMTP and FTP connections to the Internet from internal systems not authorized to send e-mail or to move files.

The firewall provides protection from various kinds of IP spoofing and routing attacks. It can also serve as the platform for IPsec. Using the tunnel mode capability, the firewall can be used to implement Virtual Private Networks (VPNs). A VPN encapsulates all the encrypted data within an IP packet.

A firewall can limit network exposure by hiding the internal network systems and information from the public Internet.

The firewall is a convenient platform for security-unrelated events such as a network address translator (which maps local addresses to Internet addresses) and has a network management function that accepts or logs Internet usage.

The firewall certainly has some negative aspects: it cannot protect against internal threats such as an employee who cooperates with an external attacker; it is also unable to protect against the transfer of virus-infected programs or files because it is impossible for it to scan all incoming files, e-mail, and messages for viruses. However, since a firewall acts as a protocol endpoint, it may use an implementation methodology designed to minimize the likelihood of bugs.

A firewall can effectively implement and control the traversal of IP multicast traffic. Some firewall mechanisms such as SOCKS are less appropriate for multicast because they are designed specifically for unicast traffic.

11.2 Firewall-Related Terminology

To design and configure a firewall, some familiarity with the basic terminology is required. It is useful for readers to understand the important terms commonly applicable to firewall technologies.

11.2.1 Bastion Host

A bastion host is a publicly accessible device for the network's security, which has a direct connection to a public network such as the Internet. The bastion host serves as a platform for any one of the three types of firewalls: packet filter, circuit-level gateway, or application-level gateway.

Bastion hosts must check all incoming and outgoing traffic and enforce the rules specified in the security policy. They must be prepared for attacks from external and possibly internal sources. They should be built with the least amount of hardware and software in order for a potential hacker to have less opportunity to overcome the firewall. Bastion hosts are armed with logging and alarm features to prevent attacks.

The bastion host's role falls into the following three common types:

- *Single-homed bastion host.* This is a device with only one network interface, normally used for an application-level gateway. The external router is configured to send all incoming data to the bastion host, and all internal clients are configured to send all outgoing data to the host. Accordingly, the host will test the data according to security guidelines.

- *Dual-homed bastion host.* This is a firewall device with at least two network interfaces. Dual-homed bastion hosts serve as application-level gateways, and as packet filters and circuit-level gateways as well. The advantage of using such hosts is that they create a complete break between the external network and the internal network. This break forces all incoming and outgoing traffic to pass through the host. The dual-homed bastion host will prevent a security break-in when a hacker tries to access internal devices.

- *Multihomed bastion host.* Single-purpose or internal bastion hosts can be classified as either single-homed or multihomed bastion hosts. The latter are used to allow the user to enforce strict security mechanisms. When the security policy requires all inbound and outbound traffic to be sent through a proxy server, a new proxy server should be created for the new streaming application. On the new proxy server, it is necessary to implement strict security mechanisms such as authentication. When multihomed bastion hosts are used as internal bastion hosts, they must reside inside the organization's internal network, normally as application gateways that receive all incoming traffic from external bastion hosts. They provide an additional level of security in case the external firewall devices are compromised. All the internal network devices are configured to communicate only with the internal bastion host.

- A tri-homed firewall connects three network segments with different network addresses. This firewall may offer some security advantages over firewalls with two interfaces. An attacker on the unprotected Internet may compromise hosts on the DMZ but still not reach any hosts on the protected internal network.

11.2.2 Proxy Server

Proxy servers are used to communicate with external servers on behalf of internal clients. A proxy service is set up and torn down in response to a client request, rather than existing on a static basis. The term proxy server typically refers to an application-level gateway, although a circuit-level gateway is also a form of proxy server. The gateway can be configured to support an application-level proxy on inbound connections and a circuit-level proxy on outbound connections. Application proxies forward packets only when a connection has been established using some known protocol. When the connection closes, a firewall using application proxies rejects individual packets, even if they contain port numbers allowed by a rule set. In contrast, circuit proxies always forward packets containing a given port number if that port number is permitted by the rule set. Thus, the key difference between application and circuit proxies is that the latter are static and will always set up a connection if the DUT/SUT's rule set allows it. Each proxy is configured to allow access only to specific host systems.

The audit log is an essential tool for detecting and terminating intruder attacks. Therefore, each proxy maintains detailed audit information by logging all traffic, each connection, and the duration of each connection.

Since a proxy module is a relatively small software package specifically designed for network security, it is easier to check such modules for security flaws.

Each proxy is independent of other proxies on the bastion host. If there is a problem with the operation of any proxy, or if future vulnerability is discovered, it is easy to replace the proxy without affecting the operation of the proxy's applications. If the support of a new service is required, the network administrator can easily install the required proxy on the bastion host.

A proxy generally performs no disk access other than to read its initial configuration file. This makes it difficult for an intruder to install Trojan horse sniffers or other dangerous files on the bastion host.

11.2.3 SOCKS

The SOCKS protocol version 4 provides for unsecured firewall traversal for TCP-based client/server applications, including HTTP, TELNET, and FTP. The new protocol extends the SOCKS version 4 model to include UDP, allows the framework to include provision for generalized strong authentication schemes, and extends the addressing scheme to encompass domain name and IPv6 addresses. The implementation of the SOCKS protocol typically involves the recompilation or relinking of TCP-based client applications so that they can use the appropriate encapsulation routines in the SOCKS library (refer to RFC 1928).

When a TCP-based client wishes to establish a connection to an object that is reachable only via a firewall, it must open a TCP connection to the appropriate SOCKS port on the SOCKS server system. The SOCKS service is conventionally located at TCP port 1080. If the connection request succeeds, the client enters negotiation for the authentication method to be used, authenticates with the chosen method, and then sends a relay request. The SOCKS server evaluates the request and either establishes the appropriate connection or denies it. In fact, SOCKS defines how to establish authenticated connections, but currently, it does not provide a clear-cut solution to the problem of encrypting the data traffic. Since the Internet at large is considered a hostile medium, encryption by using

ESP is also assumed in this scenario. An ESP transform that provides both authentication and encryption could be used, in which case the AH need not be included.

11.2.4 Choke Point

The most important aspect of firewall placement is to create choke points. A choke point is the point at which a public Internet can access the internal network. The most comprehensive and extensive monitoring tools should be configured on the choke points. Proper implementation requires that all traffic be funneled through these choke points. Since all traffic is flowing through the firewalls, security administrators, as a firewall strategy, need to create choke points to limit external access to their networks. Once these choke points have been clearly established, the firewall devices can monitor, filter, and verify all inbound and outbound traffic.

Since a choke point is installed at the firewall, a prospective hacker will go through the choke point. If the most comprehensive logging devices are installed in the firewall itself, all hacker activities can be captured. Hence, this will be detected exactly what a hacker is doing.

11.2.5 Demilitarized Zone (DMZ)

The DMZ is an expression that originates from the Korean War. It meant a strip of land forcibly kept clear of enemy soldiers. In terms of a firewall, the DMZ is a network that lies between an internal private network and the external public network. DMZ networks are sometimes called *perimeter networks*. A DMZ is used as an additional buffer to further separate the public network from the internal network.

A gateway is a machine that provides relay services to compensate for the effects of a filter. The network inhabited by the gateway is often called the *DMZ*. A gateway in the DMZ is sometimes assisted by an internal gateway. The internal filter is used to guard against the consequences of a compromised gateway, while the outside filter can be used to protect the gateway from attack.

Many firewalls support tri-homing, allowing use of a DMZ network. It is possible for a firewall to accommodate more than three interfaces, each attached to a different network segment.

11.2.6 Logging and Alarms

Logging is usually implemented at every device in the firewall, but these individual logs combine to become the entire record of user activity. Packet filters normally do not enable logging by default so as not to degrade performance. Packet filters as well as circuit-level gateways log only the most basic information. Since a choke point is installed at the firewall, a prospective hacker will go through the choke point. If so, the comprehensive logging devices will probably capture all hacker activities, including all user activities as well. The user can then tell exactly what a hacker is doing and have such information available for audit. The audit log is an essential tool for detecting and terminating intruder attacks.

Many firewalls allow the user to preconfigure responses to unacceptable activities. The firewall should alert the user by several means. The two most common actions are for the firewall to break the TCP/IP connection or to have it automatically set off alarms.

11.2.7 VPN

Some firewalls are now providing VPN services. VPNs are appropriate for any organization requiring secure external access to internal resources. All VPNs are tunnelling protocols in the sense that their information packets or payloads are encapsulated or tunnelled into the network packets. All data transmitted over a VPN is usually encrypted because an opponent with access to the Internet could eavesdrop on the data as it travels over the public network. The VPN encapsulates all the encrypted data within an IP packet. Authentication, message integrity, and encryption are very important fundamentals for implementing a VPN. Without such authentication procedures, a hacker could impersonate anyone and then gain access to the network. Message integrity is required because the packets can be altered as they travel through the Internet. Without encryption, the information may become truly public. Several methods exist to implement a VPN. Windows NT or later versions support a standard RSA connection through a VPN. Specialized firewalls or routers can be configured to establish a VPN over the Internet. New protocols such as IPsec are expected to standardize on a specific VPN solution. Several VPN protocols exist, but the Point-to-Point Tunnelling Protocol (PPTP) and IPsec are the most popular.

11.3 Types of Firewalls

As mentioned above, firewalls are classified into three common types: packet filters, circuit-level gateways, and application-level gateways. We examine each of these in turn.

11.3.1 Packet Filters

Packet filters are one of several different types of firewalls that process network traffic on a packet-by-packet basis. A packet filter's main function is to filter traffic from a remote IP host, so a router is needed to connect the internal network to the Internet. A packet filter is a device which inspects or filters each packet at a screening router for the content of IP packets. The screening router is configured to filter packets from entering or leaving the internal network, as shown in Figure 11.1. The routers can easily compare each IP address to a filter or a series of filters. The type of router used in a packet-filtering firewall is known as a *screening router*.

Packet filters typically set up a list of rules that are sequentially read line by line. Filtering rules can be applied based on source and destination IP addresses or network addresses, and TCP or UDP ports. Packet filters are read and then treated on a rule-by-rule basis. A packet filter will provide two actions, forward or discard. If the action is in the

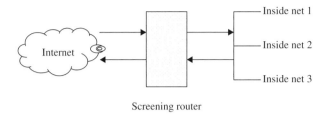

Figure 11.1 A screening router for packet filtering.

forward process, the action takes place to route the packet as normal if all conditions within the rule are met. The discard action will block all packets if the conditions in the rule are not met. Thus, a packet filter is a device that inspects each packet for predefined content. Although it does not provide an error-correcting ability, it is almost always the first line of defense. When packets are filtered at the external filter, it is usually called a *screening router*.

Since a packet filter can restrict all inbound traffic to a specific host, this restriction may prevent a hacker from being able to contact any other host within the internal network. However, the significant weakness with packet filters is that they cannot discriminate between good and bad packets. Even if a packet passes all the rules and is routed to the destination, packet filters cannot tell whether the routed packet contains good or malicious data. Another weakness of packet filters is their susceptibility to spoofing. In IP spoofing, an attacker sends packets with an incorrect source address. When this happen, replies will be sent to the apparent source address, not to the attacker. This might seem to be a problem.

Packet Filtering Rules

A packet filter applies a set of rules to each incoming IP packet and then forwards or discards the packet. The packet filter typically sets up a list of rules which may match fields in the IP or TCP header. If there is a match to one of the rules, that rule is able to determine whether to forward or discard the packet. If there is no match to any rule then two default actions (forward and discard) will be taken.

TELNET packet filtering

TELNET is a simple remote terminal access that allows a user to log onto a computer across an Internet. TELNET establishes a TCP connection and then passes keystrokes from the user's keyboard directly to the remote computer as if they had been typed on a keyboard attached to the remote machine. TELNET also carries output from the remote machine back to the user's screen. TELNET client software allows the user to specify a remote machine either by giving its domain name or IP address.

TELNET can be used to administer a UNIX machine. Windows NT does not provide a TELNET serve with the default installation, but a third-party service can be easily added. TELNET sends all user names and passwords in plaintext. Experienced hackers

Table 11.1   Telnet packet-filtering example

Rule number	Action	Source IP	Source port	Destination IP	Destination port	Protocol
1	Discard	*	23	*	*	TCP
2	Discard	*	*	*	23	TCP

can hijack a TELNET session in progress. TELNET should only be used when the user can verify the entire network connecting the client and server, not over the Internet. All TELNET traffic should be filtered at the firewall. TELNET runs on TCP port 23.

For example, to disable the ability to TELNET into internal devices from the Internet, the information listed Table 11.1 tells the router to discard any packet going to or coming from TCP port 23. TELNET for remote access application runs on TCP port 23. It runs completely in open nonencryption, with no authentication other than the user name and password that are transmitted in clear. An asterisk (*) in a field indicates any value in that particular field. The packet-filtering rule sets are executed sequentially, from top to bottom.

If a packet is passed through the filter and has a source port of 23, it will immediately be discarded. If a packet with a destination port of 23 is passed through this filter, it is discarded only after rule 2 has been applied. All other packets will be discarded.

FTP packet filtering

If the FTP service is to apply the same basic rule as applied to TELNET, the packet filter to allow or block FTP would look like Table 11.2. The FTP service is typically associated with using TCP ports 20 and 21.

One approach to handling FTP connections is explained with the following rule set. Rule 1 allows any host with the network address 192.168.10.0 to initiate a TCP session on any destination IP address on port 21. Rule 2 blocks any packet originating from any remote address with a source port of 20 and contacting a host with a network address 192.168.10.0 on any port less than 1024. Rule 3 allows any remote address that has a source port of 20 and is contacting any host with a network address of 192.168.10.0 on any port. Once a connection is set up, the ACK flag (ACK = 1) of a TCP segment is set to acknowledge segments sent from the other side. If any packet violates rule 2, it will be immediately discarded, and rule 3 will never be executed.

Table 11.2   FTP packet-filtering example

Rule number	Action	Source IP	Source port	Destination IP	Destination port	Protocol
1	Allow	192.168.10.0	*	*	21	TCP
2	Block	*	20	192.168.10.0	< 1024	TCP
3	Allow	*	20	192.168.10.0	*	TCP ACK = 1

With FTP, two TCP connections are used: a control connection to set up the file transfer and a data connection for the actual file transfer. The data connection uses a different port number to be assigned for the transfer. Remember that most servers live on low-numbered ports, but most outgoing calls tend to use higher-numbered ports, typically above 1024.

FTP is the first protocol for transferring or moving files across the Internet. Like many of the TCP/IP protocols, FTP was not designed with security in mind. It communicates with the server on two separate TCP ports 20 and 21. Each FTP server has a *command channel*, where the requests for data and directory listings are issued, and a *data channel*, over which the requested data is delivered.

FTP operates in two different modes (active and passive). In active mode, an FTP server receives commands on TCP/IP port 21 and exchanges data with the client. When a client contacts an FTP server in active mode and wants to send or receive data, the client picks an unused local TCP port between 1024 and 65 535, tells the server over the command channel, and listens for the server to connect on the chosen port. The server opens a connection from TCP port 20 to the specified port on the client machine. Once the connection is established, the data is passed across.

In passive mode, the command channel is still port 21 on the server, but the traditional data channel on port 20 is not used. Instead, when the client requests passive mode, the server picks an unused local TCP port between 1024 and 65 535 and tells the client. The client opens a connection to that port on the server. The server is listening on that port for the inbound connection from the client. Once the connection is established, the data flows across. Thus, since the client is initiating both the command and data channel connections to the server, most modern browsers use passive mode FTP for data accessing.

SMTP packet filtering

The sending and transmission of mail is the responsibility of a mail transport agent (MTA). The protocol behind nearly all MTAs is SMTP and its extension ESMTP. On the Internet, e-mail exchanges between mail servers are handled with SMTP. It is the protocol that transfers e-mail from one server to another, and it provides a basic e-mail facility for transferring messages among separate hosts. A host's SMTP server accepts mail and examines the destination IP address to decide whether to deliver the mail locally or to forward it to some other machine.

SMTP is a store/forward system, and such systems are well suited to firewall applications. SMTP receivers use TCP port 25; SMTP senders use a randomly selected port above 1023.

Most e-mail messages are addressed with hostnames instead of IP addresses, and the SMTP server uses DNS (Directory and Naming Services) to determine the matching IP address. If the same machines handle internal and external mail delivery, a hacker who can spoof DNS information may be able to cause mail that was intended for internal destinations to be delivered to an external host. A hacker who can manipulate DNS responses can redirect mail to a server under the control of the hacker. That server can then copy the mail and return it. This will introduce delays and will usually leave a trail in the log or message headers. Therefore, if it is desired to avoid situations where internal and external mail delivery are handled on the machine and internal names are resolved through DNS, it will be good practice to have the best configuration in which there is an external mail server and a separate internal mail server. The external mail server has the IP address of the internal mail server configured via a host file.

Sendmail (www.sendmail.org/) is the mailer commonly used on UNIX systems. Sendmail is very actively supported on security issues and has both an advantage and a disadvantage. Table 11.3 displays some examples of SMTP packet-filtering rule sets.

Packet filters offer their services at the network, transport, and session layers of the OSI model. Packet filters forward or deny packets based on information in each packet's header, such as the IP address or TCP port number. A packet-filtering firewall uses a rule set to determine which traffic should be forwarded and which should be blocked. Packet filters are then composed of rules that are read and treated on a rule-by-rule basis. Therefore, packet filtering is defined as the process of controlling access by examining packets based on the content of packet headers.

The following two subsections outline the specific details with relation to the circuit-level and application-level gateways for respective proxy services. Proxying provides Internet access for a single host or a small number of hosts. The proxy server evaluates requests from the client and decides which to pass on and which to disregard. If a request is approved, the proxy server talks to the real server on behalf of the client and proceeds to relay requests from the client to the real server and to relay the real server's answers back to the client. The concept of proxies is very important to firewall application because a proxy replaces the network IP address with another contingent address.

Proxies are classified into two basic forms:

- Circuit-level gateway.
- Application-level gateway.

Table 11.3 SMTP packet-filtering examples

Case	Action	Source host	Source port	Destination host	Destination port	Protocol
A	Allow	Source gateway	25	*	*	TCP
B	Allow	*	*	*	25	TCP
C	Allow	Internal host	*	*	25	TCP
D	Allow	*	25	*	*	TCP ACK flag

Case A: Connection to source SMTP port. Port 25 is for SMTP incoming. Inbound mail is allowed, but only to a gateway host.

Case B: Connection to destination SMTP port. This rule set is intended to specify that any source host can send mail to the destination. A TCP packet with a destination port 25 is routed to the SMTP server on the destination machine.

Case C: This rule set achieves the intended result that was not achieved in B. The rule takes advantage of a feature of TCP connection. This rule set states that it allows IP packets where the source IP address is one of a list of designated internal hosts and the destination TCP port 25.

Case D: This rule takes advantage of a feature of TCP connections. Once a connection is set up, the ACK flag of a TCP segment is set to acknowledge segments sent from the destination. It also allows incoming packets with a source port number of 25 that include ACK flag in the TCP segment.

Both circuit- and application-level gateways create a complete break between the internal premises network and external Internet. This break allows the firewall system to examine everything before passing it into or out of the internal network. Each of these gateways will be examined in turn in the following.

11.3.2   Circuit-Level Gateways

The circuit-level gateway represents a proxy server that statically defines what traffic will be forwarded. Circuit proxies always forward packets containing a given port number if that port number is permitted by the rule set. A circuit-leval gateway operates at the network level of the OSI model. This gateway acts as an IP address translator between the Internet and the internal system. The main advantage of a proxy server is its ability to provide network address translation (NAT). NAT hides the internal IP address from the Internet. NAT is the primary advantage of circuit-level gateways and provides security administrators with great flexibility when developing an address scheme internally. Circuit-level gateways are based on the same principles as packet filter firewalls. When the internal system sends out a series of packets, these packets appear at the circuit-level gateway where they are checked against the predetermined rules set. If the packets do not violate any rules, the gateway sends out the same packets on behalf of the internal system. The packets that appear on the Internet originate from the IP address of the gateway's external port, which is also the address that receives any replies. This process efficiently shields all internal information from the Internet. Figure 11.2 illustrates the circuit-level gateway for setting up two TCP connections.

11.3.3   Application-Level Gateways

The application-level gateway represents a proxy server, performing at the TCP/IP application level, that is set up and torn down in response to a client request, rather than existing on a static basis. Application proxies forward packets only when a connection has been established using some known protocol. When the connection closes, a firewall using application proxies rejects individual packets, even if the packets contain port numbers allowed by a rule set.

The application gateway analyses the entire message instead of individual packets when sending or receiving data. When an inside host initiates a TCP/IP connection, the application gateway receives the request and checks it against a set of rules or filters.

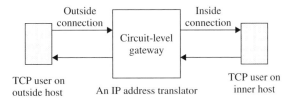

Figure 11.2   Circuit-level gateway for setting up two TCP connections.

The application gateway (or proxy server) will then initiate a TCP/IP connection with the remote server. The server will generate TCP/IP responses based on the request from the proxy server. The responses will be sent to the proxy server (application gateway) where the responses are again checked against the proxy server's filters. If the remote server's response is permitted, the proxy server will then forward the response to the inside host.

Certain transport layer protocols work better than others. For example, TCP can easily be used through a proxy server because it is a connection-based protocol, while each User Datagram Protocol (UDP) packet should be treated as an individual message because UDP is connectionless. The proxy server must analyze each UDP packet and apply it to the filters separately, which slows down the proxy process. Internet Control Message Protocol (ICMP) programs are nearly impossible to proxy because ICMP messages do not work through an application-level gateway. For example, HTTP traffic is often used in conjunction with proxy servers, but an internal host could not ping a remote host through the proxy server. Application-level gateways (proxy servers) are used as intermediate devices when routing SMTP traffic to and from the internal network and the Internet.

The main advantage of a proxy server is its ability to provide NAT for shielding the internal network from the Internet. Figure 11.3 illustrates the application-level gateway acting as a relay of the application-level traffic.

11.4 Firewall Designs

This section concerns how to implement a firewall strategy. The primary step in designing a secure firewall is obviously to prevent the firewall devices from being compromised by threats. To provide a certain level of security, the three basic firewall designs are considered: a single-homed bastion host, a dual-homed bastion host, and a screened subnet firewall. The first two options are for creating a screened host firewall, and the third option contains an additional packet-filtering router to achieve another level of security.

To achieve the most security with the least amount of effort is always desirable. When building firewall devices, the bastion host should keep the design simple with the fewest

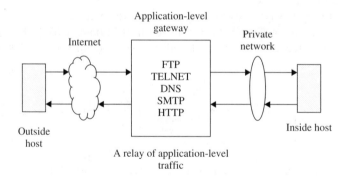

Figure 11.3 Application-level gateway for acting as a relay of application-level traffic.

possible components, both hardware and software. A bastion host is a publicly accessible device. When Internet users attempt to access resources on the Internet network, the first device they encounter is a bastion host. Fewer running services on the bastion host will give a potential hacker less opportunity to overcome the firewall. Bastion hosts must check all incoming and outgoing traffic and enforce the rules specified in the security policy. Bastion hosts are armed with logging and alarm features to prevent attacks. When creating a bastion host, it must be kept in mind that its role will help to decide what is needed and how to configure the device.

11.4.1 Screened Host Firewall (Single-Homed Bastion Host)

The first type of firewall is a screened host which uses a single-homed bastion host plus a packet-filtering router, as shown in Figure 11.4. Single-homed bastion hosts can be configured as either circuit-level or application-level gateways. When using either of these two gateways, each of which is called a *proxy server*, the bastion host can hide the configuration of the internal network.

NAT is essentially needed for developing an address scheme internally. It is a critical component of any firewall strategy. It translates the internal IP addresses to Internet Assigned Numbers Authority (IANA)-registered addresses to access the Internet. Hence, using NAT allows network administrators to use any internal IP address scheme.

The screened host firewall is designed such that all incoming and outgoing information is passed through the bastion host. The external screening router is configured to route all incoming traffic directly to the bastion host as indicated in Figure 11.4. The screening router is also configured to route outgoing traffic only if it originates from the bastion host. This kind of configuration prevents internal clients from bypassing the bastion host. Thus, the bastion host is configured to restrict unacceptable traffic and proxy acceptable traffic.

A single-homed implementation may allow a hacker to modify the router not to forward packets to the bastion host. This action would bypass the bastion host and allow the hacker directly into the network. But such a bypass usually does not happen because a network using a single-homed bastion host is normally configured to send packets only to the bastion host and not directly to the Internet.

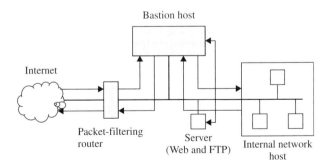

Figure 11.4 Screened host firewall system (single-homed bastion host).

11.4.2 Screened Host Firewall (Dual-Homed Bastion Host)

The configuration of the screened host firewall using a dual-homed bastion host adds significant security, compared with a single-homed bastion host. As shown in Figure 11.5, a dual-homed bastion host has two network interfaces. This firewall implementation is secure due to the fact that it creates a complete break between the internal network and the external Internet. As with the single-homed bastion, all external traffic is forwarded directly to the bastion host for processing. However, a hacker may try to subvert the bastion host and the router to bypass the firewall mechanisms. Even if a hacker could defeat either the screening router or the dual-homed bastion host, the hacker would still have to penetrate the other. Nevertheless, a dual-homed bastion host removes even this possibility. It is also possible to implement NAT for dual-homed bastion hosts.

11.4.3 Screened Subnet Firewall

The third implementation of a firewall is the screened subnet, which is also known as a *DMZ*. This firewall is the most secure one among the three implementations, simply because it uses a bastion host to support both circuit- and application-level gateways. As shown in Figure 11.6, all publicly accessible devices, including modem and server, are placed inside the DMZ. This DMZ then functions as a small isolated network positioned between the Internet and the internal network. The screened subnet firewall contains external and internal screening routers. Each is configured such that its traffic flows only to or from the bastion host. This arrangement prevents any traffic from directly traversing the DMZ subnetwork. The external screening router uses standard filtering to restrict external access to the bastion host and rejects any traffic that does not come from the bastion host. This router also uses filters to prevent attacks such as IP spoofing and source routing. The internal screening router also uses rules to prevent spoofing and source routing. Like its external counterpart, this internal router rejects incoming packets that do not originate from the bastion host and sends only outgoing packets to the bastion host.

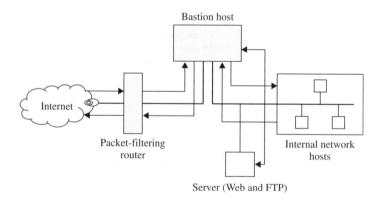

Figure 11.5 Screened host firewall system (dual-homed bastion host).

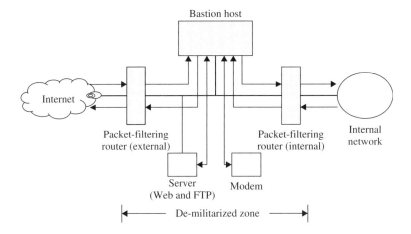

Figure 11.6 Screened subnet firewall system.

The benefits of the screened subnet firewall are based on the following facts. First, a hacker must subvert three separate tri-homed interfaces when he or she wants to access the internal network. But it is almost infeasible. Second, the internal network is effectively invisible to the Internet because all inbound/outbound packets go directly through the DMZ. This arrangement makes it impossible for a hacker to gain information about the internal systems because only the DMZ is advertized in the routing tables and other Internet information. Third, internal users cannot access the Internet without going through the bastion host because the routing information is contained within the network.

11.5 IDS Against Cyber Attacks

Recently, the information technology (IT) industry has witnessed the development of many cyber security solutions for protecting IT networks and reducing their vulnerability to cyber attacks. From firewalls to intrusion detection systems (IDSs), their security solutions have been progressed effectively in some extent for preserving data integrity, protecting data confidentiality, and maintaining data availability. This section will present a survey and comparison of Internet worm/virus detection against cyber defense.

11.5.1 Internet Worm Detection

In order to reduce the damage effectively caused by the fast propagation of malicious Internet worms, exiting IDSs should be strong enough to combat against the propagation worms, as well as to provide the security-related regulations of their monitoring networks.

A computer worm is a standalone malware program which replicates itself and spreads to other devices via network by utilizing vulnerabilities on the target devices. Therefore, a worm is a malicious piece of codes that self-progresses via network connections, exploiting security flaws in computers on the network.

Passive worms will require certain host behavior or human intervention to propagate itself until it is contacted by another host. Since the Morris worm was found in 1988, Internet worms have caused the most extensive and widespread damage of all kinds of computer attacks. Thus, an Internet worm is defined as a piece of malicious code that duplicates and propagates itself.

The life of a worm, after it is released, typically includes the following four phases:

- *Target finding.*
  Target finding is the first step in a worm's life to discover vulnerable hosts.
- *Worm transferring.*
  Transferring is referred to send a copy of the worm to the target after the victim (target) is discovered.
- *Worm activation.*
  Worm code starts to activate, duplicate, and propagate by itself until the target is discovered.
- *Worm infection.*
  Infection is the result of the worm, performing its malicious activities on the target host.

Differently from Computer Virus, the worm is a standalone program which can replicate itself through the network to spread, so it does not need to be attached. It makes the network performance weak by consuming bandwidth, increasing network traffic, or causing the denial of service (DoS). Morris worm, Blaster worm, Sasser worm, and Mydoom worm were some example of most notorious worms.

Morris worm (2 November, 1988) was written by a Cornell University student, Robert Tappan Morris, and reported to infect about 6000 major UNIX machines (damaging size is about $100 000-dollar; 10 000 000). Blaster worm (11 August, 2003) spread by exploiting BOF (buffer overflow) attack in DCOM RPC service of Windows operating system (OS), displayed some message, and rebooted the victim. Sasser worm might be the evolution of the Blast worm. Mydoom worm (26 January, 2004) was transmitted to attack Windows OS via e-mail with attachment, using user's address book. It evolved to many versions. It also affected South Korea and the United States on July 2009. Nowadays, the worm will install Trojan horse or backdoor into the PC and make the victim PC as botnet zombie.

11.5.2 Computer Virus

The Computer Virus is a kind of malicious program which can damage the victim computer and spread itself to another computer. The word "Virus" is used for most of the malicious program and it is confused with other programs, but the Virus, worm botnet, or other terms is distinct from one another.

The Virus destroys or disturbs the computer systems and user files, replicating itself via some media including hard disk, removable disc (floppy disk, USB, etc.), or network. Almost all Viruses must be attached to an executive file to be executed, because it cannot run without any action of human. Before the network is widespread, most Viruses spread via floppy disks. PC was booted from floppy disks and most of user files are also transferred by them. So the Virus in that time infected the floppy or hard disk files, or the boot sector of them. When the PC is booted or a user runs the virus attached files, the Virus uploads the malicious module and replication module to memory, sometimes it scans the victim host or files and attaches itself to them. The Virus used binary executable files, boot record, come script files, documents containing macro, and so on to transmit and infect.

11.5.3   Special Kind of Viruses

Trojan Horse

Trojan horse (or Trojan) is made to steal some information by social engineering. The term *Trojan horse* is derived from Greek mythology. The Trojan gives a cracker remote access permission like the Greek soldiers, avoiding detection of their user. It looks like some useful or helpful program, or looks like legitimate access process, but it just steal password, card number, or other useful information. The popular Trojan are Netbus, Back Orifice, Zeus, and so on.

Botnet

Botnet is the set of zombie computers connected to the Internet. Each compromised zombie computer is called as *bot*, and the botmaster, called as *C&C* (Command & Control server), controls these bots. Some Web site or e-mail has the malicious code, and if a user downloads and runs the program by curiosity or mistake, the PC is compromised by the program but the user cannot notice it. The bot has Trojan horse, backdoor, or other command tunnel, so the C&C can do anything with these bots.

Most of the bots run hidden and communicate via covert channel, so the user can hardly notice the bot. And botnet servers often liaise with each servers, the real size of bot network may be more larger than we can count.

The purpose of botnet is various, such as denial of service, STMP mail relays for sending spam mail, click fraud, steal some financial information, and login information.

There are so many kinds of botnet, and Stuxnet (June, 2010) is the most popular botnet of them. It was written to attack the industrial system with closed network, infected by USB memory. When it penetrates into one network, it spreads itself in the network and controls all of the systems. It used the system vulnerabilities such as LNK, Print Spooler, and Server Service.

Key Logger

Key logger program monitors the action of the key inputs. The key logger is of two types, software and hardware. This section concerns the software type only. It is also installed in the victim computers and logs all the strokes of keys. They save the log into some files or send the logs to the hacker by network. Key logger can steal the action of key input by kernel level, memory level, API level, packet level, and so on.

11.6   Intrusion Detections Systems

The lack of knowledge about security makes the people vulnerable to the damage due to acts of violence such as physical or electronical acts, putting on the risk the integrity of persons, companies for the loss or damage of electronic documents, physical areas, IT infrastructures, and so on.

From the point of view of the IT environment, we have to consider the implementation of IDSs to provide reliable information to network-administrator-related security of physical (Servers, Databases, Firewall, LAN, PCs, etc.) and electronic files which are considering the most valuable assets in companies.

This section presents several IDS systems which are typically deployed in IT networks.

11.6.1   Network-Based Intrusion Detection System (NIDS)

A network-based intrusion detection system (NIDS) monitors network traffic for particular network segments or devices and analyzes network, transport, and application protocols to identify suspicious activity.

NIDSs typically perform most of their analysis at the application layer such as HTTP, DNS, FTP, SMTP, and SNMP. They also analyze activity at the transport and network layers both to identify attacks at those layers and to facilitate the analysis of the application layer activity (e.g., a TCP port number may indicate which application is being used). Some NIDSs also perform limited analysis at the hardware layer.

A typical NIDS is composed of sensors, one or more management servers, multiple consoles, and optionally, one or more database servers (if the NIDS supports their use). All of these components are similar to other types of IDS technologies, except for the sensors. An NIDS sensor monitors and analyzes network activity on one or more network segments. The network interface cards that will be performing monitoring are placed into promiscuous mode, which means that they will accept all incoming packets that they see, regardless of their intended destinations. Most IDS deployments use multiple sensors, with large deployments having hundreds of sensors.

Sensors are available in two formats:

- *Appliance.* An appliance-based sensor consists of specialized hardware and sensor software. The hardware is typically optimized for sensor use, including specialized NICs and NIC drivers for efficient capture of packets, and specialized processors or other hardware components that assist in analysis. Parts or all of the IDS software might reside in firmware for increased efficiency. Appliances often use a customized, hardened OS that administrators are not intended to access directly.

- *Software Only.* Some vendors sell sensor software without an appliance. Administrators can install the software onto hosts that meet certain specifications. The sensor software might include a customized OS or it might be installed onto a standard OS just as any other application would.

Organizations should consider using management networks for their NIDS deployments whenever feasible. If an IDS is deployed without a separate management network, organizations should consider whether or not a VLAN is needed to protect the IDS communications.

In addition to choosing the appropriate network for the components, administrators also need to decide where the IDS sensors should be located. Sensors can be deployed in one of two modes:

- *Inline.* An inline sensor is deployed so that the network traffic it is monitoring must pass through it, much like the traffic flow associated with a firewall. In fact, some inline sensors are hybrid firewall/IDS devices, while others are simply IDSs. The primary motivation for deploying IDS sensors inline is to enable them to stop attacks by blocking network traffic. Inline sensors are typically placed where network firewalls and other network security devices would be placed – at the divisions between networks, such as connections with external networks and borders between different internal networks that should be segregated. Inline sensors that are not hybrid firewall/IDS devices are often deployed on the more secure side of a network division so that they have less traffic to process. Sensors can also be placed on the less secure side of a network division to provide protection for and reduce the load on the dividing device, such as a firewall.

- *Passive.* A passive sensor is deployed so that it monitors a copy of the actual network traffic; no traffic actually passes through the sensor. Passive sensors are typically deployed so that they can monitor key network locations, such as the divisions between networks, and key network segments, such as activity on a DMZ subnet. Passive sensors can monitor traffic through various methods, including the following:

  - *Spanning Port.* Many switches have a spanning port, which is a port that can see all network traffic going through the switch. Connecting a sensor to a spanning port can allow it to monitor traffic going to and from many hosts. Although this monitoring method is relatively easy and inexpensive, it can also be problematic. If a switch is configured or reconfigured incorrectly, the spanning port might not be able to see all the traffic. Another problem with spanning ports is that their use can be resource-intensive; when a switch is under heavy loads, its spanning port might not be able to see all traffic or spanning might be temporarily disabled. Also, many switches have only one spanning port, and there is often a need to have multiple technologies, such as network monitoring tools, network forensic analysis tools, and other IDS sensors, to monitor the same traffic.

  - *Network Tap.* A network tap is a direct connection between a sensor and the physical network media itself, such as a fiber-optic cable. The tap provides the sensor with a copy of all network traffic being carried by the media. Installing a tap generally involves some network downtime, and problems with a tap could cause additional downtime. Also, unlike spanning ports, which are usually already present throughout an organization, network taps need to be purchased as add-ons to the network.

  - *IDS Load Balancer.* An IDS load balancer is a device that aggregates and directs network traffic to monitoring systems, including IDS sensors. A load balancer can receive copies of network traffic from one or more spanning ports or network taps and aggregate traffic from different networks (e.g., reassemble a session that was split between two networks). The load balancer then distributes copies of the traffic to one or more listening devices, including IDS sensors, based on a set of rules configured by an administrator. The rules tell the load balancer which types of traffic to provide to each listening device.

NIDS products provide a wide variety of security capabilities. Some NIDSs offer limited information gathering capabilities, which means that they can collect information on hosts and the network activity involving those hosts. NIDSs typically perform extensive logging of data related to detected events. This data can be used to confirm the validity of alerts, investigate incidents, and correlate events between the IDS and other logging sources. NIDSs typically offer extensive and broad detection capabilities. Most products use a combination of signature-based detection, anomaly-based detection, and stateful protocol analysis techniques to perform in-depth analysis of common protocols; organizations should use NIDS products that use such a combination of techniques. The detection methods are usually tightly interwoven; for example, a stateful protocol analysis engine might parse activity into requests and responses, each of which is examined for anomalies and compared to signatures of known bad activity. Some products also use the same techniques and provide the same functionality as network behavior analysis (NBA) software.

NIDS sensors offer various prevention capabilities, including Passive only, Inline only, and both Passive and Inline.

11.6.2 Wireless Intrusion Detection System (WIDS)

A wireless intrusion detection system (WIDS) monitors wireless network traffic and analyzes its wireless networking protocols to identify suspicious activity. The typical components in a WIDS are the same as an NIDS: consoles, database servers (optional), management servers, and sensors. However, unlike an NIDS sensor, which can see all packets on the networks it monitors, a WIDS sensor works by sampling traffic because it can only monitor a single channel at a time. The longer a single channel is monitored, the more likely it is that the sensor will miss malicious activity occurring on other channels. To avoid this, sensors typically change channels frequently, so that they can monitor each channel a few times per second.

Wireless sensors are available in multiple forms. A dedicated sensor is a fixed or mobile device that performs WIDS functions but does not pass network traffic from source to destination. The other wireless sensor forms are bundled with access points (AP) or wireless switches. Because dedicated sensors can focus on detection and do not need to carry wireless traffic, they typically offer stronger detection capabilities than wireless sensors bundled with AP or wireless switches. However, dedicated sensors are often more expensive to acquire, install, and maintain than bundled sensors because bundled sensors can be installed on existing hardware, whereas dedicated sensors involve additional hardware and software. Organizations should consider both security and cost when selecting WIDS sensors.

WIDS components are typically connected to each other through a wired network. Because there should already be a strictly controlled separation between the wireless and wired networks, using either a management network or a standard network should be acceptable for WIDS components. Choosing sensor locations for a WIDS deployment is a fundamentally different problem than choosing locations for any other type of IDS sensor. If the organization uses termdefwireless local area networks (WLAN), wireless sensors should be deployed so that they monitor the range of the WLANs. Many organizations also want to deploy sensors to monitor parts of their facilities where there should be no WLAN activity, as well as channels and bands that the organization's WLANs should not use. Other considerations for selecting sensor locations include physical security, sensor range, wired network connection availability, cost, and AP and wireless switch locations.

WIDSs provide several types of security capabilities. Most can collect information on observed wireless devices and WLANs and perform extensive logging of event data. WIDSs can detect attacks, misconfigurations, and policy violations at the WLAN protocol level. Organizations should use WIDS products that use a combination of detection techniques to achieve broader and more accurate detection. Examples of events detected by WIDSs are unauthorized WLANs and WLAN devices, poorly secured WLAN devices, unusual usage patterns, the use of active wireless network scanners, denial of service attacks, and impersonation and man-in-the-middle attacks. Most WIDS sensors can also identify the physical location of a detected threat by using triangulation.

Compared to other forms of IDS, WIDS is generally more accurate; this is largely due to its limited scope (analyzing wireless networking protocols). WIDSs usually require some tuning and customization to improve their detection accuracy. The main effort is in specifying which WLANs, APs, and STAs are authorized and in entering the policy characteristics into the WIDS software. Besides reviewing tuning and customizations periodically to ensure that they are still accurate, administrators should also ensure that changes to building plans are incorporated occasionally. This is needed for accurate identification of the physical location of threats and accurate planning of sensor deployments.

Although WIDSs offer robust detection capabilities, they do have some significant limitations. WIDSs cannot detect certain types of attacks against wireless networks, such as attacks involving passive monitoring and off-line processing of wireless traffic. WIDSs are also susceptible to evasion techniques, especially those involving knowledge of a product's channel scanning scheme. Channel scanning can also impact network forensics because each sensor sees only a fraction of the activity on each channel. WIDS sensors are also susceptible to denial of service attacks and physical attacks.

WIDS sensors can offer intrusion prevention capabilities. Some sensors can instruct end points to terminate a session and prevent a new session from being established. Some sensors can instruct a switch on the wired network to block network activity for a particular wireless end point; however, this method can only block wired network communications and will not stop an end point from continuing to perform malicious actions through wireless protocols. Most IDS sensors allow administrators to specify the prevention capability configuration for each type of alert. Prevention actions can affect sensor monitoring; for example, if a sensor is transmitting signals to terminate connections, it may not be able to perform channel scanning to monitor other communications until it has completed the prevention action. To mitigate this, some sensors have two radios – one for monitoring and detection and the other for performing prevention actions. When selecting sensors, organizations should consider what prevention actions may need to be performed and how the sensor's detection capabilities could be affected by performing prevention actions.

11.6.3   Network Behavior Analysis System (NBAS)

A network behavior analysis system (NBAS) examines network traffic or statistics on network traffic to identify unusual traffic flows. NBA solutions usually have sensors and consoles, with some products also offering management servers. Some sensors are similar to NIDS sensors in that they sniff packets to monitor network activity on one or a few network segments. Other NBA sensors do not monitor the networks directly, but instead rely on network flow information provided by routers and other networking devices.

Most NBA sensors can be deployed in passive mode only, using the same connection methods (e.g., network tap, switch spanning port) as NIDSs. Passive sensors that are performing direct network monitoring should be placed so that they can monitor key network locations, such as the divisions between networks, and key network segments, such as DMZ subnets. Inline sensors are typically intended for network perimeter use, so they would be deployed in close proximity to the perimeter firewalls, often in front to limit incoming attacks that could overwhelm the firewalls.

NBA products provide a variety of security capabilities. They offer extensive information gathering capabilities, collecting detailed information on each observed host and constantly monitoring network activity for changes to this information. NBA technologies typically perform extensive logging of data related to detected events. They also typically have the capability to detect several types of malicious activity, including DoS attacks, scanning, worms, unexpected application services, and policy violations, such as a client system providing network services to other systems. Because NBA sensors work primarily by detecting significant deviations from normal behavior, they are most accurate at detecting attacks that generate large amounts of network activity in a short period of time and attacks that have unusual flow patterns. Most NBA sensors can also reconstruct a series of observed events to determine the origin of a threat.

NBA products automatically update their baselines on an ongoing basis. As a result, typically, there is not much tuning or customization to be done, other than updating firewall rule-set-like policies that most products support. A few NBA products offer limited signature customization capabilities; these are most helpful for inline sensors because they can use the signatures to find and block attacks that a firewall or router might not be capable of blocking. Besides reviewing tuning and customizations periodically to ensure that they are still accurate, administrators should also ensure that significant changes to hosts are incorporated, such as new hosts and new services. Generally, it is not feasible to automatically link NBA systems with change management systems, but administrators could review change management records regularly and adjust host inventory information in the NBA to prevent false positives.

NBA technologies have some significant limitations. They are delayed in detecting attacks because of their data sources, especially when they rely on flow data from routers and other network devices. This data is often transferred to the NBA in batches from every minute to a few times an hour. Attacks that occur quickly may not be detected until they have already disrupted or damaged systems. This delay can be avoided by using sensors that do their own packet captures and analysis; however, this is much more resource-intensive than analyzing flow data. Also, a single sensor can analyze flow data from many networks, while a single sensor can generally directly monitor only a few networks at once. Therefore, to do direct monitoring instead of using flow data, organizations might have to purchase more powerful sensors and/or more sensors.

11.6.4   Host-Based Intrusion Detection System (HIDS)

Host-based intrusion detection systems (HIDSs) monitor the characteristics of a single host and the events occurring within that host for suspicious activity. Examples of the types of characteristics an HIDS might monitor are wired and wireless network traffic, system logs, running processes, file access and modification, and system and application configuration changes. Most HIDSs have detection software known as *agents* installed on the hosts of interest. Each agent monitors activity on a single host and if prevention capabilities are enabled, also performs prevention actions. The agents transmit data to management servers. Each agent is typically designed to protect a server, a desktop or laptop, or an application service.

The network architecture for HIDS deployments is typically very simple. Because the agents are deployed to existing hosts on the organization's networks, the components usually communicate over those networks instead of using a management network. HIDS agents are most commonly deployed to critical hosts such as publicly accessible servers and servers containing sensitive information. However, because agents are available for various server and desktop/laptop OSs, as well as specific server applications, organizations could potentially deploy agents to most of their servers and desktops/laptops. Organizations should consider several criteria when selecting agent locations, including the need to analyze activity that cannot be monitored by other security controls; the cost of the agents' deployment, maintenance, and monitoring; the OSs and applications supported by the agents; the importance of each host's data or services; and the ability of the network infrastructure to support the agents' communications.

Most IDS agents alter the internal architecture of the hosts on which they are installed through shims, which are layers of code placed between existing layers of code. Although it is less intrusive to the host to perform monitoring without shims, which reduces the possibility of the IDS interfering with the host's normal operations, monitoring without shims is also generally less accurate at detecting threats and often precludes the performance of effective prevention actions.

HIDSs provide a variety of security capabilities. They typically perform extensive logging of data related to detected events and can detect several types of malicious activity. Detection techniques used include code analysis, network traffic analysis, network traffic filtering, filesystem monitoring, log analysis, and network configuration monitoring. HIDSs that use combinations of several detection techniques should generally be capable of achieving more accurate detection than products that use one or a few techniques, because each technique can monitor different characteristics of hosts. Organizations should determine which characteristics need to be monitored and select IDS products that provide adequate monitoring and analysis of those characteristics.

HIDSs usually require considerable tuning and customization; for example, many rely on observing host activity and developing baselines or profiles of expected behavior.

Others need to be configured with detailed policies that define exactly how each application on a host should behave. As the host environment changes, administrators should ensure that HIDS policies are updated to take those changes into account.

HIDSs have some significant limitations. Some detection techniques are performed only periodically, such as hourly or a few times a day, to identify events that have already happened, causing significant delay in identifying certain events. Also, many HIDSs forward their alert data to the management servers in batches a few times an hour, which can cause delays in initiating response actions. Because HIDSs run agents on the hosts being monitored, they can impact host performance because of the resources the agents consume. Installing an agent can also cause conflicts with existing host security controls, such as personal firewalls and VPN clients. Agent upgrades and some configuration changes can also necessitate rebooting the monitored hosts.

HIDSs offer various intrusion prevention capabilities; these vary based on the detection techniques used by each product. Code analysis techniques can prevent code from being executed; this can be very effective at stopping both known and previously unknown attacks. Network traffic analysis can stop incoming and outgoing network traffic containing network, transport, or application layer attacks; wireless networking protocol attacks; and the use of unauthorized applications and protocols. Network traffic filtering works as a host-based firewall and stops unauthorized access and acceptable use policy violations. Filesystem monitoring can prevent files from being accessed, modified, replaced, or deleted, which can stop malware installation and other attacks involving inappropriate file access. Other HIDS detection techniques generally do not support prevention actions because they identify events well after they have occurred.

Some HIDSs offer additional capabilities related to intrusion detection and prevention, such as enforcing restrictions on the use of removable media, detecting the activation or use of audiovisual devices, automatically hardening hosts on an ongoing basis, monitoring the status of running processes and restarting failed ones, and performing network traffic sanitization.

11.6.5 Signature-Based Systems

A signature-based IDS is based on pattern matching techniques. The IDS contains a database of patterns. Some patterns are well known by public program or domain, for example, Snort (http://www.snort.org/) and some are found by signature-based IDS companies. Using database of already found signature is much like antivirus software.

The IDS tries to match these signatures with the analyzed data. If a match is found, an alert is raised.

This approach usually yields good results in terms of few false positives. For few false positives, the Users try to select needed signatures, a task known as *tuning*, to avoid future false alerts. Tuning is performed once, and then updated from time to time. Thanks to tuning, a signature-based IDS generally generates a low rate of false alerts.

Even though the IDS make good results, it has drawbacks. First, all new attacks or polymorphic attacks, for example, zero-day attack, will go unnoticed until the responsible signatures are updated to database. It is because the IDS only tries to match signatures of its database with the attacks.

The IDS is not likely to detect even slight modifications of a known attack. Thus, attackers have a window of opportunity to gain control of the system or application under attack.

Second, the IDS need to be updated regularly, increasing the IT personnel workload and required skills.

Developing a new signature is tedious and hard work. Once an attack is made public, first, the experts need to analyze it carefully. The attack could exploit either a well-known vulnerability or a new one. The experts should develop the signature to detect the way the attack exploits a given vulnerability rather than the attack only. For example, BOF attacks could exploit different attack vectors, but those vectors will show common length. This is because attackers try to modify the attack payload without altering the attack effectiveness. So there are more chances to detect attack variations with just one signature.

A signature-based IDS raises classified alerts, for example, BOF or SQL Injection, and this classification is assigned "off-line" during the development of the signature. The importance of classification is threefold. First, Security teams can prioritize alerts without having to inspect them. Second, Security teams deploy automatic defensive countermeasures to react to certain disruptive attacks as soon as they take place, for example, dropping network traffic when a BOF is detected or changing some rules in firewall systems. Third, alerts can be correlated with each other and with other system logs, for example, firewall logs, to detect multistep attacks, that is, attacks that require the attacker to execute several actions to achieve his/her goal, and to reduce false alerts by identifying root causes. Thus, determining the attack class helps these tasks.

11.6.6 Anomaly-Based Systems

The deployment of an anomaly-based IDS typically requires expert personnel. Several parameters need to be configured, such as the duration of the training phase or the similarity metric. Every environment being different, guidelines are hard to give. Each anomaly-based IDS is also different from the others (while all signature-based IDSs work in a similar manner). Users gain little knowledge from subsequent installations; hence, deployment tasks are likely to be trial-and-error processes. Users mainly criticize three aspects of the management of current anomaly-based IDSs, each of which increases the user effort needed to run the IDS, namely,

1. An anomaly-based IDS generally raises a high number of false alerts.
2. An anomaly-based IDS usually works as a black box.
3. An anomaly-based IDS raises alerts without an exact classification.

False Alerts

Because of its statistical nature, an anomaly-based IDS is bound to raise a higher number of false alerts than a signature-based IDS. As a matter of fact, the classification of a certain input for a signature-based IDS is predetermined (for malicious inputs), while the classification of inputs for an anomaly-based IDS depends on the training data. Thus, different anomaly-based model instances could classify the same input differently.

False alerting is a well-known problem of IDSs in general and anomaly-based IDSs in particular. Security teams need to verify each raised alert; thus, an ineffective system (i.e., a system prone to raise a high number of false alerts) will require more personnel for its management.

Black Box Approach

An anomaly-based IDS carries out detection as a black box. Users have little control: quite often, they can adjust only the similarity metric used to discern licit traffic from malicious activities. Because most anomaly-based IDSs employ complex mathematical models (think of neural networks), users can neither precisely understand how the IDS engine discerns normal input nor refine the IDS model to avoid certain false alerts or to improve attack detection. Users need to spend a good deal of time to understand the inner working of the system, and they must be experts.

Lack of Attack Classification

An anomaly-based IDS raises an alert every time its model differs from the current input. The cause of the anomaly itself is unknown to the IDS. It holds little information to determine the attack class (other than the targeted IP address/TCP port and the IP source of attack). Because the model employed by the detection engine is locally built during a certain timeframe, each model is likely to be different. Hence, it is difficult to develop an off-line classifier suitable for any anomaly-based IDS instance. Because an anomaly-based IDS is supposed to detect unknown attacks, or slight modifications of well-known attacks, manual classification or the application of some heuristics are currently the only possible choices. However, manual classification is not feasible and the heuristics deliver results in a restricted context only (because the "traits" of each attack must be known before). Because alerts come unclassified, no automatic countermeasure can be activated to react to a certain threat.

Because of all the difficulties listed above for an anomaly-based IDS, a signature-based IDS is the obvious choice when users have to monitor complex systems, such as modern corporate networks, despite its inability to deal with zero-day attacks. Users can accomplish tuning more easily to avoid false alerts (thereby saving overall time), they can write their own set of signatures to detect attacks and a signature-based IDS automatically performs alert classification. An anomaly-based IDS lacks these features because researchers have mainly focused on enhancing attack detection and have not considered usability.

As a matter of fact, users could improve the usability of current anomaly-based IDSs by setting the similarity metric in such a way that the IDS generate fewer alerts. However, the detection rate would be (negatively) affected as well, thereby reducing the only advantage an anomaly-based IDS has over a signature-based IDS.

11.6.7 Evasion Techniques of IDS Systems

Attackers continuously try to find new exploits to intrude a system, while IDS system developers attempt to analyze and detect attacks. Here, four common evasions are introduced.

1. DoS
2. packet splitting
3. duplicate insertion
4. payload mutation.

Denial of Service

DoS attack intends to overwhelm network bandwidth or use system resources like CPU usage and the memory space of the system up. To do this, DoS attacks generate huge volume of network traffic and exploit the vulnerability of detection algorithm like slowing down the signature matching algorithm with modified network payload. In case of Snort, it uses backtracking for covering all possible pattern matching. The attacker could exploit this and modify the payload to take worst execution time for rule matching.

Packet Splitting

Packet splitting chops IP datagrams or TCP streams into nonoverlapping fragments known as *IP fragmentations* and *TCP segmentation*. Because of limited system resource, some IDS systems just try to match patterns to each fragmented packet without reassembling those. In this case, IDS could not detect attacking pattern in the reassembled packet. For example, an IDS may look for /bin/sh or/bin/bash to check whether the command execution code is included or not in the payload. But the attacker can deliberately split the payload with the signature into two segments, one containing /bin and the other containing /sh. If the IPS does not reassemble the two segments, it will be unable to find the signature in either segment, so the attacker can evade its detection.

Duplicate Insertion

Attackers could send duplicate or overlapping segments to IDS to confuse or neglect the content from IDS.

Because IDS system does not know network topology and victim system's OS information, if attackers send packets modifying TTL values with same sequence number, when IDS systems reassemble the packets, it gets confused with that. If IDS considered all possible combinations of packet reassembly, it gets confused as to which combination is the correct one. However, the victim system could not get all the packets sent by the attacker but the attacking packets only.

For example, an attacker sends fragmented and duplicated packets like:

1. TTL value = 30 contents="/b" sequence number = 1
2. TTL values = 13 contents "abcd" sequence number = 2
3. TTL values = 30 contents "in/b" sequence number = 2
4. TTL values = 12 contents "yep" sequence number = 3
5. TTL values = 24 contents "ash" sequence number = 3.

In this case, IDS receives all five packets and is confused how to reassemble them and hence cannot detect them as attacker packets. If hop number from attacker to victim is 20, second and fourth packets are dropped and only first, third, and fifth packets are delivered to victim system. The reassembled content is/bin/bash.

Payload Mutation

An attacker transforms malicious packet's payloads into semantically equivalent ones. Transformed packets will be different for IDS systems. Especially, the uniform resource identifier (URI) of an HTTP request could be the target of the attack. URI characters can be encoded to hexadecimal values like

Original URI: http://www.google.com?a=b

Transformed URI: http%3a%2f%2fwww.google.com%3fa=b

Attackers could transform several mutated expressions using the libwhisker library (www.wiretrip.net/rfp/ txt/whiskerids.html). This includes not only character transformation but also self-reference directory. For example, /abc/def/../def/gf/../../../bin/abc/../sh means /bin/sh.

Before trying to match patterns to URI, IDS systems need to process URL normalization (http://en.wikipedia.org/ wiki/URL_normalization) for transformed URI to original URI. IDS systems without this process will fail to detect the packet as attack.

12

SET for E-Commerce Transactions

The Secure Electronic Transaction (SET) is a protocol designed for protecting credit card transactions over the Internet. It is an industry-backed standard that was formed by MasterCard and Visa (acting as the governing body) in February 1996. To promote the SET standard throughout the payments community, advice and assistance for its development have been provided by IBM, GTE, Microsoft, Netscape, RSA, SAIC, Terisa, and Verisign.

SET relies on cryptography and X.509 v3 digital certificates to ensure message confidentiality and security. SET is the only Internet transaction protocol to provide security through authentication. It combats the risk of transaction information being altered in transit by keeping information securely encrypted at all times and by using digital certificates to verify the identity of those accessing payment details. The specifications of and ways to facilitate secure payment card transactions on the Internet are fully explored in this chapter.

12.1 Business Requirements for SET

This section describes the major business requirements for credit card transactions by means of secure payment processing over the Internet. They are listed below:

1. *Confidentiality of information* (*provide confidentiality of payment and order information*). To meet these needs, the SET protocol uses encryption. Confidentiality reduces the risk of fraud by either party to the transaction or by malicious third parties. Cardholder account and payment information should be secured as it travels across the network. It should also prevent the merchant from learning the cardholder's credit card number; this is only provided to the issuing bank. Conventional encryption by DES is used to provide confidentiality.

2. *Integrity of data* (*ensure the integrity of all transmitted data*). SET combats the risk of transaction information being altered in transit by keeping information securely encrypted at all times. That is, it guarantees that no changes in message content

Wireless Mobile Internet Security, Second Edition. Man Young Rhee.
© 2013 John Wiley & Sons, Ltd. Published 2013 by John Wiley & Sons, Ltd.

occur during transmission. Digital signatures are used to ensure integrity of payment information. RSA digital signatures, using SHA-1 hash codes, provide message integrity. Certain messages are also protected by HMAC using SHA-1.

3. *Cardholder account authentication (provide authentication that a cardholder is a legitimate customer of a branded payment card account).* Merchants need a way to verify that a cardholder is a legitimate user of a valid account number. A mechanism that links the cardholder to a specific payment card account number reduces the incidence of fraud and the overall cost of payment processing. Digital signatures and certificates are used to ensure authentication of the cardholder account. SET uses X.509 v3 digital certificates with RSA signatures for this purpose.

4. *Merchant authentication (provide authentication that a merchant can accept credit card transactions through its relationship with an acquiring financial institution).* Merchants have no way of verifying whether the cardholder is in possession of a valid payment card or has the authority to be using that card. There must be a way for the cardholder to confirm that a merchant has a relationship with a financial institution (acquirer) allowing it to accept the payment card. Cardholders also need to be able to identify merchants with whom they can securely conduct electronic commerce. SET provides for the use of digital signatures and merchant certificates to ensure authentication of the merchant. SET uses X.509 v3 digital certificates with RSA signatures for this purpose.

5. *Security techniques (ensure the use of the best security practices and system design techniques to protect all legitimate parties in an electronic commerce transaction).* SET utilizes two asymmetric key pairs for the encryption/decryption process and for the creation and verification of digital signatures. Confidentiality is ensured by the message encryption. Integrity and authentication are ensured by the use of digital signatures. Authentication is further enhanced by the use of certificates. The SET protocol utilizes cryptography to provide confidentiality of message information, ensure payment integrity, and insure identity authentication. For authentication purposes, cardholders, merchants, and acquirers will be issued with digital certificates by their sponsoring certification authorities (CAs). Thus, SET is a well-tested specification based on highly secure cryptographic algorithms and protocols.

6. *Creation of brand-new protocol (create a protocol that neither depends on transport security mechanisms nor prevents their use).* SET is an end-to-end protocol, whereas SSL provides point-to-point encryption. SET does not interfere with the use of other security mechanisms such as IPsec and SSL/TLS. Even though both technologies address the issue of security, they work in different ways and provide different levels of security. SET was specifically developed for secure payment transactions.

7. *Interoperability (facilitate and encourage interoperability among software and network providers).* SET uses specific protocols and message formats to provide interoperability. The specification must be applicable on a variety of hardware and software platforms and must not include a preference for one over another. Any cardholder with compliant software must be able to communicate with any merchant software that also meets the defined standard. In sum, it will be appropriate to introduce the TCP/IP model and Internet Protocol suite, including Electronic Payment System in Figure 12.1.

Electronic Payment System
E-cash, Mondex, Proton, Visa Cash, SET, CyberCash, CyberCoin, E-check, First Virtual

OSI model (7 layers)	TCP/IP model (4 layers)	Internet Protocol suite
Application	Application	HTTP, FTP, TELNET, SMTP, SNMP, DNS, POP, IMAP, MIME, NFS, XDR, RPC, TFTP, RIP, Gopher
Presentation		
Session	Transport	TCP, UDP
Transport		
Network	Internet	IP, ICMP, IGMP, ARP, RARP
Data Link	Network Access	Ethernet, Token-Ring, FDDI, PPP
Physical		

Figure 12.1 TCP/IP model and Internet Protocol suite. OSI, Open Systems Interconnect (seven layers); TCP/IP, Transmission Control Protocol/Internet Protocol (four layers).

12.2 SET System Participants

The participants in the SET system interactions are described in this section. A discrepancy is found between an SET transaction and a retail or mail order transaction: in a face-to-face retail transaction, electronic processing begins with the merchant or the acquirer, but in an SET transaction, the electronic processing begins with the cardholder.

- *Cardholder.* In the electronic commerce environment, consumers or corporate purchasers interact with merchants on personal computers over the Internet. A cardholder is an authorized holder of a payment card that has been issued by an issuer. In the cardholder's interactions, SET ensures that the payment card account information remains confidential.

- *Issuer.* An issuer is a financial institution (a bank) that establishes an account for a cardholder and issues the payment card. The issuer guarantees payment for authorized transactions using the payment card.

- *Merchant.* A merchant is a person or an organization that offers goods or services for sale to the cardholder. Typically, these goods or services are offered via a Web site or by e-mail. With SET, the merchant can offer its cardholders secure electronic interactions. A merchant that accepts payment cards must have a relationship with an acquirer (a financial institution).

- *Acquirer.* An acquirer is the financial institution that establishes an account with a merchant and processes payment card authorization and payments. The acquirer provides authentication to the merchant that a given card account is active and that the proposed purchase does not exceed the credit limit. The acquirer also provides electronic transfer of payments to the merchant's account. Subsequently, the acquirer is reimbursed by the issuer over some sort of payment network for electronic funds transfer (EFT).

- *Payment gateway.* A payment gateway acts as the interface between a merchant and the acquirer. It carries out payment authorization services for many card brands and performs clearing services and data capture. A payment gateway is a device operated by the acquirer or a designated third party that processes merchant payment messages (PMs), including payment instructions from cardholders. The payment gateway functions as follows: it decrypts the encoded message, authenticates all participants in a transaction, and reformats the SET message into a format compliant with the merchant's point of sale system. Note that issuers and acquirers sometimes choose to assign the processing of payment card transactions to third-party processors.

- *Certification Authority.* A CA is an entity that is trusted to issue X.509 v3 public-key certificates for cardholders, merchants, and payment gateways. The success of SET will depend on the existence of a CA infrastructure available for this purpose. The primary functions of the CA are to receive registration requests, process and approve/decline requests, and issue certificates. A financial institution may receive, process, and approve certificate requests for its cardholders or merchants and forward the information to the appropriate payment card brand(s) to issue the certificates. An independent Registration Authority (RA) that processes payment card certificate requests and forwards them to the appropriate issuer or acquirer for processing. The financial institution (issuer or acquirer) forwards approved requests to the payment card brand to issue the certificates.

Figure 12.2 illustrates the SET hierarchy which reflects the relationships between the participants in the SET system, described in the preceding paragraphs. In the SET environment, there exists a hierarchy of CAs. The SET protocol specifies a method of *trust chaining* for entity authentication. This trust chain method entails the exchange of digital certificates and verification of the public keys by validating the digital signatures of the issuing CA. As indicated in Figure 12.2, this trust chain method continues all the way up to the root CA at the top of the hierarchy.

12.3 Cryptographic Operation Principles

SET is the Internet transaction protocol providing security by ensuring confidentiality, data integrity, authentication of each party, and validation of the participant's identity. To meet these requirements, SET incorporates the following cryptographic principles:

- *Confidentiality.* This is ensured by the use of message encryption. SET relies on encryption to ensure message confidentiality. In SET, message data is encrypted with a random symmetric key which is further encrypted using the recipient's public key. The encrypted message along with this digital envelope is sent to the recipient. The recipient decrypts the digital envelope with a private key and then uses the symmetric key in order to recover the original message.

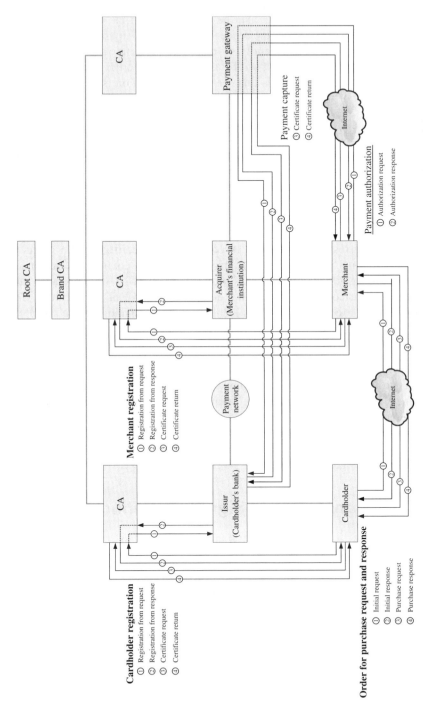

Figure 12.2   The SET transaction hierarchy indicating the relationships between the participants.

- *Integrity.* This is ensured by the use of a digital signature. Using the public/private-key pair, data encrypted with either key can be decrypted with the other. This allows the sender to encrypt a message using the sender's private key. Any recipient can determine that the message came from the sender by decrypting the message using the sender's public key. With SET, the merchant can be assured that the order it received is what the cardholder entered. SET guarantees that the order information is not altered in transit. Note that the roles of the public and private keys are reversed in the digital signature process where the private key is used to encrypt for signature and the public key is used to decrypt for verification of signature.

- *Authentication.* This is also ensured by means of a digital signature, but it is further strengthened by the use of a CA. When two parties conduct business transactions, each party wants to be sure that the other is authenticated. Before a user B accepts a message with a digital signature from a user A, B wants to be sure that the public key belongs to A. One way to secure delivery of the key is to utilize a CA to authenticate that the public key belongs to A. A CA is a trusted third party that issues digital certificates. Before it authenticates A's claims, a CA could supply a certificate that offers a high assurance of personal identity. This CA may require A to confirm his or her identity prior to issuing a certificate. Once A has provided proof of his or her identity, the CA creates a certificate containing A's name and public key. This certificate is digitally signed by the CA. It contains the CA's identification information, as well as a copy of the CA's public key. To get the most benefit, the public key of the CA should be known to as many people as possible. Thus, by trusting a single key, an entire hierarchy can be established in which one can have a high degree of trust.

The SET protocol utilizes cryptography to provide confidentiality of information, ensure payment integrity, and ensure identity authentication. For authentication purposes, cardholders, merchants, and acquirers (financial institutions) will be issued with digital certificates by their sponsoring CAs. The certificates are digital documents attesting to the binding of a public key to an individual user. They allow verification of the claim that a given public key does indeed belong to a given individual user.

12.4 Dual Signature and Signature Verification

SET introduced a new concept of digital signature called *dual signatures (DSs)*. A DS is generated by creating the message digest of two messages: order digest and payment digest. Referring to Figure 12.3, the customer takes the hash codes (message digests) of both the order message (OM) and PM by using the SHA-1 algorithm. These two hashes, h_o and h_p, are then concatenated and the hash code h of the result is taken. Finally, the customer encrypts (via RSA) the final hash code with his or her private key, K_{sc}, creating the DS. Computation of the DS is shown as follows:

$$DS = E_{K_{sc}}(h)$$

where $h = H(H(OM)\|H(PM))$

$$= H(h_o\|h_p)$$

$E_{K_{sc}}(= d_c)$ is the customer's private signature key.

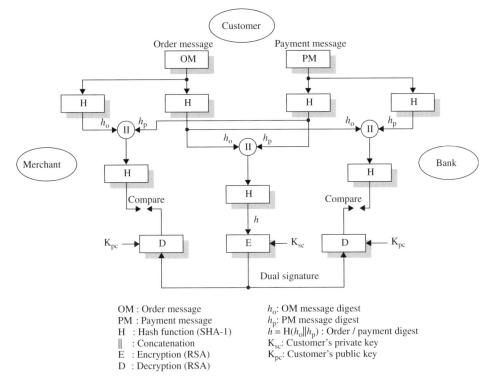

OM : Order message
PM : Payment message
H : Hash function (SHA-1)
‖ : Concatenation
E : Encryption (RSA)
D : Decryption (RSA)

h_o: OM message digest
h_p: PM message digest
$h = H(h_o\|h_p)$: Order / payment digest
K_{sc}: Customer's private key
K_{pc}: Customer's public key

Figure 12.3 Dual signature and order/payment message authentication.

Example 12.1 Computation of dual signature:
  Assume that the OM and the PM are given, respectively, as follows:

$$OM = 315a46e51283f7c647$$

$$PM = 1325f47568$$

Since SHA-1 sequentially processes blocks of 512 bits, that is, sixteen 32-bit words, the message padding must attach to the message block to ensure that a final padded message becomes a multiple of 512 bits. The 160-bit message digest can be computed from hashing the 512-bit padded message by the use of SHA-1. The padded OM and PM messages are, respectively,

Padded OM (512 bits):

315a46e5	1283f7c6	47800000	00000000
00000000	00000000	00000000	00000000
00000000	00000000	00000000	00000000
00000000	00000000	00000000	00000048

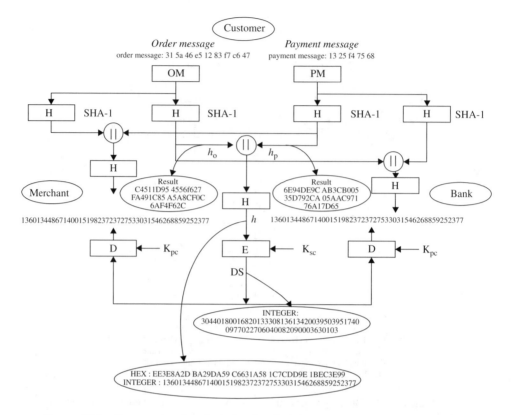

Figure 12.4   Computational analysis for the dual signature relating to Example 12.1.

Padded PM (512 bits):

1325f475	68800000	00000000	00000000
00000000	00000000	00000000	00000000
00000000	00000000	00000000	00000000
00000000	00000000	00000000	00000028

Referring to Figure 12.4, H(OM) $= h_o$ and H(PM) $= h_p$ each are obtained as follows:

h_o: c4511d95 4556f627 fa491c85 a5a8cf0c 6af4f62c (160bits)

h_p: 6e94de9c ab3cb005 35d792ca 05aac971 76a17d65 (160bits)

Concatenating these two hash codes and appending pads yields $(h_o \| h_p)$:

c4511d95	4556f627	fa491c85	a5a8cf0c
6af4f62c	6e94de9c	ab3cb005	35d792ca
05aac971	76a17d65	80000000	00000000
00000000	00000000	00000000	00000140

Taking the hash (SHA-1) of this concatenated message digests results in:

$$H(h_o\|h_p) = h$$

$$= \text{ee3e 9a3d ba2d da59 c663 1a58 1c7c dd9e 1bec 3e99} \quad \text{(hexadecimal)}$$

Transforming this resulting hash into decimal numbers yields:

$$H(h_o\|h_p) = 1360134484714001519823723727533031546268859285377 \text{(decimal)}$$

The concatenated two hashes become the input to the SHA-1 hash function. Thus, the resulting hash code h is RSA encrypted with the customer's private key $K_{sc} = d_c$ in order to obtain the DS.

To generate the public and private keys, choose two random primes, p and q, and compute the product $n = pq$. For a short example demonstration, choose $p = 47$ and $q = 73$; then $n = 3431$ and $\phi(n) = (p-1)(q-1) = 3312$. If the merchant has the customer's public key $e_c = K_{pc} = 79$ that is taken from the customer's certificate then the customer's private key d_c is computed using the extended Euclidean algorithm such that:

$$d_c \equiv e_c^{-1}(\text{mod } \phi(n))$$

$$\equiv 79^{-1}(\text{mod } 3312) \equiv 2767$$

In the digital signature process, the roles of the public and private keys are reversed, where the private key is used to encrypt (sign) and the public key is used to decrypt for verification of the signature.

To encrypt the final hash value h with d_c, first divide h into numerical blocks h_i and encrypt block after block such that:

$$\text{DS} = h_i^{d_c}(\text{mod } n)$$

This is the DS formula. Now, the DS represented in RSA-encrypted decimals can be computed as:

$$\text{DS} = 3044 \; 0180 \; 0168 \; 2013 \; 3308 \; 1361 \; 3420 \; 0395 \; 0395$$

$$1740 \; 0977 \; 0227 \; 0604 \; 0082 \; 0900 \; 0363 \; 0103$$

- *Merchant's signature verification.* Since the merchant has the customer's public key $K_{pc} = e_c = 79$, the merchant can decrypt the DS by making use of $K_{pc} = e_c$ as follows:

$$D_{K_{pc}}[\text{DS}] = \hat{h}$$

$$= 1360134484714001519823723727533031546268859285377 \quad \text{(decimal)}$$

$$= \text{ee3e 9a3d ba2d da59 c663 1a58 1c7c dd9e 1bec 3e99} \quad \text{(hexadecimal)}$$

Now assume that the merchant is in possession of the OM and the message digest for the PM $h_p = H(\text{PM})$. Then, the merchant can compute the following quantity:

$$h_M = H(H(\text{OM})\|h_p)$$

$$= \text{ee3e 9a3d ba2d da59 c663 1a58 1c7c dd9e 1bec 3e99} \quad \text{(hexadecimal)}$$

Since $h_M = \hat{h}$ is proved, the merchant has received OM and verified the signature.

- *Bank's signature verification.* Similarly, if the bank is in possession of DS, PM, the message digest h_o for OM, and the customer's public key K_{pc}, then it can compute the following quantity:

$h_B = H(h_o \| H(PM))$

$= $ ee3e 9a3d ba2d da59 c663 1a58 1c7c dd9e 1bec 3e99 (hexadecimal)

Since these two quantities are equal, $h_B = \hat{h}$, then the bank has verified the signature upon received PM.

Thus, it is verified completely that the customer has linked the OM and the PM and can prove the linkage.

12.5 Authentication and Message Integrity

When user A wishes to sign the plaintext information and send it in an encrypted message (ciphertext) to user B, the entire encryption process is as configured in Figure 12.5. The encryption/decryption processes for message integrity consist of the following steps.

1. Encryption process
   - User A sends the plaintext through a hash function to produce the message digest that is used later to test the message integrity.
   - A then encrypts the message digest with his or her private key to produce the digital signature.
   - Next, A generates a random symmetric key and uses it to encrypt the plaintext, A's signature, and a copy of A's certificate, which contains A's public key. To decrypt the plaintext later, user B will require a secure copy of this temporary symmetric key.
   - B's certificate contains a copy of his or her public key. To ensure secure transmission of the symmetric key, A encrypts it using B's public key. The encrypted key, called the *digital envelope*, is sent to B along with the encrypted message itself.
   - A sends a message to B consisting of the DES-encrypted plaintext, signature and A's public key, and the RSA-encrypted digital envelope.

2. Decryption process
   - B receives the encrypted message from A and decrypts the digital envelope with his or her private key to retrieve the symmetric key.
   - B uses the symmetric key to decrypt the encrypted message, consisting of the plaintext, A's signature, and A's public key retrieved from A's certificate.
   - B decrypts A's digital signature with A's public key that is acquired from A's certificate. This recovers the original message digest of the plaintext.
   - B runs the plaintext through the same hash function used by A and produces a new message digest of the decrypted plaintext.
   - Finally, B compares his or her message digest to the one obtained from A's digital signature. If they are exactly the same, B confirms that the message content has not

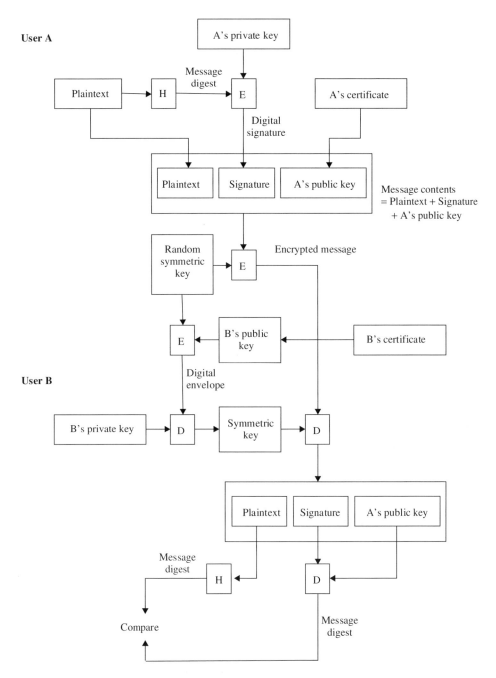

Figure 12.5 Encryption/Decryption overview for message integrity.

been altered during transmission and it was signed using A's private key. If they are not the same then the message either originated somewhere else or was altered after it was signed. In that case, B discards the message.

Example 12.2

Message Integrity Check:

User A.

Assume that the plaintext P is given as:

P = 0x135af247c613e815

The 160-bit message digest is computed from hashing the 512-bit padded plaintext by the use of SHA-1 as follows:

h = 0x8d9af6616e6063f2900833c2dcafefd1ed08f459

User A picks two random primes p = 487 and q = 229. Compute the product $n = pq$ = 111 523 and $\phi(n)$ = 110 808. Suppose the A's public key is e_A = 53 063 = 0xcf47. Then, A's private key d_A is computed using the extended Euclidean algorithm as:

$$d_A \equiv e_A^{-1}(\mathrm{mod}\ \phi(n)) \equiv 53\ 063^{-1}(\mathrm{mod}\ 110808) \equiv 71 = 0x47$$

A's private key is used to sign (encrypt) the 160-bit message digest h to produce the digital signature S_A:

S_A = 0x087760f9030ca3805ff419f4505e700cf3b18bec00d0d0cce80c9ab140dd057021a968

Now, the message contents consist of the plaintext P, signature S_A, and A's public key e_A as follows:

Message contents = 135af247c613e815087760f9030ca3805ff419f4505e700cf3b18bec

00d0d0cce80c9ab140dd057021a968cf47

A generates a random symmetric key K:

K = 0x13577ca2f8e63d79

Using this symmetric key, A encrypts the concatenated message contents as:

Encrypted message = 0x9adaff892d7c4db7f7911eacba780a6b1c6444d771f289f5a

12340aa1ccec658077f5521daddf1d78282aa96f4738426

and then sends them to user B.

User B.

User B chooses two random primes $p = 313$ and $q = 307$, which give $n = 96\,091$ and $\phi(n) = 95\,784$, respectively.

Assume that B's public key, $e_B = 109 = 0x6d$, is obtained from B's certificate. The symmetric key K is encrypted with B's public key to generate the digital envelope, which is computed as:

DE = 0x009d100c5207c1313376156091606c

B's private key d_B is computed using the extended Euclidean algorithm as:

$d_B = 7745 = 0x1e41$

The symmetric key K is recovered by decrypting the digital envelope with B's private key d_B.

K = 0x13577ca2f8e63d79

Using the recovered symmetric key, the encrypted message contents are decrypted to obtain the message contents (Plaintext + Signature + A' s public key). The message digest is computed by decrypting the recovered signature with A's public key, and it results in:

$\hat{h}$ = 0x8d9af6616e6063f2900833c2dcafefd1ed08f459

Next, the message digest is obtained using the SHA-1 hash function of the recovered plaintext. It results in the following message digest:

$\hat{h}$ = 0x8d9af6616e6063f2900833c2dcafefd1ed08f459

Thus, the MIC is completely accomplished because these two message digests are identical. Figure 12.6 gives full details of the MIC relating to this example.

12.6 Payment Processing

This section describes several transaction protocols needed to securely conduct payment processing by utilizing the cryptographic concepts introduced in Sections 12.3–12.5. The best detailed overview of SET specification appears in *Book 1: Business Description* issued in May 1997 by MasterCard and Visa. The following descriptions of secure payment processing are largely based on this book of the SET specification.

Figure 12.7 shows an overview of secure payment processing and it is worth looking at the outlines of several transaction protocols before reading the following detailed discussion.

12.6.1 Cardholder Registration

The cardholder must register with a CA before sending SET messages to the merchant. The cardholder needs a public/private-key pair for use with SET. The scenario of cardholder

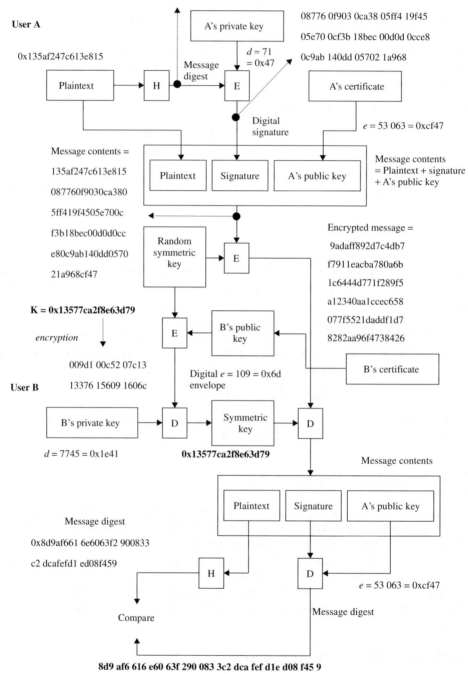

8d9 af6 616 e60 63f 290 083 3c2 dca fef d1e d08 f45 9

User A

0x135af247c613e815

08776 0f903 0ca38 05ff4 19f45

05e70 0cf3b 18bec 00d0d 0cce8

0c9ab 140dd 05702 1a968

$d = 71$
$= 0x47$

$e = 53\ 063 = 0xcf47$

Message contents =

135af247c613e815

087760f9030ca380

5ff419f4505e700c

f3b18bec00d0d0cc

e80c9ab140dd0570

21a968cf47

Message contents
= Plaintext + signature
+ A's public key

Encrypted message =

9adaff892d7c4db7

f7911eacba780a6b

1c6444d771f289f5

a12340aa1ccec658

077f5521daddf1d7

8282aa96f4738426

K = 0x13577ca2f8e63d79

encryption

009d1 00c52 07c13

13376 15609 1606c

$e = 109 = 0x6d$

User B

$d = 7745 = 0x1e41$

0x13577ca2f8e63d79

Message contents

Message digest

0x8d9af661 6e6063f2 900833

c2 dcafefd1 ed08f459

$e = 53\ 063 = 0xcf47$

8d9 af6 616 e60 63f 290 083 3c2 dca fef d1e d08 f45 9

Figure 12.6 Message integrity check relating to Example 12.2.

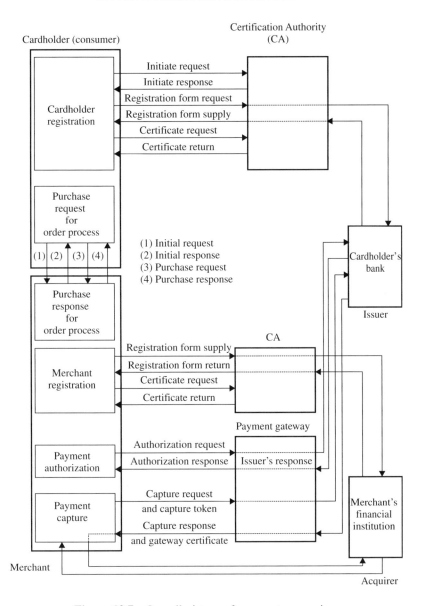

Figure 12.7 Overall picture of payment processing.

registration is described in the following.

1. Registration request/response processes

   The registration process can be started when the cardholder requests a copy of the CA certificate. When the CA receives the request, it transmits its certificate to the cardholder. The cardholder verifies the CA certificate by traversing the trust chain to the root key. The cardholder holds the CA certificate to use later during the registration process.

- The cardholder sends the *initiate request* to the CA.
- Once the initiate request is received from the cardholder, the CA generates the response and digitally signs it by generating a message digest of the response and encrypting it with the CA's private key.
- The CA sends the *initiate response* along with the CA certificate to the cardholder.
- The cardholder receives the initiate response and verifies the CA certificate by traversing the trust chain to the root key.
- The cardholder verifies the CA certificate by decrypting it with the CA's public key and comparing the result with a newly generated message digest of the initiate response.

It is worth depicting this registration process as shown in Figure 12.8.

2. Registration form process
   - The cardholder generates the registration form request.
   - The cardholder encrypts the SET message with a random symmetric key (No. 1). The DES key, along with the cardholder's account number, is then encrypted with the CA's public key.
   - The cardholder transmits the encrypted registration form request to the CA.
   - The CA decrypts the symmetric DES key (No. 1) and cardholder's account number with the CA's private key. The CA then decrypts the registration form request using the symmetric DES key (No. 1).
   - The CA determines the appropriate registration form and digitally signs it by generating a message digest of the registration form and encrypting it with the CA's private key.
   - The CA sends the registration form and the CA certificate to the cardholder.
   - The cardholder receives the registration form and verifies the CA certificate by traversing the trust chain to the root key.
   - The cardholder verifies the CA's signature by decrypting it with the CA's public key and comparing the result with a newly generated message digest of the registration form. The cardholder then completes the registration form.

The registration form process is depicted as shown Figure 12.9.

3. Certificate request/response processes
   - The cardholder generates the certificate request, including the information entered into the registration form.
   - The cardholder creates a message with request, the cardholder's public key and a newly generated symmetric key (No. 2), and digitally signs it by generating a message digest of the cardholder's private key.
   - The cardholder encrypts the message with a randomly generated symmetric key (No. 3). This symmetric key, along with the cardholder's account information, is then encrypted with the CA's public key.
   - The cardholder transmits the encrypted certificated request messages to the CA.

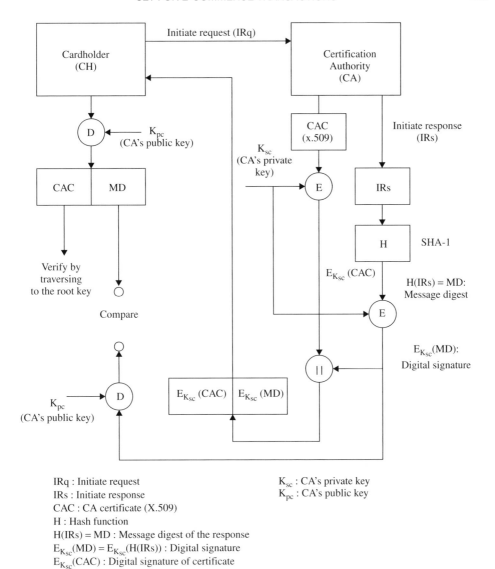

Figure 12.8 Registration request/response processes.

- The CA decrypts the No. 3 symmetric key and cardholder's account information with the CA's private key and then decrypts the certificate request using this symmetric key.
- The CA verifies the cardholder's signature by decrypting it with the cardholder's public key and comparing the result with a newly generated message digest of the certificate requested.
- The CA verifies the certificate request using the cardholder's account information and information from the registration form.

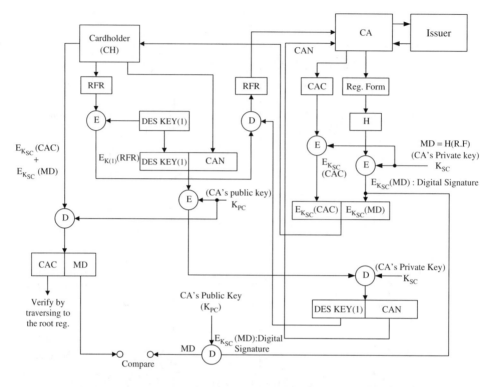

Figure 12.9   Registration form process.

- Upon verification, the CA creates the cardholder certificate, digitally signing it with the CA's private key.
- The CA generates the certificate response and digitally signs it by generating a message digest of the response and encrypting it with the CA's private key.
- The CA encrypts the certificate response with the No. 2 symmetric key from the cardholder request.
- The CA transmits the certificate response to the cardholder.
- The cardholder verifies the certificate by traversing the trust chain to the root key.
- The cardholder decrypts the response using the symmetric key (No. 2) saved from the cardholder request process.
- The cardholder verifies the CA's signature by decrypting it with the CA's public key and comparing the result with a newly generated message digest of the response.
- The cardholder stores the certificate and information from the response for future e-commerce use.

12.6.2 Merchant Registration

Merchants must register with a CA before they can receive SET payment instructions from cardholders. In order to send SET messages to the CA, the merchant must have a copy of the CA's public key which is provided in the CA certificate. The merchant also needs the registration form from the acquirer. The merchant must identify the acquirer to the CA. The merchant registration process consists of five steps as follows: (i) The merchant requests the registration form; (ii) the CA processes this request and sends the registration form; (iii) the merchant requests certificates after receiving the registration certificates; (iv) the CA creates certificates; and (v) the merchant receives certificates.

The detailed steps for the merchant registration are described in what follows.

1. Registration form process
   The registration process starts when the merchant requests the appropriate registration form.
   - The merchant sends the initiate request of the registration form to the CA. To register, the merchant fills out the registration form with information such as the merchant's name, address, and ID.
   - The CA receives the initiate request.
   - The CA selects an appropriate registration form and digitally signs it by generating a message digest of the registration form and encrypting it with the CA's private key.
   - The CA sends the registration form along with the CA certificate to the merchant.
   - The merchant receives the registration form and verifies the CA certificate by traversing the trust chain to the root key.
   - The merchant verifies the CA's signature by decrypting it with the CA's public key and comparing the result with a newly computed message digest of the registration form.
   - The merchant creates two public/private-key pairs for use with SET: key encryption and signature.

   Thus, the merchant completes the registration form. The merchant takes the registration information (name, address, and ID) and combines it with the public key in a registration message. The merchant digitally signs the registration message. Next, the merchant's software generates a random symmetric key. This random key is used to encrypt the message. The key is then encrypted into the digital envelope using the CA's public key. Finally, the merchant transmits all of these components to the CA.

2. Certificate request/create process
   The merchant starts with the certificate request. When the CA receives the merchant's request, it decrypts the digital envelope to obtain the symmetric encryption key, which it uses to decrypt the registration request.
   - The merchant generates the certificate request.
   - The merchant creates the message with request and both merchant public keys and digitally signs it by generating a message digest of the certificate request and encrypting it with the merchant's private key.

- The merchant encrypts the message with a random symmetric key (No. 1). This symmetric key, along with the merchant's account data, is then encrypted with the CA's public key.
- The merchant transmits the encrypted certificate request message to the CA.
- The CA decrypts the symmetric key (No. 1) and the merchant's account data with the CA's private key and then decrypts the message using the symmetric key (No. 1).
- The CA verifies the merchant's signature by decrypting it with the merchant's public key and comparing the result with a newly computed message digest of the certificate request.
- The CA confirms the certificate request using the merchant information.
- Upon verification, the CA creates the merchant certificate digitally signing the certificate with the CA's private key.
- The CA generates the certificate response and digitally signs it by generating a message digest of the response and encrypting it with the CA's private key.
- The CA transmits the certificate response to the merchant.
- The merchant receives the certificate response from the CA. The merchant decrypts the digital envelope to obtain the symmetric key. This key is used to decrypt the registration response containing the certificates.
- The merchant verifies the certificates by traversing the trust chain to the root key.
- The merchant verifies the CA's signature by decrypting it with the CA's public key and comparing the result with a newly computed message digest of the certificate response.
- The merchant stores the certificates and information from the response for use in future e-commerce transactions.

12.6.3 Purchase Request

The purchase request exchange should take place after the cardholder has completed browsing, selecting, and ordering. Before the end of this preliminary phase occurs, the merchant sends a completed order form to the cardholder (customer). In order to send SET messages to a merchant, the cardholder must have a copy of the certificates of the merchant and the payment gateway. The message from the cardholder indicates which payment card brand will be used for the transaction. The purchase request exchange consists of four messages: initiate request, initiate response, purchase request, and purchase response. The detailed discussions that follow describe each step fully.

1. Initiate request
   - The cardholder sends the initiate request to the merchant.
   - The merchant receives the initiate request.
   - The merchant generates the response and digitally signs it by generating a message digest of the response and encrypting it with the merchant's private key.
   - The merchant sends the response along with the merchant and payment gateway certificates to the cardholder.

2. Initiate response
   - The cardholder receives the initiate response and verifies the certificates by traversing the trust chain to the root key.
   - The cardholder verifies the merchant's signature by decrypting it with the merchant's public key and comparing the result with a newly computed message digest of the response.
   - The cardholder creates the OM using information from the shopping phase and PM. At this step, the cardholder completes payment instructions.

3. Purchase request
   - The cardholder generates a dual signature for the OM and the PM by computing the message digests of both, concatenating the two digests, computing the message digest of the result, and encrypting it using the cardholder's private key.
   - The cardholder generates a random symmetric key (No. 1) and uses it to encrypts the PM. The cardholder then encrypts his or her account number as well as the random symmetric key used to encrypt the PM in a digital envelope using the payment gateway's key.
   - The cardholder transmits the OM and the encrypted PM to the merchant.
   - The merchant verifies the cardholder certificate by traversing the trust chain to the root key.
   - The merchant verifies the cardholder's DS on the OM by decrypting it with the cardholder's public key and comparing the result with a newly computed message digest of the concatenation of the message digests of the OM and PM.
   - The merchant processes the request, including forwarding the PM to the payment gateway for authorization.

4. Purchase response
   - The merchant creates the purchase response including the merchant signature certificate and digitally signs it by generating a message digest of the purchase response and encrypting it with the merchant's private key.
   - The merchant transmits the purchase response to the cardholder.
   - If the transaction was authorized, the merchant fulfills the order to the cardholder.
   - The cardholder verifies the merchant signature certificate by traversing the trust chain to the root key.
   - The cardholder verifies the merchant's digital signature by decrypting it with the merchant's public key and comparing the result with a newly computed message digest of the purchase response.
   - The cardholder stores the purchase response.

12.6.4 Payment Authorization

During the processing of an order from a cardholder, the merchant authorizes the transaction. The authorization request and the cardholder payment instructions are then transmitted to the payment gateway.

1. Authorization request
   - The merchant creates the authorization request.
   - The merchant digitally signs an authorization request by generating a message digest of the authorization request and encrypting it with the merchant's private key.
   - The merchant encrypts the authorization request using a random symmetric key (No. 2), which in turn is encrypted with the payment gateway public key.
   - The merchant transmits the encrypted authorization request and the encrypted PM from the cardholder purchase request to the payment gateway.
   - The gateway verifies the merchant certificate by traversing the trust chain to the root key.
   - The payment gateway decrypts the digital envelope of the authorization request to obtain the symmetric encryption key (No. 2) with the gateway private key. The gateway then decrypts the authorization request using the symmetric key (No. 2).
   - The gateway verifies the merchant's digital signature by decrypting it with the merchant's public key and comparing the result with a newly computed message digest of the authorization request.
   - The gateway verifies the cardholder's certificate by traversing the trust chain to the root key.
   - The gateway decrypts the symmetric key (No. 1) and the cardholder account information with the gateway private key. It uses the symmetric key to decrypt the PM.
   - The gateway verifies the cardholder's DS on the PM by decrypting it with the cardholder's public key and comparing the result with a newly computed message digest of the concatenation of the message digest of the OM and the PM.
   - The gateway ensures consistency between the merchant's authorization request and the cardholder's PM.
   - The gateway sends the authorization request through a financial network to the cardholder's financial institution (issuer).

2. Authorization response
   - The gateway creates the authorization response message and digitally signs it by generating a message digest of the authorization response and encrypting it with the gateway's private key.
   - The gateway encrypts the authorization response with a new randomly generated symmetric key (No. 3). This key is then encrypted with the merchant's public key.
   - The gateway creates the capture token and digitally signs it by generating a message digest of the capture token and encrypting it with the gateway's private key.
   - The gateway encrypts the capture token with a new symmetric key (No. 4). This key and the cardholder account information are then encrypted with the gateway's public key.
   - The gateway transmits the encrypted authorization response to the merchant.
   - The merchant verifies the gateway certificate by traversing the trust chain to the root key.
   - The merchant decrypts the symmetric key (No. 3) with the merchant's private key and then decrypts the authorization response using the symmetric key (No. 3).

- The merchant verifies the gateway's digital signature by decrypting it with the gateway's public key and comparing the result with a newly computed message digest of the authorization response.
- The merchant stores the encrypted capture token and envelope for later capture processing.
- The merchant then completes processing of the purchase request and the cardholder's order by shipping the goods or performing the services indicated in the order.

12.6.5 Payment Capture

After completing the processing of an order from a cardholder, the merchant will request payment. The merchant generates and signs a capture request, which includes the final amount of the transaction, the transaction identifier from the OM, and other information about the transaction. A merchant's payment capture process will be described in detail in the following.

1. Capture request
   - The merchant creates the capture request.
   - The merchant embeds the merchant certificate in the capture request and digitally signs it by generating a message digest of the capture request and encrypting it with the merchant's private key.
   - The merchant encrypts the capture request with a randomly generated symmetric key (No. 5). This key is then encrypted with the payment gateway's public key.
   - The merchant transmits the encrypted capture request and encrypted capture token previously stored from the authorization response to the payment gateway.
   - The gateway verifies the merchant certificate by traversing the trust chain to the root key.
   - The gateway decrypts the symmetric key (No. 5) with the gateway's private key and then decrypts the capture request using the symmetric key (No. 5).
   - The gateway verifies the merchant's digital signature by decrypting it with the merchant's public key and comparing the result with a newly computed message digest of the capture request.
   - The gateway decrypts the symmetric key (No. 4) with the gateway's private key and then decrypts the capture token using the symmetric key (No. 4).
   - The gateway ensures consistency between the merchant's capture request and the capture token.
   - The gateway sends the capture request through a financial network to the cardholder's issuer (financial institution).

2. Capture response
   - The gateway creates the capture response message, including the gateway signature certificate, and digitally signs it by generating a message digest of the capture response and encrypting it with the gateway's private key.
   - The gateway encrypts the capture response with a newly generated symmetric key (No. 6). This key is then encrypted with the merchant's public key.

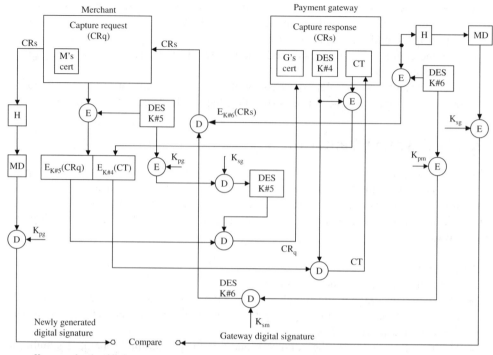

Figure 12.10   Payment capture process.

- The gateway transmits the encrypted capture response to the merchant.
- The merchant verifies the gateway certificate by traversing the trust chain to the root key.
- The merchant decrypts the symmetric key (No. 6) with the merchant's private key and then decrypts the capture response using the symmetric key (No. 6).
- The merchant verifies the gateway's digital signature by decrypting it with the gateway's public key and comparing the result with a newly generated message digest of the capture response.

Figure 12.10 shows an overview of payment capture consisting of the merchant's capture request and the gateway's capture response.

13

4G Wireless Internet Communication Technology

The mobile industry for wireless cellular services has grown at a rapid pace over the past decade. Similarly, Internet service technology has also made dramatic growth through the World Wide Web with a wire line infrastructure. Realization for complete mobile Internet technologies will become the future objectives for convergence of these technologies through multiple enhancements of both cellular mobile systems and Internet interoperability.

Flawless integration between these two wired/wireless networks will enable subscribers to not only roam worldwide but also solve the ever increasing demand for data/Internet services. However, the new technology development and service perspective of 4G systems will take many years to come. In order to keep up with this noteworthy growth in the demand for wireless broadband, new technologies and structural architectures are needed to improve system performance and network scalability greatly while significantly reducing the cost of equipment and deployment. The present concept of P2P (peer-to-peer) networking to exchange information needs to be extended to implement intelligent appliances such as a ubiquitous connectivity to the Internet services, the provision of fast broadband access technologies at more than 50 Mbps data rate, seamless global roaming, and Internet data/voice multimedia services.

The 4G system is a development initiative based on the currently deployed 2G/3G infrastructure, enabling seamless integration to emerging 4G access technologies. The path toward 4G networks should be incorporated with a number of critical trends to network integration for successful interoperability. Multiple Input Multiple Output/Orthogonal Frequency Division Multiple Access (MIMO/OFDMA)-based air interface for beyond 3G systems are called *4G systems* such as Long Term Evolution (LTE), Ultra Mobile Broadband (UMB), Mobile WiMAX (Worldwide Interoperability for Microwave Access), or Wireless Broadband (WiBro).

Wireless Mobile Internet Security, Second Edition. Man Young Rhee.
© 2013 John Wiley & Sons, Ltd. Published 2013 by John Wiley & Sons, Ltd.

13.1   Mobile WiMAX

Mobile WiMAX is one of the most promising technologies for broadband wireless communications. This new technology was adopted by the ITU as one of the IMT 2000 technologies in November 2007. Since then, Mobile WiMAX has officially become a major global cellular wireless standard along with 3GPP UMTS/HSPA and 3GPP2 CDMA2000 1xEV-DO/EV-DV.

Mobile WiMAX is a next-generation OFDM (Orthogonal Frequency Division Multiplexing)-based broadband wireless technology based on the IEEE 802.16e-2005 standards. In December 2005, the IEEE ratified 802.16e as a set of amendments to 802.16-2004. The 802.16e standard provides improved support for intercell handover, directed adjacent-cell measurement, and sleep modes to support low-power MS (mobile station) operation. On the basis of the IEEE 802.16e air interface, the Mobile WiMAX provides a broadband wireless system that enables convergence of both mobile and fixed broadband networks.

The IEEE 802.16e-2005 standard defines the physical layer (PHY) and media access control (MAC) air-link primitives between an MS and a BS (base station) for providing a mobile broadband wireless access system. This access system permits a so-called scalable deployment, for which the OFDM symbol duration and intercarrier bandwidth are defined. The Mobile WiMAX network architecture provides key functional entities, reference interfaces, and access network configuration over which a network interoperability is implemented.

The WiMAX Forum is an industry consortium promoting the IEEE 802.16 family of standards for broadband wireless access systems. The all-IP mobile network specification, conceptualized by the Network Working Group (NWG) in the WiMAX Forum, enables Mobile WiMAX to deliver a wider selection of multimedia IP services. In January 2005, the WiMAX Forum (a nonprofit organization of 414 member companies) established the NWG Group to specify an end-to-end (ETE) all-IP architecture that requires optimum operation for a broad range of IP services for the Mobile WiMAX standard. At present, more than 260 service providers are deploying fixed, portable, and mobile networks in 110 countries. In an April 2008 survey, the WiMAX Forum projected more than 93 million Mobile WiMAX users globally by 2012. A new task group for the IEEE 802.16j standard was officially established in March 2006 to amend the current IEEE 802.16e standard to support mobile multihop relay (MMR) operation in wireless broadband networks. The objective of this standard is to increase the system capacity and to broaden coverage of the current Mobile WiMAX standard.

The IEEE also plans to adopt Mobile WiMAX 2.0, called the *IEEE 802.16m*, in the later part of 2008 and could offer data rates of 100 Mbps for mobile applications and 1 Gbps for fixed uses via enhanced MIMO technology. The products of Mobile WiMAX 2.0 may be expected to appear by 2012, if adopted on schedule.

13.1.1   Mobile WiMAX Network Architecture

The Mobile WiMAX network consists of the access service network (ASN) and the connectivity service network (CSN).

Based on the IEEE 802.16e air interface, the Mobile WiMAX provides a broadband wireless system that enables convergence of both mobile and fixed broadband networks. Mobile WiMAX is the first mobile wireless system to realize enhanced spectral efficiency and system flexibility. Mobile WiMAX has adopted the following technologies: OFDMA, TDD, MIMO, AMC (Adaptive Modulation and Coding), HARQ, and all-IP, as well as mobility and security management for offering high-data-rate, low-cost, wide-area, and secured mobile multimedia services. The Mobile WiMAX network is more effective in providing multimedia data services than existing cellular wireless network (i.e., GSM/WCDMA family or CDMA2000 1xEV-DO family) and WLAN (for Wi-Fi IP-mode in small area services).

The WiMAX network architecture is designed to support a much larger coverage area than WLAN while maximizing the use of open standards and IETF protocols in a simple all-IP architecture. The WiMAX network architecture can be logically represented by a network reference model (NRM) that contains key functional entities and reference points (RPs) over which the network interoperability is defined. Mobile WiMAX represents an all-IP ETE network architecture that uses IP for ETE transport of all user data and signaling data.

The fundamental WiMAX network architecture is facilitated by key functional entities and RPs over which the network access providers (NAPs) and network service providers (NSPs) are defined. The NAP is an entity that provides WiMAX radio access infrastructure, while the NSP is an entity that provides IP connectivity.

The WiMAX architecture also enables open access to Web-based applications and enhanced Internet services, allowing operators to explore creative service offering as well as provisioning for a variety of mobile Internet business services.

1. *MS.* The MS represents a user equipment (UE) set providing wireless connectivity between a single or multiple hosts and the WiMAX network. In this context, the term *MS* is used in general to refer to both mobile and fixed device terminals.

2. *BS.* The BS is a logical network entity that primitively consists of the radio-related functions of an ASN interfacing with an MS over-the-air link corresponding to MAC and PHY specifications in IEEE 802.16 standard in order to meet the requirements defined in the WiMAX Forum System profile. To be more specific, the BS collects user terminal data via a wireless path, sends them to the ASN Gateway (GW) for the uplink (UL), and distributes the data received from the ASN-GW to the MSs for the downlink (DL). Some functions of BS include wireless access process, radio resource management (RRM), mobility support for seamless services while moving, quality of service (QoS) support for stable service, and overall equipment control.

3. *ASN.* The core elements in the ASN are the BS and ASN-GW, to be connected over IP.The ASN designates the point of entry for the MS into a WiMAX network as well as the radio access point (AP) to the MS. The ASN functions can be described as follows:
   - the 802.16 layer 2 connectivity with a WiMAX MS;
   - transfer of AAA (authentication, authorization, and accounting) messages to the subscriber's home network for authentication, authorization, and session establishment;
   - selection of the subscriber's preferred NSP;

- relay functionality for establishing layer 3 connectivity with an MS;
- RRM, QoS, and policy management;
- ASN-CSN tunneling and ASN-ASN connectivity.

ASN represents a complete set of network functions required to provide radio access to the MS, transfer AAA messages to CSN, provide preferred CSN discovery and selection, relay functionality for establishing layer 3 connectivity with MS, provide RRM, provide ASN-CSN mobility support, provide paging and location management, and provide ASN-CSN tunneling.

4. *ASN-GW.* ASN-GW is a logical entity that represents an aggregated function related to QoS, security, and mobile management for all the data connections served by its association with BSs through R6. The ASN-GW also hosts functions related to IP layer interventions with CSN through R3 as well as interactions with ASN through R4 in support of mobility. The intra-ASN RPs (R6 and R8) are only applicable to single decomposed ASN profile (ASN C), whereas the ASN-GW connects the BS with various servers and core routers in CSN, performing the routing function and thus transferring data between the BS and the CSN.

5. *Connectivity Service Network (CSN).* The CSN provides IP connectivity services to the WiMAX subscribers.The CSN elements comprise the routers, AAA proxy/servers, user databases, and GW devices for interworking.The CSN functions within logical entities across exposed RPs are listed below.
   - MS IP address and endpoint parameter allocation for user sessions
   - Internet access
   - AAA proxy/server
   - ASN-CSN tunneling support
   - Inter-CSN pipeline for roaming
   - Inter-ASN mobility
   - Subscriber billing and interoperator settlement
   - Broadcast and multicast services
   - Connectivity to WiMAX IP multimedia services
   - P2P services.

CSN is a set of network functions (listed above) that provide IP connectivity services to the subscribers. To achieve this diverse set of functions, CSN contains routers/switches, AAA and other servers, user database, and interworking GWs. CSN servers include HA (Home Agent) for the management of home addresses, the AAA server for security and accounting functions, the DNS server for conversion of IP addresses and the system names, the DHCP (Dynamic Host Configuration Protocol) server for dynamic allocation of IP, and the PCRF (Policy and Charging Rules Function) server for managing the service policy and for sending QoS setting and accounting rule information.

13.1.2   Reference Points in WiMAX Network Reference Model (NRM)

The WiMAX network architecture can be logically represented by an NRM. Each RP in the WiMAX NRM is a logical interface between different functional entities on either side of it. Different protocols associated with an RP may originate and terminate in different functional entities across that RP.

The WiMAX NRM defines the following RPs between the major functional entities.

- *R*1 (*Reference Point* 1). R1 is the RP that specifies the protocols and procedures between the MS and ASN. R1 includes the IEEE 802.16e standard that defines PHY and MAC layers air interface specifications related to control plane and management plane inter-actions.

- *R*2 (*Reference Point* 2). R2 provides the applicable protocols and procedures between the MS and CSN associated with authentication, service authorization, and IP host configuration management.

- *R*3 (*Reference Point* 3). R3 complies with control plane protocol and IP bearer plane protocol between ASN and CSN to support AAA, policy enforcement, and mobility management capabilities. It also provides necessary tunneling to transfer user data between the ASN and CSN through the ASN-GW.

- *R*4 (*Reference Point* 4). R4 offers control and bearer plane protocols between the ASNs so as to provide the MS mobility across ASNs and ASN-GWs. R4 also serves as the interoperability RP across any pair of ASNs.

- *R*5 (*Reference Point* 5). R5 arms with control plane and bearer plane protocols needed to support loaming between the CSNs, operated by the home NSP and by a visited NSP.

In addition, the intra-ASN RP (R6, R7, and R8) are defined within an ASN.

- *R*6 (*Reference Point* 6). The functionality across the ASN-GW and BS is split and signaled via R6 that includes all control and bearer plane protocols between the BS and the associated ASN-GW. The control plane consists of QoS, security- and mobility-related protocols for paging and data path establishment, and release control. The bearer plane represents the intra-ASN data path between the BS and ASN-GW.

- *R*7 (*Reference Point* 7). R7 is the optional RP, not shown in Figure 13.1. The R7 includes an optional set of control plane protocols that are applicable for AAA and policy coordination within an ASN-GW.

- *R*8 (*Reference Point* 8). R8 is the RP between BSs to ensure fast and seamless handover through direct and fast transfer of MAC context as well as data between BSs involved in handover of a certain MS. Thus, the R8 interface can facilitate the context transfer and handover optimization when the user moves from one BS to another.

Figure 13.1 illustrates the Mobile WiMAX architecture, logically demonstrated by an NRM. As briefly described earlier, the WiMAX NRM basically consists of MS, ASN, and CSN, as well as their respective interactions through RPs R1 through R8.

The Mobile WiMAX system adopts an all-IP network structure for Internet service provision. Therefore, the IP packets sent by MSs will access the Internet via the BS through the ASN-GW route. As shown in Figure 13.1, the entire WiMAX network consists of NAP (MS, BS, ASN, and ASN-GW) and NSP (CSN, various servers, and core routers).

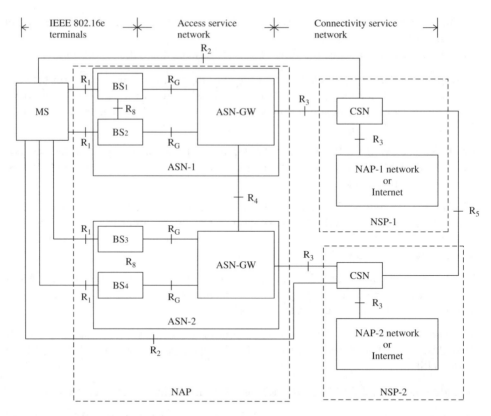

Figure 13.1 Mobile WiMAX network architecture including NRM. NAP, network access provider; NSP, network service provider; ASN, access service network; CSN, connectivity service network; ASN-GW, ASN Gateway; NRM, network reference model.

13.1.3 Key Supporting Technologies

This section covers several key technologies for supporting the mobile WiMAX system. These key supporting technologies are developed separately, combined together, and ensembled for building the Mobile WiMAX system. Each key element is introduced in the following.

Multiple Input Multiple Output (MIMO)

MIMO systems consist of multiple antenna arrays installed in both the transmitter and the receiver. The Mobile WiMAX system adopts multiple antenna technologies, that is, MIMO and beamforming (BF) to improve system throughput. The MIMO technique has drawn considerable attention to improve the channel capacity over SIMO systems by taking the spatial multiplexing effect and to increase the reliability due to the space diversity effect.

Spatial multiplexing was designed to transmit/receive multiple independent signals with the same frequency simultaneously by employing multiple transmit antennas and multiple receive antennas. First, the original signal is split into multiple bit streams before transmission at each antenna, thereby increasing significantly the transmission rate per user. At the receiver, the received data streams transmitted through the multiple transmit antennas are decoded correctly using the MIMO demultiplexer. If the channel state information (CSI) is available at the transmitter, it can help to further enhance the capacity in such a way that the total throughput can be maximized. Space diversity is devised to combine multiple data signals through an array of receive antennas that were transmitted from the source and traveled through different channels.

The arrayed antennas will also control the phases of signals to transmit toward a desired direction selectively in order to minimize the effects of interference and to combine the signal components at the desired destination in a constructive manner. This process expands the coverage and increases the throughput to the desired level that amounts to decreased interference and increased CINR (carrier-to-interference and noise ratio).

Multiple antenna technologies enable high capacities for Internet and multimedia services to mobile users, and also enhance the reliability substantially with longer transmission ranges. In practice, however, the realization of high MIMO capacity in actual radio channels is sensitive not only to the fading correlation but also to the structure of the scattering in the propagation environment. The adaptive antenna system (AAS) provides spatial division access by using multiple antennas in array. The multiple antennas used in AAS provide spatial processing gain, and the resulting antenna diversity decreases the multipath interference as well as adjacent-cell interference. As such, the AAS can help maximize the power of the desired signals while minimizing the power of the interfering signals by the directional BF effect. In fact, the two major performance degradation factors in mobile communications are the channel fading and co-channel interference.

Orthogonal Frequency Division Multiple Access (OFDMA)

OFDMA is a variation on the OFDM theme. In OFDMA, multiple access is realized by providing each user with a fraction of the available number of subcarriers. Orthogonality implies that a subcarrier is not affected by another subcarrier, which is guaranteed because all the other subcarriers take 0 value when any of the subcarriers takes the peak value (refer to Section 6.6 for a detailed description on an OFDM system). OFDMA provides multiple access by dividing the frequency band into a large number of subcarriers, which are shared by one or more users. Thus, OFDMA divides the given frequency band into multiple subcarriers, each of which is equally spaced, and modulates the user data on the subcarriers.

OFDMA has various advantages over other multiple access technologies such as FDMA, TDMA, and CDMA. FDMA was used in 1G analog communication systems, TDMA was adopted in GSM system (2G), CDMA was adopted in the cdmaOne IS-95A/B technologies (2G and 2.5G), and CDMA technologies were applied to the IMT 2000 3G systems such as WCDMA (UMTS) and CDMA2000 1x series. The development of 4G future system will rely on the MIMO/OFDMA technologies.

The OFDMA-based communication system divides the frequency band into a large number of subcarriers and processes data stream according to OFDMA communication processing in both the transmit direction and the receive direction as follows:

- *Transmit direction processing.* Input MAC data encoding/modulation (including randomization, FEC (forward error correcting), interleaving, and repetition for QPSK (quadrature phase shift keying), multiple OFDMA subcarrier mapping, inverse discrete Fourier transform (IDFT), filtering, D/A conversion, and RF transmit signal.

- *Receive direction processing.* Received RF signal, A/D conversion, filtering, DFT, subcarrier demapping, decomposition, and output data stream.

In conclusion, the principle of OFDMA is reiterated in the following. Multiple access refers to a system mechanism of sharing a communication link among multiple users. In a multiaccess communication system where transmission resources are shared among multiple users, there are two well-known techniques for RRM based on the principle of time-sharing and frequency-sharing. When FDMA is combined with OFDM, it is called *OFDMA.* The reason for sharing the radio resource in a multicarrier system is to allocate subcarriers and bit streams among users across time and frequency domains.

For the design issues of the OFDM(A) system, the following parameters are needed: the number of subcarriers, guard time, symbol duration, subcarrier spacing, modulation type per subcarrier, and the type of FEC coding. Consequently, the choice of parameters will be influenced by the system requirements such as available bandwidth, required bit rate, tolerable delay spread, and Doppler values. An OFDM signal consists of a sum of subcarriers that are modulated by using PSK or QAM (Quadrature Amplitude Modulation).

Adaptive Modulation and Coding (AMC)

AMC techniques are employed to strengthen the Mobile WiMAX system for increasing the system performance in the mobile wireless environment. For an effective operation of AMC, channel coding techniques are employed in conjunction with the modulation techniques. Coding techniques ranging from QPSK (low efficiency) to 64-QAM (high efficiency) were employed by the Mobile WiMAX to further improve the level of system capacity. Depending on the channel status, AMC should select the 64-QAM and turbo-encoding processes for the good channel condition and the QPSK modulation and convolutional coding technique for the poor channel situation. A high modulation index offers a large throughput and can be easily exploited by users near the BS where the signal-to-noise ratio (SNR) is high. For an effective operation of AMC, each MS should report its channel status to the BS for its channel estimation by using the channel quality indicator (CQI), with which type of modulation and coding techniques should be employed. The carrier-to-interference ratio (CIR) will increase as the MS moves close to the BS and decrease as the MS approaches the cell boundary.

In Mobile WiMAX, the PHY layer defines various combinations of modulation and coding rates providing a fine resolution of data rates to be used as part of link adaption. The 802.16 standard specified multiple channel coding schemes, combined with HARQ Chase.

Mobility Support

Mobile WiMAX supports mobility, inherited from the IEEE 802.16e system, that arises whenever the geographical region is divided into multiple cells and frequency spectrum is used in a repetitive manner over different cells.

The key element in mobility support is a handoff operation that refers to the converting service operation from one MS/BS wireless link to another MS/BS wireless link, maintaining continuous communication connection for providing the QoS while the MS moves from one cell to another.

There are two kinds of handover (or handoff) techniques: hard handover and soft handover. The hard handover in the Mobile WiMAX system is optimized for IP data traffic, whereas the soft handover is suitable for voice traffic in CDMA networks, but the soft handover is also used in the Mobile WiMAX system. The Mobile WiMAX operation supports the sleep mode by allowing the MS to save power from serving air interface while not in use and the idle mode by allowing the MS to be mostly idle and listen to the broadcast messages only periodically.

13.1.4 Comparison between Mobile WiMAX Network and Cellular Wireless Network

GSM/UMTS (WCDMA) family and cdmaOne IS-95/CDMA 200 1x family cellular networks differ from Mobile WiMAX network in various aspects. The Mobile WiMAX network system is founded on basis of the MIMO/OFDMA technology, whereas the cellular network systems were designed from the CDMA technology. The cellular wireless networks were initially assigned for the circuit-switched mode for voice service only and later added with the packet-switched mode for data service, whereas the Mobile WiMAX network is solely based on the IP-packet mode technology. Such a different mode of traffic operation leads to difference in architecture and results in different network configurations.

The Mobile WiMAX network used for IP traffic service consists of only two layers of BS and ASN-GW, while the circuit-switched or packet-switched cellular networks comply with four stacks of access network elements, that is, BTS (base transceiver station), RNC (Radio Network Controller), SGSN (serving GPRS Support Node), and GGSN (gateway GPRS Support Node). In fact, the simplicity of Mobile WiMAX configuration leads to many applications economically for supporting IP services much faster and more effectively than the existing cellular mobile networks.

Seamless mobility can be achieved by enabling mobile terminates to conduct seamless handovers in a transparent manner across the mobile WiMAX and 3GPP access networks. One of the challenges associated with the integration of Mobile WiMAX and 3GPP accesses is to address the various aspects of WiMAX-3GPP interworking within common network architecture, such as integrated AAA mechanisms, ETE QoS, intercellular handover, and interoperated roaming. Ongoing activities in both 3GPPs and WiMAX Forum are constantly improving and evolving the mechanisms for integration and interworking of Mobile WiMAX in 3GPP networks. This integration will be facilitated by the evolved WiMAX packet network architecture.

13.2   WiBro (Wireless Broadband)

WiBro is a Korean version based on the Mobile WiMAX technology and IEEE 802.16e standards for mobile broadband wireless access. Some technical specifications of the IEEE 802.16e standards were adopted to form the unique WiBro system having all-IP network structure tailored for Internet service provision.

The WiBro system was devised to overcome the data rate limitation of mobile phones and to add mobility to broadband Internet access. WiBro was designed to provide a variety of services for information as well as multimedia contents through the fixed line Internet access via various user terminals such as PC, PDA, and mobile phone. WiBro integrates various services in wire line Internet access and wireless mobile data to accommodate increasing demands for Internet services in stationary or mobile environments.

The WiBro system based on the Mobile WiMAX technology operates at 2.3-GHz bands with 10-MHz bandwidth and employs OFDMA for multiple access and TDD for duplexing. WiBro enables a seamless data communication even when moving at a speed of 60 km h^{-1}.

In 2004, the Ministry of Information and Communication (MIC) of the Korean Government and the TTA (the Korean standardization body) issued the basic requirements on WiBro and issued WiBro licenses to two operators, namely, KT and SK Telecom, in 2005. ETRI initiated a prototype WiBro system development in December 2004. In November 2004, Intel and LG Electronics executives agreed to ensure compatibility between WiBro and WiMAX technology. In 2005, Samsung Electronics developed the world's first commercial Mobile WiMAX system based on the 2.3-GHz band and KT deployed the WiBro network based on the 27-MHz band. In September 2005, Samsung Electronics signed a deal with Sprint Nextel Corporation to provide equipment for a WiBro trial. In June 2006, KT tried to deploy the 2.3-GHz WiBro network in the Seoul metropolitan area and launched full services of commercial WiBro in the Seoul area and its vicinity, which was the world's first WiBro deployment, in April 2007. Sprint (US), BT (UK), KDDI (JP), and TVA (BR) have or are trailing WiBro.

13.2.1   WiBro Network Architecture

Figure 13.2 depicts the architecture of the WiBro network. The WiBro network consists of MSs (MS or WiBro-AT (Access Terminal)), an ASN, and a CSN. The ASN is composed of BSs (or WiBro-AP) and ASN-GW (WiBro-PAR (packet access router)) similar to the Mobile WiMAX system. Many BSs (WiBro-APs) are located in the user side of ASN and connected to ASN-GW (or PAR), and the PAR is connected to CSN and the IP networks that contain many types of services such as AAA, HA, and network management and operations (DHCP, DNS, PCRF). CSN is connected with ASN via a router or switch.

ASN-GW (or WiBro-PAR) is the central entity of the WiBro network. It connects the CSN and multiple BSs (or WiBro-AP), which enables BSs to interwork with CSN and IP networks. Thus, ASN-GW (or WiBro-PAR) provides the functional role for transferring traffic between MSs and the external Internet. WiBro-PAR (ASN-GW) also executes the combined functions for controlling the WiBro users, mobility, and services. More specifically, the various functions of ASN-GW include handoff control, IP routing and mobility

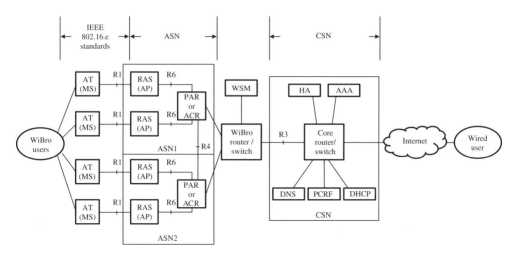

Figure 13.2 WiBro network architecture. AT, Access Terminal; AP, access point; RAS, Radio Access Station; PAR, packet access router; ACR, Access Control Router; ASN, access service network; WSM, WiMAX System Manager; HA, Home Agent; AAA, authentication, authorization, and accounting; DHCP, Dynamic Host Configuration Protocol; DNS, Domain Name Service; PCRF, Policy and Charging Rules Function; CSN, connectivity service network.

support, and user service profile for data traffic, billing service provision, information security issuer, and interworking with the HA of CSN for IP services.

WiBro is an effective communication system, providing better mobility compared to WLANs and a higher-data-rate service than cellular networks. WiBro supports all IP-based open architecture enabling easy insertion of new IP-based technology. In addition, WiBro can offer delay-sensitive applications such as VoIP (Voice over IP) as well as MoIP (Multimedia over IP). All IP-based WiBro services offer a variety of Internet contents and allow for straightforward interoperability with common data networks.

13.2.2 Key Elements in WiBro System Configuration

Before discussing key entities in the WiBro system, it is worth presenting comparative terms used between the two systems, WiMAX versus WiBro, as shown in Table 13.1.

The WiBro system consists of the following four entities: AT, RAS (Radio Access Station) (or BS), ACR (Access Control Router), and CSN. WiBro-AT is the end terminal of the network. RAS is an access entity that connects ACR and AT, the ACR connects the CSN and RAS in the ASN, and CSN is composed of various server units.

1. *WiBro Access Terminal (AT or MS).* WiBro-AT is the end user device that performs the input/output functions to process the data traffic with RAS via OFDMA, to access IP-based WiBro networks, and to perform various terminating functions, that is, wireless access to ASN, MAC processing, handover, IP-based call services, IP mobility support, radio link control, and the security provision for MS and subscribers.

Table 13.1   Comparative terms of two systems

WiMAX system	WiBro system
MS	AT
BS	AP or RAS
ASN-GW	ACR or PAR
ASN	ASN
CSN	CSN

MS, mobile station; BS, base station; ASN-GW, access service network gateway; ASN, access service network; CSN, connectivity service network; AT, Access Terminal; AP, access point; RAS, Radio Access Station; ACR, Access Control Router; PAR, packet access router.

2. *RAS (AP or BS)*. RAS is an access system that connects ACR and WiBro-AT (or MS). RAS receives subscriber data via wireless path, passes the data to the ACR for UL, and distributes the data received from the ACR to WiBro users in the DL. The RAS performs packet retransmission, packet scheduling, bandwidth allocation, ranging, and handover.

3. *ACR*. ACR connects the CSN and RAS through the RPs R_3 and R_6. The ACR performs the transferring function from data traffic between the RAS and public Internet. It connects multiple RASs to CSN and IP network and sends and receives data traffic between the Internet and MS. ACR also provides the controlling functions for WiBro user, services, and mobility. Specific ACR functions include handoff control to guarantee high mobility within its control area, IP routing and mobility support, user service information provision, billing service, and security function. In order to execute such functions, an ACR should be constructed based on the Gigabit Ethernet switch and connected to RASs based on IP.

4. *CSN*. CSN is composed of various servers, namely, the AAA server for security and accounting, the HA server for management of home address, the DNS server for conversion of IP addresses, the DHCP server for IP allocation, and the PCRF server for managing the service policy and for QoS setting. To reiterate specifically, HA accesses other networks and enables MIP users to access the Internet. HA interworks with ACR that performs a foreign agent (FA) function in mobile IPv4 environment and interworks with MS to exchange data in mobile IPv6 environment. The AAA server interfaces with the ACR and carries out subscriber AAA functions.

5. *Mobile WiMAX System Manager (WSM)*. WSM supports the management role for the network operators to supervise and maintain the ACR and RAS functionality. The interface between an ACR and a WSM is complied with the IETF standard's SNMP.

6. *RPs in WiBro System*. The interface between an ACR and the RAS in the same ASN has an RP R_6. The R_6 interface consists of a signaling plane (UDP) and a bearer plane (GRE (genetic routing encapsulation)). The interface between an ACR and another ACR in different ASNs is specified as the R_4 interface. The R_4 interface consists of a signaling plane (IP/UDP/R_4) and a bearer plane (IP/GRE).

13.2.3   System Comparison between HSDPA and WiBro

HSDPA is the most competitive system, standardized by 3GPP, evolved from WCDMA Release 5. WiBro system was launched by Korean operators at the 2.3-GHz band and 10-MHz bandwidth. Both systems use the 10-MHz bandwidth, HARQ, AMC, and QPSK, 8-PSK, and 16-QAM and 64-QAM using $R = \frac{1}{2}$ and $R = \frac{3}{4}$ turbo coding. HSDPA adopts the FDD and WCDMA schemes, while WiBro uses the TDD and OFDMA. TDD enables asymmetric radio assignment using the $2:1$ ratio for DL and UL in 6 and 3 MHz each, whereas both DL and UL use the 5-MHz bandwidth in HSDPA. Channel structure of WiBro is based on OFDMA, but HSDPA is based on WCDMA.

WiBro has 846 subcarriers and 42 symbols in a frame. WiBro's TTI (Transmission Time Interval) is 5 ms. A total of 27 symbols for DL and 15 symbols for UL are used, and 3 symbols from each link are used for control and signaling; 24 and 12 symbols are used for traffic transmission, respectively.

In WiBro system, AT (MS, handset) selects the CQI obtained by measuring CINR and PER. RAS (BS) decides the MCS (modulation and coding scheme) upon receiving CQI from WiBro-AT every 5 ms. In HSDPA system, UE (or MS) selects the CQI from measuring CINR and PER. Node B (BS) decides the MCS upon receiving CQI from UE for every HSDPA TTI of 2 ms. AT and RAS in WiBro system have the same functions as the UE and Node B in the HSDPA scheme. Hence, data traffic process in WiBro is similar to that of HSDPA despite the different channel structures. The HSDPA system of WCDMA Release 5 provides high-data-rate service on DL. HSDPA uses HS-PDSCH (high-speed packet downlink scheme channel) for flexible high-data-rate service by AMC. But WCDMA (UMTS) uses fixed MCS.

13.2.4   Key Features on WiBro Operation

To reiterate key features of the WiBro system, a few conclusive remarks are described in the following.

The UE of cellular-based system is equivalent to the AT of WiBro system. WiBro-AT provides interface to users. The WiBro system employs OFDMA for modulation and demodulation and TDD node for user multiplexing. Such a WiBro system guarantees QoS by classifying packets for priority-based transmission, that is, AT queues UL data from an application layer (L 3) before transmission.

The WiBro layer structure at the AT includes MAC layer (L 2) and a PHY layer (L 1). A UL data generated from an application layer is transmitted to the PHY layer through the MAC layer. The MAC layer controls access of the UL data to the PHY layer. The PHY layer performs radio data communication including modulation/demodulation and an RF process, while the MAC layer above the PHY layer classifies UL data packet from the application layer, stores the classified UL data packets, and transmits the stored data packet to the PHY layer for priority-based transmission according to the QoS policy.

Figure 13.3 illustrates the WiBro frame structure. The WiBro system adopts OFDMA/TDD for multiple access and duplexing. The TDD frame length is 5 ms and is segmented into a set of symbols. The frame structure is fixed as 27 symbols for the DL subframe and 15 symbols for the UL subframe. There are 864 subcarriers

Preamble (1 symbol)	MAP	DL bursts (26 symbols)	UL control (3 symbols)	UL bursts (12 symbols)

Figure 13.3   The WiBro OFDFA/TDD frame structure.

with the 1024 FFT (fast Fourier transform) sign within 10-MHz bandwidth. In WiBro specification, a subchannel is composed of a set of subcarriers. The number of subchannels is 16 in the full usage subchannel (FUSC). A bit stream carried by a subchannel during a symbol time is ended into a slot. The unit symbol time of a single subchannel is defined as a *slot*. This slot is called the *logical encoding unit*.

Referring to Figure 13.3, the DL subframe consists of 26 data bursts (symbols) and logical MAPS for DL, while the UL subframe consists of 12 data bursts and logical MAPS for UL. The MAP (maximum a posteriori) consists of the FCH (frame control header) and the downlink map (DL-MAP). The MAPs guide the SSs (subscriber stations) in decoding the following data bursts. The burst is a set of actual data slot that are allocated by a WiBro-AP (BS), for either DL or UL. Thus, SSs are responsible for which subchannel decodes data for DL and encodes for UL. The WiBro-AP (BS) should be responsible for organizing the MAPs and the data burst in every frame.

WiBro has the same MAC message format as the 802.169 format. Actually, no differences are found in their functions and MAC message formats between WiBro and 802.168 statements. The MAC message format is composed of user payload, 6-bytes fixed MAC header, optional 4 bytes CRC, and optional 12-byte encryption data.

The UL transmission is operated in the bandwidth request and grant mechanism. The SS should request a certain amount of bandwidth to the WiBro-AP (BS) first when it has some data to transfer, and the BS allocates UL bursts to the SS after scheduling decisions.

13.3 UMB (Ultra Mobile Broadband)

The UMB system is the latest member of the CDMA2000 family, a successor of the CDMA2000 1xEV-DO. It was designed for one of the next 4G mobile broadband services. UMB is designed to provide broadband connectivity to users at speeds compatible with fixed line networks such as Ethernet and Cable/DSL. UMB's IP-based mobile broadband standard enables peak download data rate of 288 Mbps with 20-MHz bandwidth and an upload data rate of 75 Mbps with 20-MHz bandwidth. UMB achieves very high data rates with high spectral efficiency and incorporates advanced communication techniques, such as lower latencies, using advanced modulation, link adaption, and HARQ for high performance with user mobility.

UMB is an IP-based OFDMA mobile broadband system designed for high-speed data rate and VoIP capability in a mobile environment that has been standardized in 3GPP2. The UMB forward link (FL) uses MIMO antennas to achieve higher system capacity and peak data rates. UMB uses OFDMA as the main modulation technique, using sophisticated control and signature mechanisms, RRM, adaptive reverse link (RL) interference management, and advanced antenna techniques such as MIMO, SDMA (space division multiple access), and BF.

The UMB solution addresses universally a large cross-section of advanced mobile broadband services by economically delivering low-rate, low-latency noise traffic at one end of the spectrum, just as efficiently as ultra-high-speed, latency-insensitive broadband data traffic at the other.

UMB provides interference management through fractional frequency reuse (FFR) for optimizing bandwidth utilization, and power control for improved coverage for edge user performance.

Coupled with the current trend toward IP-based architecture, UMB is attractive to operators seeking to future-proof their networks. Since UMB can operate in a wide range of bandwidth from 1.25 to 20 MHz, this flexibility of various deployment spectrum enables an operator to customize a UMB system for the spectrum available to the operator. Using UMB, an operator can efficiently deliver a range of multimedia and VoIP applications across multiple devices, creating exciting revenue opportunities across all the market segments.

13.3.1 Design Objectives of UMB

UMB is assigned to complement 3G deployments in order to fulfill the increasing demand efficiently for performance enhancement over existing cellular systems. At present, UMB is equipped with all the necessary features for optimal support of real-time and best effort traffic with seamless mobility.

Most important objectives for performance improvement include higher peak data rates, better spectral efficiency, lower latency, fast connection time, higher capacity, dynamic FFR, ability to seamlessly handoff between cells, and improved terminal battery life.

The UMB's high rates enable a broadband network to serve its user with consistently high data rates regardless of the user's location within the cell. Higher cell capacity can allow more simultaneous users within the network to deliver applications such as VoIP and video telephony. Higher capacity also provides higher data rates to users for both high-capacity voice and broadband data in all environments. Thus, UMB can be deployed by operators to effectively offer both VoIP and mobile broadband access. UMB delivers high capacity, increased user data rates, and lower latencies, along with an IP-based network architecture for flexible deployment. Hence, UMB is well suited to assign broadband applications with faster, more capable mobile multimedia devices.

Use of multimode devices enables seamless mobility between UMB and existing networks. A long terminal battery life will be certainly crucial for an optimal user experience.

13.3.2 Key Technologies Applicable to UMB

There are a number of key technologies to support UMB system design for improving scalability and deployment flexibility. UMB uses the following techniques to offer significant competitive advantages.

OFDMA

The multiuser version of OFDM is called *OFDMA*. UMB utilizes OFDMA technique to accommodate both the DL and UL traffic. OFDMA frame structure is equipped with separate time slots to accommodate the DL/UL. OFDMA transceivers are easily implemented using FFT and IFFT (inverse fast Fourier transform) techniques. OFDMA uses FFT to divide the allocated bandwidth into smaller orthogonal subcarriers that can then be shared among the users through parallel transmission. This division may provide tolerance to noise and multipath, while enabling more efficient use of bandwidth allocation. The ability to use different FFT sizes for OFDM modulation allows implementation across multiple bandwidth allocations. OFDMA certainly promotes spectral efficiencies with a simpler implementation when used in wider bandwidths.

OFDMA appears to be quite an attractive transmission scheme to overcome the impairment of wireless broadband channels.

Every portion of input data is transmitted on one of the available subcarriers. The signal modulation on each subcarrier is orthogonal to others and the data can be restored successfully at the receiver. All multipath signal elements can be combined without using sophisticated RAKG structures at the receiver.

OFDM is a variant of the FDM scheme in which the frequency band is divided into multiple smaller subchannels. In FDM, guard bands should be inserted when subchannelization takes place between two subchannels in order to avoid interface between them, whereas OFDM divides the frequency bandwidth into narrow orthogonal subcarriers. A subchannel is an aggregation of a number of subcarriers. These subcarriers are put together and a subchannel is created. The subcarriers include data carriers, pilot carriers, and a DC. The data carriers are used to carry data, the pilot carriers are used for channel sensing purposes, and a DC carrier is marked to the center of the channel.

MIMO Antenna Technologies

UMB incorporates a number of advanced multiple antenna techniques such as MIMO, SDMA, and BF to achieve much higher user data rates. In MIMO, the transmission implies transmitting multiple simultaneous data streams to the users. A MIMO transmission can significantly increase a user's data rates, if the user's mobile environment has a large degree of scattering and the received signal strength from the BS is strong. The key idea is that a transmitter sends multiple data streams on multiple transmit antennas and each transmitted stream goes through different paths to reach each receiver antenna. The benefit of using MIMO is that the data rate gains are achieved without allocating any additional power or bandwidth to the user. The salient features of MIMO is to offer higher throughput for a given bandwidth and higher link range for a given power value.

Cell-edge users typically do not benefit from MIMO. For cell-edge users with lower receive signal strengths, the UMB should use BF techniques to increase their data rates by focusing the transmit power from the BS in the direction of the users.

SDMA is another powerful antenna technique that allows a BS to transmit simultaneously to multiple users that are specifically separated. Thus, even though the transmissions are simultaneous, the interference caused to the users is minimal. The intersector interference caused by simultaneous transmission can be suppressed using the multiple antennas at the BS. This technique certainly increases the radio link capacity. Allowing concurrent transmissions increases the sector capacity accordingly. There is no additional terminal complexity required for SDMA. UMB will provide the necessary UL feedback channels for efficient SDMA operation because user feedback is needed in an FDD deployment scenario.

Adaptive Interference Management Techniques

UMB uses some other techniques to enhance and allow unplanned frequency as described below:

- *FFR.* FFR technique can be used to improve cell-edge performance. Users at cell-edge are put in reuse with respect to the strongest interfering sectors. Both in the RL and FL, subcarriers should be dynamically allocated to users for efficient interference control.

- *Flexband.* Flexband is a universal frequency reuse scheme, in which each sector uses an intelligent power tiring among subscribers. This method increases cell-edge data rates without affecting system capacity. Unlike traditional frequency reuse schemes that completely shut off unused subcarriers in one sector, Flexband uses all subcarriers in all sectors. In the Flexband method, users close to the BS are assigned lower power subcarriers, while those away from the BS are assigned higher power subcarriers.

- *Disjoint link support.* A terminal is allowed to select different FL and RL sectors independently, ensuring strongest link performance in both directions. Since the terminal is power-controlled by the strongest RL sector, this technique also minimizes the interference which the terminal is causing to other sectors.

- *Quasiorthogonal RL.* UMB allows scheduling of end users over the same RL subcarriers and using multiple antennas at the BS (eBS) to suppress the interferences caused by multiple transmissions on the same resources. This method is called *quasiorthogonal RL* in UMB and is used to enhance the capacity of the RL.

13.3.3 UMB IP-Based Network Architecture

UMB supports a scalable IP network architecture in which the traditional hieratic structure has been flattened by distributing the functionality of base station controller (BSC) to various other nodes. Figure 13.4 shows the UMB IP-based network architecture in which the session reference network controller (SRNC) can be implemented both in a centralized or a distributed manner.

- *eBS (evolved base station).* The UMB eBS function is similar to the combination of BTS and BSC in CDMA 200 1xEV-DO architecture. The eBS is responsible for IP-packet classification for over-the-air QoS, header compression, and user data ciphering.

- *SRNC.* SRNC performs the signaling functions for the UMB radio access network (RAN). It stores references to ongoing sessions, as well as manages paging and location information. It is also responsible for idle state management and acts as the authenticator for access authentication based on extensible authentication protocol (EAP).

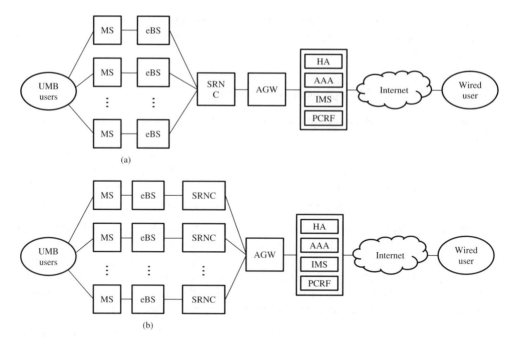

Figure 13.4 UMB IP-based network architecture. (a) Centralized scheme and (b) distributed scheme. eBS, evolved base station; SRNC, session reference network controller; AGW, Access Gateway; HA, Home Agent; AAA, authentication, authorization, and accounting; IMS, IP Multimedia Subsystem; PCRF, Policy and Charging Rules Function.

In the centralized scheme, there is a single SRNC connected to several eBSs and Access Gateway (AGW), whereas in the distributed scheme, all eBSs act as SRNCs (Figure 13.4).

- *AGW.* The AGW functionalities are similar to that of PDSN (packet data serving node) in EV-DO networks. AGW tunnels IP packets for user data to the eBS, behaves as the AAA client for accounting, and acts as the FA if applicable. AGW also acts as the QoS enforcement and intrusion detection/prevention AP.

13.3.4 Conclusive Remarks

UMB is designed to complement 3G deployments. Some of the design features that distinguish UMB from other OFDMA-based systems are lower latency handovers, tight interference control techniques, efficient signaling mechanism, and efficient tunneling mechanisms to support seamless handovers between UMB and CDMA2000 1xEV-DO. Use of multinode devices enables seamless mobility between UMB and existing networks. UMB delivers high capacity, increased user data rates, and low latencies, along with an IP-based network architecture for flexible deployments.

In November 2008, Qualcomm, a leading UMB sponsor, announced that it was ending the development of the technology, favoring LTE instead (refer to "Qualcomm halts UMB project", Reuters, November 13, 2008). It is unfortunate to hear that the UMB development project was terminated; hopefully, the UMB supporters will continue the project seeking to future-proof driven network by bringing in high-quality mobile broadband access to the mass market.

13.4   LTE (Long Term Evolution)

LTE is one of the evolved 4G technologies to succeed UMTS/HSPA (high-speed packet access) (i.e., UTRA/UTRAN) systems, supporting a range of bandwidth up to 20 MHz with peak data rate capabilities of up to 100 Mbps for DL and up to 50 Mbps for UL. Scalable bandwidths in LTE are 1.25, 1.6, 5, 10, 15, and 20 MHz. These scalable bandwidths will suit the adequate needs of different network operators that have different bandwidth allocations and also allow operators to provide different services based on the spectrum.

LTE can be a reduced number of network elements for simplifying the network architecture. LTE is a leading OFDMA-based wireless mobile broadband technology designed to provide interoperability and service continuity with existing UMTS/HSPA systems. LTE used OFDMA modulation/multiple access on the DL and SC-FDMA (single-carrier FDMA) on the UL. LTE also incorporates advanced antenna techniques such as MIMO, SDMA, and BF. LTE supports both FDD and TDD modes, allowing network operators to address all the available spectrum resources.

The LTE network supports handoffs to UMTS/HSPA and GSM/GPRS networks, providing service connectivity throughout the operator's network. LTE is ongoing to design a high-capacity mobile wireless technology for efficient delivery of real-time applications such as VoIP and various video services. High QoS can be achieved using licensed frequencies to guarantee QoS, thus reducing substantially U-plane/C-plane latency as well as round-trip delay.

13.4.1   LTE Features and Capabilities

LTE provides several key features and capabilities that enable the network operator to deliver enhanced mobile broadband services for efficient delivery of real-time applications.

- *OFDM/MIMO data transmission.* LTE uses OFDMA as its radio access technology, together with advanced antenna technologies. LTE uses the OFDM transmission technique for the DL from the enhanced BS (eNodeB) to the mobile terminal (UE). OFDM uses a large number of narrowband subcarriers for multicarrier transmission. OFDM splits data into a large number of lower-rate bit streams that are modulated onto individual subcarriers. The DL modulation schemes include QPSK, 16-QAM, and 64-QAM. Subcarrier frequencies are equally spaced by the symbol rate, and this separation makes them orthogonal by removing any cross talk.

Each user is allocated with a number of so-called resource blocks in a time and frequency dimension. The basic LTE DL physical resource can be shown by a time/frequency grid. One resource block consists of 1 slot and 12 subcarriers in the time/frequency dimension. One slot in the time element on OFDM resource blocks consists of seven symbols and slot duration time $T_s = 0.5$ ms. One resource block is composed of 1 slot in the time element ($T_s = 0.5$ ms) and 12 subcarriers in the frequency element. Therefore, 1 resource block consist of 84 ($= 7 \times 12$) resource elements. Thus, the OFDM symbols are grouped into resource blocks that have a total size of 180 kHz in the frequency domain and 0.5 ms in the time domain.

13.4.2   LTE Frame Structure

The LTE frame structure is used with FDD. As shown in Figure 13.5, the LTE frames are 10 ms in duration. These frames are divided into 10 subframes. Each subframe is 1.0 ms long and is further divided into two slots, each slot is 0.5 ms in duration. Each slot usually consists of seven OFDM symbols. The cyclic prefix (CP) is a key parameter with OFDM, which is defined as the *guard period* between symbols to combat multipath dispersion and to reduce intersymbolic interference (ISI). The CP is used repeatedly at the end of each symbol to improve signal reception.

13.4.3   LTE Time-Frequency Structure for Downlink

LTE uses OFDMA as its radio access technology, together with advanced antenna technologies. LTE utilizes the OFDM technique that uses a large number of narrowband subcarriers for multicarrier transmission. OFDMA is an excellent choice as a multiplexing scheme for the LTE DL.

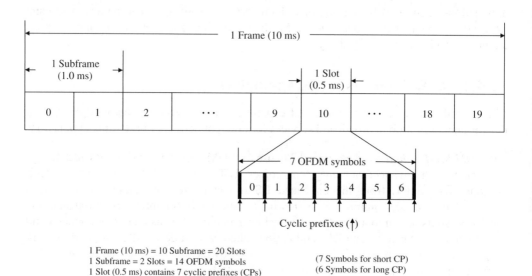

Figure 13.5   LTE frame structure.

In OFDMA, users are allocated a specific number of subcarriers for a predetermined amount of time. The total number of available subcarriers depends on the overall transmission bandwidth of the system. A physical resource block (PRB) is defined as consisting of 12 consecutive subcarriers for one slot ($T_{slot} = 0.5$ ms) in time duration. Hence, PRBs have both a time and frequency dimension. Thus, a PRB is the smallest element of resource allocation assigned by the BS scheduler.

Each user is allocated with a number of so-called resource blocks in both time and frequency domains. The basic LTE DL physical resource can be shown by aggregation of a time/frequency resource element. One resource block consists of one slot and 12 subcarriers in the time/frequency dimension. One slot in the time element on OFDM resource blocks consists of seven symbols and slot duration time $T_s = 0.5$ ms. One resource block is composed of one slot in the time element ($T_s = 0.5$ ms) and 12 subcarriers in the frequency element. Therefore, one resource block consists of 84 ($= 7 \times 12$) resource elements. Thus, the OFDM symbols are grouped into resource blocks that have a total size of 180 kHz in the frequency domain and 0.5 ms in the time domain.

As shown in Figure 13.6, the subcarriers are very tightly spaced to make efficient use of a variable bandwidth. Therefore, there is virtually no interference among adjacent subcarriers. Since spacing between the subcarriers is $\Delta F = 15$ kHz, the OFDM symbol duration time is actually $1/\Delta F + CP$. This CP is used to maintain *orthogonality* between the subcarriers. The CP length of 4.7 μs will be sufficient for handling the delay spread for most unicast scenarios.

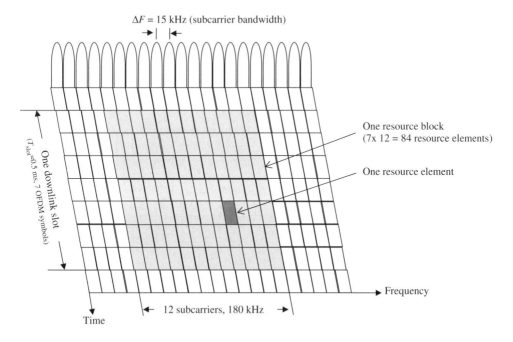

Figure 13.6 LTE downlink time-frequency resource block based on OFDM.

13.4.4   LTE SC-FDMA on Uplink

Unlike the OFDM modulation scheme used for the DL, LTE uses SC-FDMA as the basic transmission scheme for the UL. The principle advantage of SC-FDMA over conventional OFDM is a lower PAPR (peak-to-average power ratio) than would be possible using OFDMA. A low PAPR also improves coverage and the cell-edge performance. However, an optimized OFDMA implementation mitigates any issues and provides similar performance and benefits as SC-FDMA.

Figure 13.7 illustrates the LTE UL SC-FDMA signal chains, representing transmitter/receiver arrangement. SC-FDMA is well suited to the LTE UL requirements. The functional blocks in the transmit path are explained in the following list.

- *SC – constellation mapping.* This mapping process converts incoming bit stream to single-carrier symbols by means of BPSK (binary phase shift keying), QPSK, or 16-QAM techniques depending on channel conditions.

- *Serial/parallel (S/P) converter.* The S/P converter formats SC symbols into blocks for the M-point DFT (or FFT).

- *M-point DFT.* It converts SC symbol blocks into M-discrete tones.

- *Subcarrier mapping.* The subcarrier mapping maps DFT output tones to specified subcarriers for transmission. This mapping will be performed in the frequency domain. SC-FDMA systems seem to use localized subcarrier mapping for contiguous tones, rather than using distributed subcarrier mapping.

- *IDFT.* All the subcarriers for each user are the inputs for an IDFT (or FFT) that converts mapped subcarriers back into time domain for transmission.

- *CP insertion.* A CP is inserted into the composite SC-FDMA symbol to mitigate multipath effects. In order to combat against the multipath fading, IDFT reverts back to the time-domain signal for protection by a CP.

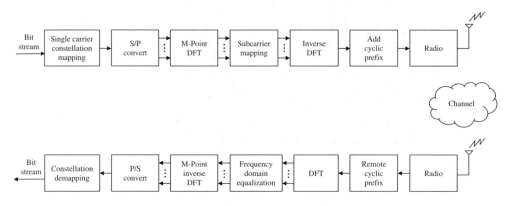

Figure 13.7   LTE uplink SC-FDMA signal chains connecting transmitter/receiver pair.

- *Radio.* The radio block converts digital signal to analog and upconverts to RF for transmission.

In the receive side chain, the process is essentially reversed.

Unlike OFDMA, the underlying SC-FDMA signal represented by the discrete subcarriers is surprisingly an SC. As a result, SC-FDMA is distinctly different from OFDM because SC-FDMA subcarriers are not independently modulated.

13.4.5 LTE Network Architecture

GSM evolution starts with GPRS before moving to EDGE and UMTS (known as *WCDMA*), and now HSPA (HSDPA/HSUPA (High-Speed Uplink Packet Access)) pairs to taking the next step to 4G LTE for achieving further substantial leaps in wireless mobile technology. The global evolution of UMTS/HSDPA Release 6 takes the step forward to Release 7 for future beyond HSPA; that is, its continued development with HSPA will lead to LTE. Technology milestone and advances in the evolution of UMTS/HSPA will continue to develop for enhancing the LTE technology.

3GPP Release 6 System is based on HSDPA and enhanced UL. The underlying LTE technology is rapidly emerging as one of the strong 4G systems and will become a major promoter in the future cellular industry. UMTS Release 6 mobile network architecture has been split between the RAN and the Core Network (CN). The RAN includes the Node B (BS) and the RNC, with each RNC managing multiple Node BSs.

The CN includes the Mobile Switching Center (MSC) for voice services and the SGSN for data services, with separate GW nodes providing access to external networks such as PSTN (public switched telephone network) and the Internet.

As shown in Figure 13.8, LTE adopts a much simpler architecture with fewer nodes and interfaces for improved performance and lower cost. The use of multiple network nodes is replaced by enhanced base stations (enhanced Node BS, or eNBS) connected to a single AGW.

In the proposed LTE architecture, the Release 6 nodes GGSN, SGSN, and RNC are merged into a single control node, that is, AGW. GGSN acts as an anchor node in the home network, while SGSN acts as an anchor node in the visiting network for handling both mobility management and session management. RNC deals with RRM, mobility management, call control, and transport network optimization. In the proposed LTE architecture, a reduced number of network nodes and interfaces along the data path will reduce the overall protocol-related processing, call setup times, latency, and interoperability cost. This change in architecture will result in a reduction in both the processing overhead and latency, as well as integrate all of the routing and interworking functions to a single AGW. Enhanced base stations (eBSs) are helping to design different spectra from 1.25 to 20 MHz, by backing with both paired (FDD) and unpaired (TDD) spectra.

The control and user planes for the UE are terminated in the AGW, which handles the CN functions provided by the GGSN and SGSN in the Release 6. The control plane protocol for UE is similar to RRC in the Release 6 for handling control of mobility and radio bearer configuration. In the user plane, the AGW handles header compression, ciphering, integrity protection, and ARQ.

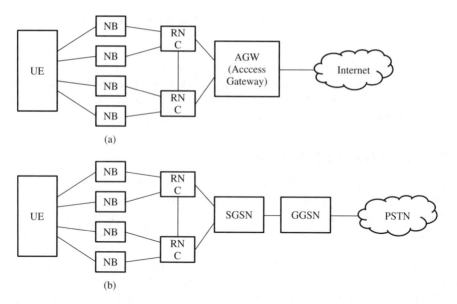

Figure 13.8 (a) LTE network architecture evolved from (b) 3G network of Release 6 architecture. UE, desktop modems, mobile phones, laptops, UMPCs (ultramobile PCs); RNC, Radio Network Controller; SGSN, serving GPRS Support Node; GGSN, gateway GPRS Support Node; eNB, enhanced Node B (enhanced base station); AGW, Access Gateway.

For more ambitious goals, the proposed LTE architecture should have the following merits:

- Complexity should be reduced due to fewer nodes and interfaces.
- All interfaces in the architecture should be IP interfaces.
- Call/bearer setup time should be reduced.
- User and control plane latency should be reduced due to fewer nodes.
- One of the benefits LTE will bring is a reduction in latency time for enhancing the behavior of time-sensitive applications such as VoIP.
- LTE is expected to deliver three to five times greater capacity than the most advanced current 3G networks.
- Placing the ARQ protocol in the AGW will provide both lossless mobility and robustness against lower-layer losses.
- Encryption and integrity protection of control plane and user plane data enable the AGW to solve a security solution at least as strong as in the Release 6.
- There should be no need for a direct eNB–eNB interface for mobility.
- A new LTE radio and CN architecture should reduce network latency, improve system performance, and provide interoperability with the existing 3GPP UMTS/HSPA and 3GPP2 CDMA2000 1xEV-DO technologies, with fewer system nodes.
- The deployment of LTE infrastructure should be as simple and cost efficient as possible.

13.4.6   Key Components Supporting LTE Design

LTE offers several important benefits and a number of features that simplify the building and management of next-generation network and uses OFDM as its radio access technology, together with advanced antenna technologies.

MIMO (Multiple Input Multiple Output)

The idea of using multiple antennas in both transmit and receive ends to increase coverage, capacity, and efficiency has been the focus of extensive research for the past 15 years. The 3GPP Release 7 specification including MIMO techniques for HSTA exploits multiple antennas at both ends of the UL and DL. LTE accommodates up to four antennas on the MS and four at the BS in a 4X4 MIMO configuration.

Various MIMO techniques include SDMA using adaptive and fixed BF and spatial multiplexing using each antenna to convey different data stream. MIMO increases peak throughput by transmitting and receiving multiple data stream within the same spectrum.

SDMA enables multiple users to send and receive data using the same time-frequency OFDM resource. In the DL, the eNode B (eNB) can transmit data simultaneously, and over the same time-frequency resource, to two users that have enough spatial separation to ensure that two data streams remain orthogonal. Similarly, on the UL, SDMA enables two users in the call to simultaneously send data to the eNB, using the same time-frequency resource. Even though the transmissions are simultaneous, the spatial separation ensures that the two data streams do not interface with each other. Allowing these concurrent transmissions increases the call capacity in both the DL and the UL.

MIMO/SDMA processing exploits spatial multiplexing, allowing different data streams to be transmitted simultaneously from different antennas, to increase the end user data rate and call capacity. When the transmitter is available for the feedback information from the receiver, MIMO enables to implement BF to increase available data rates and spectrum efficiency further.

BF increases the user data rates by effectively increasing the received signal strength at the UE. BF provides the most benefits to users in weaker signal area at the edge of the call coverage. The combined use of MIMO/OFDM will greatly improve the spectral efficiency and capacity of the LTE wireless network. Thus, MIMO/OFDM techniques will be regarded as the fundamental building blocks for all future advanced 4G wireless networks.

OFDM (Orthogonal Frequency Division Multiplexing)

OFDM(A) is an excellent choice of a multiplexing scheme for the 4G LTE DL. Currently, OFDM is an extremely successful access technology deployed in a number of wireless and wire line applications including 4G network systems such as Mobile WiMAX, WiBro, UMB, and LTE for which OFDM is widely accepted as the basis for the air interface necessary to meet the requirements for the upcoming 4G mobile networks.

LTE uses OFDM for the DL from the BS to the UE terminal. OFDM meets the LTE requirement for spectrum flexibility and enables cost-efficient solution for very wide carriers with high peak rates. For the UL, LTE uses SC-FDMA, which is to compensate for a drawback with normal OFDM.

OFDM(A) techniques employ an FFT to segment the allocated bandwidth in smaller units, which can then be shared among the users. OFDM uses a large number of narrow subcarriers for multicarrier transmission.

FFR (Fractional Frequency Reuse)

Interference control by the frequency reuse scheme is aimed to improve the cell-edge data rates and capacity across the call coverage area. Fraction of the available frequency in each sector can be reused interference in each cell and improves the cell-edge performance. Interference control can be achieved by using a fraction of the available bandwidth in each cell. FFR enables a more flexible trade-off between the overall capacity and cell-edge rate: FFR operation is derived from the eNB scheduler so that the reduction of intercell interference would be influenced by the scheduler implementation.

MBSFN (Multicast-Broadcast Single-Frequency Network)

MBMS (multimedia broadcast multicast services) are performed either in a single cell or a multicell mode. Especially, in the multicell mode, transmissions from cells are carefully synchronized to form MBSFN (Multicast-Broadcast Single-Frequency Network). LTE provides a high-capacity multicast and broadcast service using MBSFN transmission scheme.

MBSFN is an elegant application of OFDM for cellular broadcast. The operational principle is quite simple as follows: identical transmissions are broadcast from closely coordinated cells simultaneously on a common frequency, and when signals from adjacent cells arrive at the receiver, these signals are dealt with the same manner as multiple delayed signals. UE can combine the energy from multiple transmitters with no additional complexity.

All cells in the network transmit time-synchronized identical DL signals. At the UE terminal, these multiple time-synchronized transmissions appear as a single transmission with high signal strength and thus can be easily decoded. MBSFN has the flexibility to provide both vast-area broadcasts and cell-specific broadcast. The capacity benefits of the single-frequency network are highest when the same content is transmitted in all cells of the micronetwork.

13.4.7 Concluding Remarks

Following the UMTS Release 6 combined with HSUPA at the end of 2004, the evolved UTRAN (eUTRAN) has been initiated in 3GPP. The prime purpose is to develop a framework for the evolution of LTE radio access technology toward wider bandwidth, lower latency, and packet-optimized radio access technology with DL peak data rate capacity up to 100 Mbps as well as UL peak data rates up to 50 Mbps.

The LTE network architecture should be characterized by primary support for PS domain, but speech services need to be handled as VoIP calls. High-capacity voice service on LTE shall be supported by VoIP over the LTE data channels. The VoIP services are interoperable with the circuit-switched voice and VoIP services on the existing UMTS networks.

The decision has been made to adopt SC-FDMA for the UL due to superior properties in terms of UL SPAPR compared to OFDM in the DL, retaining the advantage of low PAPR when compared to OFDMA. The SC-FDMA adoption is based on the use of CP (guard interval) to allow for high-performance and low-capacity receiver implementation in the BS. For the DL, the adoption of OFDMA technology enables flexible support of different bandwidth options. This technology includes frequency domain scheduling, MIMO antenna technologies, and variable coding and modulation. Channel coding for eUTRAN is attracting optimization proposals for different error correction such as turbo coding.

LTE's high performance, integrated QoS support, and low latency allow operators to target the entire range of IP services effectively. LTE is based on a simplified IP-based network architecture that improves network latency and is designed to interoperate and ensure service continuity with existing 3GPP networks.

Although it is incomplete in coverage, the LTE specifications contain a somewhat great deal of useful information. The discussion on LTE has hopefully provided the reader with a reasonably complete description. It is regrettable that LTE specifications may not contain much detail in the conclusion remark. Works on 3GPP LTE are ongoing and specific global standard will come out by the end of 2012.

Acronyms

1xEV	1x Evolution
1xEV-DO	1x Evolution Data Only
1xEV-DV	1x Evolution Data Voice
3GPP	3rd Generation Partnership Project
ADCCP	Advanced Data Communication Control Procedures
AES	Advanced Encryption Standard (Rijndael)
AH	Authentication Header
AMPS	Advanced Mobile Phone System
ANSI	American National Standards Institute
ARP	Address Resolution Protocol
AS	Autonomous System
ASN.1	Abstract Syntax Notation One
ATM	Asynchronous Transfer Mode
BER	Basic Encoding Rules
BGP	Border Gateway Protocol
CA	Certification Authority
CBC	Cipher Block Chaining
CDMF	Commercial Data Masking Facility
CDPD	Cellular Digital Packet Data, North American Protocol
CERT	Center for Emergency Response Team
CGI	Common Gateway Interface
CIDR	Classless Inter-Domain Routing
CLNS	Connectionless Network Service
CMS	Cryptographic Message Syntax
CRC	Cyclic Redundancy Check
CRL	Certificate Revocation List
CSMA/CD	Carrier Sense Multiple Access with Collision Detection
DAC	Discretionary Access Control
DARPA	Defense Advanced Research Projects Agency
DDoS	Distributed Denial of Service
DDP	Datagram-Delivery Protocol
DER	Distinguished Encoding Rules

Wireless Mobile Internet Security, Second Edition. Man Young Rhee.
© 2013 John Wiley & Sons, Ltd. Published 2013 by John Wiley & Sons, Ltd.

DES	Data Encryption Standard
DH	Diffie–Hellman
DIT	Directory Information Tree
DMDC	DES-like Message Digest Computation
DMZ	Demilitarized Zone
DN	Distinguished Name
DNS	Domain Name Service or Domain Name System
DOI	Domain of Interpretation
DS	Dual Signature
DSA	Digital Signature Algorithm
DSS	Digital Signature Standard
DVMRP	Distance Vector Multicast Routing Protocol
EBCDIC	Extended Binary Coded Decimal Interchange Code
EC	Elliptic Curve
ECC	Elliptic Curve Cryptosystem
ECDSA	Elliptic Curve Digital Signature Algorithm
ECSD	Enhanced Circuit-Switched Data
EDGE	Enhanced Data Rate for GSM Evolution or Enhanced Data Rate for Global Evolution
EFT	Electronic Funds Transfer
ESP	Encapsulating Security Payload
ESP	Encrypted Security Payload
FDDI	Fiber Distributed Data Interface
FIPS	Federal Information Processing Standards
FTP	File Transfer Protocol
GASAPI	Generic Audit Service Application Program Interface
GPRS	General Packet Radio Service
GSM	Global System for Mobile Communications
GSSAPI	Generic Security Service Application Program Interface
HDLC	High-Level Data Link Control
HIDS	Host-Based Intrusion Detection System
HMAC	Hashed Message Authentication Codes
HMAC	Hashing Message Authentication Code
HSCSD	High-Speed Circuit-Switched Data
HSDPA	High-Speed Downlink Packet Access
HTML	Hypertext Markup Language
HTTP	Hypertext Transfer Protocol
IAB	Internet Activities Board
IAB	Internet Architecture Board
IANA	Internet Assigned Numbers Authority
ICB	International Cooperation Board
ICCB	Internet Configuration Control Board
ICMP	Internet Control Message Protocol
ICV	Integrity Check Value
IDEA	International Data Encryption Algorithm

iDEN	Integrated Digital Enhanced Network
IDS	Intrusion Detection System
IESG	Internet Engineering Steering Group
IETF	Internet Engineering Task Force
IGMP	Internet Group Management Protocol
IMAP	Internet Message Access Protocol
Inter NIC	Internet Network Information Centre
IP	Internet Protocol
IPRA	Internet Policy Registration Authority
IPsec	Internet Protocol Security
IPX	Novell Internet Packet Exchange
IRTF	Internet Research Task Force
ISAKMP	Internet Security Association Key Management Protocol
ISN	Initial Sequence Number
ISO	International Organization for Standardization
ITU-T	International Telecommunication Union-Telecommunication Section
IV	Initialization Vector
KDC	Key Distribution Center
LAN	Local Area Network
LDAP	Lightweight Directory Access Protocol
LEAF	Law Enforcement Access Field
LLC	Logical Link Control
LTE	Long Term Evolution
MAC	Media Access Control
MAC	Message Authentication Code
MBONE	Multicast Backbone
MD5	Message Digest, version 5
MIC	Message Integrity Code or Message Integrity Check
MIME	Multipurpose Internet Mail Extension
MIMO	Multiple Input Multiple Output
MOSPF	Multicast Open Shortest Path First
MSP	Message Security Protocol
MTU	Maximum Transfer Unit
NBAS	Network Behavior Analysis System
NBS	National Bureau of Standards
NCSA	National Computer Security Association
NFS	Network File System
NIC	Network Interface Card
NIDS	Network-Based Intrusion Detection System
NIST	National Institute of Standards and Technology
NMS	Network Management System
NMT	Nordic Mobile Telephone
NNTP	Network News Transfer Protocol
NSA	National Security Agency
NSAP	Network Service Access Point

NVT	Network Virtual Terminal
OFDMA	Orthogonal Frequency Division Multiple Access
OM	Order Message
ORA	Organizational Registration Authority
OSI	Open Systems Interconnect
OSPF	Open Shortest Path First
PAA	Policy Approval Authority
PCA	Policy Certification Authority
PCMCIA	Personal Computer Memory Card International Association
PCT	Private Communication Technology
PDC	Personal Digital Cellular
PEM	Privacy Enhanced Mail
PGP	Pretty Good Privacy
PKCS	Public-Key Cryptography Standards
PKC	Public-Key Certificate
PKI	Public-Key Infrastructure
PM	Payment Message
POP	Post Office Protocol
PPD	Port Protection Devices
PPP	Point-to-Point Protocol
PRBS	Pseudo-Random Binary Sequence
PSRG	Privacy and Security Research Group
QR	Quadratic Residue
RA	Registration Authority
RARP	Reverse Address Resolution Protocol
RDN	Relative Distinguished Name
RFC	Request for Comments
RIP	Routing Information Protocol
RPC	Remote Procedure Call
RSA	Rivest, Shamir, and Adleman
SA	Security Association
SAD	Security Association Database
SAGE	Security Algorithms Group of Experts
SATAN	Security Administrator Tool for Analog Network
SDLC	Synchronous Data Link Control
SEAL	Screening External Access Link
SET	Secure Electronic Transaction
SHA	Secure Hash Algorithm
SHS	Secure Hash Standard
S-HTTP	Secure HyperText Transfer Protocol
SLIP	Serial Line Internet Protocol
SMI	Structure of Management Information
S/MIME	Secure/Multipurpose Internet Mail Extension
SMTP	Simple Mail Transfer Protocol
SNMP	Simple Network Management Protocol

SNS	Social Network Services
SPD	Security Policy Database
SPE	System Packet Exchange
SPI	Security Parameter Index
SPKI	Simple Public-Key Infrastructure
SSL	Secure Sockets Layer
TACS	Total Access Communications System
TCP	Transmission Control Protocol
TFTP	Trivial File Transfer Protocol
TIS	Trusted Information System
TLS	Transport Layer Security
TS	Timestamp
UDP	User Datagram Protocol
UMB	Ultra Mobile Broadband
UMTS	Universal Mobile Telecommunication System
URI	Uniform Resource Identifier
URL	Uniform Resource Locator
VPN	Virtual Private Network
WAIS	Wide Area Information Service
WAN	Wide Area Network
WAP	Wireless Application Protocol
WiBro	Wireless Broadband
WIDS	Wireless Intrusion Detection System
WiMAX	Worldwide Interoperability for Microwave Access
WWW	World Wide Web
XOR	eXclusive OR

Bibliography

1. Aboba, B., and D. Simon, 'PPP EAP TLS Authentication Protocol', RFC 2716, October 1999.
2. Abrams, M., and H. Podell, *Computer and Network Security*. Los Alamitos, CA: IEEE Computer Society Press, 1987.
3. Adams, C., and S. Farrell, 'Internet X.509 Public Key Infrastructure Certificate Management Protocols', Internet Draft, December 2001.
4. Almquist, P., 'Type of Service in the Internet Protocol Suite', RFC 1349, July 1992.
5. Astély, D., E. Dahlman, A. Furuskär, Y. Jading, M. Lindström and S. Parkvall, 'LTE: The Evolution of Mobile Broadband', *Communications Magazine*, April 2009.
6. Atkinson, R., 'Security Architecture for the Internet Protocol', RFC 1825, August 1995.
7. Atkinson, R., 'IP Authentication Header', RFC 1826, August 1995.
8. Atkinson, R., 'IP Encapsulation Security Payload (ESP)', RFC 1827, August 1995.
9. Ballardie, A., 'Core Based Trees (CBT) Multicast Routing Architecture', RFC 2201, September 1997.
10. Bellovin, S., 'Firewall-Friendly FTP', RFC 1579, February 1994.
11. Bellovin, S., and W. Cheswick, 'Network Firewalls', *IEEE Communications Magazine*, September 1994.
12. Berners-Lee, T., and D. Connolly, 'Hypertext Markup Language – 2.0', RFC 1866, November 1995.
13. Berners-Lee, T., R. Fielding and H. Nielsen, 'Hypertext Transfer Protocol – HTTP/1.0', RFC 1945, May 1996.
14. Blakley, B., 'Architecture for Public-Key Infrastructure', Internet Draft, November 1996.
15. Boeyen, S., R. Housley, T. Howes, M. Myers and P. Richard, 'Internet Public Key Infrastructure Part 2: Operational Protocols', Internet Draft, March 1997.
16. Borenstein, N., and N. Freed, 'MIME (Multipurpose Internet Mail Extensions) Part One: Mechanisms for Specifying and Describing the Format of Internet Message Bodies', RFC 1521, September 1993.
17. Borman, D., 'TELNET Authentication: Kerberos Version 4', RFC 1411, January 1993.

Wireless Mobile Internet Security, Second Edition. Man Young Rhee.
© 2013 John Wiley & Sons, Ltd. Published 2013 by John Wiley & Sons, Ltd.

18. Borman, D., 'TELNET Authentication Option', RFC 1416, February 1993.

19. Borman, D., and C. Hedrick, 'TELNET Remote Flow Control Option', RFC 1372, October 1992.

20. Borman, D., R. Braden and V. Jacobson, 'TCP Extensions for High Performance', RFC 1323, May 1992.

21. Bradley, T., and C. Brown, 'Inverse Address Resolution Protocol', RFC 1293, June 1987.

22. Bradner, S., and A. Mankin, *IPng: Internet Protocol Next Generation*. Reading, MA: Addison-Wesley, 1996.

23. Callaghan, B., B. Pawlowski and P. Staubach, 'NFS Version 3 Protocol Specification', RFC 1813, June 1995.

24. Case, J., K. McCloghrie, M. Rose and S. Waldbusser, 'Structure of Management Information for version 2 of the Simple Network Management Protocol (SNMPv2)', RFC 1442, May 1993.

25. Case, J., K. McCloghrie, M. Rose and S. Waldbusser, 'Protocol Operations for version 2 of the Simple Network Management Protocol (SNMPv2)', RFC 1448, May 1993.

26. Case, J., K. McCloghrie, M. Rose and S. Waldbusser, 'Management Information Base for version 2 of the Simple Network Management Protocol (SNMPv2)', RFC 1907, January 1996.

27. Case, J., K. McCloghrie, M. Rose and S. Waldbusser, 'Textual Conventions for version 2 of the Simple Network Management Protocol (SNMPv2)', RFC 1903, January 1996.

28. Chapman, D., and E. Zwicky, *Building Internet Firewalls*. Sebastopol, CA: O'Reilly, 1995.

29. Cheng, P., J. Garay, A. Herzberg and H. Krawczyk, 'A Security Architecture for the Internet Protocol', *IBM Systems Journal*, 1998.

30. Cheng, P., and R. Glenn, 'Test Cases for HMAC-MD5 and HMAC-SHA-1', RFC 2202, September 1997.

31. Cheswick, W., and S. Bellovin, *Firewalls and Internet Security: Repelling the Wily Hacker*. Reading, MA: Addison-Wesley, 1994.

32. Chokhani, S., and W. Ford, 'Internet Public Key Infrastructure Part IV: Certificate Policy and Certification Practices Framework', Internet Draft, March 1997.

33. Cole, R., D. Shur and C. Villamizar, 'IP over ATM: A Framework Document', RFC 1932, April 1996.

34. Comer, D., *Internetworking with TCP/IP, Volume 1: Principles, Protocols and Architecture*. Upper Saddle River, NJ: Prentice Hall, 1995.

35. Crawford, M., 'Transmission of IPv6 Packets Over FDDI', RFC 2019, October 1996.

36. Daemen, J., and V. Rijmen, 'AES Proposal: Rijndael, AES Algorithm Submission', 3 September, 1999.

37. Davin, J., J. Galvin and K. McCloghrie, 'SNMP Security Protocols', RFC 1352, July 1992.

38. Deering, S., and R. Hinden, 'Internet Protocol, Version 6 (IPv6) Specification', RFC 1883, January 1996.

39. Deering, S., and R. Hinden, 'Internet Protocol, Version 6 (IPv6) Specification', RFC 2460, December 1998.

40. deSouza, O., and M. Rodrigues, 'Guidelines for Running OSPF Over Frame Relay Networks', RFC 1586, March 1994.

41. Dierks, T., and C. Allen, 'The TLS Protocol Version 1.0', RFC 2246, January 1999.

42. Diffie, W., and M. Hellman, 'New Directions in Cryptography', *IEEE Transactions on Information Theory*, November 1976.

43. ElGamal, T., 'A Public-Key Cryptosystem and a Signature Scheme based on Discrete Logarithms', *IEEE Transactions on Information Theory*, July 1985.

44. Ericsson, 'LTE: An Introduction', Technical white paper, June 2009.

45. Etemad, K., 'Overview of Mobile WiMAX Technology and Evolution', *IEEE Communications Magazine*, October 2008.

46. Faltstrom, P., D. Crocker and E. Fair, 'MIME Content Type for Encoded Files', RFC 1741, December 1994.

47. Farrell, S., and C. Adams, 'Internet Public Key Infrastructure Part III: Certificate Management Protocols', Internet Draft, December 1996.

48. Fielding, R., J. Gettys, J. Mogul, H. Frystyk and T. Berners-Lee, 'Hypertext Transfer Protocol – HTTP1.1', RFC 2068, January 1997.

49. Finlayson, R., 'IP Multicast and Firewalls', RFC 2588, May 1999.

50. Finlayson, R., T. Mann, J. Mogul and M. Theimer, 'Reverse Address Resolution Protocol', RFC 903, June 1984.

51. FIPS Publication ZZZ, 'Announcing the Advanced Encryption Standard (AES)', US DoC/NIST, 2001.

52. Forouzan, B. A., *TCP/IP Protocol Suite*. New York: McGraw-Hill, 2000.

53. Fox, B., and B. Gleeson, 'Virtual Private Networks Identifier', RFC 2685, September 1999.

54. Freed, N., 'Behavior of and Requirements for Internet Firewalls', RFC 2979, October 2000.

55. Freier, A. O., P. Karlton and P. C. Kocher, 'The SSL Protocol Version 3.0', Internet Draft, Netscape Communications Corporation, March 1996.

56. Fuller, V., T. Li, J. Yu and K. Varadhan, 'Classless Inter-Domain Routing (CIDR): An Address Assignment and Aggregation Strategy', RFC 1519, September 1993.

57. Galvin, J., and K. McCloghrie, 'Security Protocols for version 2 of the Simple Network Management Protocol (SNMPv2)', RFC 1446, May 1993.

58. Galvin, J., S. Murphy, S. Crocker and N. Freed, 'Security Multiparts for MIME: Multipart/Signed and Multipart/Encrypted', RFC 1847, October 1995.

59. Garfinkel, S., and G. Spafford, *Web Security & Commerce*. Cambridge, MA: O'Reilly, 1997.

60. Gasser, M., *Building a Secure Computer System*. New York: Van Nostrand Reinhold, 1988.

61. Gleeson, B., A. Lin, J. Heinanen, G. Armitage and A. Malis, 'A Framework for IP Based Virtual Private Networks', RFC 2764, February 2000.

62. Goldsmith, D., and M. Davis, 'Using Unicode with MIME', RFC 1641, July 1994.

63. Han, M., Y. Lee, S. Moon, K. Jang and D. Lee, 'Evaluation of VoIP Quality over WiBro', PAM 2008, LNCS 4979, April 2008.

64. Harkins, D., and D. Carrel, 'The Internet Key Exchange (IKE)', RFC 2409, November 1998.

65. Haskin, D., and E. Allen, 'IP Version 6 over PPP', RFC 2023, October 1996.

66. Hedrick, C., 'Routing Information Protocol', RFC 1058, June 1988.

67. Heinanen, J., 'Multiprotocol Encapsulation over ATM', RFC 1483, July 1993.

68. Hinden, R., and S. Deering, 'IP Version 6 Addressing Architecture', RFC 1884, January 1996.

69. Hinden, R., and J. Postel, 'IPv6 Testing Address Allocation', RFC 1897, January 1996.

70. Hodges, J., R. Morgan and M. Wahl, 'Lightweight Directory Access Protocol (v3): Extension for Transport Layer Security', RFC 2830, May 2000.

71. Hoffman, P., 'Enhanced Security Services for S/MIME', RFC 2634, June 1999.

72. Hoffman, P., 'SMTP Service Extension for Secure SMTP over TLS', RFC 2487, January 1999.

73. Horning, C., 'Standard for the Transmission of IP Datagrams over Ethernet Networks', RFC 894, April 1984.

74. Housley, R., W. Ford, W. Pok and D. Solo, 'Internet X.509 Public Key Infrastructure Certificate and CRL Profile', Internet Draft, September 1998.

75. Huitema, C., 'An Experiment in DNS Based IP Routing', RFC 1383, December 1992.

76. Huitema, C., *IPv6: The New Internet Protocol*. Upper Saddle River, NJ: Prentice Hall, 1998.

77. Jacobson, V., 'Compressing TCP/IP Headers for Low-Speed Serial Links', RFC 1144, February 1990.

78. Johnson, D., Menezes, A and Vanstone, S, *The Elliptic Curve Digital Signature Algorithm*, Berlin and Heidelberg: Springer-Verlag, pp. 36–63, July 2001.

79. Kantor, B., and P. Lapsley, 'Network News Transfer Protocol: A Proposed Stan-dard for the Stream-Based Transmission of News', RFC 977, February 1986.

80. Kastenholz, E., 'The Definitions of Managed Objects for the Security Protocols of the Point-to-Point Protocol', RFC 1472, June 1993.

81. Kats, D., 'A Proposed Standard for the Transmission of IP Datagrams over FDDI Networks', RFC 1188, October 1990.

82. Kent, S., and R. Atkinson, 'Security Architecture for the Internet Protocol', RFC 2401, November 1998.

83. Kent, S., and R. Atkinson, 'IP Authentication Header', RFC 2402, November 1998.

84. Kent, S., and R. Atkinson, 'IP Encapsulating Security Payload (ESP)', RFC 2406, November 1998.

85. Khare, R., and S. Lawrence, 'Upgrading to TLS Within HTTP/1.1', RFC 2817, May 2000.

86. Klensin, J., N. Freed, M. Rose, E. Stefferud and D. Crocker, 'SMTP Service Extension for 8-bit MIME transport', RFC 1652, July 1994.

87. Klensin, J., N. Freed, M. Rose, E. Stefferud and D. Crocker, 'SMTP Service Extension', RFC 1869, November 1995.

88. Koblitz, N., 'Elliptic Curves Cryptosystems', *Mathematics of Computing*, 48, pp. 203–209, 1987.

89. Koblitz, N., 'Constructing Elliptic Curves Cryptosystems in Characteristic 2', *Advances in Cryptology – Crypt '91*. Berlin and Heidelberg: Springer-Verlag, pp. 156–167, 1991.

90. Krawczyk, H., M. Bellare and R. Canetti, 'HMAC: Keyed-Hashing for Message Authentication', RFC 2104, February 1997.

91. Lai, X., and J. Massey, 'A Proposal for a New Block Encryption Standard', *Proceedings, EUROCRYPT '90*. Berlin and Heidelberg: Springer-Verlag, pp. 389–404, 1991.

92. Larmo, A., M. Lindström, M. Meyer, G. Pelletier, J. Torsner and H. Wiemann, 'The LTE Link-Layer Design', *IEEE Communications Magazine*, April 2009.

93. Laubach, M., 'Classical IP and ARP over ATM', RFC 1577, January 1994.

94. Lee, B. G., and S. Choi, *Broadband Wireless Access & Local Networks: Mobile Wimax and Wifi*. Norwood, MA: Artech House, 2008.

95. Leech, M., 'Username/Password Authentication for SOCKS V5', RFC 1929, March 1996.

96. Leech, M., M. Ganis, Y. Lee, R. Kuris, D. Koblas and L. Jones, 'SOCKS Protocol Version 5', RFC 1928, March 1996.

97. Lercier, R., and F. Morain, *Counting the Number of Points on Elliptic Curves over Finite Fields*, Lecture Notes in Computer Science, No. 921. Berlin and Heidelberg: Springer-Verlag, pp. 79–94, 1995.

98. Lloyd, B., and W. Simpson, 'PPP Authentication Protocols', RFC 1334, October 1992.

99. Lodin, S., and C. Schuba, 'Firewalls Fend Off Invasions from the Net', *IEEE Spectrum*, February 1998.

100. Lougheed, K., and Y. Rekhter, 'A Border Gateway Protocol 3 (BGP-3)', RFC 1267, October 1991.

101. Macgregor, R., C. Ezvan, L. Liguori and J. Han, 'Secure Electronic Transactions: Credit Card Payment on the Web in Theory and Practice'. IBM RedBook SG24-4978-00, 1997.

102. Madson, C., and N. Doraswamy, 'The ESP DES-CBC Cipher Algorithm With Explicit IV', RFC 2405, November 1998.

103. Madson, C., and R. Glenn, 'The Use of HMAC-MD5-96 within ESP and AH', RFC 2403, November 1998.

104. Madson, C., and R. Glenn, 'The Use of HMAC-SHA-1-96 within ESP and AH', RFC 2404, November 1998.

105. Malkin, G., 'RIP Version 2 Carrying Additional Information', RFC 1723, November 1994.

106. Malkin, G., and A. Harkin, 'TFTP Option Extension', RFC 1782, March 1995.

107. Mastercard and Visa, 'SET Secure Electronic Transaction Specification Book 1: Business Description', May 1997.

108. Maughan, D., M. Schertler, M. Schneider and J. Turner, 'Internet Security Association and Key Management Protocol (ISAKMP)', RFC 2408, November 1998.

109. McBeath, S., J. Smith, L. Chen, A. C. K. Soong and B. Hao, 'VoIP Support Using Group Resource Allocation Based On the UMB System', *IEEE Communications Magazine*, January 2008.

110. McCloghrie, K., 'An Administrative Infrastructure for SNMPv2', RFC 1910, February 1996.
111. Medvinsky, A., and M. Hur, 'Addition of Kerberos Cipher Suites to Transport Layer Security (TLS)', RFC 2712, October 1999.
112. Menezes, A. J., and S. A. Vanstone, 'Elliptic Curve Cryptosystems and their Implementation', *Journal of Cryptology*, 6, pp. 209–224, 1993.
113. Metzger, P., and W. Simpson, 'IP Authentication using Keyed MD5', RFC 1828, August 1995.
114. Metzger, P., P. Karn and W. Simpson, 'The ESP DES-CBC Transform', RFC 1829, August 1995.
115. Mockapetris, P., 'Domain Names – Implementation and Specification', RFC 1035, November 1987.
116. Mogul, J., and S. Deering, 'Path MTU Discovery', RFC 1191, November 1990.
117. Montenegro, G., and V. Gupta, 'Sun's SKIP Firewall Traversal for Mobile IP', RFC 2356, June 1998.
118. Moore, K., 'SMTP Service Extension for Delivery Status Notifications', RFC 1891, January 1996.
119. Motorola, 'Long Term Evolution (LTE): Overview of LTE Air-Interface', Technical white paper, December 2008.
120. Moy, J., 'OSPF Version 2', RFC 1583, March 1994.
121. Moy, J., 'Multicast Extensions to OSPF', RFC 1584, March 1994.
122. Myers, J., 'POP3 Authentication Command', RFC 1734, December 1994.
123. Myers, J., and M. Rose, 'Post Office Protocol – Version 3', RFC 1725, November 1994.
124. Nam, C., S. Kim and H. Lee, 'The Role of WiBro: Filling the Gaps in Mobile Broadband Technologies', *Technological Forecasting and Social Change*, 75, pp. 438–448, 2008.
125. Newman, D., 'Using TLS with IMAP, POP3 and ACAP', RFC 2595, June 1999.
126. Newman, D., 'Benchmarking Terminology for Firewall Performance', RFC 2647, August 1999.
127. Hung, N.-M, S.-D. Lin, J. Li and S. Tatesh, 'Coexistence Studies for 3GPP LTE with Other Mobile Systems', *IEEE Communications Magazine*, April 2009.
128. NIST, 'The Secure Hash Algorithm (SHA)', FIPS PUB 180-1, 1995.
129. Oppliger, R., 'Internet Security: Firewalls and Beyond', *Communications of the ACM*, May 1997.
130. Orman, H., 'The OAKLEY Key Determination Protocol', RFC 2412, November 1998.
131. Partridge, C., 'Mail Routing and the Domain System', RFC 974, January 1986.
132. Pereira, R., and R. Adams, 'The ESP CBC-Mode Cipher Algorithms', RFC 2451, November 1998.
133. Pfleeger, C., *Security in Computing*. Upper Saddle River, NJ: Prentice Hall, 1997.
134. Piper, D., 'The Internet IP Security Domain of Interpretation for ISAKMP', RFC 2407, November 1998.
135. Piscitello, D., 'FTP Operation Over Big Address Records (FOOBAR)', RFC 1639, June 1994.

136. Postel, J., 'User Datagram Protocol', RFC 768, August 1980.

137. Postel, J., 'Internet Protocol', RFC 791, September 1981.

138. Postel, J., 'Transmission Control Protocol', RFC 793, September 1981.

139. Postel, J., 'Simple Main Transfer Protocol', RFC 821, August 1982.

140. Postel, J., 'Standard for the Transmission of IP Datagrams over Experimental Ethernet networks', RFC 895, April 1984.

141. Postel, J., and J. Reynolds, 'TELNET Protocol Specification', RFC 854, May 1983.

142. Postel, J., and J. Reynolds, 'TELNET Option Specifications', RFC 855, May 1983.

143. Postel, J., and J. Reynolds, 'File Transfer Protocol', RFC 959, October 1985.

144. Postel, J., and J. Reynolds, 'Standard for the Transmission of IP Datagrams over IEEE 802 Networks', RFC 1042, February 1988.

145. Ramsdell, B., 'S/MIME Version 3 Certificate Handling', RFC 2632, June 1999.

146. Rand, D., 'PPP Reliable Transmission', RFC 1663, July 1994.

147. Rekhter, Y., 'Experience with the BGP Protocol', RFC 1268, October 1991.

148. Rekhter, Y., and P. Gross, 'Application of the Border Gateway Protocol in the Internet', RFC 1772, March 1995.

149. Rekhter, Y., and T. Li, 'An Architecture for IP Address Allocation with CIDR', RFC 1518, September 1993.

150. Rekhter, Y., and T. Li, 'A Border Gateway Protocol 4 (BGP-4)', RFC 1771, March 1995.

151. Rescorla, E., 'HTTP over TLS', RFC 2818, May 2000.

152. Rhee, M. Y., 'Message Digest Computation Using the DMDC Algorithm', Proceedings, WISA 2000, November 2000.

153. Rhee, M. Y., *Mobile Communication Systems and Security*. John Wiley & Sons (Asia) Pte Ltd, pp. 156–161, 2009.

154. Rivest, R., 'The Md5 Message-Digest Algorithm', RFC 1321, April 1992.

155. Rivest, R., *The RC5 Encryption Algorithm*. Cambridge, MA: MIT Lab. for Computer Science, 1995.

156. Rivest, R., A. Shamir, and L. Adleman, 'A Method for Obtaining Digital Signatures and Public Key Cryptosystems', *Communications of the ACM*, February 1978.

157. Rivest, R., M. J. B. Robshaw, R. Sidney and Y. L. Yin, *The RC6 Block Cipher*. Cambridge, MA: MIT Lab. for Computer Science, 1996.

158. Romao, A., 'Tools for DNS Debugging', RFC 1713, November 1994.

159. Rubin, A., D. Geer and M. Ranum, *Web Security Sourcebook*. New York: John Wiley & Sons, Inc., 1997.

160. Schnorr, C., 'Efficient Signatures for Smart Card', *Journal of Cryptology*, 1991.

161. Schoffstall, M., M. Fedor, J. Davin and J. Case, 'A Simple Network Management Protocol (SNMP)', RFC 1157, May 1990.

162. Shacham, A., R. Monsour, R. Pereira and M. Thomas, 'IP Payload Compression Protocol (IPComp)', RFC 2393, December 1998.

163. Simpson, W., 'The Point-to-Point Protocol (PPP) for the Transmission of Multiprotocol Datagrams over Point-to-Point Links', RFC 1331, May 1992.

164. Simpson, W., 'The Point-to-Point Protocol (PPP)', RFC 1661, July 1994.

165. Simpson, W., 'PPP in HDLC-like Framing', RFC 1662, July 1994.

166. Sollins, K., 'The TFTP Protocol (Revision 2)', RFC 1350, July 1992.

167. Stallings, W., *Data and Computer Communications, Fifth Edition*. Upper Saddle River, NJ: Prentice Hall, 1997.

168. Stevens, W., *TCP/IP Illustrated, Volume 1: The Protocols*. Reading, MA: Addison-Wesley, 1994.

169. Sun Microsystems, Inc., 'NFS: Network File System Protocol Specification', RFC 1094, March 1989.

170. Thayer, R., N. Doraswamy and R. Glenn, 'IP Security Document Roadmap', RFC 2411, November 1998.

171. Thomson, S., and C. Huitema, 'DNS Extensions to Support IP version 6', RFC 1886, January 1996.

172. Thomson, S., and T. Narten, 'Ipv6 Stateless Address Autoconfiguration', RFC 2462, December 1998.

173. Touch, J., 'Report on MD5 Performance', RFC 1810, June 1995.

174. Wang, M. M., and M. Dong, 'Channelization in Ultra Mobile Broadband Communication Systems: The Foward Link', Wireless Communications and Mobile Computing Conference, 2008, IWCMC '08. International, August 2008.

175. Wang, F., A. Ghosh, C. Sankaran, P. J. Fleming, F. Hsieh and S. J. Benes, 'Mobile WiMAX Systems: Performance and Evolution', *IEEE Communications Magazine*, October 2008

176. Wijnen, B., G. Carpenter, K. Curran, A. Sehgal and G. Waters, 'Simple Network Management Protocol Distributed Protocol Interface Version 2.0', RFC 1592, March 1994.

177. Yergeau, F., G. Nicol, G. Adams and M. Deurst, 'Internationalization of the Hypertext Markup Language', RFC 2070, January 1997.

178. Zweig, J., and C. Partridge, 'TCP Alternate Checksum Options', RFC 1146, March 1990.

INDEX

\# of SPIs field 317–318, 323
\# of transform 312
1xEV-DO 68
1xEV-DV 68
3DES 306, 354
3DES-CBC mode 306
3GPP 68
3GPP2 64

A5/3 75
Abstract Syntax Notation One 53, 379, 364
acceptable policy identifier 280
acceptable policy set 287–8
access control 82, 250, 270–271, 292, 350, 383–4, 440, 449
Access Control Router (ACR) 449–50
Access Gateway (AGW) 456, 468
access denied 350
access location field 281–2
access method field 281
access service network (ASN) 440
ACK flag 43, 394, 396
Acknowledgement number 44
acquirer 416–418, 420, 433
ACR: Access Control Router 449
adaptive antenna system (AAS) 445
Adaptive Interference Management Techniques 455
Adaptive Modulation and Coding (AMC) 446
ADCCP 10–11, 467
additive inverse 101–102, 106, 118–119, 241
address mapping 31
address resolution 11, 29
Address Resolution Protocol 15, 28, 467
AddRoundKey() 143, 146, 150, 152, 154

Advanced Data Communication Control 10, 467
Advanced Encryption Standard 81
AES 81
AES algorithm 136–7, 139–140, 150–151
AES key expansion 144
AES S-box 141
aggressive exchange 319
AGW (Access Gateway) 456
AH 291–4, 259, 299, 300–304, 308, 312, 317, 391
alarm 388–9, 399
American National Standards Institute 3, 251, 467
AMPS (Advanced Mobile Phone System) 64
animation 375
Anomaly-based Systems 411
ANSI 3
ANSI X3.66 10
ANSI X9.30 CRL format 251
ANSI X9.30 standard 251
anticlogging 309
anticlogging token 309
antireplay service 291, 302–3
anycast address 35
AP: Access Point 449–450
Apple-Talk Datagram-Delivery Protocol 52
application layer 9, 12–14, 41–4, 46, 52, 337, 388, 404, 410, 451
application proxy 390, 397
application/pgp-signature protocol 378
application-level gateway 387, 389–390, 392, 396–400
application-level proxy 390
Armor checksum 358, 360
Armor tail 358, 360

ARP 13, 15, 27–30, 61
ARP reply 28–31
ARP spoofing 61
ARPANET 1
AS 54, 467
ASCII 12, 52, 175, 345, 358, 372, 374–7, 381
ASCII Armor 257, 353, 357–9, 377
   Armor headers 310–311
   Armor head line 358–9
ASCII character 48–9, 52, 309, 357, 376
ASCII-Armored data 359
ASN.1 53, 257, 283, 379, 381, 467
ASN-Gateway (ASN-GW) 441–2
asymmetric key pairs 416
asynchronous modem link 4
Asynchronous Transfer Mode 5, 467
AT: Access Terminal 448
ATM 5, 388, 467
ATM network 5, 388
attenuation 6
attribute 48, 270–272, 274
Attribute certificate 380
audit 343
audit log 390, 392
authentication 33, 35, 38, 39, 341, 344, 355
Authentication Header 291
authentication only exchange 319
Auth-Only SA 324
Authority information access extension 281
authority key identifier extension 276
authorization request 435–6
authorization response 436–7
Autonomous System 54

Back Orifice 61
backbone 54
bank's signature verification 424
Base 64 encoding 376–7, 381–2
base exchange 318
Base Station (BS) 440
basic constraints extension 276, 279–88
Basic Encoding Rule 379
basic path validation algorithm 287
bastion host 389–390, 398–401
beamforming (BF) 444
BER 379
BGP 8, 54, 55
bit stream 9, 452
bitwise parallel 178
Black box approach 412

bogus packet 306, 308
Border Gateway Protocol 8, 54–5
Botnet 403
Bourne Shell 48
branded payment card account
broadcast 416
broadcast-type protocol 44
browser 47
brute-force attack 84, 95, 214
buffer 11
Buffer overflow 59
bugs 388

CA 249, 251, 254, 256, 258, 261–4, 266,
   268–270, 219, 229
CA certificate 279, 350, 429–430, 433
CA name 289
CA's private key 273, 285, 430–434
CA's public key 263, 266, 269, 271, 420, 430,
   432, 433–4
CA's signature 273, 430, 432–4
cache table 29–31
cache-control module 31
CAD/CAM 12
capture request 473, 438
capture response 437–8
capture token 436–7
cardholder 416–417, 420, 429–430
cardholder account 415–416
cardholder account authentication 416
cardholder certificate 432, 435
cardholder credit card number 415
cardholder payment instruction 435
cardholder purchase request 436
cardholder registration 427
cardholder's account information 436
cardholder's account number 430
cardholder's issuer 437
cardholder's private key 430, 435
cardholder's public key 430–431, 435–6
cardholder's signature 431
Carrier Sense Multiple Access with Collision
   Detection 2
carrier-to-interference ratio (CIR) 446
caching 47
CAST-128 354–5
CBC mode 97, 306, 327
CDMA 2000 1x EV (1x Evolution) 74
CDMA 2000 1x EV-DO (1x Evolution Data
   Only) 74

CDMA 2000 1x EV-DV (1x Evolution Data Voice) 74
CDMA 64
CDMA cellular system 181
cdmaOne IS-95B 69
CDPD (Cellular Digital Packet Data) 65
cell 5, 66
certificate authority 315, 322
certificate authority field 315
certificate data field 314–315, 322
certificate encoding field 314–315
certificate path validation 271
certificate payload 314–315, 322
certificate policies extension 278, 288
certificate policy identifier 288
certificate request 332, 335–6
certificate request message 335–6
certificate request payload 315, 322
certificate response 432, 434
Certificate Revocation List 282, 314, 380
certificate revocation request 259, 266
certificate verify message 350
Certification Authority 261, 418
certification path 249, 268, 274
certification path constrain 271
certification path constraints extension 279
certification path length constraint 279
certification path validation 271, 287
certification revocation signature 365
CGI 48–9
chain of certificate 267
chain of trust 249, 264–5
change cipher spec message 326–7, 331, 337–8, 351
channel quality indicator (CQI) 446
channel state information (CSI) 445
Cheapernet 8
checksum 40, 44
choke point 387, 391
CIDR 31
CINR (carrier-to-interference and noise ratio) 445
cipher 135–6, 143, 146
cipher key 135, 140, 152
Cipher-Block Chaining mode 97, 327
ciphertext 75, 82, 91, 100, 103, 128, 208, 215, 304
circuit proxy 390, 397
circuit-level gateway 389–390, 396–7

classless addressing 31
Classless Interdomain Routing 31
client certificate message 336–7
client key exchange message 336–7
client socket address 41
ClientHello.random 333, 339
CLNS 52
Cloaked URL 61
closure alert 332, 349
CMS 379–81
coaxial cable 2
code bits 43
codepoint 18
column-wise permutation 164
command channel 395
Common Gateway Interface 48
community operation 1
compressed message 330, 356
compression algorithm 328, 335, 338, 351, 356, 372
compression (zip) 257, 353
Computer Virus 402
concatenation 167, 170, 187–8, 224, 335
confidentiality 33, 38, 97, 257, 270, 291–2, 294, 299, 302, 353–5, 357, 377, 383–4, 415–416, 418, 420
congestion 7, 17, 35
congruence 220–221
connecting devices 5
   bridge 7
   gateway 8, 14
   repeater 6
   router 7, 8
   switch 5
connection reset 43
connectionless delivery 33
connectionless integrity 291–2
Connectionless Network Service 52
connectionless protocol 16, 44
connection-oriented cell switching network 5
connectivity service network (CSN) 440, 442
constrained subtree 288
constrained subtree state variable 288
constraint 275, 279–280, 289
content tag 362
contiguous mask 26
contiguous string 26
cookie 309–310
coprime 208

Core Network (CN) 461
CRC 4, 357, 360, 452
credit card transaction 415–416
credit limit 417
critical extended key usage field 281
CRL 251, 261, 266, 277, 282–5, 379–380
CRL distribution points extension 281
CRL entry extensions 285
  certificate issuer 286
  Greenwich Mean Time (Zulu) 286
  hold instruction code 286
  invalidity date 286
  reason code 286
CRL extensions 284
  authority key identifier extension 284
  CRL number field 284
  delta CRL indicator 284
  issuer alternative name extension 284
  issuing distributing point 285–6
CRL sign bit 277
Cross-site request forgery 60
Cross-site scripting 60
cryptographic checksum 201
cryptographic message syntax 379–380
CSMA/CD 2, 3, 8
CSRF 60
current read state 326
current write state 326
CWTS 68
cyclic prefix (CP) 458
Cyclic prefix insertion 460
Cyclic Redundancy Check 4, 357

DARPA 1
data capture 418
data channel 395
data compression 12
data confidentiality 33
data content type 381
data diffusion 173
data encryption bit 277
Data Encryption Standard 81, 161, 269
data expansion function 345
data formatting 12
data integrity 6, 33, 38, 195, 296, 299, 325, 353
data link control protocol 10
data link layer 4, 9–10, 27, 29, 44, 388
data origin authentication 195, 277, 291–2, 296
DDOS 58
DDP 52

decimation process 168
decipher only bit 277
decode error 350
decrypt-error 350
decryption 38, 82, 84, 91, 100, 109, 117
decryption key 106, 208
decryption key sub-blocks 106
decryption failed 349
Defense Advanced Research Project Agency 1
delete payload 317–318, 323–4
delta CRL indicator 251, 284
DeMilitarized Zone 387, 391
demultiplexing 20, 46
dequeue 30
DER 379
DES 81–2, 84, 86, 89, 91, 114
DES-CBC 97, 296
DES-like Message Digest Computation 161
destination address 3, 7, 11, 37, 39, 41, 294
destination extension 39
destination host 18, 22, 27, 40, 45, 320
destination IP address 21, 27, 29, 295, 300,
  394–5
destination physical address 27, 29
destination port number 42
DH-DSS 337
DH-RSA 337
differential cryptanalysis 81, 103
differentiated services 18
Diffie-Hellman key exchange scheme 381
Diffie-Hellman parameters 335
diffusion 78, 101, 124
digested-data content type 382
digital certificate 415–416, 418, 420
digital envelope 253, 382, 418, 424, 427, 433–6
digital signature 161–2, 201, 203, 207, 215,
  249, 251, 253, 256–8, 277–8, 316, 353,
  355–6, 372, 377–8, 380–381, 416, 418,
  420, 423–4
Digital Signature Algorithm 188, 203, 227, 256
digital signature bit 277
Digital Signature Standard 227, 335–6
direct delivery 27
Directory Access Protocol 250, 282
Directory Information Tree 269
discrete logarithm 203–204, 227, 230, 239, 244
Disjoint Link Support 455
Distance-vector Multicast Routing Protocol 55
distance-vector routing 54–5

Distinguished Encoding Rule 379
DIT 269
DMDC 161–2, 176
DMZ network 387, 391
DNS 14, 23, 47, 53, 269, 395–6, 404, 442, 448–450
DNS name 278, 280, 284
DOI 292–4, 306, 309, 311–314, 316–24
Domain Name System 23, 53
Domain Naming Service 14–15
Domain of Interpretation 292
Doubling 230, 237
doubling point 232
downlink map (DL-MAP) 452
DSA 188, 203, 227–8, 240, 244, 256–8, 266, 282, 356, 369
DSS 227, 253, 256, 335, 337, 356, 380
dual ring 3
dual signature 257, 420, 435
dual-homed bastion host 389, 398, 400
duplicate insertion 413
DUT/SUT rule set 390
DVMRP 55
dynamic mapping 27–8, 31
dynamic table 8

eavesdropping 38, 325
EBCDIC 12
eBS (evolved Base Station) 455
EC 230, 232–42, 244–64
EC domain parameter 244–5
ECC 203, 230, 239–240
ECDSA 240, 244–5
ECDSA signature generation 244
ECDSA (Elliptic Curve Digital Signature Algorithm) 240
ECSD (Enhanced Circuit-Switched Data) 69
EDE mode 96–7
EDGE (Enhanced Data rate for GSM Evolution) 69
EDI address 278
electronic commerce 1, 161, 416–417
electronic funds transfer (EFT) 257, 418
Electronic Funds Transfer 257, 418
ElGamal 203
ElGamal authentication scheme 219
ElGamal encryption algorithm 216–217, 239
ElGamal public-key cryptosystem 239
ElGamal signature algorithm 216
elliptic curve 230, 239

Elliptic Curve Cryptosystem 230, 239
Elliptic Curve Cryptosystem (ECC) 230, 239
Elliptic Curve Digital Signature Algorithm 240
Elliptic curves 230
Email spoofing 61
Encapsulating Security Payload 97, 291
encapsulation 44, 46, 381, 383, 390, 450
encapsulation protocol 387–8
encipher only bit 277
Encrypt-Decrypt-Encrypt mode 96
encrypted certificate request message 434
encrypted-data content type 382
encryption 34, 58, 67, 99, 104, 107, 161
encryption key 100–102, 208, 380, 382
encryption key sub-blocks 106
end-entity certificate 287–9
ending tag 48
end-to-end protocol 416
enveloped-data content type 381–2
ephemeral port 45, 49
error alert 349
error control 10, 13, 40, 46
error reporting message 40
ESMTP 395
ESP 35, 38, 291–5, 299, 301–6, 308, 391
ESP header 295, 301, 304, 306
ESP payload data 307
ESP trailer 305
ESP transport mode 305–6
ESP tunnel mode 305–6
Ethernet 2–3, 7–9, 11, 22, 28–9
ETSI 64–5
Eudora 50
Euler's criterion 235
Euler's formula 208
Euler's totient function 208, 215
excluded subtree 288
excluded subtree state variable 288
Executor 61
expanded key table 109–110, 112, 114
expiration timer 54
explicit policy identifier 288
explicit policy state variable 288
Exploit 59
Export_restriction 350
extended Euclidean algorithm 138, 210, 215, 218, 220, 423, 426–7
extended key usage field 280
exterior routing 54

external bastion host 389
external Data Representation 50
external mail server 396
external screening router 399–400

f8 75
Facebook 56
False alerts 411
Fast Ethernet 8
fatal handshake failure alert 336
FCH (frame control header) 452
FDDI 2–3, 7, 13
FEC 446
Federal Information Processing Standards 188
Fermat's theorem 222, 235
FFR (Fractional Frequency Reuse) 464
FI function 75–6, 79
Fiber Distributed Data Interface 2–3
fiber-optic cable 2–3, 5, 405
File Transfer Protocol 14, 22, 49
File Transfer Protocol server 22, 49
finished message 337–8, 350–351
finished-label 351
finite field 139, 203–4, 206, 219–220, 230,
        237, 239, 244
FIPS 81–2, 188, 356
firewall 394, 387–92, 394–9, 400–401, 404–5,
        408, 410–411
FL function 76–8
flag field 4, 19
Flexband 455
flow control 4–5, 10–11, 13, 44
flow label 35, 40
FO function 75–8
follower 56
FOMA 70
forward link (FL) 452
four MD5 nonlinear functions 171, 178
Fractional Frequency Reuse (FFR) 453
fragment size 19
fragmentation 33, 35, 328
fragmentation module 7
fragmentation offset field 19
frame 3–4, 7
frame fragmentation 11
Frame Relay 4–5
FTP 14, 22–3, 44, 46–7, 49–50, 388, 390,
        394–5, 398, 404
FTP active mode 395
FTP packet filtering 394

FTP passive mode 395
full usage subchannel (FUSC) 452
full-duplex service 41, 43

garbage collection timer 54
gateway authorization request 375
gateway authorization response 436
gateway capture response 436
gateway digital signature 438
gateway private key 436
gateway public key 436
gateway signature certificate 437
gateway certificate 434, 436, 438
GEA3 75
generic certification of a user ID and public-key
        365
generic payload header 311–318, 320
GetBulk operation 53
GetNext operation 53
GIF 375–7
Gopher 47
GPRS (General Packet Radio Service) 69
Graphics Interchange Format 357
Groupon 57
GSM (Global System for Mobile
        Communications) 65
guard period 458

hacker 51, 61, 389, 391–3, 395, 399, 400–401,
        403
HARQ 446
hash code 11, 162, 168, 176, 188, 204, 227,
        245, 254, 337, 355, 364, 420
hash function 161–2, 195, 227–8, 253, 258,
        266, 282, 296, 308, 315, 322, 345, 352,
        370, 377, 424
hash payload 315, 322
Hashed Message Authentication Code 296
HDLC 10–11
header length 16
header 5–6, 16
header checksum 20, 40
header length field (HLEN) 16
heterogeneous platform 1
hexadecimal colon 34
hierarchical tree structure 264
higher-numbered port 395
High-level Data Link Control 10
HMAC 195–6, 189–99, 200–201, 296
HMAC-MD5 198, 297–8, 308, 314, 342–3

HMAC-SHA-1 198, 200, 298–9, 308, 341–2
hop count 7–8, 54
hop limit 39–40
Host-based Intrusion Detection System (HIDS) 409
hostid 23–6, 31, 33–4
HSCSD(High-Speed Circuit-Switched Data) 69
HSDPA (High Speed Downlink Packet Access) 73
HTML 47–8, 51
HTML tag 48
HTTP 14, 44, 47, 62, 291, 325, 372, 387, 390, 398, 404
HTTP GET command 48
HyperText Markup Language 47
HyperText Transfer Protocol 14, 44, 47, 325

IA5String 281
IAB 1–2, 250
IANA 22, 280, 300, 303–4,
IANA-registered address 399
ICB 1
ICCB 1
ICMP 11, 13–15, 19–20, 40, 295, 301, 305, 398
ICMP error message 19
ICV 299, 301, 304, 306–8
IDEA 81, 99–100, 306–7, 330, 354
IDEA decryption 106
IDEA encryption 101
IDEA encryption key 101
iDEN (Integrated Digital Enhanced Network) 66
identification field 19
identification payload 313–314, 321
identity authentication 275, 416, 420
identity protection exchange 319
Identity Theft 59
IDFT 446
IEEE token ring 13
IESG 1–2
IETF 1–2, 18, 32–3, 51, 250, 257, 291–2, 314, 325, 353, 441, 450
IGMP 15, 20, 41, 55
IKE 291, 299, 303, 308
image scanning 12
IMAP 14, 50–52, 417
i-mode 67
IMT 2000 68
inbound traffic 393

indirect delivery 27
Inform operation 53
information acquisition 1
inhibited policy mapping 289
inhibited policy-mapping field 280
initial policy identifier 287
initial policy set 288
Initialization Vector 97, 303, 327
initiate request 430, 433–4
initiate response 430, 434–5
initiator and responder cookie pair 312, 316–317
Inner CBC 97
inner IP header 295, 301, 306
inner padding 195, 296
input module 29–31
input-byte array 136
inside signature 382–4
insufficient-security 350
integer multiplication 124
integrated-salted S2K 370–371
integrity 377, 382, 384, 420
Integrity Check Value 299
inter symbolic interference (ISI) 458
interdomain routing 31
interior routing 54
internal bastion host 389
internal mail server 396
internal screening router 400
internal-error 350
International Cooperation Board 1
International Data Encryption Algorithm 81, 99–100, 306
International Organization for Standardization 9
Internet Activities Board 1–2, 250
Internet Architecture Board 2
Internet Assigned Numbers Authority 22, 280, 300, 399
Internet Configuration Control Board 1
Internet Control Message Protocol 11, 40, 295, 398
Internet Draft 310, 325
Internet Engineering Steering Group 1
Internet Engineering Task Force 1, 18, 250, 325
Internet Group Management Protocol 15, 41
Internet Key Exchange 291
Internet layer 13–15
Internet Lightweight Directory Access 250
Internet Mail Access Protocol (IMAP) 14, 51

Internet Message Access Protocol 14, 50–51
Internet Protocol 4, 15
Internet Protocol next generation 33
Internet Protocol version 4 (IPv4) 16
Internet Request for Comments 250
Internet Research Task Force 1, 250
Internet Society 2
Internet transaction protocol 415, 418
Internet Worm Detection 401
interoperability 249, 278, 282, 292, 294,
    379–380, 416, 439–41, 443, 449, 457,
    461–2
intranet 387–8
intrusion detection system(IDS) 401
inverse cipher 135–6, 150, 152, 158, 160
Inverse DFT 460
inverse S-box 150, 154
InvMixColumns() 150–151, 155
InvShiftRows() 150, 154
InvSubBytes() 150, 154
IP 11, 13, 15
IP address 22–4, 26, 28–9, 42
IP address class 23, 24
IP address translator 397
IP addressing 21–2, 24
IP authentication header 299
IP datagram 16–18, 20–21, 25, 27–9, 40–42,
    44–5, 291, 307
IP destination address 294
IP header 19–21, 42, 44–5, 295, 299
IP header option 21
IP host 392
IP multicast traffic 388
IP packet 7, 27, 29, 40, 295, 301, 306, 388,
    392–3, 447, 455
IP router 7
IP routing 27, 50, 448, 450
IP security document roadmap 292
IP source address 39, 309, 388, 393
IP spoofing 61
IP spoofing 61, 388, 393, 400
IP subnet 388
ipad 195–200, 296–9, 341–4
IPng 17, 32–3
IPsec AH 300, 312, 317
IPsec AH Format 300
    authentication data 301
    next header 300
    payload length 300

sequence number 300
SPI 300
IPsec DOI 311, 314, 317
IPsec ESP format 303
    authentication data 304
    next header 304
    pad length 304
    padding 304
    payload data 303
    sequence number 303
    SPI 303
IPv4 17, 31–40, 295, 301, 305–6
IPv4 addressing 33
IPv4 context 301, 305
IPv4 datagram 17–18
IPv4 header 16, 301
IPv6 17, 31–5, 37, 39
IPv6 addressing 33
IPv6 context 301, 306
IPv6 extension headers 35, 295
IPv6 header 35
IPv6 header format 33
IPv6 packet format 35
IPX 52
IRTF 1, 250
ISAKMP 291, 294, 308–9
ISAKMP exchanges 310, 318–319
ISAKMP header 309–310, 312, 316–317, 320,
    323
    exchange type 309–310, 318, 320, 323
    flags field 310, 320
authentication only bit 310–311, 323
commit bit 310–311
    encryption bit 310, 323
    initiator cookie 310
    length 311
    major version 310
    message ID 311
    minor version 310
    next payload 311
    responder cookie 310
ISAKMP message 309, 311–312, 314–316,
    319–20
ISAKMP payload 261, 265, 272
ISAKMP payload processing 319
    authentication only bit 323
    certificate payload processing 322
    certificate request payload processing 322
    delete payload processing 323

general message processing 319
generic payload header processing 320
hash payload processing 322
identification data field 321
identification payload processing 321
ISAKMP header processing 320
key exchange payload processing 321
nonce payload processing 323
notification payload processing 323
notify message type 323
proposal payload processing 321
security association payload processing320
signature data field 323
signature payload processing 322
transform payload processing 321
ISAKMP SA 311, 319, 324
ISAKMP SPI 312, 317
ISO 9
ISO Latin-5 359
issuer 251, 272
issuer alternative name extension 279, 282, 284
issuer domain policy 278
ITU-T 3, 68, 251, 271, 274, 279
IV 97

Java 47, 49, 71
Joint Photographic Experts Group 375
JPEG 375

KASUMI 75, 78, 80, 481
key agreement bit 277
key attribute information 271, 274
key certificate signing bit 277
key encryption bit 277
key exchange method 335, 337–8, 352
key exchange payload 311, 313, 318, 321
key expansion algorithm 109–110, 114
key expansion routine 135, 140, 142
key generation scheme 162, 171
key identifier field 276
Key logger 403
key material packet 367
   key packet variant 367
   public-key packet format 368
   secret-key packet format 369
key revocation signature 365
key schedule algorithm 124
key usage extension 276–7, 280
Keyed-hashing Message Authentication Code
   340

Keylogging 61
KGCORE 75
known-plaintext attack 95
Korn Shell 48

Lack of attack classification 412
LAN 2, 41
LAP-B 11
LDAP 250–251
legal issue 260
Legendre symbol 235
Link Access Procedure, Balanced 11
link activation 10
link address 22
link deactivation 10
Linkedin 57
link-state routing 54
literal data packet 364, 366
LLC 27
Local Area Network 2
log name 49
Logging 388–91, 399
logging service 388
logging strategy 388
logical address 22, 27–8
logical function 189
Logical Link Control 27
logical network addressing 11
Long Term Evolution (LTE) 439, 457
low-numbered port 395
LUCIFER 81
LZ77 356–7
LZFG 356
LZSS 356

MAC (Media Access Control) 27, 440
MAC (Message Authentication Code) 296
magic constants 110, 124
mail order transaction 417
Mail Transport Agent 395
MAP (maximum a posteriori) 452
masking 26
masking pattern 26
master secret 327, 338–9, 350–352
Maximum Transfer Unit 7
MBMS (Multimedia broadcast multicast
   services) 464
MBONE 41
MBSFN (Multicast-Broadcast Single Frequency
   Network) 464

MD5 100, 161, 176–7, 181, 253
MD5 algorithm 226
Media Access Control 27
merchant 415–418
merchant account data 434
Merchant Authentication 416
merchant authorization request 436
merchant capture request 438
merchant payment capture 437
merchant private key 433–8
merchant public key 434–7
merchant registration 433
merchant signature 435
merchant's point of sale system 418
merchant signature certificate 435
Message Authentication Code 296
message confidentiality 188, 415, 418
message contents 426
message digest 161–2, 166–170, 181, 188,
    191, 195, 214, 225–8, 430–438
message forgery 325
message integrity 251, 377, 392, 424
message integrity check 377, 426
message padding 188, 421
Message Security Protocol 251
metric 8
MIC 378, 427
MIME 52, 257, 353, 360, 372–3
MIME security content type 377
MixColumns() 143, 146
MMS 65
mobile multihop relay (MMR) 440
Mobile Station (MS) 440
mobile station registration 162
Mobile Switching Center (MSC) 461
Mobile WiMAX 440
Mobile WiMAX System Manager (WSM) 450
modem 9, 65, 400
modular reduction 138
MOSAIC 256
MOSPF 55
Motion Picture Experts Group 375
MPEG 375
MPI 366, 368
M-point DFT 460
MSP 251
MTU 7, 19, 37–8
MTU table 7
multicast 22

multicast address 23, 35, 41
multicast backbone 41
multicast host 23
Multicast Open Shortest Path First 55
multicast routers 41
multihomed bastion host 389
multihomed host 27
multipart/encrypted content type 375, 377
multipart/signed content type 374, 377–8
Multiple Input Multiple-Output (MIMO) 444
multiplexing 46
multiplicative identity 138
multiplicative inverse 102, 106, 208
Multipurpose Internet Mail Extension 52

name constraints extension 279
name subtree 280
Directory and Naming Services 395
NAT 397–400
National Bureau of Standards 81
National Institute of Standard and Technology
    81
National Security Agency 81
NBS 81
Netbus 61
netid 22–5, 31–4
network access layer 13, 22
network access providers (NAPs) 441
network address resolution 11
network address translator 388
network behavior analysis (NBA) 406
Network Behavior Analysis System (NBAS)
    408
Network File System 49
network interface card 22, 404
network layer 4–5, 11, 27, 41, 388, 404
network layer protocol 15, 32, 41
network management function 12, 388
Network Management System 52
network reference model (NRM) 442
network service providers (NSPs) 441
Network Virtual Terminal 56, 372
Network Working Group (NWG) 440
Network-based Intrusion Detection System
    (NIDS) 404
Newhall ring 13
next header 35, 37, 40
    authentication 38
    encrypted security payload 35
    fragmentation 37

hop-by-hop option 35–6
    security parameter index 38, 294
    source routing 21, 36–7, 39
NFS 49
NIC 22, 404
NIST 81
NMS 52
NMT (Nordic Mobile Telephone) 64
Node B 461
nonce data field 316
nonce payload 309, 316, 323
nonces 309, 316
nonrepudiation bit 277
no-renegotiation 350
notification payload 316–317, 320–323
Novell Internet Packet Exchange 52
NSA 81
NVT 56, 372
NVT ASCII data 373

Oakley key determination protocol 291, 308
object identifier 258
octet-stream 375, 377–8
OFDMA 453
offset value 19
OID 258, 266, 275, 278, 280–282
opad 196, 199, 297, 299, 341–2
OpenPGP message format 377
Open Shortest Path First protocol 8, 54
Open System Interconnect model 4
OpenPGP digital signature 378
options 21, 44
ORA 249, 258, 261–4, 266, 270
order digest 257, 420
order information 420
order message 420
Organisational Registration Authority 249, 258,
    261–4, 266, 270
Orthogonal Frequency Division Multiple
    Access (OFDMA) 445
orthogonality 459
OSI model 4
OSPF 8, 54
outbound traffic 388–9, 391
Outer CBC 98
outer IP header 295, 301, 306
outer padding 195, 296
output module 29
output-feedback mode 75
outside signature 383–4

PAA 249, 258, 261, 266, 271
packet 1–2, 6
packet 361
packet filter 387, 389, 392–4
packet filtering router 398–9
packet filtering rule 394, 396
packet header 361, 396
packet length 363
packet mode terminal 4
packet sniffer 262
packet splitting 413
packet tag 361
packet-by-packet basis 392
packet-filtering firewall 392, 396
packet-switching network 5
packet-switching protocol 4
padded message 188–9
PAR: Packet Access Router 448
parity bit 82, 97
passphrase 370
Password cracking 60
path validation algorithm 287, 289
path validation module 289
Path-vector routing 55
payload length 40, 300, 311–314
Payload mutation 413–414
payment authorization 418, 435
payment authorization service 418
payment capture 437
payment card 416–417
payment card account 416–417
payment card authorization 417
payment card brand 418, 434
payment card certificate 418
payment card transaction 418
payment digest 257, 420
payment gateway 418, 434–8
payment gateway public key 436
payment gateway's key 435
payment integrity 257, 416, 420
payment message 418
payment processing 257, 415–416, 427
PCA 249, 258–62, 264, 266, 270–271, 289
PCA name 289
PCS 64
PCMCIA card 270, 330
PDA 57
PDC (Personal Digital Cellular) 67
peer-to-peer communication 12

PEM CRL format 251
pending read state 326
pending write state 326
perimeter 387
perimeter network 391
periodic timer 54
Perl 49
PGP 13–14, 95, 100, 256–7, 264, 291, 313, 353–9
PGP 5.x 368, 371
PGP 5.x key 366–7
PGP packet structure 363
    message packet 363–4
    session key packet 363, 365–6
    signature packet 363–6
phase 1 exchange 311
P-hash 345
Phishing 58
physical address 7, 11, 22, 27–31
physical layer 4–6, 9, 27, 44
physical resource block (PRB) 459
PKI 249–51, 258–9, 261
PKIX 266, 269, 271
plaintext 82, 304, 306, 327, 330, 359, 364, 370, 393, 424
P-MD5 378
point at infinity 230, 235–8
point-to-point encryption 416
Point-to-Point Protocol 4
Point-to-Point Tunnelling Protocol 392
Policy Approval Authority 249, 258
Policy Certification Authority 249, 260
policy constraints extension 280, 288
policy mapping extension 278
policy-mapping state variable 288
polynomial modulo 137
POP3 50–52, 373
port number 41–2, 44–5
Post Office Protocol 14
Post Office Protocol version 3 50–51
PostScript 375
PPP 4
PPP frame 4
PPTP 392
PRBS state transition function 171
precedence 17–18
premaster secret 337–340, 352
preoutput block 91–2, 94–5
presentation layer 11

Pretty Good Privacy 14, 100, 256, 353
PRF 344
primary ring 3
prime factor 222, 227
Prime factorization 215
prime field 230, 239
Prime Field 230
prime number 203, 208, 215, 227, 258
primitive element 203, 237, 239, 258
priority 35, 40
Privacy Enhanced Mail (PEM) 14, 251
Privacy Infringement 59
private key 208, 212, 214–215, 222, 227, 270, 278, 282, 284, 286, 352, 370, 381
private-key usage period extension 278
Procedure 271
proposal # field 312
proposal payload 312
proposal-id field 312
Protocol (PPTP) 392
Protocol 312–313, 317–24, 325, 327–8, 331–4, 339, 344, 349–350, 372–380
protocol suite 12
protocol-id field 312, 317, 321, 323
protocol-version 350
proxy ARP 29
proxy module 390
proxy server 47, 51, 389–390, 396–9
pseudocode 110, 114, 143, 147
pseudo-random binary sequence 171
pseudo-random function 344
P-SHA-1 345
PSK (Phase Shift Keying) 446
public key 208, 213–215, 222, 227, 249
public-key algorithm 203, 230, 239, 355, 366–9, 371
public-key certificate 249–250, 261–2, 418
public-key Infrastructure 249
public-key packet 365, 368–9, 379
purchase request 434–5
purchase response 434–5

QAM (Quadrature Amplitude Modulation) 446
QPSK 446
quadratic nonresidue 234–5
quadratic residue 234–5
Quasi Orthogonal Reverse Link 455
query message 40–41
queue 29–30
Quoted-printable 375–6, 378

RA 263, 418
Radio Access Network (RAN) 461
Radio Access System (RAS) 449
Radio Network Controller (RNC) 461
radio resource management (RRM) 452
radix-64 conversion 257, 353, 357
radix-64 encoding 357–8, 360–361
Random symmetric key 424, 430, 433–6
RARP 13, 15, 27–8, 31
RAS: Radio Access Station 449
RC5 decryption algorithm 117
RC5 encryption algorithm 108, 114
RC6 decryption algorithm 128
RC6 encryption algorithm 125
Rcon[i] 141, 143
RDN 270
Receive Direction Processing 446
recompressed message 356
record route option 21, 39
record-overflow 350
reference points (RPs) 441–3, 450
Registration Authority 418
registration form 431, 433
registration form process 430, 433
registration form request 430
registration information (name, address and ID) 433
registration request 433
registration request process 418, 429
registration response process 418, 429
relative distinguished name 270
relatively prime 204, 208, 217, 219, 222
remote access 61
Remote Login 12, 44, 56
remote server 50, 62, 398
repository 250, 266, 269
required explicit policy field 280
Reserve Address Resolution Protocol 15, 31
resource sharing 14
reverse link (RL) 452
RFC 250
Rijndael algorithm 82, 135
RIP 8, 54–5
RIPEMD-160 296
Rlogin 44, 55–6, 388
root CA 249, 264, 287, 336, 418
Rootkit 60
RotWord() 140
round constant word array 141

round key 75, 86, 91, 124, 126
router 7–8, 17–18, 20–22, 25–9, 294–5, 392, 399, 408, 448
Routing Information Protocol 8, 54
routing module 7
routing table 7–8, 27–8, 34, 54–5, 401
row/column-wise permutations 164
row-wise permutation 164
RPC 50
RSA encryption algorithm 208, 258
RSA public-key cryptosystem 207, 212
RSA signature scheme 212

S/MIME 14, 95, 257, 291, 353, 372–3, 379, 382
S/MIME version 3 agents 379
S2K specifier 369–370
SA 291, 294–95, 300–301, 303–4, 306, 308–13, 318–24
SA attributes field 313
SAD 294–5
salted S2K 370–371
S-box 76, 78–82, 101, 114, 143
SC–constellation mapping 460
Schnorr's authentication algorithm 222
Schnorr's public-key cryptosystem 222
Schnorr's signature algorithm 224
Schoolbus 61
screened host firewall 398, 399–400
screened subnet firewall 398, 400–401
screening router 392–3, 399–400
SDLC 11
SDMA (Space Division Multiple Access) 452–3
secondary ring 3
secret key parameter 114
secret-key packet 368–370
Secure Electronic Transaction 257, 415
Secure Hash Algorithm 188, 214, 356
Secure Hash Standard 188
Multipurpose Internet Mail Extension (S/MIME) 14, 353
secure payment processing 415, 427
secure payment transaction 416
Secure Sockets Layer version 3 325
Secure/Multipurpose Internet Mail Extension 14, 353
Security Association 291, 294
Security Association Database 294–5
security association payload 311

security gateway 294–5, 301, 306, 387
security multiparts 377
security option 21
Security Parameter Index 38, 294
Security Policy Database 292, 294
security protocol identifier 294
self-signed certificate 287–9
sendmail 50–51, 396
sequence number 43, 284, 300, 303, 327, 413
Serial/parallel converter 460
server certificate 335
server hello done message 335–7
server key exchange message 335
server socket address 41
Serverhello.random 333, 339
serving GPRS Support Node (SGSN) 462
Session hijacking 62
session layer 11, 396
session state 326
   cipher spec 327
   compression method 326
   is resumable 327
   master secret 327
   peer certificate 326
   session identifier 326
SET 257, 271, 415–21
SET payment instruction 433
SHA 188–190, 225, 228
SHA primitive functions 190
SHA-1 161, 188–91, 214, 253, 258, 266
SHA-1 algorithm 195, 335, 420
shared secret data 188
ShiftRows() 143, 145, 150
SHS 188
Signaling System #7 9
signature payload 316, 318, 323
Signature-based Systems 410
signed-data content type 381
Silencer 61
Simple Mail Transfer Protocol 44, 50, 372
Simple Network Management Protocol 14, 46, 52
single-homed bastion host 389, 398–400
sliding window protocol 44
Smart Phone 57
Smart TV 57
SMI 52–3
SMS 65

SMTP 14, 47, 50–52, 61, 257, 353, 372–3, 376–7, 388, 395, 404
SMTP packet filtering 395–6
SMTP server 50–51, 395
SNMP 14, 46, 52–3, 404
socket address 41
socket pair 42
SOCKS 387–8, 390
SOCKS port 390
SOCKS protocol version 4 390
SOCKS server 390
source address 39, 388, 393
source host 18–19, 40, 45
source IP address 16, 19, 21, 40
source port number 42, 45
source routing 21, 33, 36, 39, 400
source routing option 21
Spam 59
spatial division multiple access (SDMA) 463
SPD 292, 294–5
SPE 11
SPI 38–9, 294–5, 300, 303, 308, 312, 317, 321, 323–4
SPI field 38–9, 294, 303, 312, 317–318
SPI size 312, 317–318, 323
Spoofing attack 61
SQL Injection 411
SRNC (Session Reference Network Controller) 455
SS7 9
SSD 188
SSL Alert Protocol 331
   bad-certificate 332
   bad-record-mac 332
   certificate-expired 332
   certificate-revoked 332
certificate-unknown 332
   close-notify 332
   decompression-failure 332
   illegal-parameter 332
   no-certificate 332
   unexpected-message 332
   unsupported certificate 332
SSL Change Cipher Spec Protocol 331
   change cipher spec message 331
   current state 331
   padding state 331
SSL connection 327
   client write key 327

client write MAC secret 327
initialisation vectors 327
sequence numbers 327
server and client random 327
server write key 327
server write MAC secret 327
SSL Handshake Protocol 327, 332
  cipher suites 334
  client hello 334
  client hello message 332, 334
  client version 334
  ClientHello.cipher-suite 334
  ClientHello.compression-method 334
  ClientHello.session-id 334
  compression method 334
  handshake failure alert 334
  hello request 334
  server hello message 332, 334–5
  server version 334
  session ID 334
SSL Record Protocol 325, 327–330
  appended SSL record header 330
  compression and decompression 328
  Fragmentation 328
  MAC 329
SSL session 326
SSLv3 325
SSLv3 protocol 339
SSL/TLS 271, 416
stand-alone signature 365
starting tag 48
state 136, 140, 143, 146, 150–151
state array 136, 143, 146–7
static mapping 27–8
static table 8
Striker 61
string-to-key (S2K) 369
Structure of Management Information 52
stub link 55
SubBytes() 143, 147
Subcarrier mapping 460
subject alternative name 275, 279, 287
subject directory attributes extension 279
subject distinguished name 269
subject domain policy 278
subject identification information 271
subject key identifier 275–6
subject key identifier extension 276
subkey 76, 78–9, 100, 102, 365, 367

subkey binding signature 365
subnet 24, 26
subnet addressing 26
subnetid 24–5
subnetting 24–6, 33
SubWord() 141
Sun's Remote Procedure Call 50
supernetting 24, 26, 33
swapped output 175–6
swapping operation 103
switching mechanisms 5
  circuit switching 5–6
  message switching 5–6
  packet switching 5–6
symmetric block cipher 81–2, 109, 135
Synchronous Data Link Control 11
syntax selection 12
System Packet Exchange 11

TACS (Total Access Communications System) 64
tampering 270, 325
Target finding 402
TCP 11, 14–16, 20, 35, 41–4
TCP data 42–3
TCP header 42–4, 393
TCP packet format 42
TCP port 20 395
TCP port 21 23
TCP port 23 23, 394
TCP port 25 51, 395
TCP port 395
TCP port number 388, 396, 404
TCP segment 17, 38, 42–3, 328, 392
TCP/IP four-layer model 12
TCP/IP protocol 11
TDMA-136 66
TELNET 22–23, 44, 47, 55–6, 393–4
TELNET packet filtering 393
Telnet server 23, 56
TFTP 23, 49–50
Thicknet 8
Thinnet 8
TIA 64
Time to live (TTL) 20
timestamp option 21
timestamp signature 365
TLS certificate verify message 350
TLS change cipher spec message 351
TLS finished message 351

TLS handshake protocol 349
TLS handshake-message 351
TLS master-secret 352
TLS premaster-secret 352
TLS record layer 349
TLS record protocol 351
TLS server hello message 352
TLS v1 325, 339–340
TLS v1 protocol 339
token 2
Token Ring 2–3, 7, 13, 22
TOS field 17–18
trace 235
traffic control 11
transaction protocol 415, 427
transform # field 313
transform payload 312–313, 321
transform-id field 313, 321
Transmission Control Protocol 15, 41
Transmit Direction Processing 446
transparent data 338
transport layer 5, 11, 13–14
Transport Layer Protocols 41
Transport Layer Security version 1 325
transport mode 38, 295, 301, 304–6
transport mode SA 295
tri-homed firewall 389
tri-homing 387, 391
triple DES 81, 95–6, 306–7
triple DES-EDE mode 97
triple wrapped message 382–5
triple wrapping 384
Trivial File Transfer Protocol 23, 49
Trojan horse 403
Trojan horse 51, 60–61, 403
Trojan horse sniffer 390
truncated (TR) 78
trust chain to the root key 430, 432–8
trust chaining 418
truth table 178, 190
TTL 20, 39–41, 413
tunnel mode 295, 299, 301,
    303–7, 388
tunnel mode SA 295
tunneling protocol 392
Point-to-Point Tunneling 392
Twisted Ethernet 8
twisted-pair cable 2
Twitter 56

two-key cryptosystem 215
Type of service (ToS) 17

UDP 14–15, 17, 20, 35, 41–2, 44–7, 52, 299,
    301, 305, 309, 390, 398, 450
UDP header 44
    Destination port number 45
    ephemeral port number 45
    pseudoheader 46
    source port number 45
    UDP checksum 46
    UDP length 45
    universal port number 45
UDP packet 44
UDP port 309, 392
UMB (Ultra Mobile Broadband) 452
UMTS (Universal Mobile Telecommunication
    System) 73
uncompressed message 356
unicast 22, 27–8, 35
unicast address 35, 294
uniform resource identifier 278, 414
universal addressing system 22
unknown-ca 350
URG flag 43
urgent pointer 43
URI 278, 281–2, 284, 414
URL 47, 61–2, 281, 414
user authentication 249, 251, 253
User Datagram Protocol 14, 44, 398
user key 131, 133
user-canceled 350
UTF-8 359

v3 key fingerprint 368
v4 key fingerprint 364
variable number of rounds 109
variable-length secret key 109
VCI 5
vendor ID payload 318
version 2 packet 368
version 3 packet 368
version 35
version 4 packet 368
version field (VER) 17, 274, 369
Virtual Channel Identifier 5
Virtual Path Identifier 5
Virtual Private Network 264, 388
Virus 58
virus 58, 388, 401–3, 481

virus-infected programs or files 388
VPI 5
VPN 264, 388, 392
VPN protocol 392

WAN 2–4, 22, 24
WAP (Wireless Application Protocol) 67
Web page 47–8
Web server 47–8
Web traffic 47
Website 1
Website spoofing 61
WiBro (Wireless Broadband) 448
WiBro Access Terminal (AT) 449
Wide Area Network 2–3
windows NT 51, 392
window scale factor 44
window size 43
Wireless Intrusion Detection System (WIDS) 406
word size 109
World Wide Web 14–15, 44, 47, 439
Worm 58
Worm activation 402
Worm infection 402
Worm transferring 402
WWW 47

X.25 4–5, 11
X.400 51
X.500 directory 250–251, 271, 285
X.500 name 269–70, 272–3
X.509 AC 380
X.509 certificate 251, 266, 271, 315
X.509 certificate format 251, 271
X.509 certificate format 251, 271
  certification path constraint 271, 275
  extensions related to CRL 275
  issuer 273
  issuer unique identifier 273
  issuer's signature 273

key and policy information 275–6
serial number 272
signature algorithm 272–3
subject and issuer attribute 275, 278
subject name 273
subject public-key information 273
subject unique identifier 273
validity period 272
version number 272
X.509 CRL format 251, 282
X.509 Public-Key Infrastructure 273
X.509 v1 certificate 271, 273
X.509 v2 certificate 270, 273
X.509 v2 CRL format 251, 282, 284–5
  issuer name field 282
  UTC Time, Generalised Time 283
  X.509 distinguished name 283
  X.509 type name 282
  next update field 283
  revoked certificates field 284
  signature field 283
  algorithm identifier 282
hash functions–MD5 and SHA-1 282
signature algorithm–RSA and DSA 282
  this update field 283
  this update field 283
  issue date of CRL 283
  version field (optional) 282
X.509 v3 certificate 251, 271, 274, 282, 326, 335
X.509 v3 certificate format 275
X.509 v3 public-key certificate 418
XDR 50
Xerox Wire 8
XSRF 60
XSS 60
xtime() 138

Zero-Extension (ZE) 78
ZIP algorithm 354, 356
Zombie 403